Springer Series On Environmental Management

Springer Series on Environmental Management

Volumes published since 1989

The Professional Practice of Environmental Management (1989)
R.S. Dorney and L. Dorney (eds.)

Chemical in the Aquatic Environment: Advanced Hazard Assessment (1989)
L. Landner (ed.)

Inorganic Contaminants of Surface Water: Research and Monitoring Priorities (1991)
J.W. Moore

Chernobyl: A Policy Response Study (1991)
B. Segerstahl (ed.)

Long-Term Consequences of Disasters: The Reconstruction of Friuli, Italy, in Its International Context, 1976–1988 (1991)
R. Geipel

Food Web Management: A Case Study of Lake Mendota (1992)
J.F. Kitchell (ed.)

Restoration and Recovery of an Industrial Region: Progress in Restoring the Smelter-Damaged Landscape near Sudbury, Canada (1995)
J.M. Gunn (ed.)

Limnological and Engineering Analysis of a Polluted Urban Lake: Prelude to Environmental Management of Onondaga Lake, New York (1996)
S.W. Effler (ed.)

Assessment and Management of Plant Invasions (1997)
J.O. Luken and J.W. Thieret (eds.)

Marine Debris: Sources, Impacts, and Solutions (1997)
J.M. Coe and D.B. Rogers (eds.)

Environmental Problem Solving: Psychosocial Barriers to Adaptive Change (1999)
A. Miller

Rural Planning from an Environmental Systems Perspective (1999)
F.B. Golley and J. Bellot (eds.)

Wildlife Study Design (2001)
M.L. Morrison, W.M. Block, M.D. Strickland, and W.L. Kendall

Selenium Assessment in Aquatic Ecosystems: A Guide for Hazard Evaluation and Water Quality Criteria (2002)
A.D. Lemly

Quantifying Environmental Impact Assessments Using Fuzzy Logic (2005)
R.B. Shephard

Changing Land Use Patterns in the Coastal Zone: Managing Environmental Quality in Rapidly Developing Regions (2006)
G.S. Kleppel, M.R. DeVoe, and M.V. Rawson (eds.)

The Longleaf Pine Ecosystem: Ecology, Silviculture, and Restoration (2006)
S. Jose, E.J. Tokela, and D.L. Miller (eds.)

Linking Restoration and Ecological Succession (2007)
L.R. Walker, J. Walker, and R.J. Hobbs (eds.)

Margaret M. Carreiro
Yong-Chang Song
Jianguo Wu
Editors

Ecology, Planning, and Management of Urban Forests

International Perspectives

Springer

Editors
Margaret M. Carreiro
Department of Biology
University of Louisville
Louisville, KY 40292
USA
m.carreiro@louisville.edu

Yong-Chang Song
East China Normal University
Shanghai 200062
China
ycsong@des.ecnu.edu.cn

Jianguo Wu
School of Life Sciences
Arizona State University
Tempe, AZ 85287
USA
Jingle.Wu@asu.edu

Series Editors
Bruce N. Anderson
Planreal Australasia
Keilor, Victoria 3036
Australia
bnanderson@compuserve.com

Robert W. Howarth
Program in Biogeochemistry and Environmental Change
Cornell University; Corson Hall
Ithaca, NY 14853
USA
rwh2@cornell.edu

Lawrence R. Walker
Department of Biological Sciences
University of Nevada Las Vegas
Las Vegas, NV 89154
USA
walker@unlv.nevada.edu

ISBN: 978-0-387-71424-0 e-ISBN: 978-0-387-71425-7

Library of Congress Control Number: 2007929477

Cover illustration: The aerial photo of Louisville, Kentucky, USA, a City of Parks along the banks of the Ohio River, was taken by Keith Mountain, Department of Geography and Geosciences, University of Louisville.

Printed on acid-free paper

9 8 7 6 5 4 3 2 1

springer.com

Preface

Urbanization, an inevitable consequence of human social development, is occurring rapidly and is global in scope. Urbanization has brought measurable benefits to human societies, such as concentrated populations and labor forces that facilitate large-scale production of goods and services, extensive transportation systems that foster trade and economic development, advanced communication and information technologies that bolster education and scientific enterprises, health care and public facilities and services, job opportunities, and cultural diversity—all resulting in higher overall living standards. However, urbanization has also resulted in a number of negative impacts on the environment, including encroachment on farmland and natural habitats, increase in impervious surfaces, reduction in native biodiversity, enormous and concentrated consumption of energy and resources that result in equally large production of waste and pollution, and isolation of humans from nature. How can we take better advantage of the benefits and minimize the negative impacts of urbanization? This is a critical question that must be addressed in the development of modern cities around the world. Ecological cities ("eco-cities") represent a new approach to meeting this challenge.

Urban forests play a fundamentally important role in building ecological cities, because they improve the environmental quality of the urban environment and the aesthetics of urban landscapes. Thus, in many developed and developing countries, the evolution of urban forestry has been recognized as an essential means of maintaining urban ecosystem health, improving human living conditions, fostering a harmonious human–nature relationship, and ultimately achieving urban sustainability. Shanghai, as one of the largest megacities in the world, has been searching for planning and design principles for building an ecologically sound metropolitan region, and large-scale development of urban forestry is under way. Thus, it was quite appropriate and timely that the International Symposium on Urban Forestry and Eco-Cities was held in Shanghai (September 19–23, 2002). The symposium was organized by East China Normal University, the Shanghai Municipal Agricultural Commission, and the Shanghai Agriculture and Forestry Bureau, with support from the Forestry Bureau of the People's Republic of China, the Chinese Academy of Forestry, the Forestry Society of China, the Shanghai Foreign Affairs Office, the Shanghai Municipal Construction and Management Commission, the Shanghai Planning Commission, the Shanghai Environmental Protection Bureau, the Shanghai Urban Planning Administration Bureau, and the Shanghai Landscape Administration Bureau. Scientists, practitioners, and policy makers from Asia, Europe, and North America participated both in the symposium and subsequent site visits throughout the Shanghai region to observe and offer comment on the urban forestry programs of several cities and towns. This book has evolved from the presentations and discussions held at this meeting.

East China Normal University
Shanghai, China
September 2006

Yong-Chang Song

Email: ycsong@des.ecnu.edu.cn (or contact his English-speaking colleague, Dr. Junxiang Li, at: jxli@des.ecnu.edu.cn)

Note: Although all of the illustrations in the book are reproduced in black-and-white, the original color files can be accessed by using the following link: http://springer.com/978-0-387-71424-0.

Acknowledgments

We wish to thank East China Normal University (ECNU), the Shanghai Municipal Agricultural Commission, and the Shanghai Agriculture and Forestry Bureau for organizing the International Symposium on Urban Forestry and Eco-Cities, which inspired this book. We also would like to convey our special appreciation to the many student volunteers and staff at ECNU who made foreign visitors at this symposium feel welcome, oriented, and comfortable during their stay in Shanghai. We are also grateful to the organizers of the field trips to different urban parks, gardens, plantations, and forest remnants within Shanghai and its surrounding region. Being able to see these sites and the work being done there firsthand, and to have discussions with the community leaders who are managing them, greatly enriched the experiences and knowledge shared during the symposium.

We are indebted to the many authors who contributed their knowledge, expertise, and time so that we could produce this book. Their patience and cooperation during the revision and editorial process were much appreciated. Many reviewers (listed below) also contributed to the quality of the manuscripts through their insightful suggestions and constructive critique. Thanks to James Baxter, Jürgen Breuste, Loren Byrne, Nancy Golubiewski, Glenn Guntenspergen, Gordon Heisler, Wei Ji, Faith Kostel-Hughes, Chris Martin, Jennifer Mattei, Joseph McBride, Mark McDonnell, Janet Morrison, Laura Musacchio, Jari Niemalä, Paul Nolan, David Nowak, Richard Pouyat, Hai Ren, George Robinson, Reuben Rose-Redwood, Weijun Shen, Mark Smale, Tara Trammell, Christopher Tripler, Paige Warren, Duning Xiao, and Wei-xing Zhu. A special thanks to Keith Mountain for taking the aerial photo of Louisville for the book cover. We also wish to thank Janet Slobodien, Ann Avouris, and Herman Makler of the editorial staff at Springer, and Geethalakshmi Srinivasan of SPi Publisher Services for their help, advice, and patience throughout the process of writing, editing, and completing this book.

Contents

Contributors

Margaret M. Carreiro
University of Louisville, Department of Biology, Louisville, KY, USA

Sophia Shuang Chen
Nanjing Institute of Geography and Limnology, Chinese Academy of Sciences, Nanjing, China

Wendy Y. Chen
University of Hong Kong, Department of Geography, Hong Kong, China

Y.C. Cho
Seoul Women's University, Faculty of Environment and Life Sciences, Seoul, Korea

Liang-Jun Da
East China Normal University, Department of Environmental Science, Shanghai, China; and Shanghai Key Laboratory for Ecology of Urbanization Processes and Eco-Restoration, Shanghai, China

Bin Du
Tsinghua University, Department of Environmental Science and Engineering, Beijing, China

Jun Gao
East China Normal University, Department of Environmental Science, Shanghai, China

Peng Gong
University of California-Berkeley, Department of Environmental Science, Policy and Management; Berkeley, CA, USA

Henning Haeupler
Ruhr-University, Department of Special Botany, Working Group Geobotany, Bochum, Germany

Volker Heidt
University of Mainz, Department of Geography, Mainz, Germany

Chenglin Huang
Anhui Agricultural University, Institute of Forestry and Landscape, Hefei, China

Michael Jestaedt
Forest Administration of Hessia, Germany

C.Y. Jim
The University of Hong Kong, Department of Geography, Hong Kong, China

Keith Jones
Forestry Commission, Northwest England, UK

Nerys Jones
National Urban Forestry Unit, UK

W.A. Kenney
University of Toronto, Faculty of Forestry, Toronto, ON, Canada

J. James Kielbaso
Michigan State University, Department of Forestry, East Lansing, MI, USA

A.N. Lee
Seoul Women's University, Faculty of Environment and Life Sciences, Seoul, Korea

Chang Seok Lee
Seoul Women's University, Faculty of Environment and Life Sciences, Seoul, Korea

Junxiang Li
East China Normal University, Department of Environmental Science, Shanghai, China

Akira Miyawaki
Japanese Center for International Studies in Ecology (JISE), Yokohama, Japan

Marco Neef
University of Mainz, Department of Geography, Mainz, Germany

Rolf Schulzke
Hessian Ministry of the Environment, Rural Development and Consumer Protection, Wiesbaden, Germany; and WWF International, Caucasus Programme Office, Tbilisi, Georgia

Giacomo Secco
Department of Geography, University of Padua, Padua, Italy

Guojun Song
Renmin University, Environmental Institute of China, Beijing, China

Yong-Chang Song
East China Normal University, Department of Environmental Science, Shanghai, China; and Shanghai Key Laboratory for Ecology of Urbanization Processes and Eco-Restoration, Shanghai, China

Sebastian Stoll
Regional Council of Kassel, Forestry Division, Kassel, Germany

Yujie Wang
East China Normal University, Department of Environmental Science, Shanghai, China

Zongguo Wen
Tsinghua University, Department of Environmental Science and Engineering, Beijing, China

Rüdiger Wittig
Johann Wolfgang Goethe-University, Institute for Ecology, Evolution and Diversity; Department of Ecology and Geobotany, Frankfurt, Germany

Jianguo (Jingle) Wu
Arizona State University, School of Life Sciences and Global Institute of Sustainability, Tempe, AZ, USA

Wenyou Wu
Anhui Agricultural University, Institute of Forestry and Landscape, Hefei, China

Zemin Wu
Anhui Agricultural University, Institute of Forestry and Landscape, Hefei, China

Jun Yang
Department of Landscape Architecture and Horticulture, Temple University, Ambler, PA, USA

Kunmin Zhang
State Environmental Protection Administration, Beijing, China

Shaoujie Zhang
Anhui Agricultural University, Institute of Forestry and Landscape, Hefei, China

Jinxing Zhou
Chinese Academy of Forestry Science, Institute of Forestry Science, Beijing, China

Wayne C. Zipperer
U.S.D.A. Forest Service, University of Florida, Gainesville, FL, USA

Grazia Zulian
Department of Geography, University of Padua, Padua, Italy

I
Perspectives and Approaches in Urban Forestry

1
Introduction: The Growth of Cities and Urban Forestry

Margaret M. Carreiro

Background

In the last 100 years, there have been two very dramatic changes in human society. First, our global population quadrupled to its present 6.3 billion, and second, we have become an urban species (United Nations Department of Economic and Social Affairs/Population Division, 2004). In 1900, 86% of humanity lived in the rural countryside, interacting directly with the natural world on a daily basis. However, as of 2006 over 50% of us now live at densities greater than 625 people per square kilometer in cities containing more than 100,000 individuals. United Nations predictions hold that by 2050 nearly two thirds of the estimated world population of 9 billion people will live in cities. This demographic shift has occurred at an uneven pace throughout the world, with Western industrialized nations having experienced it earlier, accounting for the fact that today about 75% to 80% of Europeans and North Americans are already city dwellers. Therefore, the lion's share (90%) of the increase in urbanized humanity over the next few decades will occur in the developing nations, particularly those in Asia. Moreover, the number of large cities throughout the world is increasing rapidly. In 1900, there were only 19 cities with a million or more inhabitants. Today, there are over 400, with 564 projected by 2015 (United Nations Department of Economic and Social Affairs/Population Division, 2004). There now are 19 megacities, with 10 million or more people each, whereas just 20 years ago there were only eight.

Many of these cities are struggling to provide basic services such as drinking water, waste removal, sanitation, and shelter for their people. Other cities are not experiencing such crises as acutely, but nevertheless suffer chronically from vastly unequal allocation of resources and services to their citizens, hotter mesoclimates (urban heat-island effect), flooding, poor air and water quality, and intermittent water shortages that portend more serious scarcity in the future (Hinrichsen, 2002; Shao et al., 2006; Yang and Pang, 2006; Zhao et al., 2006). Cities, and therefore the majority of humanity, are becoming increasingly stressed by environmental and social factors that negatively impact our physical and psychological well-being. As the economist, John Kenneth Galbraith, stated: "The test of the quality of life in an advanced economic society is now largely in the quality of urban life. Romance

M.M. Carreiro et al. (eds.), *Ecology, Planning, and Management of Urban Forests: International Perspectives.*

may still belong to the countryside—but the present reality of life abides in the city" (quoted in Miller, 2002).

For the last 50 years there has been a growing realization that the solutions to most of these problems reside in making cities more efficient in their consumption of energy and materials and disposal of waste products, and in altering patterns of urban development to reduce the amount of impervious "gray" infrastructure (e.g., buildings and roads) and to increase the amount of "green" infrastructure, particularly trees. This realization has been expressed in the concepts of the eco-cities movement, adopted by many environmentalists and urban designers throughout the world (Register, 2002). How do we create cities that are more ecologically sustainable and resilient to fluctuations in internal and external environmental forces, and that provide healthier conditions so that people can not only exist but also thrive? For solutions to such challenges to urban quality of life, urban designers, planners, and managers are reexamining in more detail the many benefits and services we derive from natural and semi-natural habitats in and around cities.

This renewed appreciation for incorporating more nature into urban design has not occurred overnight, but has been evolving for over a century. Since the middle of the 19th century, our instinctive need for including plants in our cities resulted in several beautification movements in the United States and Europe that created public parks and gardens in many cities (Schmid, 1975; Konijnendijk et al., 2006). Landscape architects of that time considered these public green spaces important places for relaxation and recreation, and for increasing contact with nature, so obviously missing in the hard, paved "deserts" we had constructed for ourselves. These architects also advocated the creation of tree-lined avenues that added aesthetic grace and gentility to our residential areas. It is in these beautification movements that we can trace the beginnings of present-day urban forests in Europe and North America. However, by the late 1970s some forestry professionals in the U.S. and Canada realized that trees and other woody vegetation in cities provide more than social amenities (Konijnendijk et al., 2006). While there are costs associated with planting and managing urban trees, they also supply us with many environmental and economic benefits, including improving air and water quality, reducing noise pollution, controlling floods, preventing soil erosion, cooling the urban heat island, reducing the energy required to cool and heat buildings, increasing real estate values, and in some cases increasing the supply of drinking water in ex-urban areas just beyond cities. Indeed, about 50 years ago the U.S. government decided that these "forests" were distinctive enough in their purpose and requirements so that a new division within the U.S. Forest Service, Urban Forestry, should be formed to study and manage trees and other woody vegetation in cities.

Providing Urban Forestry with its own political identity in turn stimulated the growth of the science of urban forestry and arboriculture in many universities throughout the world. After decades of research in many nations, we now possess a large body of scientific knowledge on the ecological, economic, and social roles of trees, woodlands, and other green spaces in and near cities. Professionals in different countries have also learned which planning and management strategies have worked best in their specific regions for the acquisition, restoration and maintenance of woodlands and other urban green spaces. There is now a pressing need to share, compare, and consolidate the diverse knowledge gained from studies conducted in different cities and countries, so that we can identify and improve those approaches and methods that may work in other different geographic, environmental and sociopolitical settings. The next logical step is to extend this knowledge and apply it to problems at increasingly larger spatial scales than heretofore considered. This involves addressing landscape and regional scale questions dealing with optimizing the spatial patterns of green infrastructure for defined functional priorities according to each city's needs and in the context of its immediate suburban and rural surroundings. The city's lifelines do not abruptly end at its municipal boundary, hence the need for a more holistic approach to redesigning our urban areas, linking them with their rural environment, and to considering the roles of urban forestry in such efforts.

Urban forestry is now poised to "go global" with the overt and ambitious intent of using trees and natural to semi-natural habitat patches for ameliorating the negative environmental effects of urbanization, and for contributing to the monumental, long-term mission of creating more livable, ecologically sustainable eco-cities. With this in mind, leading scientists in China took the initiative to invite urban forestry experts from approximately a dozen countries to share their insights and knowledge during the International Symposium on Urban Forestry and Eco-Cities held in Shanghai, China in September 2002. This book contains contributions from experts in Asia, Europe, and North America, most of whom attended this conference to share the diverse plans and studies in urban forestry occurring in their respective countries. Contributions from our Asian colleagues, particularly the Chinese, provide Westerners with an astonishing alternative view of the larger spatial extent and differing circumstances under which they are boldly rebuilding their urban forest infrastructure in world-class megacities like Shanghai and Beijing. However, this book is not simply about planning and design, but about science and management. It contains studies and perspectives on urban forests from a broad array of basic and applied scientific disciplines including ecosystem ecology, biogeochemistry, landscape ecology, plant community ecology, geography, and social science. In addition, we hope that readers will profit from the diversity of international perspectives and case studies contributed by academic and governmental experts in management, planning, and restoration. Examples of how science has infused practice and how practice has informed science are plentiful in this book. In addition, we hope that the studies provided will help motivate more scientists, planners, and managers to work together and to adopt a broader landscape ecology approach to urban forestry, and, in so doing, better address the pressing needs for improving the quality of life in their respective cities.

Scope of Book

This book provides multicultural and multidisciplinary perspectives and information on the roles, planning, management, and restoration of urban forests from experts in Asia, North America, and Europe. The book is divided into three parts: Part I,

Perspectives and Approaches in Urban Forestry; Part II, Planning, Managing, and Restoring Urban Forests; and Part III, Synthesis and Directions for Future Research, Planning, and Implementation.

Part I focuses on the broad array of approaches to the study and management of urban forests, reflective of the varied histories and national contexts of these cities. In Chapter 2, Wu describes an overarching, sociobiological model for the study of urban ecosystems (also see Wu, 2006), and challenges us to enlarge the scope of urban forestry by placing it into a landscape ecological and regional framework required if the urban forestry is to contribute to the creation of eco-cities. In Chapter 3, Wittig describes the conceptual foundation for defining and establishing an eco-city, and in Chapter 4, Zhang and colleagues compare and evaluate different indicators for monitoring progress in improving the ecological soundness of cities and quality of life for their citizens.

Since urban forests incur maintenance costs, an accounting of their ecological and social benefits in monetary terms is also urgently needed for decision makers to balance trade-offs of particular urban development scenarios involving green space and trees.

To promote such understanding, Chen and Jim in Chapter 5 review the literature on the ecological services provided by urban green infrastructure, and Heidt and Neef in Chapter 6 discuss how urban green space can ameliorate the negative socioeconomic impacts of the heat-island effect and air pollution specifically. In Chapter 7, Zipperer explores how the practice of urban forestry, often involving decision making at the individual tree level, can benefit by applying an ecosystem perspective, and thereby improve the benefits and reduce the costs of urban forest maintenance. In Chapter 8, Nerys Jones summarizes the history of urban forestry in Great Britain and its legacy in remediating the negative impacts of industrialization on vegetation and people, and describes the management approaches that have met with best success in that country.

In Chapter 9, Jim shifts our attention to developing nations, like China, and focuses on diverse, but systematic, planning strategies for maximizing greening along a gradient from densely settled compact cities to peri-urban areas. In addition, he suggests ways to overcome institutional inertia concerning urban greening, particularly in developing nations. We continue this focus on urban forestry in China in Chapter 10, in which Song and colleagues provide a timeline for the recent and rapid development of the field of urban ecology and forestry in China, and offer a detailed view of the current status of ecological planning in Shanghai. This chapter includes an overview of Shanghai's plans for improving urban green infrastructure and water quality and its socioeconomic plans for starting and maintaining eco-communities and eco-industrial parks as steps toward developing circular economies and achieving local and regional sustainability goals. This chapter ends with recommendations for the types of urban ecological studies that would be most profitable for applied and basic scientists to pursue so that relationships between nature and people can be "harmonized."

While most of the chapters in this book concentrate on the management and modification of entire urban landscapes, Chapter 11 by Carreiro describes how one approach (the urban–rural gradient approach) can be used to study how urban and suburban land use might affect the biogeochemical and ecological functioning of a single habitat type, natural forest remnants, that may exist in urban and urbanizing landscapes. Part I ends with Chapter 12, in which Miyawaki imparts insights from decades of experience in restoring and constructing natural and semi-natural forest habitats in cities throughout east Asia. His reforestation experiments have demonstrated that developing partnerships among citizens, private and government agencies, and scientists in these restoration efforts is critical to the long-term success of these constructed habitats as well as the promotion of the eco-cities movement in Asia.

Part II emphasizes case studies in planning, managing, and restoring urban forests in cities throughout the world. In Chapter 13, Keith Jones describes a Geographic Information Systems (GIS)-based, tool (the Public Benefit Recording System) for identifying and prioritizing areas of degraded land for woodland restoration that will maximize socioeconomic and environmental benefits. The Public Benefit Recording System has been used successfully to guide strategic planning and restoration efforts in urban forestry in North West England and provides a model that is adaptable to any city throughout the world. In Chapter 14, Li and colleagues discuss the ecological roles performed by various types of landscape corridors in the Shanghai metropolitan area. In addition they describe efforts there in planning and implementing a multifunctional green corridor network to improve air and water quality, and to promote biological conservation and nature appreciation by the city's people.

Chapters 15 to 19 focus on management issues that have arisen in urban forests of different types in different countries. In Chapter 15, Kielbaso reviews survey trends on tree stocking, tree values, and societal benefits in urban forests in the United States. A checklist of criteria for evaluating the soundness of a city management program in urban forestry is also included. In Chapter 16, Chen and Jim report findings of their study of tree communities comprising the urban forest of Nanjing, China. They discuss the relationships of tree species density, composition, size structure, and performance to residential neighborhoods, industrial areas, roadsides, garden parks, and institutional sites. To improve management of the city's urban forest, they also conducted a management survey for various agencies involved in tree planting and maintenance. This permitted an assessment of the quality of urban tree care and allowed them to identify responsibility and communication gaps between agencies and between agencies and the landscaping industry that would improve the quality of Nanjing's urban forestry program. In Chapter 17, Wu and colleagues describe the community structure and distribution of urban trees and forest patches in Hefei, China. In addition they also provide information on ecosystem-level measures of forest structure such as tree biomass distribution, leaf area, and leaf area indices, and finish the chapter with recommendations for urban forest improvement in Hefei.

In Chapter 18, we shift to an exploration of the issue of multipurpose forestry in Germany. Schulzke and Stoll provide a brief historical review of traditions and laws that affected the growth and maintenance of forests surrounding cities in the German state of Hesse. Initially managed to provide timber as a commodity, these forests are now being managed for watershed protection and recreation as well. The authors suggest strategies for reducing conflicts arising from multipurpose forestry, and discuss how these peri-urban forests can connect a city and its rural surroundings into a more integrated system. Based on his experiences in Germany, Jestaedt in Chapter 19 follows with detailed recommendations for managing forests in and near cities primarily for public recreation and provides information on specialized planning needs, management challenges, and silviculture. In Chapter 20, Secco and Zulian combine geographic, landscape ecological, and sociological perspectives and approaches in evaluating the benefits of parks to city inhabitants. Their method and evolving model are being tested and implemented in cities and towns in Italy to improve planning for public green spaces by attempting to optimize the distribution and allocation of park resources relative to neighborhood demography, distance measures, and use of parks by people.

Chapters 21 and 22 describe strategies and tools for managing urban forests. Kenney describes in Chapter 21 an improved approach for urban forest managers who wish to determine the current structure and distribution of their urban forests. Kenney demonstrates that the three-dimensional leaf area index (LAI) and potential leaf area index (PLAI) are more informative than two-dimensional canopy cover as a measure of urban forest mass and applies his approach to the city of Toronto, Canada. The use of LAI and PLAI can be linked more directly to many social benefits derived from urban forests, such air pollution filtration, carbon sequestration, and thermal buffering, and should assist planners and managers in improving the protection and growth of urban forest resources. In Chapter 22, Yang and colleagues describe their use of satellite remote sensing imagery from Landsat 5 and 7 to determine the spatial and temporal changes in urban vegetation cover that occurred in Beijing, China, between 1991 and 2002, a period of accelerated growth in this city. In addition to providing technical information on the use of this tool to identify hot spots of vegetational change, they also describe how the misuse of simple indices, such as total vegetation cover data, by decision makers appears to be linked with a decline in urban vegetation in inner portions of the city where human density and need for green space is greatest. Remote sensing studies that examine spatial and temporal vegetation change can therefore constitute an informational feedback mechanism so that decision makers can try to redress unintended shifts in allocation of green cover that are not consistent with larger-scale policy goals of achieving ecologically sustainable cities.

Part II concludes with four case studies involving restoration experiments in urban forestry from Germany, Korea, and China (Chapters 23 to 26). Due to its large coal deposits, the Ruhr Valley has long been a region of intense industrial activity in Germany. Coal-mine abandonment over time has therefore created a landscape of mine waste heaps quite close to densely populated areas. In Chapter 23, Haeupler reports on vegetational studies that have explored the potential for these heaps to become revegetated so as to support greater biological diversity, recreational activities, and nature appreciation. By using sites along a chronosequence of abandonment, Haeupler assesses the relative success these reclamation efforts have had thus far in accelerating plant successional processes. In Korea, Lee and colleagues have also been attempting to restore forests in habitats degraded by nearby industrial complexes. In Chapter 24, they describe their rationale and approach for identifying native tree species that are most tolerant of the polluted and altered soil conditions near these complexes. The results of their restoration trials lead them to suggest that success occurs when the conditions and regional characteristics of each site are considered, and they argue against a uniform formulaic approach to restoration, especially in polluted locations. In Chapter 25, Lee and colleagues provide restoration plan recommendations for the city of Seoul, Korea, based on their mapping and sampling studies of soil properties and vegetation. In Chapter 26, Da and Song describe their adaptation of the Miyawaki afforestation method (described in Chapter 11) in accelerating the growth of near-natural forest patches containing the potential natural vegetation for the region throughout Shanghai, China. Their efforts are part of larger plans for improving environmental quality of life in the city, and moving Shanghai closer toward developing in a more sustainable manner.

In Part III, Chapter 27, the editors provide a synthesis and suggest directions for future research in urban forestry by comparing and contrasting varied national needs and perspectives described in the book. Unifying themes that emerge from these diverse contributions are emphasized.

References

Hinrichsen, D., Salem, R., and Blackburn, R. (2002) Meeting the Urban Challenge: Population Reports, Series M. Report No. 16. The Johns Hopkins Bloomberg School of Public Health, Population Information Program, Baltimore, MD.

Konijnendijk, C.C., Ricard, R.M., Kenney, A., and Randrup, T.B. (2006). Defining urban forestry—a comparative perspective of North America and Europe. Urban Forestry & Urban Greening 4:93–103.

Miller, G.T. (2002) Living in the Environment, 12th ed. Brooks/Cole, Pacific Grove, CA.

Register, R. (2002) Ecocities: Building Cities in Balance with Nature. Berkeley Hills Books, Berkeley, CA.

Shao, M., Tang, X.Y., Zhang, Y.H., and Li, W.J. (2006) City clusters in China: air and surface water pollution. Frontiers in Ecology and the Environment 4:353–361.

Schmid, J.A. (1975). Urban Vegetation. Report No. 161. Department of Geography, University of Chicago, Chicago, IL.

United Nations Department of Economic and Social Affairs/Population Division. (2004) World Population Prospects: The 2004 Revision, Volume III: Analytical Report. United Nations, New York.

Wu, J. (2006) Landscape ecology, cross-disciplinarity, and sustainability science. Landscape Ecology 21:1–4.

Yang, X.L., and Pang, J.W. (2006) Implementing China's "Water Agenda 21". Frontiers in Ecology and the Environment 4:362–368.

Zhao, S.Q., Da, L.J., Tang, Z.Y., Fang, H.J., Song, K., and Fang, J.Y. (2006) Ecological consequences of rapid urban expansion: Shanghai, China. Frontiers in Ecology and the Environment 4:341–346.

2
Toward a Landscape Ecology of Cities: Beyond Buildings, Trees, and Urban Forests

Jianguo (Jingle) Wu

Human population growth and urbanization are two dominant demographic trends in our time (Brown, 2001). World population has continued to grow exponentially for the past several decades, and reached 6.2 billion in 2002, with a current annual increase rate of almost 80 million (Earth Policy Institute, 2002). The proportion of the total world population that is urban was only a few percent in the 1800s, but it increased to 14% by 1900, rapidly jumped to about 30% in 1950 (Platt, 1994a; Wu and Overton, 2002), and is passing 50% now. Evidently, as the world's human population has increased exponentially, so has the proportion of people living in cities (Fig. 2.1). It has been projected that 60% of the world's population will reside in urban areas by 2025 (Platt, 1994a). In 1800, there was only one city, Beijing, in the entire world that had more than a million people; 326 such cities existed 200 years later (Brown, 2001). The urban population is growing three times faster than the rural population (Nilsson et al., 1999), and we are now witnessing a historically unprecedented and monumental, global-scale, rural-to-urban transition. To quote Lester Brown (2001), "For the first time, we will be an urban species!"

At a more regional scale, urban people already account for more than two thirds of the European population today. In the United States, 74% of the population resided in urban areas in 1989, and this number will increase to more than 80% by 2025 (Pickett et al., 2001). The historical record so far has shown that both the number of mega-cities as well as the number of urban dwellers have increased much faster in developing countries than in developed countries. For example, nearly 40% of the population of the Asia-Pacific region is now urban, and the region contains 13 of the 25 largest cities of the world. It has been estimated that by 2015 about 903 million people in Asia will live in cities with a population of over one million people (cf. Wu and Overton, 2002). While the world's urban population is projected to rise to 60% by 2025, nearly half of these people will reside in the Asia-Pacific region. Undoubtedly, urbanization will continue to have significant impact on the environment as well as on economic, social, and political processes at local, regional, and global scales.

Urbanization has profoundly transformed many natural landscapes throughout the world, and contributed significantly to the current crisis of biodiversity loss and deterioration of ecosystem services. Although cities cover less than 2% of the

M.M. Carreiro et al. (eds.), *Ecology, Planning, and Management of Urban Forests: International Perspectives.*

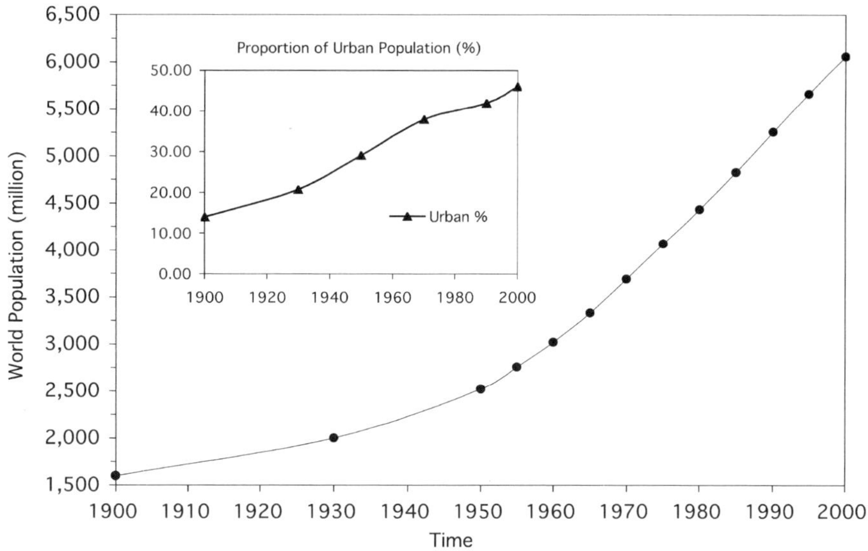

Fig. 2.1 Increase in the total world population and the proportion of the urban dwellers in the 20th century (1900–2000). Data were from United Nations (2001), Platt (1994a), and World Resources Institute (1998)

earth's land surface, they account for 78% of carbon emissions, 60% of residential water use, and 76% of the wood used for industrial purposes (Brown, 2001). About half of the world's nitrogen fixation is mediated by humans (Galloway, 1998), and the ecological impacts of urbanization in terms of biodiversity, biogeochemistry, and ecosystem services go far beyond the city limits. Also, rapid urbanization since the 1990s has been accompanied by a proliferation of slums and dysfunctional neighborhoods with high health risks, especially in most developing countries. High rates of urbanization and industrialization have increased the demands for land, water, and energy, and resulted in expanding transportation networks that constitute a key accelerating factor in economic growth as well as environmental degradation. Urbanization in many countries has resulted in air and water pollution, loss of productive agricultural land, loss and fragmentation of species habitats, overextraction of groundwater resources, and deforestation as a consequence of increased demand for construction timber. The most serious air pollution problems often occur in urban areas. A survey by the World Health Organization (WHO) and United Nations Environment Program found that the levels of suspended particulate matter (SPM) in 10 of the 11 cities they examined were two times higher than WHO's guidelines for protecting human health. It is important to realize that the ecological influences of cities go far beyond the space they occupy. Urban ecological footprints can be enormous because of their huge demands for energy, food, and other resources, and the regional and global impacts of their wastes and emissions on soil, air, and water (Wackernagel and Rees, 1996; Rees, 1997; Luck et al., 2001; Wu and Overton, 2002). For example, London's population consumes some 55,000 gallons of fuel and some 6600 tons of food, and emits 160,000 tons of carbon dioxide (CO_2) every single day. Such consumption requires a land base 12.5 times the size of London to support its population (Beatley, 2000). Vancouver's ecological footprint was estimated as being 180 times that of its city size (cf. Collins et al., 2000).

Clearly, cities are places where people are most concentrated, and where environmental problems are most devastating. Although there are apparently a myriad of political, socioeconomic, and environmental causes and consequences of urban problems, it is certain that to alleviate these problems our cities must be designed, planned, and managed in a more ecologically sound manner. Up until now, urbanization has, for the most part, increasingly isolated humans from nature through artifacts and technology. But it is clear that if an agreeable human quality of life is to be sustained in urban systems, then the ecological state of its natural components must be improved and harmony between people and nature must be set as a goal. In short, sustainable cities are most likely to be ecologically sound cities—eco-cities. To achieve the ecological integrity of cities, urban forests and other types of green spaces are critically important, and they must be explicitly and adequately considered in the design, planning, and management of urban systems. This chapter reviews some of the changing perspectives and approaches in urban ecology, and outlines several key concepts and principles in landscape ecology that are relevant to the research and practice of urban forestry and the development of eco-cities.

Urban Forests and Their Values

The urban forest usually refers to all woody plants in and around the city, including street trees, yard trees, park trees, and planted or remnant forest stands (Miller, 1997; Helms, 1998; Konijnendijk, 1999). Many studies have documented that urban forests may have a number of ecological/environmental, economic, and sociocultural benefits. For example, urban forests can improve air quality by absorbing particulates and pollutants (e.g., ozone, chlorine, sulfur dioxide, nitrogen dioxide, fluorine), sequester atmospheric CO_2, reduce soil erosion and purify water, serve as habitats for plants and animals, alleviate noise pollution, moderate local/regional climate to save energy consumption (i.e., reducing urban temperature in summer and heat loss in winter), increase real estate values, improve neighborhood and landscape aesthetics, and enhance the psychological well-being of urbanites (Burch and Grove, 1993; Platt et al., 1994; Miller, 1997; Kennedy et al., 1998; Nilsson et al., 1999).

Some of the ecological and socioeconomic values of urban forests are quite impressive, and may even sound astounding to traditional ecologists. For example, according to a report by the United States Department of Agriculture's (USDA) Center for Urban Forest Research (USDA/CUFR, 2002), parking lot trees in Davis, California, reduced the surface temperatures of asphalt by as much as 20°C (36°F), and cabin temperatures of vehicles by over 26.1°C (47°F). The parking lot trees in Sacramento, California, with an overall 8.1% effective shade area, generated annual benefits of $700,000/year, and increasing the shade to 50% will boost the benefits to $4 million/year (McPherson et al., 1999; McPherson, 2001; USDA/

CUFR, 2002). Data from 31 California cities showed that air temperature was warming due to the urban heat-island effect at a rate of 0.4°C (0.72°F) per decade since 1965 (Akbari et al., 1992), while the increase rate of downtown temperatures for the entire United States has varied from 0.14° to 1.1°C (0.25° to 2°F) per decade since the 1950s (McPherson, 1994). This urban warming had direct economic and energy use consequences. McPherson (1994) estimated that about 3% to 8% of electric demand in the U.S. was used to compensate for the urban heat-island effect. A cost-benefit analysis of energy-efficient landscaping with trees in Tucson, Arizona, estimated that the net benefits for planting 500,000 trees was $236.5 million for a 40-year planning horizon; computer simulations projected that an additional 100 million mature trees in U.S. cities could save 30 billion kilowatt-hours of energy for heating and cooling, and consequently reduce CO_2 emissions by as much as 8 billion kilograms (8 million metric tons) per year (cf. McPherson, 1994).

Urban forest benefits are not just economic. The following classic example demonstrates the psychological and health-improvement value of urban forests. Ulrich (1984) examined the records for 1972 to 1981 for recovery of 46 patients after gallbladder surgery in a suburban Pennsylvania hospital to determine whether a window view with or without trees might have any restorative influences. The results showed that the 23 patients who could view a small stand of deciduous trees from their room windows had significantly shorter hospital stays, received fewer negative evaluative comments in nurses' notes, and took fewer painkillers than the other 23 who had windows facing a brown brick wall. Wilson (1984) and Kellert and Wilson (1993) argued that people, when isolated from nature, will suffer psychologically, which may lead to a measurable decline in well-being—the biophilia hypothesis. Other empirical studies corroborate this hypothesis (Roszak et al., 1995; Brown, 2001). Given all these measurable social and economic benefits, urban forests (and all urban green spaces) should be properly maintained, planned, and managed. However, all the ecological and socioeconomic functions have not been well studied by scientists, and are not well known to the public. Consequently, municipal budget allocations to green space and urban forestry are often smaller than needed for their maintenance.

To enhance more integrative research and promote values of urban forestry, it is necessary to broaden the concept of urban forestry. Urban forestry is closely related to "community forestry" and "social forestry" (Miller, 1997; Nilsson et al., 1999). Traditionally, the study of urban forests has focused primarily on local-scale and applied issues (Konijnendijk, 1999), and urban forests are often managed as individual trees instead of from the perspective of a whole forest ecosystem (University of Florida/Institute of Food and Agricultural Sciences, 2001). However, since any urban environment is extremely heterogeneous in space and dynamic in time, and since areas containing urban trees and forest patches are often geographically fragmented, an urban forest may be most appropriately treated as a landscape that consists of a variety of changing and interacting patches of different shape, size, and history. Urban trees and forests are integral parts of this urban landscape—a dynamic patch mosaic system. As a science of the relationship between spatial heterogeneity and ecological processes, therefore, landscape ecology provides many useful concepts and principles for urban planning and design in general and for urban forestry in particular, as will be explained below.

Changing Perspectives in Urban Ecology

A major goal of urban ecology is to understand the relationship between the spatiotemporal patterns of urbanization and ecological processes. Thus, the study of urban morphology and its evolution is critically important. As early as 1825, the German economist von Thünen asserted that the urban morphology of an isolated city would be characterized by concentric economic rings (e.g., business, residential, industrial, agriculture), as dictated by simple cost-benefit relations (the principle of marginal spatial utility; cf. Portugali, 2000). Von Thünen's work laid an important foundation for the theory of urban development, including the concentric zone theory and the central place theory, which depict cities as more or less concentric or symmetric structures with one or more central business districts (CBDs). In contrast with the concentric-ring models, the sector theory allows for corridors or wedges of industrialization due to the influence of transportation networks. The multiple nuclei theory recognizes the multiple centers of specialized activities (e.g., finance, industry, commerce, residence) and describes an asymmetric patch mosaic pattern. These theories of urban forms are commonly found in textbooks in social sciences, and represent the exceptions rather than the norm when applied to real cities. In particular, the concentric zone theory, the sector theory, and the multiple nuclei theory were developed based primarily on studies of American cities (Chicago, San Francisco, and Boston, respectively) several decades ago, and thus they are less applicable to cities in other countries or even to most young American cities (Thio, 1989).

Cities may differ drastically in their architectural appearance and environmental settings, but one commonality is that the diversity and spatial arrangement of their landscape elements undoubtedly affect and are affected by physical, ecological, and socioeconomic processes within and beyond their boundaries. Ecologists have long studied the effects of spatial pattern of urbanization on ecological processes (Stearns and Montag, 1974; Sukopp, 1990, 1998; Loucks, 1994; Breuste et al., 1998; Zipperer et al., 2000). In fact, urban ecological studies date back several decades ago when botanists, notably of the Berlin school of urban ecology (Sukopp, 1990, 1998), documented the spatial distribution of plants in and around cities. In contrast, the Chicago school of urban ecology defined the field as the study of the relationships between people and their urban environment by applying concepts developed in plant and animal ecology, most prominent of which are concepts of dominance, competition, invasion, and succession (Thio, 1989). Apparently, this view of urban ecology is a subdiscipline of social or human ecology and focuses more on people rather than on biological organisms and their organization within cities.

Based on the degree of emphasis and reliance on biological ecology as well as conceptual and methodological frameworks, I distinguish five urban ecological approaches (Fig. 2.2). These approaches are essentially developed from three broad

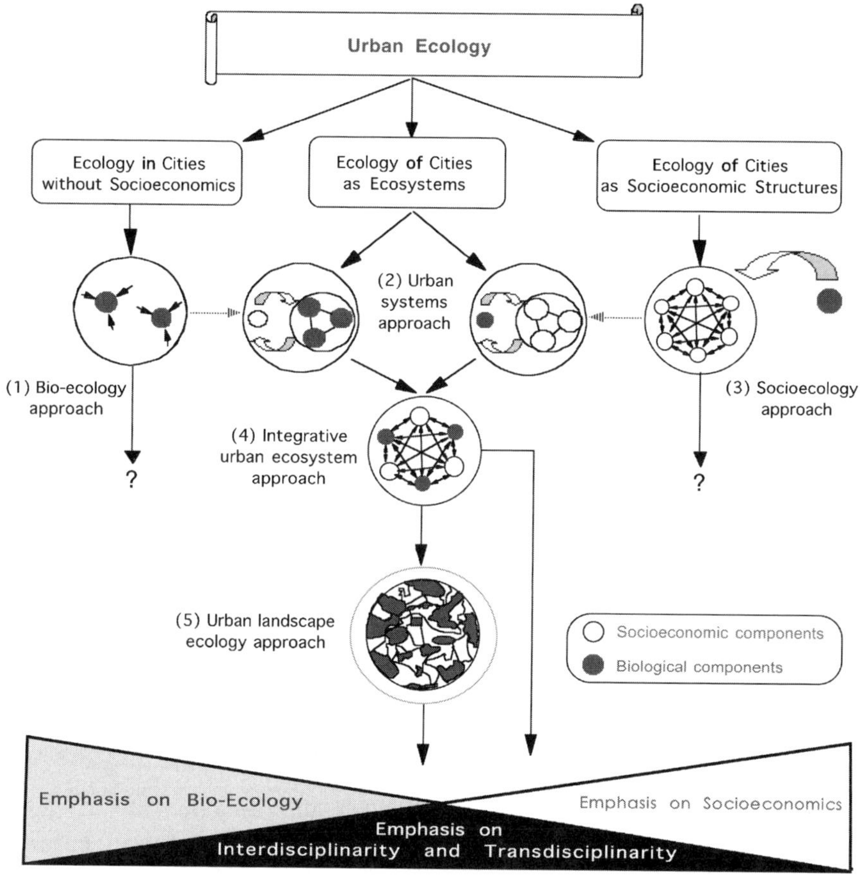

Fig. 2.2 Development of different perspectives in urban ecology. In general, there has been an evolution of perspectives from the ecology *in* cities to the ecology *of* cities, from isolated organismal to landscape studies, and from disciplinary investigations to interdisciplinary integration. See text for more detail

perspectives on urban ecology: ecology in cities (the first approach), ecology of cities as socioeconomic structures (the second approach), and ecology of cities as ecosystems (the third to fifth approach). The first approach focuses solely on the ecology of plants and animals living in urban areas, assuming that this can be accomplished without explicitly considering socioeconomic causes and consequences. This approach leads to what may be called the bio-ecology perspective (Fig. 2.2). In sharp contrast, the second approach treats cities as socioeconomic structures or organizations. It tackles complex urban social and economic patterns and processes by applying some concepts and principles from biological ecology, while, ironically, biological organisms and their associations (populations and communities) within cities are overlooked. This approach leads to the so-called socioecology perspective (Fig. 2.2). Obviously, both of these approaches capture only certain components of the urban system, but neither of them singly is adequate to understand the city as a society–nature interactive system where components affect each other.

The third approach considers the city as an urban system that is composed of both socioeconomic and biological components (Fig. 2.2). While this approach seems to combine some of the elements in the previous two approaches, it is characterized mainly by the systems methodology that emphasizes causal relations, feedback, and various interactions among system components. This urban systems perspective focuses either on socioeconomic dynamics (e.g., Forrester, 1969) or ecological processes (e.g., Stearns and Montag, 1974). Although both ecological and socioeconomic components are recognized here, they are not well balanced and integrated. Further integration between the bioecology and socioecology perspectives and between human ecology and ecosystem ecology has led to the fourth approach, the integrative urban ecosystem approach (Fig. 2.2). An example of this is Zev Naveh's total human ecosystem (Naveh and Lieberman, 1984). This is really an urban ecosystem perspective in that it treats both the biological and socioeconomic components of the city as equally important and in an integrative rather than divisive manner (also see Pickett et al., 1997). Finally, over the past two decades with the acutely growing awareness of the importance of considering spatial heterogeneity and its ecological consequences for understanding system processes, the urban landscape ecology approach has emerged (Fig. 2.2). This landscape approach emphasizes not only the diversity and interactions of the biological and socioeconomic components of the city, but also the spatial pattern of these elements and their ecological consequences from the scale of small patches to that of the entire urban landscape, and to the regional context in which the city resides (Pickett et al., 1997; Zipperer et al., 2000; Luck and Wu, 2002; Wu and David, 2002). Several contrasting characteristics of these different perspectives and associated approaches are summarized in Table 2.1.

Urban planning and design also seem to have experienced a paradigm shift in the past one-and-a-half centuries. For example, Platt (1994b) provided a lucid discussion on how the concepts of open space in North American cities have evolved in relation to urban design and planning. The "Picturesque Rurality" favored "the establishment of large, lavishly planted urban parks," but "put less emphasis on functional utility than on aesthetic effect through landscape design and horticulture"; the "City Beautiful" monumentalism "emphasized large, geometric plazas embellished with fountains, statuary, and formal landscaping;" the "Garden City" notion advocated having open spaces of different forms (e.g., practical community parks and individual garden plots) as major elements of the city and throughout the core of the city (Platt, 1994b). Although the City Beautiful and Garden City were among the most influential paradigms in urban design and planning, it is evident that modern urban designing and planning principles have moved beyond an initial focus on city form and human interests. Efforts by urban planners, designers, and architects to combine urban morphology with ecological functioning and efforts by ecologists to integrate the "ecology in cities" with

Table 2.1 Different perspectives on urban ecology, corresponding research approaches, and their major characteristics

Perspectives on urban ecology	Ecology in cities without socioeconomics	Ecology of cities as ecosystems	Ecology of cities as socioeconomic structures
Approaches to studying urban ecology	• Bioecology approach	• Urban systems approach • Integrative urban ecosystem approach • Urban landscape ecology approach	• Socioecology approach
Major characteristics	• Urban areas disturbed as environment	• Cities as unique ecosystems	• Cities as socioeconomic systems
	• Basic ecology in urban environment	• Humans as integral components of landscape systems	• Humans as the primary or the only system components
	• Humans as disturbance agents	• Consideration of both ecological and socioeconomic patterns and processes	• Ecological principles and methods used only as metaphors
	• Spatiotemporal patterns of organisms and human influences	• Problem-solving and solution-driven research	• Dominated by methodologies developed in social sciences
	• Non–solution-driven research	• Strong interdisciplinary interactions between natural and social sciences	• Little cross-disciplinary interactions between natural and social sciences
	• Little cross-disciplinary interactions between natural and social sciences		

Note: See Fig. 2.2 for a schematic representation of how these different perspectives and approaches evolve and relate to each other.

socioeconomic patterns and processes have brought both sides much closer to a common perspective—a landscape ecological perspective of cities.

In the next section, I shall discuss the major elements of landscape ecology and explore how landscape ecological principles may be used for improving the research and practice of urban forestry.

A Landscape Ecology Perspective on Cities

What Is Landscape Ecology?

Landscape ecology is the science and art of studying and influencing the spatial pattern of landscapes and its ecological consequences. The "science" of landscape ecology provides the theoretical basis for understanding the formation, dynamics, and ecological effects of spatial heterogeneity, and the relationship between landscape pattern and ecological and socioeconomic processes over different scales in space and time. The "art" of landscape ecology reflects the humanistic perspectives necessary for integrating biophysical and socioeconomic and cultural components within the landscape in general, and landscape design, planning, and management in particular. The term *landscape ecology* was coined by Carl Troll (1939), a German geographer. Before the early 1980s, landscape ecology was essentially a regional applied science, practiced mainly in Europe and focusing on land planning and human–ecosystem interactions (Naveh and Lieberman, 1984). The globalization of landscape ecology started with a series of publications in North America (Forman and Godron, 1986; Moss, 1988; Turner, 1989; Turner and Gardner, 1991). In the past two decades landscape ecology has experienced unprecedented rapid development in both theory and applications, and established itself as both a field of study and a new ecological paradigm (Wu and Loucks, 1995; Wu, 2000).

Based on the views of a group of leading landscape ecologists, Wu and Hobbs (2002) summarized six key issues that define the scope of landscape ecology: (1) interdisciplinarity or transdisciplinarity, (2) integration between basic research and applications, (3) conceptual and theoretical development, (4) education and training, (5) international scholarly communication and collaborations, and (6) outreach and communication with the public and decision makers. The terms of *interdisciplinarity* and *transdisciplinarity* have been defined variously in the literature, but I find the definitions summarized by Tress et al. (2005) both clear and satisfactory. Interdisciplinary research involves multiple disciplines that have close cross-boundary interactions to achieve a common goal based on a concerted framework, thus producing integrative knowledge that cannot be obtained from disciplinary studies. On the other hand, transdisciplinary research involves both cross-disciplinary interactions and participation from nonacademic stakeholders or governmental agencies guided by a common goal, thus producing integrative new knowledge and uniting science with society (Tress et al., 2005).

The six key issues are all related to each other, and may be important to sciences other than landscape ecology. But the emphasis on beyond-bioscience interdisciplinarity and real-world problem solving is one of the several characteristics distinguishing landscape ecology from the traditional bioecological disciplines such as population or community ecology. Because the structure and functioning of landscapes are influenced by a myriad of physical, biological, socioeconomic, cultural, and political forces, the ecology of landscapes must be interdisciplinary. This is necessary for landscape ecology to provide the scientific basis for resource management, land use planning, biodiversity conservation, and other broad-scale environmental issues. The same group of landscape ecologists also identified a list of top research topics in the field: (1) ecological flows in landscape mosaics; (2) causes, processes, and consequences of land use and land cover change; (3) nonlinear dynamics and landscape complexity; (4) scaling and uncertainty analysis; (5) methodological development; (6) relating landscape metrics to ecological processes; (7) integrating humans and their activities into landscape ecology; (8) optimization of landscape pattern; (9) landscape conservation and sustainability; and (10) data acquisition and accuracy assessment (Wu and Hobbs, 2002).

In essence, landscape ecology is a highly interdisciplinary field of study that focuses on spatial patterning of landscape elements and its relationships to ecological processes on different scales in space and time. No matter which aspects of the landscape one may concentrate on, be they biophysical, socioeconomic, or both, the landscape ecological paradigm helps bring the phenomena into perspective by integrating pattern, process, scale, and hierarchy. The key issues and research topics seem equally relevant to the science and practice of urban forestry and ecological cities. In particular, I suggest that the several principles discussed below may be used to guide the planning, managing, and design of urban forests and eco-cities.

Landscape Ecological Principles for Urban Forestry and Eco-Cities

Hierarchy Theory of Landscapes

Landscapes are nested hierarchical systems in both structure and function (Miller, 1978; Haigh, 1987; Urban et al., 1987; Wu and Loucks, 1995; Wu, 1999; Bessey, 2002). A hierarchy or hierarchical system can broadly be defined as a partial ordering of interactive entities (Simon, 1973). In hierarchical systems, higher levels are characterized by slower and larger entities (or low-frequency events), and lower levels are characterized by faster and smaller entities (or high-frequency events). The upper level exerts constraints (e.g., as boundary conditions) to the lower level, whereas the lower provides initiating conditions to the upper (Wu, 1999). Hierarchy theory suggests that when one studies a phenomenon at a particular hierarchical level (the focal level, often denoted as level 0), the mechanistic understanding comes from the next lower level (level −1), whereas the significance of that phenomenon can only be revealed at the next higher level (level +1).

The urban forest clearly forms a nested spatial hierarchy: individuals trees, tree corridors (e.g., trees along streets and roads), and networks (e.g., trees around parking lots, residential and urban blocks), patches of different shape and size (e.g., trees as aggregates in parks or remnant or planted forest fragments), and the entire urban forest in and around the city that also includes other types of green spaces (e.g., lawns, golf courses, and shrub communities). Clearly, urban forest planning, management, and designing should not stop at the urban fringe. This hierarchical view suggests that a fuller understanding and appreciation of urban forests can be gained by considering them at multiple scales. We need to see the trees, the forest, the corridors, the patches, the urban landscape, and the regional context, as well as understanding the hierarchical linkages among all of them!

Pattern-Process Principle

An important principle in landscape ecology is that the spatial pattern affects and is affected by ecological processes, and that the relationship between pattern and process is scale dependent. Here "pattern" includes both the composition (e.g., the number and abundance of land cover types) and configuration (e.g., the shape and spatial arrangement of landscape elements) of the landscape. "Scale" refers to the grain size (e.g., the spatial or temporal resolution of an observation set) or the extent (e.g., the total area or time duration of a study). The role of scale is ultimately important for understanding the relationship between pattern and process. If the spatial pattern changes much more slowly than the process under consideration (e.g., regional topography versus population dynamics of an animal species), the pattern-process relationship is mostly one directional: pattern affects process. However, when pattern and process are within the same spatial domain and operate on similar time scales, the pattern-process relationship is interactive. For example, the fine-scale spatial pattern of species composition and biomass in a grassland is interactive with the grazing process by cattle. The pattern affects the grazing behavior, and grazing immediately modifies the pattern and creates new patterns. Of course, in the case of overgrazing, the pattern can be totally destroyed, and a relatively homogeneous degraded or even desertified land is left behind.

The pattern-process principle certainly has implications for urban forestry and eco-cities. For example, the large-scale patterns of geomorphology, hydrology, and socioeconomic factors in an urban area set constraints on ecological processes, and thus determine where urban forests may be best maintained or planted, but local soil conditions are more likely to determine how well individual trees grow. For a variety of ecological and socioeconomic purposes, it is not only the diversity and the total amount of urban trees and forests that are important, but also the shape and spatial arrangement of individual trees and forest patches. In addition, the planning and designing of urban forests and the city as a whole must consider the multiple and sometimes conflicting ecological and socioeconomic purposes at different scales.

Landscape Connectivity

Landscape connectivity refers to the degree of connectedness among landscape elements (patches, corridors, and matrix) of the same or similar type (e.g., forest habitats, lakes, or rivers). Landscape connectivity includes both structural and functional components. Structural connectivity measures how spatially connected landscape elements are, whereas functional connectivity measures how connected an ecological process (e.g., dispersal, nutrient dynamics) is in space over a certain time scale. Clearly, landscape connectivity is dependent on both the scale of observation and ecological processes under consideration. Even for the same landscape, its connectivity may vary radically when different processes are considered (e.g., beetle movement, bird flying, seed dispersal, fire spread). With the accelerating human dominance of the earth system, landscapes have been increasingly fragmented, and wildlife habitats have been reduced in the total amount and disconnected in spatial pattern. Thus, a central question in conservation biology and landscape ecology is how landscape connectivity of habitats affects biodiversity and ecosystem processes.

Landscape connectivity is closely related to the structural and functional attributes of corridors and networks (Forman, 1995). Corridors are linear landscape elements that may function as habitats (e.g., riparian ecosystems, vegetated corridors), conduits (e.g., vegetated strips, roads), filters/barriers (e.g., windbreaks, roads), sources (areas that give off materials), or sinks (areas that receive materials). Corridors of the same or similar types interconnect to form a network, whose functionality is determined by network density (the amount or abundance of corridors), network connectivity (the degree to which all corridors are connected), and network circuitry (the degree to which loops or circuits are present in the net) (Forman and Godron, 1986; Forman, 1995). In general, corridors are undoubtedly important landscape elements. But the exact role of corridors of a particular type can only be understood with respect to the species or ecological process under consideration and, again, these will change with scale. In the past decade, the concept of landscape connectivity in terms of corridors and networks has increasingly been applied in nature conservation, resource management, and land-use planning (Noss, 1987; Cook, 1991; Cook and van Lier, 1994; Poiani et al., 2000; Opdam et al., 2001).

Percolation theory has been particularly useful for understanding landscape connectivity both structurally and functionally (Gardner et al., 1987, 1992; With and Crist, 1995). Percolation theory is the basis for studying the flow of liquids through material aggregates. In the context of landscape ecology, percolation may refer to the spread of any process through connected structural elements across the landscape. The most intriguing feature of percolation theory is the existence of a critical density of landscape components at which landscape function abruptly changes (Green, 1994; Turner et al., 2001). For example, a model landscape in which habitat and nonhabitat pixels are randomly distributed essentially has no clusters spanning across the entire landscape before the total percent habitat cover reaches the critical density or percolation threshold of $P_c = 59.28\%$. However, once the threshold is approached or exceeded, the probability of forming spanning clusters jumps to 100%, implying that much of the landscape is functionally connected (Green, 1994; Turner et al., 2001). Thus, percolation theory suggests that there are connectivity thresholds that significantly influence the flows of energy, materials, and organisms across the landscape mosaics of various kinds. Empirical studies have shown that real landscapes, most of which are clumped, often have a lower critical density value than the theoretical one predicted by percolation theory, and that landscape connectivity is a function of both the structural interconnectedness and the behavioral or dynamic features of the phenomenon.

How does this knowledge inform our thinking about urban forests? Urban forests typically contain many scattered individual trees, narrow strips, and small patches. Simply put, they are highly fragmented and often geographically disconnected. To enhance the benefits that can be derived from their ecological and socioeconomic functions, it is important to maintain a proper degree of connectivity among the different components of the urban forest across a range of spatial scales. At the same time, it is important to bear in mind that increased connectivity may also promote the spread of exotic species, epidemics, and disturbances such as fires. Overall, the concepts and knowledge of connectivity, corridors, networks, and percolation thresholds developed in landscape ecology may be useful for planning, managing, and designing urban forests as well as eco-cities.

Metapopulation Theory

In fragmented landscapes, biological populations live in geographically distributed habitat patches. A metapopulation is a system of such local populations spatially separated by unsuitable environments but still functionally and genetically connected by dispersal. Thus, metapopulations integrate the structurally nested habitat hierarchy with functionally dynamic population processes. Two salient characteristics of metapopulations are frequent local species extinction at the habitat patch level and species recolonization at the habit patch mosaic (or landscape) level. Metapopulation theory predicts that species that are locally unstable can still persist at the landscape (or regional) scale if the connectivity among habitat patches is beyond some threshold value (Opdam, 1991; Wu et al., 1993). How exactly the spatial pattern of habitat patches and corridors affects the local extinction, regional recolonization, and eventually persistence of species is a central question of metapopulation dynamics (Hanski and Gilpin, 1997; Hanski, 1999; Opdam et al., 2001).

As mentioned earlier, urban forests are a hierarchical patch dynamic system (Wu and Loucks, 1995; Wu, 1999), and may be viewed as a metapopulation when the focus is on the population dynamics and species persistence of trees in vegetated habitats. This metapopulation view becomes even more appropriate and necessary when animal species are considered. Conceptually, this is a special case of the more general hierarchical perspective of urban forestry outlined above (see Hierarchy Theory of Landscapes).

Landscape Self-Organizing Complexity

Landscapes are complex spatial systems in which heterogeneity, nonlinearity, and contingency are the norm. Findings in the sciences of complexity and nonlinear dynamics suggest that spatially extended complex systems like landscapes are often self-organizing (Perez-Trejo, 1993; Lobo and Schuler, 1997; Aber et al., 1999; Phillips, 1999). Self-organization is the capacity of complex systems to develop and change internal structures spontaneously and adaptively in order to cope with or manipulate their environment (Cilliers, 1998; Levin, 1999). Self-organizing systems tend to increase their complexity in time, and are replete with emergent properties, phase transitions, and threshold behaviors. Several inferences have emerged from this self-organizing complexity perspective: (1) local interactions play a critical role in the formation of regional and global patterns, while large-scale factors set constraints; (2) the exact behavior of complex systems is inherently unpredictable; (3) the traditional system stability based on homeostatic equilibrium is unachievable; and (4) system metastability (or nonequilibrium resilience) is determined primarily by the system's internal diversity, flexibility, and adaptability in response to unpredictable environmental changes.

Cities and urban landscapes are prototypical examples of self-organizing complex systems that have a large number of diverse components interacting nonlinearly (Portugali, 2000). It is extremely difficult or impossible to precisely predict the ecological and socioeconomic future of such systems no matter how much information we have on them—a view that completely defies the traditional Newtonian determinism. But this does not mean that we cannot understand or even influence their dynamics. Urban forests are a part of the self-organizing and complex urban landscape, and their structure, function, and interactions with other landscape components will affect the landscape's behavior. As such, planning and design should aim to increase the entire system's ability to cope with environmental uncertainties and extreme events (e.g., floods, fires, and epidemics that are intensified by humans). Equally important is the realization that humans are also an affected component of the complex system, not just a source of disturbance. As the most active, and sometimes most powerful, agents in urban landscapes, we have important roles to play in shaping their dynamics. We cannot precisely predict the urban future, but we can certainly influence it through our actions.

Aggregate-with-Outliers Principle

Forman (1995) proposed a landscape planning principle, the aggregate-with-outliers principle, which states that "one should aggregate land uses, yet maintain corridors and small patches of nature throughout developed areas, as well as outliers of human activity spatially arranged along major boundaries." This principle accommodates several important landscape ecological attributes. In particular, intentional aggregation of large patches of natural vegetation protects aquifers and stream networks, provides habitats for large-home-range species and interior-requiring species, and maintains a more natural disturbance regime and a high degree of landscape connectivity. Landscapes with patches of variable sizes provide habitats for a range of species from specialists to generalists. While vegetated corridors can enhance species movements and landscape connectivity, the overall multiple-scale, heterogeneous planning promotes "risk spreading," genetic variation, and multipurpose socioeconomic activities (Forman, 1995). In addition, Dramstad et al. (1996) illustrated 55 more specific landscape ecology principles for landscape architecture and land-use planning. The preferred characteristics of patches, edges/boundaries, corridors/connectivity, and landscape mosaics are discussed for the purpose of conserving biodiversity, an increasingly important goal from the point of view of urban planners or policy makers (see Dramstad et al., 1996, for specific examples).

Discussion and Conclusion

We are witnessing a moment in human history when, for the first time, the majority of the global human population lives in urban areas. The plethora of environmental and socioeconomic problems that challenge most cities throughout the world suggests that our cities, in general, need to be designed, planned, and managed better so as to become more ecologically and socioeconomically sustainable. Indeed, a new urbanism has been called for, which is based fundamentally on promoting the ecological relevance and limits of urban design and planning (Beatley, 2000). Urban forestry is an important part of this endeavor. Urban trees and forests often form a hierarchy of patches from isolated individuals to networks of corridors and to relatively large and contiguous patches (which are not always managed by the same municipal or governmental agencies and departments—fragmented patches run by fragmented often undercommunicating agencies). Urban forests may function as an air/water purifier, a temperature modulator or energy saver, a soil stabilizer, a wildlife habitat, a noise barrier, a landscape beautifier, a real estate value booster, and even a psychological comforter! However, despite their large-scale ecological roles, urban forests have traditionally been studied and managed largely at local, rather than regional, scales.

From a landscape ecological perspective, in planning and designing urban forests and eco-cities, we must consider various levels of nested contexts and expand our thinking (1) beyond "trees" to consider their connections and interactions with higher levels of vegetation aggregates, such as forest patches, corridors, and networks; (2) beyond "forests" to consider how forest patches interact with other land-use/cover types in space and time within urban areas; (3) beyond "urban" to take into account the regional environmental context of the city and its influence on forested habitats; (4) beyond "science" (in the classic and narrow sense) to develop an interdisciplinary landscape ecology of cities that integrates science with planning, designing, and management practices; (5) beyond "now" to plan for long-term environmental and socioeconomic sustainability; and (6) beyond framing our thinking in terms of "homeostatic stability" so we can build "cities of resilience" that are capable of coping with surprises generated by the nonlinear interactions originating from inside and unpredictable environmental changes from outside the city.

To achieve these goals, I have argued that urban forests need to be viewed as an integral part of the urban landscape—a dynamic patch mosaic system. As such, a landscape ecological perspective is needed for urban forestry. Specifically, several principles can be used to guide the practice of urban forestry and planning, including hierarchy theory of landscapes, the pattern-process principle, landscape connectivity, metapopulation theory, landscape self-organizing complexity, and the aggregate-with-outliers principle. Of course, landscape ecology is only one of a number of ecological, environmental, and social sciences that are relevant to urban forestry and the realization of eco-cities. But I argue that the perspectives provided by landscape ecology provide a spatially explicit, interdisciplinary framework through which pattern and process within and across cityscapes can be related. They also facilitate the communications among scientists, practitioners, policy makers, and the public because concepts of pattern and process, connectivity and functionality, and hierarchical components and linkages are essential for both research in and the practice of urban forestry.

Acknowledgments This chapter is based on an invited presentation at the International Symposium on Urban Forestry and Eco-Cities, Shanghai, September 16–22, 2002. I would like to thank Margaret Carreiro and anonymous reviewers for helpful comments on the manuscript of this chapter. My research in urban landscape ecology has been supported by grants from U.S. Environmental Protection Agency (EPA) (Science to Achieve Results Program, R827676-01-0) and U.S. National Science Foundation (NSF) (DEB 97-14833/CAP-LTER and BCS-0508002).

References

Aber, J.D., Bugmann, H.K.M., Kabat, K.P., et al. (1999) Hydrological and biogeochemical processes in complex landscapes—What is the role of temporal and spatial ecosystem dynamics? In: Tenhunen, J.D., and Kabat, P. (eds.) Integrating Hydrology, Ecosystem Dynamics, and Biogeochemistry in Complex Landscapes. Wiley, Chichester, pp. 335–355.

Akbari, H., Davis, S., Dorsano, S., Huang, J., and Winnett, S. (1992) Cooling Our Communities: A Guidebook on Tree Planting and Light-Colored Surfacing. U.S. Environmental Protection Agency, Washington, DC.

Beatley, T. (2000) Green Urbanism: Learning from European Cities. Island Press, Washington, D.C.

Bessey, K.M. (2002) Structure and dynamics in an urban landscape: toward a multi-scaled view. Ecosystems 5:360–375.

Breuste, J., Feldmann, H., and Uhlmann, O. (eds.) (1998) Urban Ecology. Springer, Berlin.

Brown, L.R. (2001) Eco-Economy: Building an Economy for the Earth. W.W. Norton, New York.

Burch, W.R., and Grove, J.M. (1993) People, trees and participation in the urban frontier. Unasylva 44:19–27.

Cilliers, P. (1998) Complexity and Postmodernism: Understanding Complex Systems. Routledge, New York.

Collins, J.P., Kinzig, A., Grimm, N.B., et al. (2000) A new urban ecology. American Scientist 88:416–425.

Cook, E.A. (1991) Urban landscape networks: an ecological planning framework. Landscape Research 16:8–15.

Cook, E.A., and van Lier, H.N. (eds.) (1994) Landscape Planning and Ecological Networks. Elsevier, Amsterdam.

Dramstad, W.E., Olson, J.D., and Forman, R.T.T. (1996) Landscape Ecology Principles in Landscape Architecture and Land-Use Planning. Harvard University Graduate School of Design/Island Press, Cambridge, MA.

Earth Policy Institute. (2002) Key indicators. http://www.earth-policy.org/Indicators.

Forman, R.T.T. (1995) Land Mosaics: The Ecology of Landscapes and Regions. Cambridge University Press, Cambridge, England.

Forman, R.T.T., and Godron, M. (1986) Landscape Ecology. Wiley, New York.

Forrester, J. (1969) Urban Dynamics. Productivity Press, Portland.

Galloway, J.N. (1998) The global nitrogen cycle: changes and consequences. Environmental Pollution 102:15–24.

Gardner, R.H., Milne, B.T., Turner, M.G., and O'Neill, R.V. (1987) Neutral models for the analysis of broad-scale landscape pattern. Landscape Ecology 1:19–28.

Gardner, R.H., Turner, M.G., Dale, V.H., and O'Neill, R.V. (1992) A percolation model of ecological flows. In: Hansen, A.J., and Castri, F.D. (eds.) Landscape Boundaries: Consequences for Biotic Diversity and Ecological Flows. Springer-Verlag, New York, pp. 259–269.

Green, D.G. (1994) Connectivity and complexity in landscapes and ecosystems. Pacific Conservation Biology 1:194–200.

Haigh, M.J. (1987) The holon: hierarchy theory and landscape research. Catena Supplement 10:181–192.

Hanski, I. (ed.) (1999) Metapopulation Ecology. Oxford University Press, New York.

Hanski, I.A., and Gilpin, M.E. (eds.) (1997) Metapopulation Biology: Ecology, Genetics, and Evolution. Academic Press, San Diego.

Helms, J. (ed.) (1998) Dictionary of Forestry. Society of American Foresters, Bethesda, MD.

Kellert, S.R., and Wilson, E.O. (eds.) (1993) The Biophilia Hypothesis. Island Press, Washington, DC.

Kennedy, J.J., Dombeck, M.P., and Koch, N.E. (1998) Values, beliefs and management of public forests in the Western World at the close of the twentieth century. Unasylva 49:16–26.

Konijnendijk, C.D. (1999) Urban woodland and policies and concepts in Europe: between tradition and innovation. In: Proceedings of International Conference on Community Forestry, December 7–8, 1999, London. http://www.communityforest.org.uk/tpsn.html.

Levin, S.A. (1999) Fragile Dominion: Complexity and the Commons. Perseus Books, Reading, PA.

Lobo, J., and Schuler, R.E. (1997) Efficient organization, urban hierarchies and landscape criteria. In: Schweitzer, F. (ed.) Self-Organization of Complex Structures: From Individual to Collective Dynamics. Gordon and Breach Science Publishers, Amsterdam, pp. 547–558.

Loucks, O.L. (1994). Sustainability in urban ecosystems: beyond an object of study. In: Platt, R.H., Rowntree, R.A., and Muick, P.C. (eds.) The Ecological City. University of Massachusetts Press, Amherst, MA, pp. 49–65.

Luck, M., Jenerette, G.D., Wu, J., and Grimm, N.B. (2001) The urban funnel model and the spatially heterogeneous ecological footprint. Ecosystems 4:782–796.

Luck, M., and Wu, J. (2002) A gradient analysis of urban landscape pattern: A case study from the Phoenix metropolitan region, Arizona, USA. Landscape Ecology 17:327–339.

McPherson, E.G. (1994) Cooling urban heat islands with sustainable landscapes. In: Platt, R.H., Rowntree, R.A., and Muick, P.C. (eds.) The Ecological City: Preserving and Restoring Urban Biodiversity. University of Massachusetts Press, Amherst, MA, pp. 151–171.

McPherson, E.G. (2001) Sacramento's parking lot shading ordinance: environmental and economic costs of compliance. Landscape and Urban Planning 57:105–123.

McPherson, E.G., Simpson, J.R., Peper, P.J., and Xiao, Q. (1999) Benefit-cost analysis of Modesto's municipal urban forest. Journal of Arboriculture 25:235–248.

Miller, D.H. (1978) The factor of scale: ecosystem, landscape mosaic, and region. In: Hammond, K.A., Macinio, G., and Fairchild, W.B. (eds.) Sourcebook on the Environment: A Guide to the Literature. University of Chicago Press, Chicago, pp. 63–88.

Miller, R. (1997) Urban Forestry: Planning and Managing Urban Greenspaces, 2nd ed. Prentice Hall, Englewood Cliffs, NJ.

Moss, M.R. (ed.) (1988) Landscape Ecology and Management. Proceedings of the First Symposium of the Canadian Society of Landscape Ecology and Management, May 1987. Polyscience, Montreal.

Naveh, Z., and Lieberman, A.S. (1984) Landscape Ecology: Theory and Application. Springer-Verlag, New York.

Nilsson, K., Konijnendijk, C.D., and Randrup, T.B. (1999) Urban forestry: Where people meet trees. In: Proceedings of International Conference on Community Forestry, December 7–8, 1999, London. http://www.communityforest.org.uk/tpsn.html.

Noss, R.F. (1987) Corridors in real landscapes: a reply to Simberloff and Cox. Conservation Biology 1:159–164.

Opdam, P. (1991) Metapopulation theory and habitat fragmentation: a review of Holarctic breeding bird studies. Landscape Ecology 5:93–106.

Opdam, P., Foppen, R., and Vos, C. (2001) Bridging the gap between ecology and spatial planning in landscaping ecology. Landscape Ecology 16:767–779.

Perez-Trejo, F. (1993) Landscape response units: Process-based self-organizing systems. In: Haines-Young, R., Green, D.R., and Cousins, S. (eds.) Landscape Ecology and Geographic Information Systems. Taylor & Francis, London, pp. 87–98.

Phillips, J.D. (1999) Divergence, convergence, and self-organization in landscapes. Annals of the Association of American Geographers 89:466–488.

Pickett, S.T.A., Burch, W.R., Dalton, S.E., Foresman, T.W., Grove, J.M. and Rowntree, R. (1997) A conceptual framework for the study of human ecosystems in urban areas. Urban Ecosystems 1:185–199.

Pickett, S.T.A., Cadenasso, M.L., Grove, J.M., et al. (2001) Urban ecological systems: linking terrestrial ecological, physical, and socioeconomic components of metropolitan areas. Annual Review of Ecology and Systematics 32:127–157.

Platt, R.H., Rowntree, R.A., and Muick, P.C. (eds.) (1994) The Ecological City: Preserving and Restoring Urban Biodiversity. University of Massachusetts Press, Amherst, MA.

Platt, R.H. (1994a) The ecological city: Introduction and overview. In: Platt, R.H., Rowntree, R.A., and Muick, P.C. (eds.) The Ecological City: Preserving and Restoring Urban Biodiversity. University of Massachusetts Press, Amherst, MA, pp. 1–17.

Platt, R.H. (1994b) From commons to commons: evolving concepts of open space in North American cities. In: Platt, R.H., Rowntree, R.A., and Muick, P.C. (eds.) The Ecological City: Preserving and Restoring Urban Biodiversity. University of Massachusetts Press, Amherst, MA, pp. 21–39.

Poiani, K.A., Richter, B.D., Anderson, M.G., and Richter, H.E. (2000) Biodiversity conservation at multiple scales: Functional sites, landscapes, and networks. BioScience 50:133–146.

Portugali, J. (2000) Self-Organization and the City. Springer, Berlin.

Rees, W.E. (1997) Urban ecosystems: the human dimension. Urban Ecosystems 1:63–75.

Roszak, T., Gomes, M., and Kanner, A. (eds.) (1995) Restoring the Earth, Healing the Mind. Sierra Club Books, San Francisco.

Simon, H.A. (1973) The organization of complex systems. In: Pattee, H.H. (ed.) Hierarchy Theory: The Challenge of Complex Systems. George Braziller, New York, pp. 1–27.

Stearns, F., and Montag, T. (eds.) (1974) The Urban Ecosystem: A Holistic Approach. Dowden, Hutchinson & Ross, Stroudsburg, PA.

Sukopp, H. (1990) Urban ecology and its application in Europe. In: Sukopp, H., Hejny, S., and Kowarik, I. (eds.) Urban Ecology: Plants and Plant Communities in Urban Environments. SPB Academic Publishing B.V., The Hague, Netherlands, pp. 2–22.

Sukopp, H. (1998) Urban ecology—scientific and practical aspects. In: Breuste, J., Feldmann, H., and Uhlmann, O. (eds.) Urban Ecology. Springer, Berlin, pp. 3–16.

Thio, A. (1989) Sociology: An Introduction, 2nd ed. Harper & Row, Cambridge, MA.

Tress, G., Tress, B., and Fry, G. (2005) Clarifying integrative research concepts in landscape ecology. Landscape Ecology 20:479–493.

Troll, C. (1939) Luftbildplan and okologische bodenforschung. Zeitschraft der Gesellschaft fur Erdkunde Zu Berlin 7–8:241–298.

Turner, M.G. (1989) Landscape ecology: the effect of pattern on process. Annual Review of Ecology and Systematics 20:171–197.

Turner, M.G., and Gardner, R.H. (eds.) (1991) Quantitative Methods in Landscape Ecology. Springer-Verlag, New York.

Turner, M.G., Gardner, R.H., and O'Neill, R.V. (2001) Landscape Ecology in Theory and Practice: Pattern and Process. Springer-Verlag, New York.

Ulrich, R.S. (1984) View through a window may influence recovery from surgery. Science 224:420–421.

United Nations. (2001) World Population Prospects: The 2000 Revision. United Nations, New York.

University of Florida (UF)/Institute of Food and Agricultural Sciences (IFAS). (2001) Restoring the Urban Forest Ecosystem. School of Forestry Resources and Conservation, Florida Cooperative Extension Service, Institute of Food and Agricultural Sciences, University of Florida, Gainsville. http://edis.ifas.ufl.edu/MENU_FR:CIR1266.

Urban, D.L., O'Neill, R.V., and Shugart, H.H. (1987) Landscape ecology: a hierarchical perspective can help scientists understand spatial patterns. BioScience 37:119–127.

U.S. Department of Agriculture (USDA)/Center for Urban Forest Research (CUFR). (2002) Where are all the cool parking lots? Center for Urban Forest Research, Pacific Southwest Research Station, USDA Forest Service, Davis, CA. http://cufr.ucdavis.edu/.

Wackernagel, M., and Rees, W.E. (1996) Our Ecological Footprint: Reducing Human Impact on the Earth. New Society Publishers, British Columbia, Canada.

Wilson, E.O. (1984) Biophilia. Harvard University Press, Cambridge, MA.

With, K.A., and Crist, T.O. (1995) Critical thresholds in species responses to landscape structure. Ecology 76(8):2446–2459.

World Resources Institute. (1998) 1998–99 World Resources—A Guide to the Global Environment. Oxford University Press, New York.

Wu, J., and David, J.L. (2002) A spatially explicit hierarchical approach to modeling complex ecological systems: theory and applications. Ecological Modelling 153:7–26.

Wu, J., and Loucks, O.L. (1995) From balance-of-nature to hierarchical patch dynamics: a paradigm shift in ecology. Quarterly Review of Biology 70:439–466.

Wu, J., Vankat, J.L., and Barlas, Y. (1993) Effects of patch connectivity and arrangement on animal metapopulation dynamics: a simulation study. Ecological Modelling 65:221–254.

Wu, J.G. (1999) Hierarchy and scaling: extrapolating information along a scaling ladder. Canadian Journal of Remote Sensing 25:367–380.

Wu, J.G. (2000) Landscape Ecology: Pattern, Process, Scale and Hierarchy. Higher Education Press, Beijing.

Wu, J.G., and Hobbs, R. (2002) Key issues and research priorities in landscape ecology: an idiosyncratic synthesis. Landscape Ecology 17:355–365.

Wu, J.G., and Overton, C. (2002) Asian ecology: pressing problems and research challenges. Bulletin of Ecological Society of America 83:189–194.

Zipperer, W.C., Wu, J., Pouyat, R.V., and Pickett, S.T.A. (2000) The application of ecological principles to urban and urbanizing landscapes. Ecological Applications 10:685–688.

3
Principles for Guiding Eco-City Development

Rüdiger Wittig

Before a town may rightly be termed an "eco-city," it must first meet a number of requirements, which can be summarized in one sentence: To the greatest possible extent, an eco-city should function in the same way as a natural ecosystem. However, a comparison between urban and natural ecosystems reveals that in general this goal is currently far from being realized. Cities and natural ecosystems differ greatly in their energy sources, in the origin of nutrients and other materials used, and in patterns of waste disposal and material cycling. To establish criteria for an eco-city, we must first remember what a typical city looks like. Bearing the characteristics of a typical city in mind, one can approach the question of the conditions that have to be fulfilled before a town or city may rightly be termed an eco-city. However, it is recognized that socioeconomic conditions do constrain the ability of cities to mimic nature in terms of energy flow and material cycling. Nonetheless, to clarify these requirements it is useful to compare and contrast the characteristics and functions of cities and natural ecosystems.

Characteristics of a City

The most important characteristics of a city can be enumerated as follows:

- High building density
- High proportion of sealed surfaces (pavement, buildings)
- Great importation of fossil fuels for energy
- Great importation of nutrients (food), building materials, goods
- Concentration of diverse industries
- High levels of trade and commerce
- Dense vehicular traffic
- Many entertainment venues and many cultural institutions
- High waste production
- Contamination of air, water, and soils
- Light pollution
- Noise pollution

M.M. Carreiro et al. (eds.), *Ecology, Planning, and Management of Urban Forests: International Perspectives.*

These qualities result in a city having a great impact on its surroundings. These characteristics also create large differences between cities and natural ecosystems.

Differences Between Cities and Natural Ecosystems

As shown in Table 3.1, cities and natural ecosystems differ greatly in the following respects:

- Their main energy sources
- The origin of matter
- The composition of their surfaces and vertical structures
- The direction of energy and material flows
- The methods of waste disposal

Most of these contrasts contribute to large differences in the direction of energy and material flows. While in a natural terrestrial ecosystem a great amount of recycling takes place (Jordan, 1982), in a city almost no material cycling can be detected (closed loops for material reusage). The almost complete absence of internal matter cycling within cities was shown in ecosystem studies carried out in Vienna in the 1990s (Punz et al., 1996), and earlier in Brussels (Duvigneaud and Denayer-de Smet, 1975) and Hong Kong (Boyden et al., 1981). Particularly impressive were the results from Vienna, which showed that an area 1336 times the size of Vienna was needed to provide for the energy consumed by his city (Fussenegger et al., 1995).

Steps Toward and Demands on an Eco-City

To convert a typical city into an eco-city, the differences shown in Table 3.1 must be minimized, so that the ecological footprint of the city (Wackernagel and Rees, 1996) is reduced. Wittig et al. (1998) established five principles to guide decision makers wishing to move their city toward greater ecological sustainability (see also Wittig et al., 1995):

1. Media that support life (soil, water, air) must be protected.
2. Energy consumption must be reduced.
3. Material use should be reduced and materials recycling increased.
4. The amount and kind of nature in the city must be enhanced through conservation and restoration activities.
5. A rich variety of spatial structure and space must be provided.

Achieving any one of these principles often allows a city to meet at least two or three others simultaneously, as explained in the following examples. Reduction in material flow (principle No. 3) results in less traffic, and thus a reduction in energy consumption (principle No. 2). Less energy consumption in turn results in fewer

Table 3.1 Important ecological differences between cities and natural terrestrial ecosystems

Feature	Cities	Natural terrestrial ecosystems
Main energy source(s)	Fossil fuels (coal, petroleum, natural gas), nuclear power	Solar
Origin of nutrients and other materials	Mostly external	Mostly internal
Surface and structure	Surface sealed; vertical structure dominated by artificial hard surfaces	Permeable soil surfaces; structure dominated by vegetation in most cases
Flow of energy and material	High input-output systems for materials; almost no internal cycling of matter	High percentage of matter recycled within the system
Waste disposal	Large number of dumping sites often far from the city	Nutrient export low

emissions and thus less air, water, and soil pollution (principle No. 1). Conservation and promotion of nature (principle No. 4) promotes more unsealed soils (i.e., protection of life media such as soil [principle No. 1]) and also contributes to there being more variety in urban structure and space (principle No. 5).

Urban forests, be they trees planted in dense groups or in linear fashion along streets, highways, and residential areas, planted woodland parks, and true forest remnants within a city can all contribute greatly to the realization of the above-mentioned five principles. The economic, ecological, and social advantages and benefits of urban forests have been thoroughly discussed in earlier studies (Rowntree, 1986; Oke, 1989; Kuhn et al., 1998; Simpson, 1998) and are also discussed in this volume (Chapters 5 and 6), and therefore are not discussed fully here. However, to list a few, trees promote greater biological diversity at all city scales, and reduce air and noise pollution, local flooding, the urban heat island at the whole city scale via evapotranspiration, and at a more local scale reduce the energy requirements for cooling of buildings directly receiving their shade. Of course, other types of vegetation in cities not classified as "forests" also make important contributions to a city's goal of developing into an eco-city. In particular, roof and wall greening can play important roles in the thermal control of buildings and thus lower building energy costs (Höschele and Schmidt, 1974; Harazono et al., 1990/91; Kuttler, 2005).

While much of the vegetation in cities is planted and managed, natural vegetation or biotopes also exist within many densely populated cities. The long-term maintenance of these valuable natural systems is sometimes in doubt. However, the example of Frankfurt am Main (Wittig, 2002), which contains six nature reserves totaling an area of 127.5 hectares, shows that it is possible to successfully maintain natural reserves, even though they are surrounded by densely built-up areas. For example, Wittig and Nawrath (2002) identified Frankfurt as particularly important for conserving endangered dry meadows: 10% to 25% of the total area of these endangered habitat types still existing in the Federal State of Hesse are situated in the nature reserves of the city of Frankfurt (Table 3.2).

Table 3.2 The importance of a city (Frankfurt am Main) in providing habitat for the conservation of rough meadows in the Federal State of Hesse, Germany

Endangered plant community		Area (hectares)		Percentage of the Hessian
Common name	Scientific name	State of Hesse	Frankfurt	areas situated in Frankfurt
Thrift-rough meadow	Diantho-Armerietum	20	2	10%
Silvergrass dune vegetation	Spergulo-Corynephoretum	35	4	11%
Bromegrass mesoxerophytic meadow	Mesobrometum	40	10	25%

Source: Wittig and Nawrath (2002) after Gregor (1992).

The Role of Socioeconomic Conditions

While discussing urban ecology and urban forestry from the points of view of ecologists, foresters, city planners, and landscape architects, one should not forget that the realization of an eco-city is not only dependent on input from people in these disciplines, but also highly influenced by socioeconomic conditions and the attitudes of the population. From this point of view, the requirements laid down in the Charter of European Cities and Towns Towards Sustainability (1994) must be regarded in addition to the purely ecological principles mentioned above. The charter states that the following factors should be promoted:

- A general change in values toward a realistic environmental awareness by both businesses and households
- A sustainable economic development, in particular a significant reduction of energy consumption and a shift from nonrenewable to renewable energy sources
- An effective use of all political and technical instruments and tools available for promoting an ecosystem approach to urban management

Thus far, awards and rankings have been used as tools for providing an incentive for cities to achieve the goals of the charter. That is why the European Sustainable City Award was created. For this award only those cities that have developed a common vision for local sustainable development with input from a wide variety of members and sectors (stakeholders) of the local community can apply. In 2004–05 the first detailed report on city quality of life in the United States was produced and included indicators of sustainability, programs, policies, and performance (Sustain Lane, 2005). Twenty-five U.S. cities were chosen for an evaluation of their relative levels of sustainability. Two cities (San Francisco and Portland) were identified as "sustainability leaders," ten cities were classified as "moving toward sustainability," and seven were placed in the category "mixed sustainability progress." The end of the queue was represented by four cities in the category of "sustainability laggard," and two cities (Houston and Detroit) where in the category "sustainability is in danger."

Conclusion

Today the average city differs greatly in structure and function from natural ecosystems and hence is a long way from becoming sustainable. To achieve the status of an eco-city, these differences have to be minimized. Principles for achieving such a minimization exist as goals. When trying to meet these principles, however, the important role of socioeconomic conditions should not be forgotten. The Aarlborg Charter represents the determination of many European cities to working toward greater sustainability and can serve as a model for other cities considering the path of evolving into eco-cities.

References

Boyden, S., Millar, S., Newcombe, K., and O'Neill, B. (1981) The Ecology of a City and Its People. Australia National University Press, Canberra.

Charter of European Cities and Towns Towards Sustainability, Aalborg. (1994) Internet-document. http://www.aalborg.dk/Borgerportal/Engelsk/Aalborg+Charter.htm.

Duvigneaud, P., and Denayer-de Smet, S. (eds.) (1975) L' Ecosystème Urbain—Application à l'Agglomération bruxelloise. Publication de Colloque International organisé par L'Agglomération de Bruxelles 14 et 15 septembre 1974, Bruxelles.

Fussenegger, K., Dörflinger, A.N., Maier, R., and Punz, W. (1995) Der Flächengebrauch einer Großstadt am Beispiel Wien. Verhandlungen Zoologisch Botanische Gesellschaft Österreich 132:233–249.

Gregor, T. (1992) Hessische Magerrasen. Botanik Naturschutz Hessen, Beiheft 4:50–64.

Harazono, Y., Teraoka, S., Nakase, I., and Ikeda, H. (1990/91) Effects of rooftop vegetation using artificial substrates on the urban climate and the thermal load of buildings. Energy and Buildings 15–16:435–442.

Höschele, K., and Schmidt, H. (1974) Klimatische Wirkungen einer Dachbegrünung. Garten und Landschaft 6: 334–337.

Jordan, C.F. (1982). The nutrient balance of an Amazonian Rain Forest. Ecology 63(3):647–654.

Kuhn, A., Ballach, H.-J., and Wittig, R. (1998) Vegetation as a Sink for PAH in Urban Regions. In: Breuste, J., Feldmann, H., and Uhlmann, O. (eds.) Urban Ecology. Springer, Berlin, Heidelberg, New York.

Kuttler, W. (2005) Stadtklima. In: Hupfer, P., and Kuttler, W. (eds.) Witterung und Klima. 11. Aufl. Teubner Verlag, Stuttgart pp. 371–432.

Oke, T. (1989) The micrometeorology of the urban forest. Philosophical Transactions of the Royal Society of London, [B] 324:335–349.

Punz, W., Maier, R., Hiertz, P., and Dörflinger, A.N. (1996) Der Energie- und Stoffhaushalt Wiens. Verhandlungen Zoologisch Botanische Gesellschaft Österreich 133:27–39.

Rowntree, R.A. (ed.) (1986) Ecology of the urban forest part II: function. Urban Ecology 9:227–434.

Simpson, J.R. (1998) Urban forest impacts on regional cooling and heating energy use: Sacramento county case study. J Arboriculture 24(4):201–214.

Sustain Lane. (2005) http://www.sustainlane.com/cityindex/citypage/3/SustainLane+US+City+Rankings.html

Wackernagel, M., and Rees, W. (1996) Our Ecological Footprint: Reducing Human Impact on the Earth. New Society Publishers, Gabriola Island, BC.

Wittig, R. (ed.) (2002) Die Naturschutzgebiete in Frankfurt am Main. Geobotanische Kolloquien.

Wittig, R., and Nawrath, S. (2002) Die Frankfurter Naturschutzgebiete: Situation, Effizienz, Probleme, Handlungsbedarf und Chancen. Geobot. Kolloq. 17:3–12.

Wittig, R., Breuste, J., Finke, L., et al. (1995) Wie soll die aus ökologischer Sicht ideale Stadt aussehen? Forderungen der Ökologie an die Stadt der Zukunft. Z. Ökologie u. Naturschutz 4:157–161.

Wittig, R., Sukopp, H., and Breuste, J. (1998) Ökologische Stadtplanung. In: Stadtökologie, 2nd ed. Fischer, Stuttgart, Jena, Lübeck, Ulm., pp. 401–432.

4
A Multiple-Indicators Approach to Monitoring Urban Sustainable Development

Kunmin Zhang, Zongguo Wen, Bin Du, and Guojun Song

In June 1972, the United Nations Conference on the Human Environment (Stockholm) issued the Declaration on the Human Environment, which proposed the following: "Planning must be applied to human settlements and urbanization with a view to avoiding adverse effects on the environment and obtaining maximum social, economic and environmental benefits for all." Since then, the construction of eco-cities to promote sustainable development gradually became a hot topic worldwide. The concept of an eco-city originated from the Garden City design of Howard in 1903 (Huang and Chen, 2002). In 1975, United States environmentalist and urban designer Richard Register and others created the organization Eco-City Builders (Register, 1996, 2002). During the past two decades, great progress has been made in the theory of the eco-city due to the efforts of many scholars, such as Ma Shijun in China; Oleg Yanitsky, a former Soviet ecologist; and David Engwicht, an Australian community activist. More recently, a number of pioneer eco-cities have appeared, including Bangalore in India, Curitiba and Santos in Brazil, Whyalla in Australia, Whitaker in New Zealand, Copenhagen in Denmark, Alberta in Canada, and, in the United States, Berkeley, California, Cleveland, Ohio, and Portland, Oregon (Zhang et al., 2003a).

Much attention has also been paid to the eco-city concept and movement in China. Many cities have begun eco-city planning, research, or construction (Luo and Zeng, 1999; Zhang et al., 2000; Zhang and Wen, 2001; Zhang, 2001), including Yichun in Jiangxi Province, Ma'anshan in Anhui Province, Yantai and Rizhao in Shandong Province, Yangzhou in Jiangsu Province, and Shanghai and Guangzhou.

Many international organizations (e.g., the United Nations Environment Program, the United Nations Center for Human Settlements, the Asian and the Pacific Economic Cooperation) have also conducted projects and research on urban sustainable development. However, so far there has not been any commonly accepted definition of an eco-city, not to mention an established and internationally acknowledged eco-city (Zhanget al., 2003a; Wen et al., 2005a). Through academic research and case study descriptions, it has become recognized that the construction of a sustainable eco-city is a step-by-step process. An eco-city is closely linked to the sustainability of the local-to-national state of the economy, society, and the environment, and can be gradually established through the efforts of its stakeholders. As cities adopt policies intended to promote sustainability, the development of

M.M. Carreiro et al. (eds.), *Ecology, Planning, and Management of Urban Forests: International Perspectives.*

indicators to monitor progress will be an important component of the informational feedback needed to accomplish such a transformation.

Indicator Functions

An eco-city is a complex system intended to harmonize economic development, social advancement, and ecological conservation, a city where the flow of materials, energy, information, population, and currency is highly efficient (Zhang, 2001; Gao et al., 2001; Zhang and Wen, 2001). The eco-city is required to develop within the bounds of its larger ecosystem in order to realize sustainable development. Consequently, we can conclude that sustainable development becomes an obvious characteristic of an eco-city.

In order that an eco-city should not remain an abstract concept, it is essential to construct a set of common indicators to monitor and compare the structure and function of different cities contemporaneously and over time. These indicators should provide at least the following (Zhang et al., 2003a): (1) explanatory tools to translate the concepts of sustainable development into practical terms; (2) pilot tools to assist in making policy choices that promote sustainable development and provide directional guidance for decision makers when facing alternative policies; and (3) performance assessment tools to decide how effective efforts to meet sustainable development goals and objectives have been.

Fifteen years after the United Nations Conference on Environment and Development (UNCED) in 1992, there are still no indicators for measuring sustainable development that have been globally accepted. Currently, research on indicators for sustainable development at national or regional levels is still underway, while research on urban ecologically sustainable development remains at a conceptual level. Despite all the challenges, this issue inevitably needs to be addressed. This chapter discusses four Chinese cities as comparative case studies for applying and developing sustainability indicators: Suzhou and Yangzhou in Jiangsu Province, Ningbo city in Zhejiang province, and Guangzhou in Guangdong province. All four cities are located in southeast China and were chosen because of their advanced state of development, which could provide us with a 10-year series of data to explore. In addition, each of their local governments has made commitments to eco-city building as an important development goal.

Progress in Developing Indicators and Evaluation Models for Sustainability

International Experiences in Monitoring Sustainability

Since 1992, many international organizations, nongovernmental organizations, and some countries have conducted research and proposed their indicators (Wen, 2005), such as the "drive-state-response" framework by the United Nations Commission on Sustainable Development (UNCSD), the Framework Indicators of Sustainable

Development (FISD) by the United Nations Statistic Department and the Scientific Committee on Problems of the Environment (SCOPE). At the same time, indicators have been proposed at different levels, including international, national, local, and departmental. The representative frameworks are Alberta, Canada, at a local level and the Netherlands at the national level. At the city and community levels, many countries such as the United States, United Kingdom, Denmark, and Norway have built different representative indicators for the "sustainable city" or "green city." For example, Seattle has constructed a set of indicators for a "sustainable Seattle" with a similar methodology to that of the Urban Ecologically Sustainable Development Indicators (UESDIs) described in this chapter.

Several researchers (Moffat et al., 1999; Zhang et al., 2003a) have tried some new approaches, for example, using a comprehensive evaluation method to develop a "substitutable index" to accommodate different aspects of sustainable development and to compensate for factors that traditional economic indices overlook. This method has been widely applied (Zhang et al., 2003b). Usually these indices are highly integrated to transform a certain complex system into a numerical value that decision makers can more easily understand. Several typical indices include (1) the Human Development Index (Murray, 1991); (2) the Sustainable Process Index (Moser, 1994); (3) the Social Progress Index (Desai, 1993); (4) the Index for Sustainable Economic Welfare (Daly et al., 1994); and (5) Material Input per Service Unit (Schmidt-Bleek, 1994).

Some departments and research institutes in China have started to study indicators of sustainable development and have made some progress on multi-methodology and application (Administrative Center for China's Agenda, 2000; Zhang et al., 2003b; Wen et al., 2005b), but no indicator has been widely accepted. Most indicators at the urban level emphasize only environmental issues without including other dimensions of sustainable development. The city of Guangzhou proposed a set of indicators for an eco-city in 2001 that included indicators for eco-city urban planning and assessment. The State Environmental Protection Administration (SEPA) set down Quantitative Indicators of Urban Synthetically Environmental Harness in 1989 and the National Checking Indicators of Model City of Environmental Protection in 1997 (Task Group of Indicators System on Sustainable Development, 1999). In July 2002, SEPA issued the National Checking Indicators of a Beautiful Town. Furthermore, some cities, such as Yichun, Nanning, Beijing, Yangzhou, and Rizhao, have carried out some studies and practices evaluating urban sustainability.

Review of the Methodology on Indicators and Evaluation Models for Sustainability

Current indicators and evaluation models for sustainable development have been primarily classified into three methodological groups: system engineering, monetary evaluation, and biophysical. These indicators and models monitor sustainability through multifaceted concepts. However, weaknesses, of both a theoretical and a practical nature, exist with all of the indicators, as described below. It is important to understand these limitations when using the indicators (Zhang et al. 2003b; Wen, 2005).

System Engineering Methodology

Examples include indicators developed by the United Nations (2001), the Barometer of Sustainability designed by Prescott (1995), the Environmental Sustainability Index of the Global Leaders of Tomorrow Environmental Task Force (2002), and the Indicators System of China's Sustainability, proposed by the research group on sustainable development of the Chinese Academy of Science (1999). These indicators generally use the comprehensive evaluation method to monitor and analyze progress on sustainable development. The primary challenge of this methodology lies in data aggregation and integration of indicators, but the scales and dimensions of these data are largely different. This type of indicator is difficult to put into practice because of its complex framework.

Monetary Evaluation

This type of methodology is usually developed to amend the System of National Accounting (SNA), and includes such indicators as the Genuine Saving (Dixon and Hamilton, 1997), Green Gross Domestic Product (GDP), the Index for Sustainable Economic Welfare (Daly et al., 1994), and the Genuine Progress Indicator (Cobb et al., 1995; Hamilton, 1999; Wen et al., 2006). However, there are some limitations, such as (1) the price cannot truly reflect the scarcity of a natural resource because of the unavailable externality; (2) it is difficult to determine the discount rate, and the resources are not reversible after depletion; and (3) natural capital and human-made capital cannot be substituted and supplemented for each other. In addition, from the standpoint of strong sustainability, indicators based on monetary evaluation as weak sustainability cannot reflect real social development (Wen et al., 2006).

Biophysical Methodology

The Ecological Footprint, designed by Wackernagel and Rees (1996) and Rees (2000), is a representative model of this approach. Comparing the calculated results from the eco-footprint model with the ecological service capacity provided by natural capital, one can estimate the gap between ecological carrying capacity and human activities under certain conditions. This model can reflect only the effect of economic policy on the environment, and overlooks other important influencing factors caused by land utilization, such as land degradation resulting from urbanization, pollution, and erosion.

Measuring Sustainability and Designing an Indicators Framework

As mentioned above, it is very difficult for any single set of indicators to tell us all we need to know about urban sustainability, since sustainability itself is such a multifaceted concept (Zhang et al., 2003a; Wen et al., 2005b). Even when studying the same time period in a city, different indicators that evaluate sustainability may arrive at conflicting conclusions. Therefore, additional models that measure sustainability from a multidisciplinary perspective and using different assessment methodologies are valuable for obtaining a more complete picture of urban sustainability (Zhang et al., 2003b). In addition, the indicator should be as simple as possible so that communication among the general public, the decision makers, and the media can be efficient and clear.

Here we use a group of five models to assess progress toward the development of a sustainable eco-city (Fig. 4.1), including (1) the index of Approximate Environmental-Adjusted Net Domestic Product (AEANDP), (2) the Genuine Saving Rate (GSR), (3) the Eco-Footprint (EF), (4) the Index of Sustainable Economic Welfare (ISEW), and (5) the Genuine Progress Indicator (GPI). These five indicators are designed in three dimensions from weak (e.g., AEANDP, GSR, ISEW, and GPI) to strong sustainability (e.g., EF), which try to monitor and capture factors encompassing sustainability more thoroughly. Similar to the research in Scotland done by Moffatt and others (Moffatt, 1996; Moffatt et al., 1999), we found that different indicators provide different results.

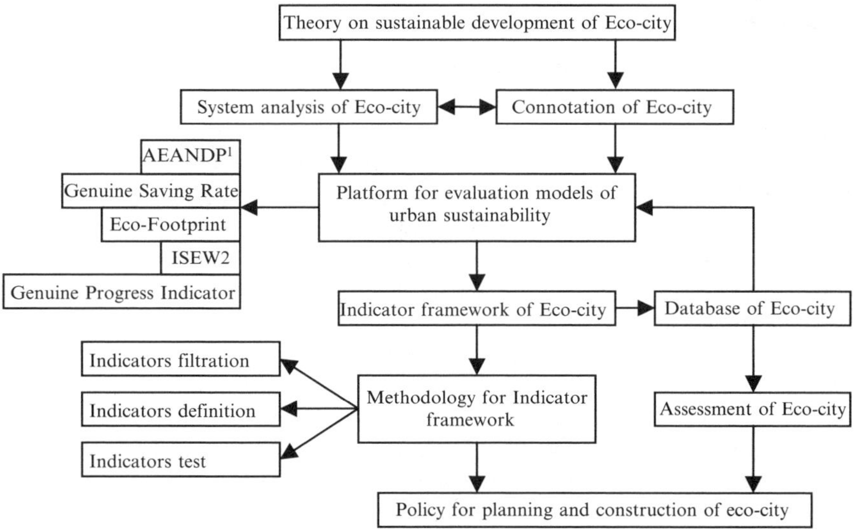

Fig. 4.1 Methodological framework for research on urban sustainability. AEANDP, Approximate Environmental-Adjusted Net Domestic Product. ISEW, Index of Sustainable Economic Welfare

Table 4.1 Models to assess sustainability trends

Models	Unit	Characteristics
Genuine Saving Rate (GSR)	Percentage	Test of "weak sustainability": if national income can make savings growth greater than depreciation of human-made and natural capital, development is considered sustainable.
AEANDP	Currency	Amends traditional accounting system of national economy to attach a price to natural resources and deduct economic loss due to environmental pollution. Continuously negative AEANDP indicates a decrease in urban sustainability.
Eco-footprint (EF)	Land area	Test of "strong sustainability": calculates two indicators at a particular population and economic scale: (1) the biologically productive area needed to maintain resource consumption and waste absorption; (2) biologically productive area that a region can provide. Judges whether the regional productive consumption activity is within the carrying capacity of the local or defined ecosystem by comparing the two indicators.
ISEW	Currency	Reflects the sustainable economic welfare and living quality of members of the whole society.
Genuine Progress Indicator (GPI)	Currency	Includes societal, economic and environmental factors, and respectively calculates their benefits and costs to measure local sustainability.

The interdisciplinary framework shown in Table 4.1 and Figure 4.1 can be used to monitor different facets of urban sustainability. For example, the GSR index focuses on the need to reinvest in all forms of capital, whereas the ecological index (EF) focuses on energy and matter requirements needed to maintain a city. Developing and using indicators that are sensitive enough to detect the important interactions among social, economic, and ecological systems are an essential component of sustainable development policy, because it is very hard to choose among alternative policies and implement them without the ability to predict their impacts on multiple aspects of sustainability. In considering the calculated results for a real city in a case study, we proposed a framework of indicators for ecological sustainability that can probably be adapted to any specific city. We established indicators (Urban Ecologically Sustainable Development Indicators [UESDI]) according to system engineering principles in the designed framework. Our research used the five methodologies mentioned above in four case studies (Suzhou city and Yangzhou city in Jiangsu Province, Ningbo city in Zhejiang Province, and Guangzhou city in Guangdong Province) to measure their individual sustainability and to make comparisons among them. The case cities are rapidly growing cities in the coastal region of China.

Methodology in Our Case Study

Our study assessed the sustainability of these four cities over a 10-year period (1991 to 2000) using the five models in Table 4.1. Sustainability trends for each city over the 10 years were determined using each model individually and using an integrated index, the UESDI. This summary index was based on a total of 35 indicators and 74 variables (see Construction of Our Indicators Framework [the UESDI], below), which were analyzed using principal component analysis to obtain the integrated index. Because all models have a specific policy impact, they are easy for stakeholders to understand and accept, and the results can be used for comparisons over time (temporal trend for a single city) and across regions (between different cities). The five models and UESDI can then be used to measure sustainable development in real decision-making contexts.

Assessment of Urban Sustainability: Explanations of Each Model

Approximate Environmental-Adjusted Net Domestic Product

The AEANDP takes the standard macroeconomic measure of net domestic product and deducts from it the value of depreciation in major elements of natural capital and changes in pollution flows. The AEANDP increases when there are increases in natural capital stock or in technological advancement to improve the efficiency of natural capital. On the other hand, the sustainability level of social income decreases when natural capital stock decreases. The AEANDP in our study was the residual value calculated by deducting from the gross domestic product (GDP) the depreciation of human-made capital and environmental resources, and economic loss due to environmental pollution.

Genuine Saving Rate

In 1995, the World Bank proposed a rough estimation of the genuine saving rate (GSR) in the report entitled "Monitoring Environmental Progress," and accepted the rate of genuine saving in the GDP as a new indicator for measuring the current status of and the potential for national economic development after deterioration of natural resources and economic loss due to environmental pollution were subtracted from the GDP. The policy implications of the GSR are that continuous reductions in the growth of genuine savings result in a reduction of wealth.

Calculation of the GSR in our study follows that of the World Bank (Hamilton, 1999; Zhang et al., 2003a): GDP is calculated by subtracting gross consumption, then adding investment in education; the result is the traditional standard of state wealth accumulation—the gross savings. When depreciation of produced capital is subtracted from gross savings, the value of net savings is obtained. Net savings is closer to the concept of sustainability, but it still only takes into account human-produced capital, not natural capital. Net savings subtracts consumption of natural resources and expense of pollution from gross saving, and deducts the economic loss caused by long-term environmental damage caused by such factors as emissions of CO_2 and ozone-depleting substances to obtain the genuine savings of cities in this case study. Finally, the GSR is calculated by dividing the GDP by genuine savings.

Eco-Footprint

The ecological footprint (EF) method, developed by Wackernagel and Rees (1996) is a newer method for measuring sustainable development by assessing the ecological impacts of regions, nations, or cities. The EF measures the area of the earth's surface needed to provide natural resources and pollutant absorption services for people at different population and economic scales. The EF aggregates human impact on the biosphere into one number: the biologically productive land occupied exclusively for a given human activity. The EF of any defined population (from a single individual, household, city, region, or nation) is the area of biologically productive land and water area occupied exclusively to produce the resources and assimilate wastes generated by that population, using the prevailing technology (Rees, 2000). By comparing the EF with the area of land available, it is relatively easy to judge whether the regional productive consumption activity is within the carrying capacity of local ecosystems (Haberl et al., 2001). When the local carrying capacity is less than the eco-footprint, then an ecological deficit is identified. Redefining Progress (2002) calculated the EFs and ecological capacities of 152 nations in the world. The results showed that the global EF covered 13.7 billion hectares (ha) in 1999, or 2.3 global ha per person (a global hectare is 1 hectare of average biological productivity), while the global ecological carrying capacity was about 11.4 billion ha. Therefore, human consumption of natural resources that year overshot the earth's biological capacity by about 20%.

Index of Sustainable Economic Welfare

The ISEW is a sociopolitical index originally developed in the USA by Daly and Cobb (1989) to measure human quality of life. We used the following equation to calculate the ISEW in this study:

ISEW = (Personal consumers' expenditure + Nondefense expenditures + Capital balance) – (Defense expenditures + Losses due to pollution + Depletion of natural capital).

Genuine Progress Indicator

The GPI was introduced by Redefining Progress in 1995 to amend the conventional GDP accounting system (Wen et al., 2004a). This indicator is an improvement over the ISEW in that it includes allowances for fairness in the existing distribution of income, as well as additional measures of environmental degradation, defense expenditures, and unpaid work (Wen et al., 2004b, 2005c). The GPI model begins with personal expenditure, adjusted for some factors (such as the income distribution index), then adds certain new accounting values (for example, the merit of housework, child care, home repairs, and gardening), and deducts certain additional costs (such as economic loss from crime) as well as those from pollution.

Construction of Our Indicators Framework (the UESDI)

In our analysis of ecological, economic, and societal characteristics of the four cities in our case study based on results of the five evaluation models, we set up the UESDI with an array of variables. This framework includes five subsystems: the resources support system (R), the societal support system (S), the economic support system (E), the environmental support system (A), and the institutional support system (I). The number of indicators in different subsystems differs from city to city. Furthermore, the variation in the number of different indicators in this framework is appropriate and diverse. The UESDI framework to monitor urban sustainability in our study includes a total of 35 indicators and 74 variables. The steps entailed in our comprehensive assessment approach are as follows (Zhang et al., 2003b):

1. Collect and process raw data for the years 1996 to 2000 for the five subsystems in the UESDI mentioned above.
2. Normalize the data needed for indicators of all the subsystems.
3. Use the principal component analysis approach (Hu and He, 2000) to derive a value that reflects the level of aggregated development (L) for each of the five subsystems singly.
4. Integrate the L values of the five subsystems with average weighting, and then analyze the results to assess their level of sustainable development.

Then we can examine the trends of sustainability during the 10-year period with the comprehensive evaluation method for indicators. Finally, a pentagon radar chart is used to visualize and communicate the degree of integrated development level and harmonization (equity) among the five support subsystems. The next section presents the results of models of GSR and EF in the four cities.

Results of This Case Study

Sustainability Assessment Using the AEANDP and GSR Models

The GDP and GDP per capita of all the case cities grew rapidly during the study period (Table 4.2). This trend can be used as a variable to compare with the monetary indicator of sustainability in this chapter. The following is a description of the results of the AEANDP and GSR model assessments: Trends for the AEANDP in all four cities increased between 1991 and 2001, but at different rates (Fig. 4.2), with Guangzhou rising most rapidly and Yangzhou most slowly.

Table 4.2 Total gross domestic product (GDP) and GDP per capita of the four case study cities

	Ningbo		Guangzhou		Yangzhou		Suzhou	
Year	GDP (billion yuan)	GDP per capita (thousand yuan)	GDP (billion yuan)	GDP per capita (thousand yuan)	GDP (billion yuan)	GDP per capita (thousand yuan)	GDP (billion yuan)	GDP per capita (thousand yuan)
1991	16.99	3.30	38.67	6.42	18.06	1.94	25.31	4.49
1992	21.31	4.12	51.07	8.34	23.79	2.55	35.97	6.34
1993	31.64	6.09	74.08	11.88	33.06	3.54	52.60	9.24
1994	46.35	8.86	97.62	15.32	47.04	5.02	72.09	12.62
1995	60.93	11.58	124.31	19.22	60.50	6.44	90.31	15.76
1996	79.59	15.01	144.49	22.02	35.11	7.89	100.21	17.46
1997	89.74	16.83	164.63	24.70	37.67	8.44	113.26	19.70
1998	97.34	18.19	184.16	27.32	40.16	8.99	125.00	21.73
1999	104.17	19.35	205.67	30.03	42.70	9.54	135.84	23.57
2000	117.58	21.74	237.59	33.91	47.21	10.48	154.07	26.65
2001	131.06	24.10	268.58	37.69	51.00	11.21	176.03	30.32

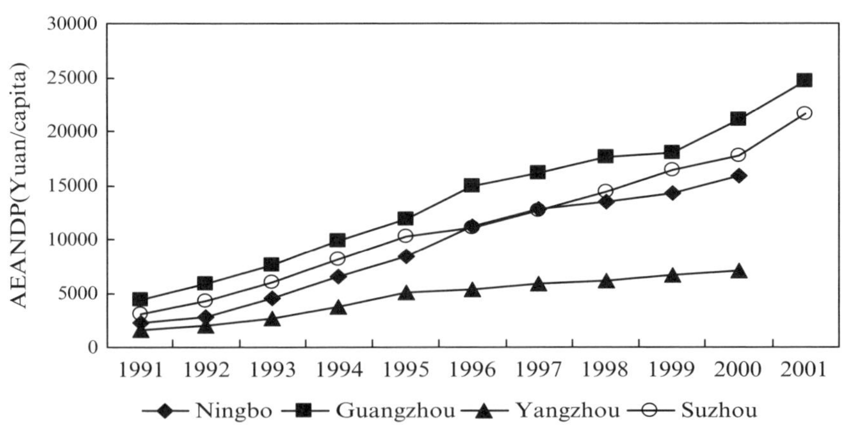

Fig. 4.2 Per capita trends in AEANDP in yuans for four Chinese cities from 1991 to 2001

For Ningbo the GDP growth rate even when normalized to 1990 values exceeded 30% in 1996, but then began to slow in 1997 (Table 4.2). In 2000, the population of Ningbo was 5.41 million, and the GDP per capita reached $2626. The results for Ningbo show that from 1991 to 2000 its GDP increased from 17 billion to 117.6 billion yuan. Judging by the trends in the AEANDP to GDP ratio from 1991 to 2000 (about 70% to 74%), we can conclude that Ningbo is moving toward sustainable development during the study period.

We again use Ningbo city as an example for examining trends in the GSR. In 2000, its GDP was 117.6 billion yuan. After gross consumption (45.1 billion yuan) and goods and services exports (21.8 billion yuan) are subtracted from the GDP, and education investment (3.2 billion yuan) added, the resulting gross savings is 37.2 billion yuan. After depreciation of produced capital (15.4 billion yuan) is subtracted from the gross savings, a net savings of 21.8 billion yuan is obtained. This net savings value is a better indicator of sustainability than is the gross savings value, but still only considers human-produced capital. Once the costs of natural resource depletion, pollution damage, and economic loss caused by long-term environmental influences (such as emission of CO_2 and consumption of ozone depleting substances) are deducted, the result is a genuine savings of 8.9 billion yuan. Finally, the GSR for the year 2000 (7.6%) is derived by dividing genuine savings by the GDP.

Our analysis shows that GSR in Ningbo city grew quickly from 1991 to 1995 (Table 4.2). Since 1996, the GSR began to decline due possibly to its declining GDP growth rate, its foreign debts, and its growing depletion of natural resources and environmental pollution damage. This declining GSR trend mirrors a decreasing trend in Ningbo's rate of development, even though its GSR is still positive. If the trend continues, Ningbo's GSR may become negative in the near future. So if sustainable development is a priority, these negative GSR trends must be stabilized or reversed.

The changing trends in GSR (Table 4.3) reveal not only the quality of urban development but also the level of urban sustainability for these four cities.

Table 4.3 Comparison of Genuine Savings Rate (% of GDP) trends for four cities in China

Year	Ningbo	Guangzhou	Yangzhou	Suzhou
1991	7.5	27.1	6.1	21.2
1992	12.1	28.8	9.5	25.1
1993	16.4	26.9	13.5	25.7
1994	24.1	25.1	14.4	26.0
1995	21.3	22.6	13.3	26.7
1996	23.6	22.0	12.3	24.9
1997	13.1	15.5	12.0	25.1
1998	10.8	16.4	12.6	28.4
1999	10.8	11.2	12.8	28.6
2000	7.6	12.6	10.6	28.0
2001	–	13.4	–	27.6

Among them, Suzhou has the highest GSR, with an average rate of 23.6%, which shows a strong potential for sustainable development in future. The higher domestic saving and lower personal consumption in Suzhou compared to the other four cities resulted in its attaining the highest GSR. Guangzhou has the second highest GSR, with an average annual rate of 18.9%. However, recent trends for Guangzhou are declining in spite of its rapid economic development and improvement in environmental quality, because its economic activity greatly relies on consumption of nonrenewable natural resources (mainly coal and oil). Ningbo is third with an average annual GSR of 14.7% from 1991 to 2000. However, its GSR began to fall quickly after 1996 because of a deficiency in gross investment and an accelerated depreciation of fixed assets. Finally, the average annual GSR for Yangzhou over the 10-year period was 11.5%, the lowest among these cities. This was primarily caused by the higher depreciation of fixed capital and depletion of natural resources, as well as the long-term damage due to CO_2 emissions.

Ecological Footprint Model Assessments

Figure 4.3 shows the temporal trends in the per capita ecological footprint (hectares per person) for each city from 1991 to 2001. Total EF per capita in all four cities has increased quickly since 1991 because of the vast consumption of materials and resources, especially in Ningbo, whose EF per capita grew from 2.59 ha to 4.98 ha per person from 1991 to 2001. The per capita EF for these four cities exceeded the average for China (1.8 ha) (Redefining Progress, 2002) and were above the average of other developing countries like India (1.1 ha). Although the EF for these four cities is smaller than those of many developed

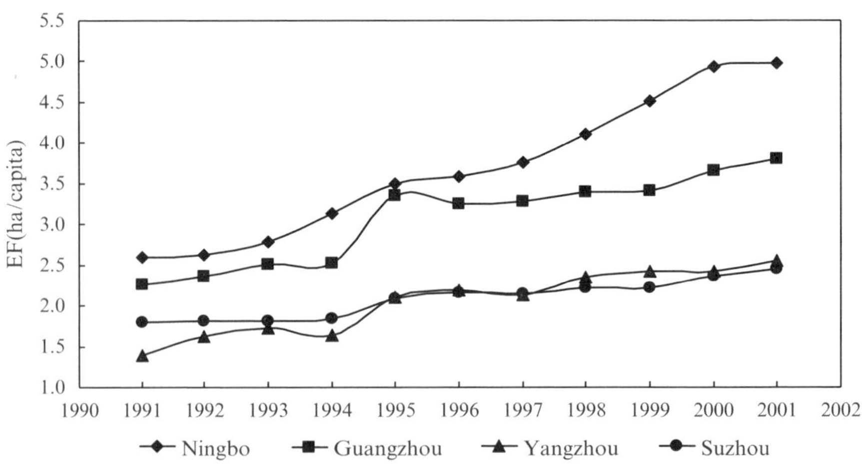

Fig. 4.3 Per capita ecological footprint (EF) for four Chinese cities from 1991 to 2001

countries, they can still be considered high in relation to their economic level. High population density and low per capita ecological carrying capacity (ECC) led to large ecological conflicts in these cities. Our estimates of ecological carrying capacity per person can be arranged in order as follows: Ningbo (0.40 ha), Suzhou (0.38 ha), Yangzhou (0.35 ha), and Guangzhou (0.29 ha). These values are smaller than the average for China (0.89 ha), and far less than the global average (2.3 ha). The ecological conflicts per capita of these cities (defined as the value remaining when ECC is subtracted from EF) are all more than 2 ha in 2001. In descending order the ecological conflict values were Ningbo (4.5 ha), Guangzhou (3.3 ha), Yangzhou and Suzhou (2.0 ha), demonstrating that all four cites are ecological unsustainable.

Index of Sustainable Economic Welfare and Genuine Progress Indicator Model Assessments

We used the ISEW and GPI as to describe sustainability status and trends using sociopolitical criteria. Trends for ISEW (Fig. 4.4) and GPI (Fig. 4.5) increased for all four cities from 1991 to 2001, but at varying rates. The ISEW rate increase for Ningbo was the most rapid, reflecting a greater than 500% rise in its per capita GDP and average ISEW growth rate of 24% during that period. However, Ningbo's per capita ISEW increase was significantly slower than that of its per capita GDP. Ningbo's ISEW peaked in 1996 and then declined by 2001. As a result, the difference between its per capita GDP and ISEW is more noticeable in later years. Comparing the growth rate of Ningbo's per capita ISEW and per capita GDP over the period 1992 to 2000, we found that growth rate in per capita ISEW is lower than the growth rate of per capita GDP for all years except 1993. One potential reason

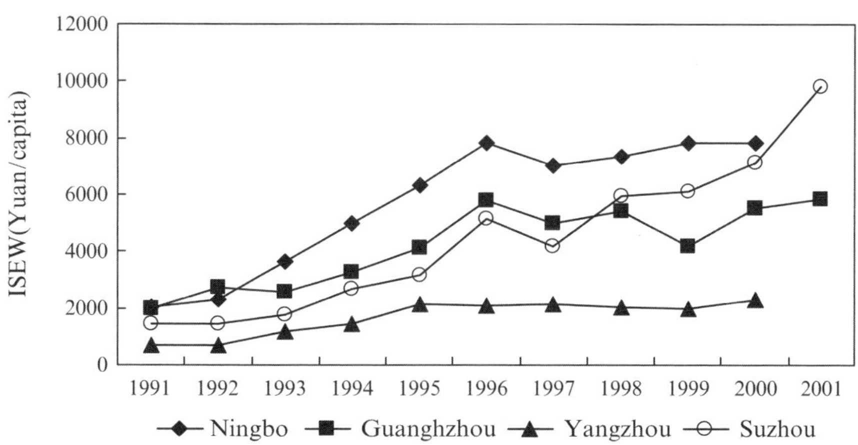

Fig. 4.4 Trends in per capita Index of Sustainable Economic Welfare (ISEW) for four Chinese cities from 1991 to 2001

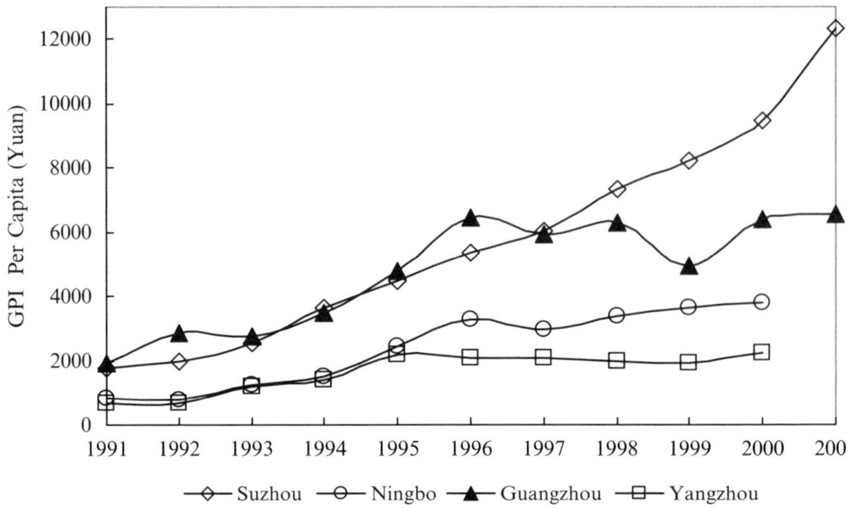

Fig. 4.5 Per capita Genuine Progress Indicator (GPI) for four Chinese cities from 1991 to 2001

for this pattern could be that the negative components of human welfare (e.g., loss of wetlands and costs of water pollution) have been growing much faster than consumption expenditure and the positive components of human welfare. In general, it seems that Ningbo's current production processes are threatening the future welfare of its citizens. The GPI analysis shows similar trends as the ISEW in that from 1991 to 2000 the per capita GDP increased from 3315 to 21735 yuan, but the per capita GPI only increased from 750 to 3514 yuan. Although per capita GPI did increase, it did so more slowly than the growth in per capita GDP, and therefore the gap between the two indices increased over this period. The GPI results for the other cities (Fig. 4.5) show that Suzhou's GPI was increasing rapidly, while that of Yangzhou improved slowly before 1995 but stagnated after that. The trend for Ningbo was only slightly greater than that of Yangzhou.

Urban Ecologically Sustainable Development Indicators Model Assessments

The methodology that we used in this part of the study is a comprehensive assessment approach, and we focus our in-depth explanation on Ningbo. We collected and processed raw data for the five subsystems (resource, economic, institutional, social, and environmental supports) used to calculate the UESDI. We then used principal component analysis to derive a value (L) that reflects the level of aggregated development for the city for each year for each of the support systems (Fig. 4.6). Due to many difficulties in evaluating the

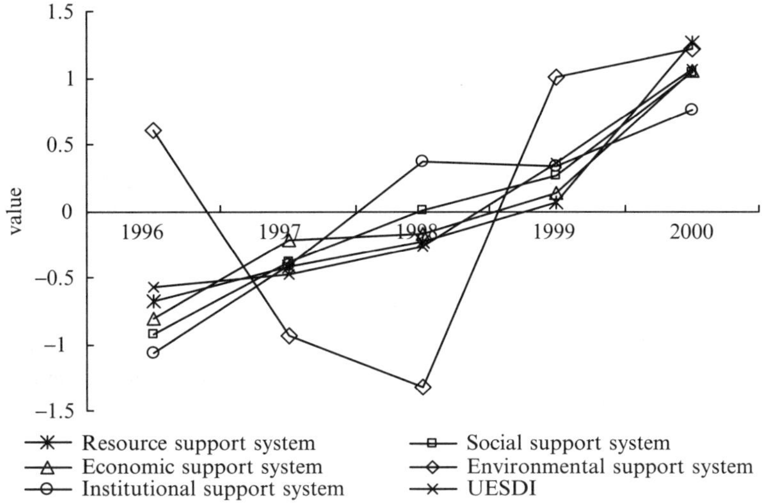

Fig. 4.6 Analysis of changes over time in the L-value (y-axis) for the five support systems and the Urban Ecologically Sustainable Development Indicators (UESDIs) for Ningbo city. The L-value is derived from many variables using principal components analysis

contribution of each subsystem, in this analysis we subjectively assigned an average weighting factor to each of the five subsystems to get one integrated index (L). We can examine sustainability trends over the past 10 years based on the changing L-values of the UESDI. Finally, the pentagon radar chart (Fig. 4.7) reflects and compares the integrated development level and the degree of coordination or correlation among the five support subsystems for Ningbo in 1996 and again in 1999. For example, during that interval the economic support system grew both in absolute terms and in proportion to other support systems, whereas the social support system did not keep up proportionately with economic growth.

Integrated Assessment of Urban Sustainability for Ningbo

Based on the results of the above five models and to make a more comprehensive analysis, we will continue to use Ningbo city as an example to offer our summary assessment of its urban sustainability. Trends in sustainability for Ningbo from 1991 to 2001 were obtained from each model and are shown together in Table 4.4. One model (EF) indicates that Ningbo is developing unsustainably, while two models (UESDI and GSR) indicate that its sustainability is marginal. However, use of the other three models could allow one to argue that Ningbo's development trajectory is sustainable. Therefore, we found that a set of scientific indicators, rather than

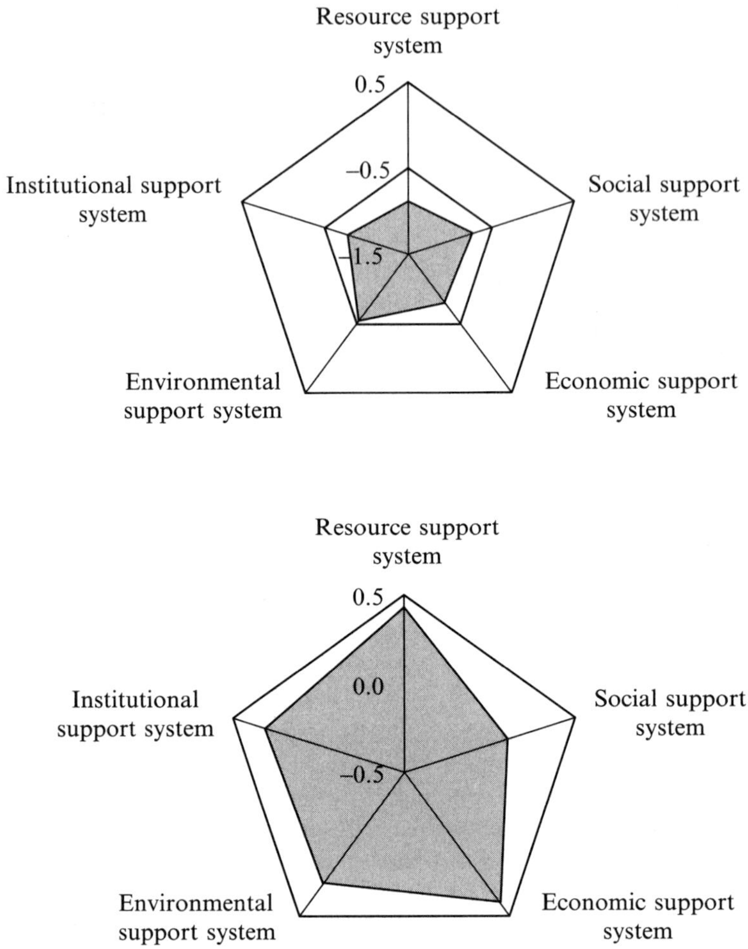

Fig. 4.7 The value of different support systems in Ningbo in 1996 (top) and 1999 (bottom)

Table 4.4 Sustainability assessments for Ningbo based on five indicators

Indicator	Model	Measure	Sustainable	Marginal	Unsustainable
Economic	AEANDP	Currency units	Yes	–	–
Economic	GSR	% of GDP		Yes	–
Biophysical	EF	Land area/capita		–	Yes
Sociopolitical	GPI	Currency units	Yes	–	–
Sociopolitical	ISEW	Currency units	Yes	–	–
Integrated	UESDI	Unit less		Yes	

dependence on one indicator, is very necessary for decision makers to monitor and assess ecologically sustainable development of cities. Local governments are more and more interested in setting their cities on a sustainable path. Sustainable development indicators can be a key mechanism for encouraging progress in this direction.

Conclusion

Based on the results derived from the five models and the UESDI for measuring urban sustainability for four Chinese cities, we have arrived at some key conclusions: (1) although indicators represent the status of a city or cities during the same study period, results from the methods or models differ from one another; therefore (2) using a single indicator to measure sustainable development is incomplete and not recommended; (3) each indicator has its own biases both in theory and in practice, so it would be more reasonable and innovative to measure sustainability from multidisciplinary standpoints; (4) indicators should be simplified as much as possible so that policy makers, the pubic, and the media can communicate effectively with each other. However, multiple models from different disciplines should be adopted to monitor progress on sustainable development. This approach would help decision makers grasp the comprehensive situation of urban sustainability from multidimensions.

Through the use of this case study with these five models and UESDI on urban sustainability, our research highlights that it is necessary to summarize the advantages and weaknesses of indicators and evaluation models on urban sustainable development systematically. In addition, we need to pay attention to individual characteristics of each city when comparing among cities within and between countries. We feel that it is possible to construct practicable and normative indicators through stepwise statistical computation methods to monitor and evaluate the states and trends in ecologically sustainable development for cities. These indicators can provide valuable decision-making support for municipal governments and administrative departments at different levels.

References

Administrative Center for China's Agenda 21. (2000) The Proceedings of the International Workshop on Sustainable Development Indicators. Heilongjiang People Press, Harbin, China, pp. 68–74.

Chinese Academy of Science. (1999) The Strategy Report of China's Sustainable Development in 2000. Science Press, Beijing, pp. 235–244.

Cobb, C., Halstead, T., and Rowe, J. (1995) If the GDP is up, why is America down? Atlantic Monthly, October:59–78.

Daly, H.E., and Cobb, J.B. (1989) For the Common Good. Beacon Press, Boston.

Daly, H.E., Cobb, J.B., Jr., and Cobb, C.W. (1994) For the Common Good: Redirecting the Economy toward Community, the Environment, and a Sustainable Future. Beacon Press, Boston.

Desai, M. (1993) Income and alternative measures of well-being. In: Westendorff, D.G., and Ghai, D. (eds.) Monitoring Social Progress in the 1990s. Averbury (published for UNRISD), Brookfield, WI, pp. 23–40.

Dixon, J. and Hamilton K. (1997) *Expanding the Measure of Wealth—Indicators of Environmentally Sustainable Development*. Environmental Department, The World Bank, Washington, D.C. pp. 19–30.

Gao, D., Huang, Y., and Li, H. (2001) Preliminary research on eco-city. China Sustainable Development 4: 8–13.

Global Leaders of Tomorrow Environment Task Force. (2002) Environmental Sustainability Index. World Economic Forum Annual Meeting, 2002, Davos, Switzerland, pp. 4–53.

Haberl, H., Erb, K.H., and Krausmann, F. (2001) How to calculate and interpret ecological footprints for long periods of time: the case of Austria 1926–1995. Ecological Economics 38:25–45.

Hamilton, C. (1999) The genuine progress indicator methodological developments and results from Australia. Ecological Economics 30:13–28.
Hu, Y., and He, S. (2000) Comprehensive Assessment Methods. Science Press, Beijing.
Huang, G., and Chen, Y. (2002) The Theory and Planning Methods for Eco-City. Science Press, Beijing.
Luo, S., and Zeng, Z. (1999) Review on the study on indicators framework for sustainable development. Human and Geography 14(4):55–59.
Moffatt, I. (1996) Sustainable Development: Principles, Analysis and Policies. The Parthenon Publishing Group, New York, London, pp. 50–169.
Moffatt, I., Hanley, N., Wilson, M., and Faichney, R. (1999) Time series analysis of indicators of sustainability for Scotland, 1980–1993. Ecological Economics 28:55–73.
Moser F., (1994), Proceedings of the International Symposium: Evaluation Criteria for a Sustainable Economy. EFB Event No. 90, Inst. of Chem. Eng., University of Technology, Graz, April 6–7.
Murray, C.J.L. (1991) Development data constraints and the Human Development Index. UNRISD, Geneva, p. 27.
Prescott, A.R. (1995) The barometer of sustainability: a method of assessing progress towards sustainable societies. International Union for the Conservation of Nature and Natural Resources and PADATA, Gland, Switzerland, and Victoria, BC.
Redefining Progress. (2002) Ecological footprints and ecological capacities of 152 nations—the 1996 update. Oakland, CA. http://www.rprogress.org.
Rees, W. (2000) Eco-footprint analysis: merits and brickbats. Ecological Economics 32:371–374.
Register, R. (1996) The Eco-City Movement—Deep History, Movement of Opportunity. Village Wisdom/ Future Cities: The Third International Eco-City and Ecovillage Conference, Oakland, CA, pp. 26–29.
Register, R. (2002) Ecocities: Building Cities in Balance with Nature. Berkeley Hills Books, CA.
Schmidt-Bleek, F. (1994) The Fossil Makers. Factor 10 and More. Birkhauser, Basel, Boston, and Berlin.
Task Group of Indicators System on Sustainable Development. (1999) Manual of Study on Urban Environmentally Sustainable Development in China. China Environmental Science Press, Beijing.
United Nations. (2001) Indicators of Sustainable Development: Guidelines and Methodologies. Division for Sustainable Development, Department of Economic and Social Affairs, United Nations.
Wackernagel, M., and Rees, W. (1996) Our ecological footprint. New Society Publishers, Gabriola, BC, and Philadelphia, PA.
Wen, Z. (2005) Capital Extension Methodology: The Simulation Study into Policy Alternatives Towards Sustainable Development. Dissertation, Tsinghua University.
Wen, Z., Zhang, K., Du, J., and Du, B. (2004a) Defects of the GDP accounts system and the amendatory approaches: methodology and case study. Journal of China University of Geosciences (Social Sciences Edition) 4(3):43–46.
Wen, Z., Zhang, K., Chen W., et al. (2004b) Genuine progress indicator methodology and the case study. China Soft Science 8:145–151.
Wen, Z., Zhang, K., Huang, L., et al. (2005a) Genuine saving rate: an integrated indicator to measure urban sustainable development towards an eco-city. International Journal of Sustainable Development and World Ecology 12:1–13.
Wen, Z., Zhang, K., Du J., et al. (2005b) Methodology study on monitoring regional sustainability. Economic Geography 25(1):26–32.
Wen, Z., Zhang K., and Li W. (2006) Case study on the genuine progress indicator methodology to measure economic welfare in China. Ecological Economics 2007, 63:463–475.
Zhang, K. (2001) Policies and Actions on Sustainable Development in China (English version). China Environmental Science Press, Beijing. (Chinese version published by same press in 2004.)
Zhang, K., He, X., and Wen, Z. (2000) The study on urban environmentally sustainable development indicators in China. China Population, Resources and Environment 10(2):54–59.
Zhang, K., He, X., and Wen, Z. (2003a) Study of indicators of urban environmentally sustainable development in China. International Journal of Sustainable Development 6(2):170–182.
Zhang, K., and Wen, Z. (2001) Progress in indicators of urban ecologically sustainable development. Urban Environment and Urban Ecology 4(6):1–4.
Zhang, K., Wen, Z., Du, B., and Song, G. (2003b) Evaluation and indicators on Eco-city. Chemical Industry Press, Beijing.

5
Assessment and Valuation of the Ecosystem Services Provided by Urban Forests

Wendy Y. Chen and C.Y. Jim

Urban forests, composed of trees and other vegetation, are integral parts of urban ecosystems. Whether planted intentionally or left by default, urban forests appeared even in the earliest settlements. In urban areas, the constituent greenery provides a broad range of benefits, including opportunities for residents to have daily contact with nature, and to enjoy attractive landscapes and recreational activities (Grey and Deneke, 1986; Rowntree, 1986; Ulrich, 1986; Dwyer et al., 1992; Miller, 1997; Bolund and Hunhammar, 1999; Tyrväinen and Miettinen, 2000). In addition, vegetation in cities moderates microclimate extremes and reduces regional pollution (Botkin and Beveridge, 1997; Whitford et al., 2001). They contribute to an improved quality of urban life in many ways, even though these functions are often taken for granted by the public and some city authorities.

The environmental benefits and natural functions provided by urban forests can be interpreted as ecosystem services, which are defined as benefits that the human population can derive, directly or indirectly, from ecosystem functions (Costanza et al., 1997). The urban population must rely mainly on services derived from external ecosystems, such as food and energy. However, the diversified benefits generated by urban forests, which are limited in comparison with the amount of imported ecosystem services, could be more instrumental in solving local environmental problems. They could significantly improve the quality of urban life, and play a paramount role in stabilizing and sustaining urban ecosystems (Daily, 1997; Bolund and Hunhammar, 1999; Jensen et al., 2000). Such ecosystem services, however, are not very tangible and generally not well understood or appreciated. Recent studies have generated a wealth of scientific information on the magnitude of their benefits. A more direct interpretation of these benefits for laypersons could promote their preservation and enhancement.

A useful approach is to quantify these natural services, and then follow with a valuation of these nonmarket and noncommodity goods. The results could be translated into the universal language of monetary units, and be compared with alternatives to facilitate decision making related to natural resources or the environment. Valuation is inseparable from the choices and decisions humans have to make about ecosystems (Bingham et al., 1995; Costanza et al., 1997; Barbier et al., 1998; Costanza, 2000). Some studies have attempted to quantify the ecosystem services generated by urban

M.M. Carreiro et al. (eds.), *Ecology, Planning, and Management of Urban Forests: International Perspectives.*

forests, such as recreational opportunities (Tyrväinen and Miettinen, 2000; Tyrväinen, 2001; Jim and Chen, 2006a), carbon dioxide sequestration and carbon storage (Nowak, 1993, 1994a; McPherson, 1998; Brack, 2002; Nowak and Crane, 2002), air pollutant removal (Nowak, 1994b; Beckett et al., 1998; Nowak and Dwyer, 2000; Akbari et al., 2001; Brack, 2002), microclimate regulation (Heisler et al., 1994; McPherson et al., 1997; Akbari, 2002), and rainwater retention (Sanders, 1986; Nowak and Dwyer, 2000; McPherson and Simpson, 2002). These studies provide an objective, scientific, and convincing basis for the planning and management of urban forests (Tyrväinen, 2001; Nowak et al., 2002a,b).

This chapter reviews studies on ecosystem services provided by urban forests, and provides an overall assessment of the status of the science. Case studies have provided a rich array of specific and objective data to verify in concrete terms many benefits of urban greenery, some of which were in the past mainly advocated as broad-brush interpretations, assumptions, and postulates. Particular attention is given to empirical studies of benefits generated by urban forests, including the identification of ecosystem services, the intrinsic value embodied in such services, and the methods to assess their value. The extensive findings from assessing ecosystem services should provide a firm basis for proceeding to the next logical step of valuation. Associated potential applications are also discussed.

Ecosystem Services Provided by Urban Forests

Different approaches have been adopted to categorize the diversified benefits of urban forests (Grey and Deneke, 1986; Phillips, 1993; Miller, 1997). Although most services are indirect and intangible, they play important roles in the sustainable operation of local ecosystems, and contribute notably to the welfare of urban society. These services have long been recognized, and a large body of literature has attempted to identify and quantify them.

Biomass Functions

Plants in urban forests act as primary producers to absorb carbon dioxide and generate oxygen through photosynthesis. The annual rate of O_2 release and CO_2 sequestration depends on photosynthetic capacity of plants, which in turn depends on species composition and age structure of urban vegetation (Rowntree and Nowak, 1991; Nowak, 1993, 1994a; McPherson, 1994a, 1998; Moll and Kollin, 1996; Whitford et al., 2001; Brack, 2002; Nowak and Crane, 2002). It was estimated that an acre (0.405 hectare [ha]) of tree cover in Brooklyn, New York, could generate a net value of approximately 2.8 ton (2.856 metric ton [t]) of oxygen per year (excluding the effect of tree decomposition) based on field data of tree density and trunk diameter at breast height (DBH) measurements (Nowak et al., 2002b). This amount could satisfy the annual oxygen consumption of 14 people. However, it is debatable whether oxygen produced by urban forests is important, since there are many sources of oxygen and plenty of oxygen in the atmosphere. However, urban forests do contribute oxygen to the atmosphere, and, together with the gaseous and particulate air pollution removed by vegetation, they can improve the quality of urban air (Guan and Chen, 2003).

Vegetation can directly and indirectly reduce atmospheric CO_2, a greenhouse gas associated with an increased risk of global warming (Schneider, 1989; Nowak, 1993, 1994a; Moll and Kollin, 1996). Trees store carbon and actively sequester it during growth. Trees also cool the city by shading and evapotranspiration, which reduces the demand for air conditioning, thereby avoiding CO_2 emissions associated with electric power generation (Heisler, 1986a; Akbari and Taha, 1992; Nowak, 1994a; McPherson, 1998; Akbari, 2002). Compared with other plant life forms, trees have a larger biomass, higher leaf area index, and longer life span, and are more effective in retaining carbon and cooling the air.

Usually tree cover is used to estimate the storage rate of CO_2 (Dorney et al., 1984; Nowak, 1993; McPherson, 1994b). Radial trunk growth data have been used to calculate annual carbon sequestration (Nowak, 1994a; Jo and McPherson, 1995, 2001; McPherson, 1998). The capacity of trees to capture carbon has varied. The average amount of carbon stored per tree in the city of Sacramento is 2343 kg, compared to 336 kg in Oakland and 756 kg in Chicago. The annual CO_2 uptake ranged from 35 to 43, 22 to 36, and 1.02 to 48 kg per tree, respectively, in Sacramento, Chicago, and Brooklyn (Nowak, 1993, 1994a; McPherson, 1998; Nowak et al., 2002b). The amount of CO_2 emissions avoided depends on the degree of air temperature decrease by the citywide tree canopy cover. Urban carbon emissions could be decreased by 0.2% to 3.8% at 11% tree cover, and 3.2% to 3.9% at 33% cover (Jo and McPherson, 2001). Regional climate and the fuel composition used to generate electricity can influence potential CO_2 emission avoidance. Vegetation could significantly reduce CO_2 emissions in regions with a long cold season and that use coal as the primary fuel (McPherson, 1998).

Tree species, age, health condition, weather, and environmental conditions could influence the amount of CO_2 uptake and carbon storage. For a given city, selecting species with large final dimensions, providing good growing sites and conditions, keeping them strong and vigorous, and permitting them to reach their biological potentials in terms of size and physiology could raise the cost-effectiveness of the urban forest in terms of carbon sequestration and carbon emissions avoidance.

Environmental Benefits

Environmental benefits are key services provided by urban forests, including pollutant absorption and filtration, microclimate regulation, noise reduction, and rainwater retention. Such functions, however, are not easy to comprehend by the average layperson because of their intangible characteristics. Human activities have introduced many pollutants into cities, such as sulfur dioxide gas, nitrogen oxide gases, particulates, and noise. These contaminants have created major environmental and public health problems in many cities (Bolund and Hunhammar, 1999). The heat-island effect makes cities hot and uncomfortable places to live and work. Billions of dollars must be spent annually to lower the elevated temperature in attempts to shift the bioclimatic regime into the comfort zone inside buildings. Various studies have demonstrated that urban forests can effectively mitigate these problems.

Air Pollutant Abatement

Air pollution is a rather pervasive and serious problem in urban areas, especially its potential to damage human health (Nowak, 1994b; Beckett et al., 1998). Air pollution also may induce other problems, such as damage of vegetation and human-made materials, visibility reduction, and acidic deposition (Fenger et al., 1998; Kojima and Lovei, 2001). In modern cities, the major air pollutants include sulfur dioxide (SO_2), nitrogen oxides (NO_x), ozone (O_3), and particulates usually expressed as fine particulate matter with a diameter less that 10 μm (PM_{10}). SO_2 and NO_x mostly result from stationary fossil fuel combustion sources and automobiles; O_3 is formed through chemical reactions involving NO_x and volatile organic compounds (Nowak, 1994b; Fenger et al., 1998); PM_{10} is associated mainly with road-transport emissions (Chow et al., 1996; Samaras and Sorensen, 1998). Other human activities can also produce PM_{10}, such as combustion of fossil fuels in power stations, industrial processes, construction, and chemical reactions involving gaseous pollutants (Nowak, 1994b; Beckett et al., 1998).

Air pollutants are removed from the atmosphere by trees in urban green spaces mainly through dry deposition, a mechanism by which gaseous and particulate pollutants are transported to and absorbed into plants mainly through their surfaces (Smith, 1990; McPherson, 1998; Lovett et al., 2000; Fowler, 2002; see also Chapters 1 and 11). The effectiveness of this ecosystem service varies by plant species, canopy area, type and characteristics of air pollutants, and local meteorological environment (Sehmel, 1980; Smith, 1990; Zhou, 1993; Nowak, 1994b; Fowler, 2002). Gaseous pollutants could be absorbed into plant tissues through the stomata together with CO_2 in the process of photosynthesis, and together with O_2 in respiration. After entering the plant, transfer and assimilation could fix the pollutants in the tissues. Inside the plant, SO_2 and NO_2 would react with water on inner-leaf cell walls to form sulfurous and sulfuric acids, and nitrous and nitric acids (Legge and Krupa, 2002). These acids may further react with other food compounds to be transported to different parts of the plant (Smith, 1990; Nowak, 1994; Li and Xu, 2002).

Leaves, branches, stems, and associated surface structures (e.g., pubescence on leaves) could trap particles that are later washed off by precipitation (Smith, 1990). In addition, tree transpiration can increase air humidity, thus aiding settlement of airborne particulates (Grey and Deneke, 1986). Urban tree canopies are more effective in capturing particles than other vegetation types due to their greater surface roughness (Manning and Feder, 1980), which increases turbulent deposition and impaction processes by inducing localized increases in wind speed (Croxford et al., 1996; Beckett, et al., 1998). Variations in the structure and micro-roughness of a leaf surface affect patterns of PM_{10} deposition (Burkhardt et al., 1995; Beckett et al., 1998). In addition, the location of trees in relation to pollutant sources is important. Particulate interception by vegetation would be considerably higher near their sources (Impens and Delcarte, 1979; Spitsyna and Skripal'shchikova, 1991).

The ecosystem service of urban forests in removing air pollutants has been quantified and calculated (Nowak, 1994b; Taha, 1996, 1997; Beckett et al., 1998; Nowak et al., 1998, 2000; Rosenfeld et al., 1998; Scott et al., 1998; McPherson et al., 1999; Akbari et al., 2001; Akbari, 2002). For instance, for land covered by street trees in St. Louis, particle removal reached about 2.75 lb/acre/day (3.08 kg/ha/day) (DeSanto et al., 1976b). For other locations covered by trees in the same city, the removal rate was 1.3 to 3.9 lb/acre/day (1.4 to 4.4 kg/ha/day) for particles, 1.2 to 3.6 lb/acre/day (1.3 to 4.0 kg/ha/day) for NO_x, 20.2 to 66.3 lb/acre/day (22.7 to 74.4 kg/ha/day) for SO_2, and 30.9 to 99.5 lb/acre/day (34.7 to 111.6 kg/ha/day) for O_3 (DeSanto et al., 1976a).

In Chicago, many factors, including aerodynamic roughness, atmospheric stability, pollutant concentration, solar radiation, temperature, turbulence, wind velocity, particle size, gaseous chemical activity and solubility, and vegetative characteristics, were taken into consideration in estimating the dry deposition rate of air pollutants (Nowak, 1994b). The removal rate by Chicago's urban forests in 1991 was estimated to be 0.7 kg/ha/year for CO, 2.1 kg/ha/year for SO_2, 2.4 kg/ha/year for NO_2, 5.5 kg/ha/year for PM_{10}, and 6.0 kg/ha/year for O_3. Removal occurred mainly during the in-leaf season, and total removal was up to 87.5 kg/ha/year. In Frankfurt, Germany, a street with trees had 3000 dust particles per liter of air, whereas streets without trees in the same neighborhood had 10,000 to 30,000 particulates per liter of air (Mink and Witter, 1982).

Microclimate Amelioration

Urban areas are well known to be warmer than the surrounding countryside by an average of 0.5° to 1.5°C in temperate latitudes (Hutchison and Taylor, 1983; Grey and Deneke, 1986; Oke, 1989; Grimmond and Oke, 1995; Yokohari et al., 2001; Akbari, 2002), and up to 3°C in tropical areas (Tso, 1996). The elevated air temperature that defines these urban heat islands is often accompanied by reduced relative humidity. Both changes in urban microclimate can make city centers and other densely built-up areas uncomfortably hot for humans. To adjust the indoor microclimate artificially to the comfort zone, large amounts of energy must be consumed (Akbari et al., 2001). That urban forests could ameliorate microclimate has been strongly perceived by citizens in Guangzhou city in China (Jim and Chen, 2006b).

The heat island is intensified by the lack of vegetation and the common occurrence of dark surfaces in urban areas (Grey and Deneke, 1986; Akbari et al., 1990, 2001; Grimmond and Oke, 1995; Tso, 1996; Whitford et al., 2001; Akbari, 2002).

Common building materials such as asphalt, concrete, steel, and glass are ineffective insulators; thus, the absorbed heat is readily conducted and transmitted. These materials also have high thermal capacities that store large amounts of heat energy during the daylight hours, helping to raise daytime air temperature. At night, the stored heat is readily dissipated, raising the temperature in the surrounding air and contributing to the increase in the frequency of hot nights in cities. The concrete canyon phenomenon traps air and reduces air movement, thus also trapping heat within the urban matrix. Unlike these dark artificial surfaces, only about 20% of incident solar energy falling on a leaf is re-radiated. Therefore, green plants could significantly reduce the amount of re-radiated long-wave radiation in cities (Peck and Associates, 1999).

Urban forests can effectively modify the microclimate and improve thermal comfort in the summer through three mechanisms (Oke, 1989; Akbari et al., 1990; Taha et al., 1997; Nowak et al., 2000; Simpson, 1998; Luvall et al., 2000; Gómez et al., 2001). First, appropriately situated trees can prevent some solar radiation from striking buildings, thus reducing initial heating and heat storage, and reducing energy used to cool buildings (Heisler, 1986b; Simpson and McPherson, 1996). Trees on the west side of buildings are the most valuable, followed by the east, and then the south in the Northern Hemisphere and the north in the Southern Hemisphere. Their efficiency may change in relation to geographical conditions (Parker, 1983). Deciduous trees are particularly beneficial for their ability to admit solar radiation during the winter, while blocking it during the summer (Pitt et al., 1979; Akbari, 2002).

Second, urban forests can act as windbreaks to modify the ambient conditions around buildings. Scattered trees planted throughout a neighborhood were found to increase surface roughness, thereby reducing wind speeds (Heisler, 1990). The effectiveness of windbreaks depends on tree height, width, length, and permeability (Robinette, 1972; Pitt et al., 1979). Lower wind speed could reduce penetration of outside air into indoor space, which could be beneficial during both the heating and cooling seasons. Overall, trees and other vegetation can lower outdoor temperature in summer, and reduce heat loss in winter (Heisler, 1986a; Akbari and Taha, 1992; Heisler et al., 1994; McPherson, 1994d; Akbari, 2002).

Third, urban forests could lower summer air temperatures through evapotranspiration (Liu, 1998; Akbari, 2002). An average mature tree can transfer up to 100 gallons (about 378.5 kg) of water into the atmosphere through transpiration in a hot summer day to cool the ambient air (Kramer and Kozlowski, 1960; Kozlowski and Pallardy, 1997). Transpired water from leaf surfaces can cool the air because latent heat of vaporization from the ambient air is absorbed to convert liquid water into water vapor. The physical process of evaporation from soil surfaces associated with trees similarly contributes to air cooling. Evapotranspiration during the summer from an area with good urban forest cover can notably decrease air temperature and increase relative humidity (Meier, 1990/91; Barradas, 2000; Akbari, 2002), generating an "oasis effect" in the urban fabric. In such an environment, people will feel more comfortable and buildings will consume less cooling energy (Heisler, 1986a, 1990; McPherson, 1994a,b,d; Laverne and Lewis, 1996; Simpson, 1998; Akbari, 2002).

Yokohari et al. (2001) used an internal boundary layer (IBL) model to measure the cooling impact of paddy fields on summer air temperature in a residential area in Tokyo. The study demonstrated that the cooling effect of urban vegetation could extend approximately 150 meters (m) into surrounding residential areas. To reap the benefits of such an extended cooling effect, Yokohari et al. suggested a wider distribution of urban forests throughout urban areas. They also proposed that streets should be aligned parallel to prevailing winds during hot summers and open onto green spaces as far as possible.

The ecosystem services of trees can be valuated in relation to air temperature regulation and associated energy savings for cooling and heating. Some studies documented the differences in cooling energy use between houses on landscaped and non-landscaped sites. Parker (1983) measured the cooling energy consumption of a mobile trailer in Miami, Florida, and found properly located trees and shrubs reduced electricity use for air conditioning by as much as 50% (Akbari et al., 2001). In summer of 1992, Akbari et al. (1997) monitored peak-power and cooling energy savings by shade trees next to two houses in Sacramento, California. They found the shading and microclimate effects of trees yielded seasonal cooling energy savings of 30%, corresponding to an average of 4kWh/day. Simpson (1998) estimated that tree shade reduced cooling load by 12% in a residential location in Sacramento County. In Chicago, shade from a large street tree located to the west of a typical brick residence can reduce annual air conditioning energy use by 2% to 7% (McPherson et al., 1997).

A few more studies have focused on the quantification of the evapotranspiration and wind-shielding effects through computer simulations. Heisler (1986a,b, 1990) investigated the impact of trees in reducing wind speed and the impact of tree location around a house on energy use. Akbari and Taha (1992) used Heisler's data to simulate energy use of typical houses in cold climates. They found that in cold climates, a 30% increase in urban tree cover can reduce winter heating energy use by 10%, and evergreen trees planted on the north side of buildings can effectively protect the buildings from the cold north wind.

Noise Reduction

Noise may be potential sources of physical and psychological stress to humans. Unwanted sound is widely engendered in cities because of their high concentration of people and machinery. Generally, noise levels in excess of 70 decibels___ (dBA) are perceived as annoying. Sound attenuation, involving absorption, deflection, reflection, refraction, and masking, take place over distance, but can be absorbed over shorter distances by the use of barriers. Both living vegetation and artificial structures help to dampen noise levels (Farnham and Beimborn, 2003).

Urban forests can be used as effective noise attenuators. High frequencies, which are the most bothersome, could be absorbed, deflected, or refracted by leaves, twigs, and branches of trees and shrubs of urban forests with proper design (Aylor, 1972; Miller, 1997). Trees also mask noise by generating pleasant sounds as wind moves tree leaves or as birds sing in the canopy (Miller, 1997). Some studies suggest that when planted with enough width and density, vegetation can noticeably reduce noise. For noise reduction, trees with dense crowns, and shrubs should be planted close to the noise sources. A 30-m-wide tree belt combined with soft ground surfaces can reduce loud noise by 50% or more (6 to 10 dBA) (Miller, 1997; Nowak and Dwyer, 2000). Even narrow vegetation belts with sufficient branch and foliage density and strategically placed could be rather effective (Harris and Cohn, 1985). Reduction of 3 to 5 dBA can be achieved by even a 3-m-wide belt of dense trees and shrubs with dense foliage at their base. However, room for such vegetative belts is often unavailable in compact city neighborhoods. Thus vegetation in urban areas could be more effectively employed to screen noise at the source rather than abating noise at recipient sites (Anderson et al., 1984; Lorenzo et al., 2000; Fang and Ling, 2003).

Rainwater Retention

In urban ecosystems, most surfaces are occupied by impermeable structures and surfaces such as buildings and roads, with occasional vegetation and soil cover. Such an anthropogenic land surface substantially modifies the pathways and behaviors of the hydrological cycle in cities (Driver and Troutman, 1989; White, 2002). Due to a low coverage by vegetation and unsealed soil surfaces in cities, rainfall interception and evaporation of intercepted water are reduced. With less permeable area, rain infiltration into soil, and hence soil moisture storage and groundwater recharge, are also significantly less (Whitford et al., 2001). The bulk of the rainwater has to run off from the impermeable land surface, and heavy investment in storm water drainage systems is needed to avoid water accumulation and flooding.

Urban forests and underlying soil play important roles in rainwater retention and runoff avoidance. Subsequently, the urban forests release the retained water gradually into the environment, thus stabilizing discharge into rivers (Kato et al., 1997). The combination of diversion and delaying effects on surface channel discharge can suppress and postpone peak flows, lighten the load on storm water drains (Baines, 2000; Nowak and Dwyer, 2000), and reduce the likelihood of flooding. The potential retention capability of urban forests is related to vegetation type and degree of impervious cover. This capacity could be augmented by installing at strategic locations custom-built detention ponds with permeable vegetated bottoms.

Rainwater retention by urban vegetation can reduce the size and density of drains needed in a city, and hence the costs of constructing and maintaining a city's drainage infrastructure (Grimmond et al., 1994; Kato et al., 1997; Xiao et al., 1998; Baines, 2000; Nowak and Dwyer, 2000; Girling and Kellett, 2002). The cost savings of avoiding flood damages could be factored into the assessment of this ecosystem service. The ancillary benefits of enhanced recharging of the groundwater, and associated environmental and practical implications, could also be considered.

Findings from hydrologic simulations indicate that different amounts of existing tree-canopy cover reduce urban storm water by 4% to 8%, and that a modest increase in tree cover can further decrease runoff. For example, Sanders (1986) found that in Dayton, Ohio, an existing 22% tree canopy cover reduced potential runoff by 7% for an intensive storm, and an increase in canopy to 29% would reduce runoff by nearly 12%. Consequently, the cost of storing and discharging storm water for Dayton would be notably eased if more forest cover existed.

Recreation and Aesthetic Services

For the general public, recreational possibilities and aesthetic enjoyment may be the most readily appreciated benefits of urban forests (Smardon, 1988; Baines, 2000; Lorenzo et al., 2000; Tyrväinen and Miettinen, 2000; McPherson and Simpson, 2002; Jim and Chen, 2006a). Vegetation softens the urban hardscape to create a more aesthetically pleasing landscape. Vegetation creates different colors, shapes, dimensions, textures, sounds, and feels, and these attributes vary infinitely with season, time of day, and weather conditions (Miller, 1997). In addition, vegetation used subtly as a screen and buffer plays an important role in blocking incompatible or undesirable views, channeling people's sight toward beautiful views, and furnishing a natural frame for scenery (Brush et al., 1979; Smardon, 1988; Miller, 1997).

Vegetation is key to making cities pleasant and livable. Using engineering and landscape skills, and integrating grass, shrubs and trees, urban forests can be created as landscaped civic spaces where people can gather and mingle (Millard, 2000). Where appropriate, urban forests are supplemented with playgrounds and sports fields in a comprehensive recreational-plus–green space system. Recreational use of urban forests may include a diverse range of passive and active pursuits, including sitting to relax, reading, sunbathing, listening to or playing music, playing with friends or children, children climbing and hiding, picnicking or eating, and watching and feeding wildlife or birds (Dwyer et al., 1992; Liu, 1998).

Urban forests usually are positive symbols of landscape beauty, although residents' cultural and educational background might affect their preference for species, design styles, and their use of the spaces (Schroeder and Anderson, 1984; Kent, 1993; Oguz, 2000; Tyrväinen et al., 2003; Todorova et al., 2004). Clear views with low-density understory vegetation are associated with increased pleasure and are preferred by visitors (Hull and Harvey, 1989; Tyrväinen et al., 2003). Through various natural attributes, urban forests provide residents with contrasts and diversions to the monotonous, and even harsh, indoor and outdoor conditions that dominate cities (Jim, 1987; Smardon, 1988; Miller, 1997).

Other Ecosystem Services

In addition to the benefits identified above, urban forests provide other ecosystem services that meet the criteria of having economic value, contribute to societal wealth, and are scarce in supply. However, these services might go unrecognized by many urban residents due to their intangible character.

Health and Psychological Services

Urban forests, as principal surrogates for "wilder" nature in cities, provide tranquil and healthy environment for stressed residents (Schroeder and Anderson, 1984; Davey Resource Group, 1993; Ulrich, 1999; Hunter, 2001). Schroeder (1986) found that the common feelings recalled by visitors to the Morton Arboretum (Chicago, Illinois) included peacefulness, serenity, and tranquility. People's positive feeling toward parklands is believed to be connected with our human evolutionary link with nature and hence biophilia, because natural places are associated with ancient survival values of supplying food and water (Wilson, 1999). There is also a close association between attractive urban forests and their therapeutic value especially for people who are ill (Ulrich, 1986, 1999; Todorova et al., 2004). Patients recovering in a hospital ward with high-caliber green views through the windows demanded less pain relief medication and nursing attention, and recovered at a faster rate (Ulrich, 1984). Horticultural gardens have been used in developing programs for therapeutic purposes (Smardon, 1988; Marcus and Barnes, 1999; Jackson, 2003). Natural settings can also reduce mental fatigue and aggression (Kuo and Sullivan, 2001b). Green surroundings in residential neighborhoods tend to reduce the incidence of crimes and reduce the fear of crime (Kuo and Sullivan, 2001a).

Wildlife Habitats

Urban forests provide habitats for a multitude of wildlife, such as birds, mammals, insects, reptiles, and amphibians (Rowntree, 1986; Davey Resource Group, 1993; Adams, 1994; Bradley, 1995; Miller, 1997; Dunster, 1998) that enhance a site's attractiveness and aesthetic enjoyment (Matthews et al., 1988; Nilon et al., 1995). Biotic richness, abundance, and composition are tied to vegetation quality, fragmentation, and other urban forest characteristics (Rowntree, 1986; Nilon et al., 1995; Bolger et al., 1997; Keefe and Giuliano, 2004; Brennan and Schnell, 2005). In a recent study of bird assemblages in Melbourne, Australia, remnant parks with more complex biomass structure and species composition hosted more birds (higher abundance and species diversity) than street vegetation (White et al., 2005).

Biodiversity Conservation

It has been recognized that urban ecosystems can play an important role in biodiversity conservation (MacDonald, 1996; Schiller and Horn, 1997; van den Berg et al., 1998; Maurer et al., 2000; Savard et al., 2000; Lëfvenhaft et al., 2002; Gyllin and

Grahn, 2005), for both floral and faunal species. Although in urban areas, traditional nature conservation seems to be only marginally possible for some practical reasons, scientific planning and a clear understanding of biodiversity among residents would help protect biodiversity in urban ecosystems (Jensen et al., 2000; Lëfvenhaft et al., 2002).

Education and Sites for Scientific Research

Exposure to the diversity of urban nature stimulates the senses and provides an informal mode of outdoor classroom education. Such serendipitous learning opportunities are seldom conveyed by traditional classroom education (Jim, 1987). Green views and green spaces near homes and schools can positively influence the behavior and performance of children. Greenery has been found to contribute to improved concentration and self-discipline, fewer afflictions from attention deficit disorder, fewer behavioral problems, lower likelihood of truancy, and better scholastic achievement (Taylor et al., 2001a,b). Urban forests also provide sites for conducting scientific research, which can contribute to a more thorough and holistic understanding of urban ecosystems in general (McDonnell et al., 1997; also see Chapters 1, 11, and 22).

Valuation of Ecosystem Services Provided by Urban Forests

Ecosystem services provided by urban forests are obviously important to human life and the sustainability of urban ecosystems. The quantification and valuation of these ecosystem services provide information and insight for humans to contemplate alternatives and make decisions relating to urban forests. Some ecosystem services, such as food, timber, and other forest products, can be traded in conventional commodity markets for which the valuing technique is straightforwardly based on common economic principles. Other types discussed above, defined as public goods with positive externalities, demand unconventional valuation to address their nonmarket and noncommodity traits. Many external market techniques for their monetary valuation have been developed (Heal, 2000b). These primarily include use of the hedonic price index (discussed below), replacement cost and travel cost methods (based on actual transactions), and the contingent valuation method (based on hypothetical transactions) (O'Connor and Spash, 1999; Farber et al., 2002, 2006; Freeman, 2003).

The monetary value of ecosystem services provides useful information for benefit–cost analyses or natural resource damage assessments (Shechter and Freeman, 1994; Aldred, 1997; Toman, 1998), for which valuing is a necessary step (Heal, 2000b). Heated debates and some controversies have evolved regarding the choice of valuation methodology (Bromley, 1990; Galbraith, 1992; Common, 1995; Foster, 1997; Grove-White, 1997; McFadden, 1999; Ludwig, 2000).

Concerns about cost-effectiveness are a common and even key consideration in the formulation of scientifically based, environmental management policies (Spash, 1997). This suggests the need to provide workable and more generally acceptable methods for valuing ecosystem services. Recent developments in this field are promising and convincing.

Valuation Methods

With origins in economics, ecology, sociopsychology, and other related disciplines, several methods have been developed to measure the value of diversified ecosystem services (Bingham et al., 1995). Commonly, these techniques fall into four categories: conventional market approaches, household production functions, hedonic pricing, and experimental methods (survey or contingent valuation methods) (Pearce, 1993).

Conventional market approaches offer a basis for valuing ecosystem services that can be transacted in markets. They include market price (or shadow price), tax (or subsidy), tax quota (Randall, 1987, 1991; Heal, 2000a,b), opportunity cost, replacement cost, and averting expenditure (Garrod and Willis, 1992, 1999; Pearce and Warford, 1993; Stern, 1999; Starrett, 2000). The incentives embodied in some of these methods could help to strike a balance between efficient production and distribution of ecosystem services. For instance, the common implementation of the carbon tax furnishes a feasible means of managing human interactions with the earth's ecological base, working effectively to control carbon dioxide emission through realizing the social cost of damaging an environmental asset (Heal, 2000b).

Household production functions are based on the notion that marketed (and nonmarketed) ecosystem services are demanded as intermediaries in a household's consumption process (Smith, 1991). Consumers often gain utility not directly from the goods that they purchase, but instead they transform the goods by a household production function into something that they value. For example, the consumers purchase flour and eggs, and the uses some time and their labor to produce a cake. The consumers did not really want the flour or sugar, but they purchased them so that they could produce the cake that they actually wanted. The focus on household production functions helps to identify the potential links between marketed and nonmarketed ecosystem services. A household allocates some of its available labor, time, and possibly income to an activity that is affected in some way by ecosystem services generally recognized as "environmental quality" (i.e., the state of the environment or the goods and services it provides). The household therefore combines its labor, environmental quality, and other goods to "produce" goods or a service, but only for its own consumption and welfare (i.e., for household utility). By determining how changes in environmental quality influence this household production function and thus the welfare of the household, it is possible to value these changes. For example, the protection of watershed forests could provide an ecosystem service of mitigating droughts, the value of which could be derived through measuring the household savings in water collection costs in relation to household water consumption. Approximate methods could then be developed to gauge the value of provision of ecosystem services by identifying relevant parameters in the indirect utility function (Smith, 1991). Based on the assumption of either a substitute or a complementary relationship between the ecosystem services and marketed commodities consumed by household, household behaviors could be modeled, such as the time allocation model (travel cost model) for recreation (Loomis, 1987; Maille and Mendelsohn, 1993; Riera, 2000), household labor allocation model for water and food collection (Pattanayak, 2004), and averting behavior models (discrete choice model) that account for the health and welfare impacts of pollution (Smith, 1991; National Research Council, 2005).

Hedonic pricing deals with market-priced goods where a certain component or aspect that is particularly pleasing to consumers could be assigned a price index to reflect its contribution to the total value of the goods in question (Lancaster, 1966; Palmquist, 1991; Sheppard, 1999; Gatto and De Leo, 2000; Heal, 2000b). For example, the value of a garden in a property is a component of the house's total real estate market price, and the garden's value can be quantified separately. People usually are willing to pay more for a beautiful view, if two houses are identical except for that view. The extra portion of the transaction price can be estimated as the value of aesthetic service generated by the garden or the pleasant view, although in practice many attributes of a house, such as size, quality, and neighborhood, could jointly influence the price. Some statistical models, such as linear (parametric, semiparametric, and nonparametric), semilogarithmic, double logarithmic, and Box-Cox transformation, have been developed to separate the part of the variation in prices that is attributed to each characteristic (Palmquist, 1991; Haab and McConnell, 2002; Freeman, 2003; Malpezzi, 2003).

Experimental methods, mainly the contingent valuation method and more recently, contingent choice models, have been adopted to capture non-use values of ecosystem services. A carefully designed sampling of representative individuals in a community is required to select the respondents. They are then probed for their preferences for ecosystem services by answering questions about hypothetical choices. The findings of the survey will be extrapolated to the population as a whole. Based on the method of soliciting responses, two approaches can be identified. In the willingness-to-pay approach, respondents are asked how much they would be willing to pay to ensure a welfare gain from the change in the provision of ecosystem services. The alternative is the willingness-to-accept approach, in which respondents are asked how much they would be willing to accept to endure a welfare loss from a reduced provision of the services (Mitchell and Carson, 1989; Carson, 1991; Hoevenagel, 1994; Gatto and De Leo, 2000). The contingent valuation method has been widely applied to value nonmarket ecosystem services provided by natural resources and is increasingly accepted as a valuation method (McConnell and Walls, 2005).

There is an extensive literature on the application of the above valuation methods and the assessment of their advantages and weaknesses (e.g., Farber et al., 2002).

The choice of a suitable approach is dependent on the ecosystem services and the socioeconomic profile of the community in question.

Empirical Valuation of Urban Forests

Various approaches have been employed to assess the value of ecosystem services generated by urban forests, such as replacement cost (McPherson, 1994c; Price, 2003); hedonic pricing (More et al., 1988; Tyrväinen, 1997; McPherson et al., 1999; Tyrväinen and Miettinen, 2000; McPherson and Simpson, 2002; Price, 2003; Kim and Wells, 2005; Jim and Chen, 2006), externality cost (Nowak, 1994b; Hall, 1997; McPherson and Simpson, 2002), travel cost (Dwyer et al., 1983; Grandstaff and Dixon, 1986; Price, 2003), and contingent valuation (Dwyer et al., 1989; Tyrväinen and Väänänen, 1998; Kwak et al., 2003; Jim and Chen, 2006a).

The extensive literature on urban forest value has focused on North American cities. In a study of Chicago's urban forest ecosystem, energy savings, air-pollution mitigation, carbon dioxide sequestration and avoided carbon emissions, avoided runoff, and other benefits associated with trees can outweigh planting and maintenance costs. The benefits were valued at $59 million per year, whereas cost was $21 million, for a net present value of $38 million or $402 per tree planted (30 years, 7% discount rate, 95,000 trees planted). A benefit–cost ratio of 2.83 indicates that the value of projected benefits is nearly three times the value of projected costs (McPherson, 1994c; McPherson et al., 1997). Tree location would affect the benefits derived. It was suggested that the benefit–cost ratios were the largest for trees in residential yards and public housing sites. In this study, the value of energy savings included net heating savings in winter and cooling savings in summer. For Chicago this was estimated using Chicago weather data and a utility price of $0.12/kWh for electric power and $5.00 per million British thermal units (MBtu) for natural gas.

For the value of air quality improvement in the same study, the traditional costs of pollution control were used. These were $1307/ton for PM_{10}, $490/ton for O_3, $4412/ton for NO_2, $1634/ton for SO_2, and $920/ton for CO (1 short U.S. ton = 0.907 metric tons). For the value related to carbon dioxide (carbon dioxide sequestered and avoided), traditional costs of control ($0.011/lb) were used and carbon emission rates were $0.11 lb/kWh (KWh = kilowatt hour) and $29.9 lb/MBtu (1 lb = 0.454 kg). For the value of hydrologic benefits, typical retention/detention costs for storm water control ($0.02/gal) were applied for the ecosystem service of runoff avoided and potable water cost ($0.00175/gal) was used for avoided power plant water consumption (1 U.S. gallon = 3.785 liters). For the valuation of other benefits, which might comprise aesthetic value, improved health, wildlife value, and social empowerment, replacement cost was used. Specific values of this case study (McPherson, 1994c) are given in Table 5.1.

In a comparison of urban forests in two California cities (McPherson and Simpson, 2002), the average annual value was found to be $53.17/tree (total $4.8 million) in

Table 5.1 Annual value of ecosystem services generated by tree plantings in Chicago by location (30 year analysis, 7% discount rate, in thousands of U.S. dollars)

Benefit	Park	Yard	Street	Highway	Housing	Total
Energy						
Shade	233	984	1,184	91	75	2,567
ET cooling	340	1,296	1,676	135	105	3,552
Wind reduction	1,479	5,648	7,302	586	457	15,472
Subtotal	2,052	7,928	10,162	812	637	21,591
Air quality						
PM_{10}	8	11	11	2	1	33
O_3	1	2	1	0	0	4
NO_2	8	19	18	2	2	49
SO_2	8	23	21	2	2	56
CO	1	1	1	0	0	3
Subtotal	26	56	52	6	5	145
Carbon dioxide						
Sequestered	37	65	82	12	5	201
Avoided	92	359	465	37	27	980
Subtotal	129	424	547	49	32	1,181
Hydrologic						
Runoff avoided	46	170	494	24	15	749
Saved at power plant	6	26	32	3	2	69
Subtotal	52	196	526	27	17	818
Other benefits[a]	8,242	11,854	12,262	1,926	923	35,207
Total	10,501	20,458	23,549	2,820	1,614	58,942

ET denotes the combination of evaporation and transpiration.The total row was obtained by summing the three subtotal values and the "Other benefits" value.

[a] Other benefits theoretically represented the value of nonmarket benefits such as aesthetic value, improved health, wildlife value, and social empowerment, which were not calculated separately in this study.

Source: McPherson (1994c).

Modesto and \$83.39/tree (total \$2.3 million) in Santa Monica. The valuation included summer energy savings, carbon dioxide sequestration, air quality improvement, storm water retention, aesthetic contributions, and others. The average value per tree was anticipated to increase with tree age and size. This total value outweighs the annual expenditure on urban forests, which in Modesto were \$2.6 million and \$1.5 million in Santa Monica (Table 5.2).

Some studies focus on special ecosystem services rather than a holistic analysis of urban forest benefits. The meteorological impact of large-scale tree-planting programs has been analyzed in selected United States metropolitan areas: Atlanta, Chicago, Dallas, Houston, Los Angeles, Miami, New York, Philadelphia, Phoenix, and Washington, DC. Model simulations showed that trees could cool these cities on average by 0.3° to 1°C, and by up to 3°C in some locations within a city having big trees and a long duration of shading. For most cities, total (direct and indirect) annual energy savings were \$10 to \$35/year/100 m^2 of roof area in residential and commercial zones (Akbari et al., 2001; Akbari, 2002).

Table 5.2 Annual benefits and costs of urban forests in two American cities in California

Benefit and cost		Modesto	Santa Monica
Benefit	Energy saving	1,000,560	147,534
	CO_2 sequestration	312,920	48,974
	Air quality improvement	538,106	147,682
	Storm water retention	616,139	110,784
	Aesthetic/others	2,380,415	1,894,758
	Total value	4,848,140	2,349,732
Program expenditure	Planting	167,062	22,900
	Pruning	1,202,252	863,380
	Removals	342,896	49,500
	Other	186,722	73,764
	Administration	315,572	102,404
	Subtotal	2,214,504	1,111,948
Nonprogram expenditure	Hardscape repair	297,586	271,344
	Leaf clean-up	106,426	27,808
	Claims and legal	68,000	132,900
	Subtotal	472,012	432,052
Revenue		63,132	–
Net expenditure		2,623,384	1,544,000
Net benefit		2,224,756	805,732
Benefit–cost ratio		1.85	1.52

Source: McPherson and Simpson (2002).

The energy saving potential of urban green spaces in three U.S. cities—Baton Rouge, Sacramento, and Salt Lake City—have also been investigated (Konopacki and Akbari, 2000; Akbari et al., 2001; Akbari, 2002). Several scenarios of strategically placing trees around a building for maximum impacts were considered. A three-dimensional meteorological model was built to simulate the potential impact of trees on ambient cooling benefits and to calculate the energy savings for each region. For the three cities above, a net annual savings in energy expenditure of $6.3 million, $12.8 million, and $1.5 million, respectively, was calculated.

The ecosystem services of amenity and recreation provided by urban forests have been measured using the hedonic pricing method. Tyrväinen and Miettinen (2000) reported that in Salo, Finland, buyers were willing to pay 4.9% more to obtain a dwelling with a forest view. In addition, an increase of 1 km to a green space was found to reduce the house price by 5.9% (reflecting a reduction in recreational opportunity). By applying this hedonic pricing method, the greater the percentage of forested land in the housing district, the higher the house price. The total value of urban forests (only those reflected by housing market, including recreational and aesthetic services) in the study area is 3.84 million Euro dollars (about $4.22 million).

In another study using the contingent valuation method in the same Finnish town, most respondents were willing to pay 31 to 76 FIM/month (about $6 to $14/

month) for the use of urban forests for recreational benefits. Different management quality and venue location might be valued differently by residents (Tyrväinen, 2001). Jim and Chen (2006a) assessed the value of recreation and amenity services provided by urban forests in Guangzhou in south China by the contingent valuation method. They found that aggregate willingness-to-pay by residents was 547.09 million RMB per year (about $66.23 million), which was six times higher than the annual maintenance expenditure for the city's green spaces. The expressed payment level was significantly and positively associated with income (one of the weaknesses of this method is that poor people cannot be considered to the same extent as those with higher incomes). The responses mean that residents consider urban forests to be superior "goods." This first study of an Asian metropolis provides interesting results on human responses to green space in a developing city, and the findings are largely comparable to those obtained in American, European, and Asian cities (e.g., Tyrväinen and Vaananen, 1998; Kleiber, 2001; Kramer et al., 2002; Kwak et al., 2003).

Computer simulations have been used to develop models of urban forest effects (the UFORE model of Nowak et al., 1998). In many urban areas in America, CityGreen software has been used to assess the main ecosystem services generated by urban forests and their value (American Forests, 2002). The computational formulas for CityGreen are based on case studies of urban forests in North American cities (for a critique of CityGreen, see Longcore et al., 2004). In these studies involving both UFORE and CityGreen, ecosystem services included were air pollutant removal, storm water retention, energy savings attributed to urban trees, stored and sequestered carbon, and avoided carbon emissions. For air pollutant removal, in which ozone, sulfur dioxide, nitrogen dioxide, PM_{10}, and carbon monoxide are included, the amount removed is based on dry-deposition on trees during the on-leaf growing season (following the UFORE model developed by Nowak and Crane [2000] based on data collected in 50 U.S. cities), and the value is calculated based on local externality costs set by state public service commissions. Storm water retention ability is based on the TR-55 model for simulating urban hydrology for small watersheds developed by the U.S. Natural Resources Conservation Service (Soil Conservation Service, 1986). CityGreen calculates storm water runoff volume, peak flow, and time of concentration and percentage change under different land cover scenarios. Calculation of the monetary value of these ecosystem services is based on the local average cost of constructing detention basins of a size needed to hold the excess runoff. For energy savings for residential buildings, an annual average of $11.00 per home is adopted, based on American Forests' analysis of existing tree canopy on one- or two-story single-family detached homes. Biomass function uses UFORE model parameters to calculate carbon stored in and annual carbon sequestered by trees. Avoided emission of carbon dioxide is estimated according to kilowatt-hour savings in the energy module, multiplied by U.S. Energy Information Administration data for state-level fuel sources used in electricity production. For biomass functions, only the capacity is given, and no monetary value is calculated. The results of some case studies are summarized in Table 5.3.

Table 5.3 Results of U.S. case studies of the quantitative assessment and valuation of urban forest annual benefits using the CityGreen software

Study area	Tree canopy (year of study) (% cover)	Pollutant removal		Carbon dioxide			Stormwater retention Capacity Value		Summer energy savings (million US$)
		Capacity (million lb)	Value (million US$)	Stored (million ton)	Sequestered (thousand ton)	Avoided (million ton)	Capacity (billion ft^3)	Value (billion US$)	
Houston	30% (1999)	83	208	37.5	138	10.8	2.4	1.3	1.86
Willamette	24% (2000)	178	419	73	563	0.1	10.1	20.2	–
Charlottesville	41% (2000)	230	567	0.9	7.2	–	5	10	–
Roanoke	35% (1997)	14	40.5	9	41	–	1	2	–
Union City	33% (1996)	0.14	0.32	–	–	–	6.9×10^{-3}	0.01	75×10^{-3}
Chattanooga	16.5% (1996)	5.3	12.8	2.4	4		0.38	0.76	–
Atlanta	26% (1996)	19	47	8.3	58	0.7	1.18	2.36	2.8

Source: American Forests. http://americanforests.org/resources/rea

These empirical studies indicate that the value of various ecosystem services provided by urban forests is very high, and often greatly exceeds the cost of tree planting and maintenance. Longer-term public benefits could be raised by increasing tree cover, by planting the right kinds of trees in proper locations, and by providing sound tree management (McPherson, 1994c; Nowak and Dwyer, 2000).

Application of Ecosystem Services and Their Value in the Management of Urban Forests

Prescribing set standards for the amount of urban forest cover needed to generate a certain amount and type of ecosystem service for residents is difficult, because the sustainability of urban ecosystems disproportionately relies on materials and ecosystem services imported from areas lying outside their boundaries (Bolund and Hunhammar, 1999), such as carbon dioxide sequestration and freshwater supply. However, it is possible to modify urban forests to yield ecosystem services in specific locations through judicious allocation and replacement, for example, by planting trees in a given compass direction relative to a house to improve energy savings. Likewise, the value of ecosystem services provided by urban forests could be maximized by configuring the pattern and structure of vegetation to suit the unique character and need of each landscape situation (Bradley, 1995).

A market basis for sustaining urban development relies on the observation that many ecosystem functions provide direct services or values to human economic endeavors, and that such services could be explicitly included in the valuation of proposed development projects. Sustainability, therefore, could be linked to the study and measurement of these ecosystem functions and services. Incorporated as an integral component of development planning, urban forest distribution could be better adjusted to realize ecosystem services with high and accreting value. In essence, the valuation of ecosystem services can contribute to the goals of sustainable development of cities (eco-cities), and to enhance conventional development planning that often favors economic considerations at the expense of natural ones.

A bridge could be built to link urban ecology with economics through the identification of ecosystem services, that depend on the interaction of ecological factors within urban ecosystems and gives them a monetary value that can be tied to a city's economy. Better information on the economic importance of urban forests is crucially needed if we are to sustain urban forests and conserve natural capital in cities. Although it is still difficult to capture all ecosystem services into conventional, market-based economic analyses, urban planning that encompasses the wide range of benefits and values provided by urban forests could help to create special landscapes in a multi-functional, productive, and sustainable way (de Groot, 2006). Therefore, a realization of the worth of urban vegetation together with construction of more resource-efficient city structures and designs could advance our goal of creating workable eco-cities that align with the spirit of smart growth (Gatrell and Jensen, 2002).

Urban forest budgets are often deficient compared with allocations to other municipal services and infrastructures (Konijnendijk, 1997; Miller, 1997). This is often partly attributed to inadequate understanding of urban forest benefits to human society besides those involving their routine ornamental role. Thus financial constraints could result in poor management of urban forests, leading to destabilizing feedback that diminishes their functions, reduces their usage, and erodes people's confidence in their usefulness. Monetary valuation provides a scientific, objective, and convincing message that urban forests can contribute to societal wealth and quality of urban life. Valuation of an urban forest's services could more persuasively justify financial support for enhancing the planning and management of urban forests to sustain these benefits. Well-defined economic incentives, such as removal or damage charges (Nowak et al., 2002a) and subsidization to establish and improve urban forests, could also be based on appropriate valuation of urban forests. Systematic planning and management of urban forests could then be realized, and in turn their socioeconomic value could be more fully be translated, appreciated, and expanded.

Cost-benefit analysis is traditionally applied as a planning and decision-making instrument. By providing clear and true reflection of urban forest benefits and associated values, better insights could be nurtured to inform debates and decisions on the trade-offs involving introducing new urban forests with respect to alternative land-use options. A relatively low priority is commonly accorded to urban forests in the policy-making process involving development. This biased attitude is partly attributed to the insufficient understanding of urban forest functions and the failure to express these functions in monetary terms (Konijnendijk, 1997; Miller, 1997; Tyrväinen, 1997; Jansson and Nohrstedt, 2001; Ekins, 2003). For many land owners and developers, other forms of land use, such as commercial, housing, and industrial uses, are more beneficial than urban forests. But for the general public, healthy and sustainable urban ecosystems are more important for achieving a higher quality of urban life. The trade-off between local short-term economic benefits and long-term sustainability of urban ecosystems has always been a critical issue that needs to be more emphatically addressed in human development. The valuation of ecosystem services provides a methodology and an instrument to compare and contrast alternative options in universal monetary units. Thus urban forests could be given an equal footing and equal treatment in the intense contest for use of scarce urban land.

Transmitting information on urban forest values is pivotal if we are to gain wide public recognition of their importance (McPherson and Simpson, 2002; Tyrväinen et al., 2003). Ecosystem services and estimation of their monetary values could be used as constructive media for engaging the public, planners, policy makers, and managers in urban forest projects. As urban forest benefits are largely intangible and not easily perceived by ordinary citizens, various public and formal educational programs are necessary to convey accurate knowledge. Awareness that urban forests can contribute to societal wealth and health is essential to shift public attitudes from apathy to support (Davey Resource Group, 1993). Clear understanding of relevant benefits could equip and encourage residents to participate in urban forest development and conservation projects (Jepson and Canney, 2003). Without explicit recognition of urban forest functions, human activities might intentionally or inadvertently degrade forest benefits (Jansson and Nohrstedt, 2001) with little forethought and afterthought. In a world of increasing democratization, the public wants the chance to participate in different stages of urban development, from institution and planning to implementation and management. Supported by sufficient evidence of the wide array of contributions to the community, it is possible to rally strong public support for urban forests.

Conclusion

Since the publication of the benchmark work of Costanza et al. (1997), a large body of research has documented the description, identification, analysis, and valuation of ecosystem services generated by various biomes and specific natural ecosystems. Relatively few researchers, however, have ventured outside of remnant natural ecosystems into the human- dominated realm, namely urban ecosystems (Daily and Ehrlich, 1999). It may be difficult for urban residents to appreciate and support conservation of nature in remote areas, if they do not understand nature encountered in their everyday city life. Without a sound knowledge of the diversified ecosystem services generated by urban forests, attempts to improve paradigms for the functioning of urban ecosystems would be incomplete.

Urban forests, one of the most active, complex, and dynamic natural components in urban ecosystems, supply a varied range of services to sustain the system. These include carbon dioxide sequestration, oxygen release, carbon sinks, air pollutant removal, microclimate regulation, noise reduction, rainwater retention, recreation, aesthetic enjoyment, health and psychological services, wildlife habitats, biodiversity conservation, education, and scientific research opportunities. These ecosystem services efficiently mitigate some negative impacts that invariably accompany urbanization, such as air pollution and the heat island. Urban societies desperately desire to find cost-effective solutions to such vexing and aggravating environmental problems. Expanding and improving urban forests provide promises for a natural and sustainable answer.

Some ecosystem services have been quantified and valuated as monetary units to facilitate benefit–cost analyses, to inform public policies, and to integrate urban forests into projects for the enhancement of urban sustainability (Tyrväinen, 2001). Convincing results have shown that urban forest values can always outweigh their maintenance costs, resulting in a high benefit–cost ratio. The annual surplus values could easily amortize the installment costs in a short time, with a handsome profit to be reaped in the long term. Many factors can affect the absolute value of ecosystem services generated by urban forests, such as geographical location of the city, species composition and structure of urban forests, and the choice of valuation methodology. By analyzing the preferred ecosystem services in a certain site and the relationship between function and value, efficient allocation and configuration of urban forests could be designed to maximize ecological functions and contributions to societal wealth and sustainability.

Difficulties and uncertainties should be acknowledged in the quantification and valuation of ecosystem services provided by urban forests, especially transforming ecological understanding of ecosystem functions and services into economically relevant terms (Koomen et al., 2005; National Research Council, 2005), and these issues offer a fertile ground for further studies. More in-depth research is necessary to explore how urban forests provide ecosystem services through complicated interactions among ecological elements within urban ecosystems. In addition, we could study the variables that should be included in the quantification models to improve their accuracy and predictive power. Further refinement of valuation techniques and econometric models could help to overcome the bottleneck of translating ecosystem services more precisely into economic values. Methods could be developed to conduct studies with broader applicability.

In a relentlessly urbanizing world, there is increasing demand and pressure to maximize the use of scarce urban land for development. Fortunately, increasing awareness of the positive roles urban forests play in maintaining environmental and human health counterbalances these forces. The continual rise in human knowledge, affluence, leisure time, and mobility will no doubt augment the demand to protect and enhance nature in cities. Unfortunately, in many countries, urban forest budgets are declining, a trend that is diametrically opposed to the call and need for better and more urban nature (Konijnendijk, 1997; Ekins, 2003). A comprehensive understanding of urban forest values is of crucial importance for promoting better policies and greater support for such natural treasures (McPherson, 1994c; Jensen et al., 2000; Nowak and Crane, 2000; Nowak et al., 2002a).

Acknowledgments The authors would like to express gratitude to the research grant support kindly provided by the Committee on Research and Conference Grants of the University of Hong Kong, the Urban China Research Network Small Grant Award of the Urban China Research Network at the University of Albany, and the John Z. Duling Research Grant of the International Society of Arboriculture TREE Fund.

References

Adams, L.W. (1994) In our own backyard: Conserving urban wildlife. Journal of Forestry 92:24–25.

Akbari, H. (2002) Shade trees reduce building energy use and CO_2 emissions from power plants. Environmental Pollution 116:S119–S126.

Akbari, H., Bretz, S.E., Taha, H.G., Kurn, D.M., and Hanford, J. (1997) Peak power and cooling energy savings of high-albedo roofs. Energy and Buildings 25:117–126.

Akbari, H., Pomerantz, M., and Taha, H. (2001) Cool surfaces and shade trees to reduce energy use and improve air quality in urban areas. Solar Energy 70:295–310.

Akbari, H., Rosenfeld, A.H., and Taha, H. (1990) Summer heat islands, urban trees, and white surfaces. ASHRAE Transactions 96:1381–1388.

Akbari, H., and Taha, H. (1992) The impact of trees and white surfaces on residential heating and cooling energy use in four Canadian cities. Energy 17:141–149.

Aldred, J. (1997) Existence value, moral commitments and in-kind valuation. In: Foster, J. (ed.) Valuing Nature: Economics, Ethics and Environment. Routledge, New York, pp. 155–169.

Anderson, L.M., Mulligan, B.E., and Goodman, L.S. (1984) Effects of vegetation on human response to sound. Journal of Arboriculture 10:45–49.

American Forests (2002) CITYgreen: Calculating the Value of Nature, Version 5.0. American Forests, Washington, DC.

Aylor, D.E. (1972) Noise reduction by vegetation and ground. Journal of Acoustical Society of America 51:197–205.

Baines, C. (2000) A forest of other issues. Landscape Design 294:46–47.

Barbier, E.B., Markandya, A., and Pearce, D.W. (1998) Environmental sustainability and cost-benefit analysis. In: Barbier, E.B. (ed.) The Economics of Environment and Development: Selected Essays. Edward Elgar, Northampton, MA, pp. 54–64.

Barradas, V.L. (2000) Energy balance and transpiration in an urban tree hedgerow in Mexico City. Urban Ecosystems 4:55–67.

Beckett, K.B., Freer-Smith, P.H., and Taylor, G. (1998) Urban woodlands: their role in reducing the effects of particulate pollution. Environmental Pollution 99:347–360.

Bingham, G., Bishop, R., Brody, M., et al. (1995) Issues in ecosystem valuation: improving information for decision making. Ecological Economics 14:73–90.

Bolger, D.T., Alberts, A.C., Sauvajot, R.M., et al. (1997) Response of rodents to habitat fragmentation in Costal Southern California. Ecological Applications 7:552–563.

Bolund, P., and Hunhammar, S. (1999) Ecosystem services in urban areas. Ecological Economics 29:293–301.

Botkin, D.B., and Beveridge, C.E. (1997) Cities as environments. Urban Ecosystems 1:3–19.

Brack, C.L. (2002) Pollution mitigation and carbon sequestration in an urban forest. Environmental Pollution 116:195–200.

Bradley, G. (1995) Integrating multidisciplinary perspectives. In: Bradley, G. (ed.) Urban Forest Landscapes: Integrating Multidisciplinary Perspectives. University of Washington Press, Seattle, WA, pp. 3–11.

Brennan, S., and Schnell, G. (2005) Relationship between bird abundances and landscape characteristics: the influence of scale. Environmental Monitoring and Assessment 105:209–228.

Bromley, D.W. (1990) The ideology of efficiency: searching for a theory of policy analysis. Journal of Environmental Economics and Management 19:86–107.

Brush, R.O., Williamson, D.N., and Fabos, J.G. (1979) Visual screening potential of forest vegetation. Urban Ecology 4:207–216.

Burkhardt, J., Peters, K., and Crossley, A. (1995) The presence of structural surface waxes on coniferous needles affects the pattern of dry deposition of fine particles. Journal of Experimental Botany 46:823–831.

Carson, R.T. (1991) Constructed markets. In: Braden, J.B., and Kolstad, C.D. (eds.) Measuring the Demand for Environmental Quality. Elsevier Science, Amsterdam, pp. 121–162.

Chow, J.C., Watson, J.G., and Lowenthal, D.H. (1996) Sources and chemistry of PM_{10} aerosol in Santa Barbara County, CA. Atmospheric Environment 30:1489–1499.

Common, M. (1995) Sustainability and Policy: Limits to Economics. Cambridge University Press, Cambridge, England.

Costanza, R., d'Arge, R., de Groot, R., et al. (1997) The value of the world's ecosystem services and natural capital. Nature 387:253–260.

Costanza, R. (2000) Social goals and the valuation of ecosystem services. Ecosystems 3:4–10.

Croxford, B., Penn, A., and Hillier, B. (1996) Spatial distribution of urban pollution: civilizing urban traffic. Science of the Total Environment 190:3–9.

Daily, G.C. (ed.) (1997) Nature's Services: Societal Dependence on Natural Ecosystems. Island Press, Washington, DC.

Daily, G.C., and Enrlich, P.R. (1999) Managing earth's ecosystems: an interdisciplinary challenge. Ecosystems 2:277–280.

Davey Resource Group. (1993) Consolidating and Communicating Urban Forest Benefits. International Society of Arboriculture Research Trust, Savoy, IL.

de Groot, R. (2006) Function-analysis and valuation as a tool to assess land use conflicts in planning for sustainable, multi-functional landscapes. Landscape and Urban Planning 75:175–186.
DeSanto, R.S., Glaser, R.A., McMillen, W.P., MacGregor, K.A., and Miller, J.A. (1976a) Open Space as an Air Resource Management Measure, Vol. II: Design Criteria, EPA-450/3–76–028b, U.S. Environmental Protection Agency.
DeSanto, R.S., MacGregor, K.A., McMillen, W.P., and Glaser, R.A. (1976b) Open Space as an Air Resource Management Measure, Vol. III: Demonstration Plan (St. Louis, MO.), EPA-450/3–76–028c, U.S. Environmental Protection Agency.
Dorney, J.R., Guntenspergen, G.R., Keough, J.R., and Stearns, F. (1984) Composition and structure of an urban woody plant community. Urban Ecology 8:69–90.
Driver, N., and Troutman, B. (1989) Regression models for estimating urban storm run-off quality and quantity in the United States. Journal of Hydrology 109:221–236.
Dunster, J.A. (1998) The role of aborists in providing wildlife habitat and landscape linkages throughout the urban forest. Journal of Arboriculture 24:160–167.
Dwyer, J.F., McPherson, E.G., Schroeder, H.W., and Rowntree, R.A. (1992) Assessing the benefits and costs of the urban forest. Journal of Arboriculture 18:227–234.
Dwyer, J.F., Peterson, G., and Darragh, A. (1983) Estimating the value of urban forests using the travel cost method. Journal of Arboriculture 9:182–185.
Dwyer, J.F., Schroeder, H.W., Louviere, J.J., and Anderson, D.H. (1989) Urbanites willingness to pay for trees and forests in recreation areas. Journal of Arboriculture 15:247–252.
Ekins, P. (2003) Identifying critical natural capital: conclusions about critical natural capital. Ecological Economics 44:277–292.
Fang, C.F., and Ling, D.L. (2003) Investigation of the noise reduction provided by tree belts. Landscape and Urban Planning 63:187–195.
Farber, S., Costanza, R., Childers, D.L., et al. (2006) Linking ecology and economics for ecosystem management. BioScience 56:121–133.
Farber, S.C., Costanza, R., and Wilson, M.A. (2002) Economic and ecological concepts for valuing ecosystem services. Ecological Economics 41:375–392.
Farnham, J., and Beimborn, E. (2003) Noise Barrier Design Guidelines. Centre for Urban Transportation Studies, University of Wisconsin-Milwaukee, Milwaukee, WN. http://www.uwm.edu/Dept/CUTS/noise/noiseb.htm
Fenger, J., Hertel, O., and Palmgren, F. (eds.) (1998) Urban Air Pollution: European Aspects. Kluwer Academic, Dordrecht.
Foster, J. (1997) Environmental value and the scope of economics. In: Foster J. (ed.) Valuing Nature? Economics, Ethics and Environment. Routledge, London, pp. 1–17.
Freeman, M.A. (2003) The Measurement of Environmental and Resource Values: Theory and Methods, 2nd ed. Resource for the Future, Washington, DC.
Freer-Smith, P.H., Holloway, S., and Goodman, A. (1997) The uptake of particulates by an urban woodland: site description and particulate composition. Environmental Pollution 95:27–35.
Galbraith, J.K. (1992) The Culture of Contentment. Houghton Mifflin, Boston.
Garrod, G., and Willis, K.G. (1992) The environmental economic impact of woodland: a two-stage hedonic price model of the amenity value of forestry in Britain. Applied Economics 24:715–728.
Garrod, G., and Willis, K.G. (1999) Economic Valuation of the Environment: Methods and Case Studies. Edward Elgar, Cheltenham, UK.
Gatto, M., and De Leo, G.A. (2000) Pricing biodiversity and ecosystem services: the never-ending story. Bioscience 50:347–355.
Gatrell, J.D., and Jensen, R.R. (2002) Growth through greening: developing and assessing alternative economic development programmes. Applied Geography 22:331–350.
Girling, C., and Kellett, R. (2002) Comparing stormwater impacts and costs on three neighbourhood plan types. Landscape Journal 21:100–109.
Gómez, F., Tamarit, N., and Jabaloyes, J. (2001) Green zones, bioclimatics studies and human comfort in the future development of urban planning. Landscape and Urban Planning 55:151–161.

Grandstaff, S., and Dixon, J.A. (1986) Evaluation of Lumpince park in Bangkok, Thailand. In: Dixon, J.A., and Hufschmidt, M.M. (eds.) Economic Valuation Techniques for the Environment: A Case Study Workbook. Johns Hopkins University Press, Baltimore, pp. 121–140.

Grey, G.W., and Deneke, F.J. (1986) Urban Forestry, 2nd ed. John Wiley and Sons, New York.

Grimmond, C.S.B., and Oke, T.R. (1995) Comparison of heat fluxes from summertime observations in the suburbs of four North American cities. Journal of Applied Meteorology 34:873–889.

Grimmond, S., Souch, C., Grant, R., and Heisler, G. (1994) Local scale energy and water exchange in a Chicago neighborhood. In: McPherson, E.G., Nowak, D.J., and Rowntree, R.A. (eds.) Chicago's Urban Forest Ecosystem: Results of the Chicago Urban Forest Climate Project. General Technical Report NE-186, US Department of Agriculture, Forest Service, Radnor, PA, pp. 41–62.

Grove-White, R. (1997) The environmental "valuation" controversy. In: Foster, J. (ed.) Valuing Nature: Economics, Ethics and Environment. Routledge, London, pp. 21–31.

Guan, D.S., and Chen, Y.J. (2003) Roles of urban vegetation on balance of carbon and oxygen in Guangzhou, China. Journal of Environmental Sciences 15:155–159.

Gyllin, M., and Grahn, P. (2005) A semantic model for assessing the experience of urban biodiversity. Urban Forestry and Urban Greening 3:149–161.

Haab, T.C., and McConnell, K.E. (2002) Valuing Environmental and Natural Resources: the Econometrics of Non-market Valuation. Edward Elgar, Cheltenham, UK.

Hall, D.C. (1997) Albedo and vegetation demand-side management options for warm climates. Ecological Economics 24:31–45.

Harris, R.A., and Cohn, L.F. (1985) Use of vegetation for abatement of highway traffic noise. ASCE Journal of Urban Planning and Development 111:34–48.

Heal, G. (2000a) Valuing ecosystem services. Ecosystems 3:24–30.

Heal, G. (2000b) Nature and the Marketplace: Capturing the Value of Ecosystem Services. Island Press, Washington, DC.

Heisler, G.M. (1986a) Energy savings with trees. Journal of Arboriculture 12:113–125.

Heisler, G.M. (1986b) Effects of individual trees on the solar radiation climate of small buildings. Urban Ecology 9:337–359.

Heisler, G.M. (1990) Mean wind speed below building height in residential neighborhoods with different tree densities. ASHRAE Transactions 96:1389–1396.

Heisler, G.M., Grimmond, S., Grant, R.H., and Souch, C. (1994) Investigation of the Chicago's urban forests on wind and air temperature within residential neighborhoods. In: McPherson, E.G., Nowak, D.J., and Rowntree, R.A. (eds.) Chicago's Urban Forest Ecosystem: Results of the Chicago Urban Forest Climate Project. General Technical Report NE-186, US Department of Agriculture, Forest Service, Radnor, PA, pp. 19–40.

Hoevenagel, R. (1994) An assessment of the contingent valuation method. In: Pethig, R. (ed.) Valuing the Environment: Methodological and Measurement Issues. Kluwer Academic, Dordrecht, pp. 195–227.

Hull, R.B., and Harvey, A. (1989) Explaining the emotion people experience in suburban parks. Environment and Behaviour 21:323–345.

Hunter, I.R. (2001) What do people want from urban forestry? The European experience. Urban Ecosystems 5:277–284.

Hutchison, B.A., and Taylor, F.G. (1983) Energy conservation mechanisms and potentials of landscape design to ameliorate building microclimates. Landscape Journal 2(1):19–39.

Impens, R.A., and Delcarte, E. (1979) Survey of urban trees in Brussels, Belgium. Journal of Arboriculture 5:169–176.

Jackson, L.E. (2003) The relationship of urban design to human health and condition. Landscape and Urban Planning 64:190–200.

Jansson, Å., and Nohrstedt, P. (2001) Carbon sinks and human freshwater dependence in Stockholm county. Ecological Economics 32:351–355.

Jensen, M.B., Persson, B., Guldager, S., Reeh, U., and Nilsson, K. (2000) Green structure and sustainability—developing a tool for local planning. Landscape and Urban Planning 52:117–133.

Jepson, P., and Canney, S. (2003) Values-led conservation. Global Ecology and Biogeography 12:271–274.
Jim, C.Y. (1987) Urban trees in Hong Kong: benefits and constraints. The Arboricultural Journal 11:145–164.
Jim, C.Y., and Chen, W.Y. (2006a) Recreation-amenity use and contingent valuation of urban greenspaces in Guangzhou, China. Landscape and Urban Planning 75:81–96.
Jim, C.Y., and Chen, W.Y. (2006b) Perception and attitude of residents toward urban green spaces in Guangzhou (China). Environmental Management 38(3):338–349.
Jim, C.Y., and Chen, W.Y. (2006c) Impacts of urban environmental elements on residential housing prices in Guangzhou (China). Landscape and Urban Planning 78:422–434.
Jo, H.K., and McPherson, E.G. (1995) Carbon storage and flux in urban residential greenspace. Journal of Environmental Management 45:109–133.
Jo, H.K., and McPherson, E.G. (2001) Indirect carbon reduction by residential vegetation and planting strategies in Chicago, USA. Journal of Environmental Management 61:165–177.
Kato, Y., Yokohari, M., and Brown, R.D. (1997) Integration and visualization of the ecological value of rural landscape in maintaining the physical environment of Japan. Landscape and Urban Planning 39:69–82.
Kent, R.L. (1993) Attributes, features and reasons for enjoyment of scenic routes, a comparison of experts, residents, and citizens. Landscape Research 18:92–102.
Keefe, E.M., and Giuliano, W.M. (2004) Effects of forest structure on the distribution of southern flying squirrels (Glaucomys volans) in urban parks. Urban Ecosystems 7:55–64.
Kim, Y.S., and Wells, A. (2005) The impact of forest density on property values. Journal of Forestry 103:146–151.
Kleiber, O. (2001) Valuation of recreational benefits and visitor conflicts in an urban forest. Fifth International Conference of the International Society for Ecological Economics (ISEE), September 26–29, 2001, Moscow, Russia.
Kojima, M., and Lovei, M. (2001) Urban Air Quality Management: Coordinating Transport, Environment, and Energy Policies in Developing Countries. World Bank Technical Paper No. 508, Washington, DC.
Konijnendijk, C.C. (1997) Urban Forestry: Overview and Analysis of European Forest Policies. European Forest Institute Working Paper 12, Joensuu, Finland.
Konopacki, S., and Akbari, H. (2000) Energy savings calculation for urban heat island mitigation strategies in Baton Rouge, Sacramento and Salt Lake City. Proceedings of the 2000 ACEEE Summer Study on Energy Efficiency in Buildings, Vol. 9, Pacific Grove, CA.
Koomen, E., Dekkers, J., Koetse, M., Rietveld, P., and Scholten, H. (2005) Valuation of metropolitan open space, presenting the research framework. The 45th Congress of the European Regional Science Association, August 23–27, 2005, Amsterdam, The Netherlands.
Kozlowski, T.T., and Pallardy, S.G. (1997) Physiology of Woody Plants, 2nd ed. Academic Press, San Diego.
Kramer, P.J., and Kozlowski, T.T. (1960) Physiology of Trees. McGraw-Hill, New York.
Kramer, R.A., Holmes, T.P., and Haefele, M. (2002) Using contingent valuation to estimate the value of forest ecosystem protection. In: Sills, E.O., and Abt, K.L. (eds.) Forests in a Market Economy, Kluwer Academic, Dordrecht, The Netherlands, pp. 303–320.
Kuo, F.E., and Sullivan, W.C. (2001a) Environment and crime in the inner city: Does vegetation reduce crime? Environment and Behavior 33:343–367.
Kuo, F.E., and Sullivan, W.C. (2001b) Aggression and violence in the inner city: Effects of environment via mental fatigue. Environment and Behavior 33:543–571.
Kwak, S.J., Yoo, S.H., and Han, S.Y. (2003) Estimating the public's value for urban forest in the Seoul metropolitan area of Korea: a contingent valuation study. Urban Studies 40:2207–2221.
Lancaster, K.J. (1966) A new approach to consumer theory. Journal of Political Economy 74:132–157.

Laverne, R.J., and Lewis, G. (1996) The effect of vegetation on residential energy use in Ann Arbor, Michigan. Journal of Arboriculture 22:234–243.
Lëfvenhaft, K., Björn, C., and Ihse, M. (2002) Biotope patterns in urban areas: a conceptual model integrating biodiversity issues in spatial planning. Landscape and Urban Planning 58:223–240.
Li, W., and Xu, W. (2002) Effects of sulphur dioxide on enzyme activity of selected plants. In: He, X., and Ning, Z. (eds.) Advances in Urban Forest Ecology. China Forestry Publishing House, Beijing, pp. 199–201 (in Chinese).
Liu, H.T. (1998) Urban Forestry in China: A Biogeographical Study in Guangzhou City. PhD dissertation, University of Hong Kong, Hong Kong.
Longcore, T., Li, C., and Wilson, J.P. (2004) Applicability of CityGreen urban ecosystem analysis software to a densely built urban neighborhood. Urban Geography 25:173–186.
Loomis, J.B. (1987) The economic value of instream flow: methodology and benefit estimates for optimum flows. Journal of Environmental Management 24:169–179.
Lorenzo, A.B., Blanche, C.A., Qi, Y., and Guidry, M.M. (2000) Assessing residents' willingness to pay to preserve the community urban forest: a small-city case study. Journal of Arboriculture 26:319–325.
Lovett, G.M., Traynor, M.M., Pouyat, R.V., Carreiro, M.M., Zhu, W., and Baxter, J.W. (2000) Atmospheric deposition to oak forests along an urban-rural gradient. Environmental Science and Technology 34:4294–4300.
Ludwig, D. (2000) Limitations of economic valuation of ecosystems. Ecosystems 3:31–35.
Luvall, J.C., Quattrochi, D.A., Rickman, D., and Estes, M. (2000) Urban forests cool hot cities. Journal of Forestry 98:149–150.
MacDonald, L. (1996) Global problems local solutions: measuring the value of the urban forest. American Forests 102:26–29.
Maille, P., and Mendelsohn, R. (1993) Valuing ecotourism in Madagascar. Journal of Environmental Management 38:213–218.
Malpezzi, S. (2003) Hedonic pricing models: a selective and applied review. In: O'Sullivan, T., and Gibb, K. (eds.) Housing Economics and Public Policy. Blackwell, Oxford, UK, pp. 67–89.
Manning, W.J., and Feder, W.A. (1980) Biomonitoring Air Pollutants with Plants. Applied Science, London.
Marcus, C.C., and Barnes, M. (1999) (eds.) Healing Gardens: Therapeutic Benefits and Design Recommendations. John Wiley and Sons, New York.
Matthews, M.J., O'Connor, S., and Cole, R. (1988) Database for the New York State urban wildlife habitat inventory. Landscape and Urban Planning 15:23–37.
Maurer, U., Peschel, T., Schmitz, S. (2000) The flora of selected urban land-use types in Berlin and Potsdam with regard to nature conservation in cities. Landscape and Urban Planning 46:209–215.
McConnell, V., and Walls, M. (2005) The Value of Open Space: Evidence from Studies of Nonmarket Benefits. Resources for the Future, Washington, DC.
McDonnell, M.J., Pickett, S., Pouyat, R., et al. (1997) Ecosystem processes along urban-to-rural gradients. Urban Ecosystems 1:21–36.
McFadden, D. (1999) Rationality for economists? Journal of Risk and Uncertainty 19:73–105.
McPherson, E.G. (1994a) Cooling urban heat islands with sustainable landscapes. In: Platt, R.H., Rowntree, R.A., and Muick, P.C. (eds.) The Ecological City: Preserving and Restoring Urban Biodiversity. University of Massachusetts Press, Boston, pp. 151–171.
McPherson, E.G. (1994b) Using urban forests for energy efficiency and carbon storage. Journal of Forestry 92:36–41.
McPherson, E.G. (1994c) Benefits and costs of tree planting and care in Chicago. In: McPherson, E.G., Nowak, D.J., and Rowntree, R.A. (eds.) Chicago's Urban Forest Ecosystem: Results of the Chicago Urban Forest Climate Project. General Technical Report NE-186, US Department of Agriculture, Forest Service, Radnor, PA, pp. 115–134.

McPherson, E.G. (1994d) Energy-saving potential of trees in Chicago. In: McPherson, E.G., Nowak, D.J., and Rowntree, R.A. (eds.) Chicago's Urban Forest Ecosystem: Results of the Chicago Urban Forest Climate Project. General Technical Report NE-186, US Department of Agriculture, Forest Service, Radnor, PA, pp. 95–113.

McPherson, E.G. (1998) Atmospheric carbon dioxide reduction by Sacramento's urban forest. Journal of Arboriculture 24:215–223.

McPherson, E.G., Nowak, D., Heisler, G., et al. (1997) Quantifying urban forest structure, function, and value: the Chicago Urban Forest Climate Project. Urban Ecosystems 1:49–61.

McPherson, E.G., and Simpson, J.R. (2002) A comparison of municipal forest benefits and costs in Modesto and Santa Monica, California, USA. Urban Forestry and Urban Greening 1:61–74.

McPherson, E.G., Simpson, J.R., Peper, P.J., and Xiao, Q. (1999) Benefit-cost analysis of Modesto's municipal urban forest. Journal of Arboriculture 25:235–248.

Meier, A. (1990/91) Strategic landscaping and air-conditioning savings: a literature review. Energy and Buildings 15–16:479–486.

Miller, R.W. (1997) Urban Forestry: Planning and Managing Urban Greenspaces, 2nd ed. Prentice-Hall, Englewood Cliffs, NJ.

Millard, A. (2000) The potential role of natural colonization as a design tool for urban forestry—a pilot study. Landscape and Urban Planning 52:173–179.

Mink, G., and Witter, G. (1982) Haeuser mit Gruenem Pelz, Ein Handbuch zur Hausbegruenung. Verlag Dieter Fricke GmbH, Frankfurt.

Mitchell, R.C., and Carson, R.T. (1989) Using Surveys to Value Public Goods: The Contingent Valuation Method, Resources for the Future, Washington, DC.

Moll, G., and Kollin, C. (1996) Urban trees and carbon. American Forests 102(3):26–28.

More, T.A., Stevens, T., and Allen, P.G. (1988) Valuation of urban parks. Landscape and Urban Planning 15:139–152.

National Research Council. (2005) Valuing Ecosystem Services: Toward Better Environmental Decision-Making. Committee on Assessing and Valuing the Services of Aquatic and Related Terrestrial Ecosystems, Water Science and Technology Board, Division on Earth and Life Studies, National Research Council. National Academies Press, Washington, DC.

Nilon, C.H., Long, C.N., and Zipperer, W.C. (1995) Effects of wildland development on forest bird communities. Landscape and Urban Planning 32:81–92.

Nowak, D.J. (1993) Atmospheric carbon reduction by urban trees. Journal of Environmental Management 37:207–217.

Nowak, D.J. (1994a) Atmospheric carbon dioxide reduction by Chicago's urban forest. In: McPherson, E.G., Nowak, D.J., and Rowntree, R.A. (eds.) Chicago's Urban Forest Ecosystem: Results of the Chicago Urban Forest Climate Project. General Technical Report NE-186, US Department of Agriculture, Forest Service, Radnor, PA, pp. 83–94.

Nowak, D.J. (1994b) Air pollution removal by Chicago's urban forest. In: McPherson, E.G., Nowak, D.J., and Rowntree, R.A. (eds.) Chicago's Urban Forest Ecosystem: Results of the Chicago Urban Forest Climate Project. General Technical Report NE-186, US Department of Agriculture, Forest Service, Radnor, PA, pp. 63–81.

Nowak, D.J., Civerolo, J.C., Rao, S.T., Sistla, G., Luley, C.J., and Crane, D.E. (2000) A modeling study of the impact of urban trees on ozone. Atmospheric Environment 34:1601–1613.

Nowak, D.J., and Crane, D.E. (2000) The Urban Forest Effects (UFORE) Model: quantifying urban forest structure and functions. In: Hansen, M., and Burk, T. (eds.) Proceedings Integrated Tools for Natural Resources Inventories in the 21st Century. IUFRO Conference, 16–20 August 1998, Boise, ID. General Technical Report NC-212, U.S. Department of Agriculture, Forest Service, North Central Research Station, St. Paul, MN, pp. 714–720.

Nowak, D.J., and Crane, D.E. (2002) Carbon storage and sequestration by urban trees in the USA. Environmental Pollution 116:381–389.

Nowak, D.J., Crane, D.E., and Dwyer, J.F. (2002a) Compensatory value of urban trees in the United States. Journal of Arboriculture 28:194–199.

Nowak, D.J., Crane, D.E., Stevens, J.C., Ibarra, M. (2002b) Brooklyn's Urban Forest. General Technical Report NE-290, US Department of Agriculture, Forest Service, Northeastern Research Station, Newtown Square, PA.
Nowak, D.J., and Dwyer, J.F. (2000) Understanding the benefits and costs of urban forest ecosystems. In: Kuser, J.E. (ed.) Handbook of Urban and Community Forestry in the Northeast. Kluwer Academic/Plenum, New York, pp. 11–25.
Nowak, D.J., McHale, P.J., Ibarra, M., Crane, D., Stecens, J., and Luley, C. (1998) Modeling the effects of urban vegetation on air pollution. In: Gryning, S., and Chaumerliac, N. (eds.) Air Pollution Modeling and its Application XII, Plenum, New York, pp. 399–407.
O'Connor, M., and Spash, C. (1999) Valuation and the Environment: Theory, Method and Practice. Edward Elgar, Northampton, MA.
Oguz, D. (2000) User surveys of Ankara's urban parks. Landscape and Urban Planning 52:165–171.
Oke, T.R (1989) The micrometeorology of the urban forest. Philosophical Transactions of the Royal Society of London Series B, Biological Sciences 324:335–348.
Palmquist, R.B. (1991) Hedonic methods. In: Braden, J.B., and Kolstad, C.D. (eds.) Measuring the Demand for Environmental Quality. North-Holland, Amsterdam, The Netherlands, pp. 77–120.
Parker, J.H. (1983) Landscaping to reduce the energy use in cooling buildings. Journal of Forestry 81:82–84.
Pattanayak, S.K. (2004) Valuing watershed services: concepts and empirics from Southeast Asia. Agriculture, Ecosystems and Environment 104:171–184.
Pearce, D.W. (1993) Economic Value and the Natural World. Earthscan, London.
Pearce, D.W., and Warford, J.J. (1993) World Without End: Economics, Environment, and Sustainable Development. Oxford University Press, Oxford, England.
Peck & Associates (1999) Green roofs for healthy cities. Peck & Associates, Toronto. http://www.peck.ca.
Phillips, L.E. (1993) Urban Trees: A Guide for Selection, Maintenance, and Master Planning. McGraw-Hill, New York.
Pitt, D., Soergell II, K., and Zube, E. (1979) Trees in the city. In: Laurie, I.G. (ed.) Nature in Cities. John Wiley and Sons, New York, pp. 205–229.
Price, C. (2003) Quantifying the aesthetic benefits of urban forestry. Urban Forestry and Urban Greening 1:123–133.
Randall, A. (1987) Resource Economics: An Economic Approach to Natural Resource and Environmental Policy, 2nd ed. John Wiley and Sons, New York.
Randall, A. (1991) Total and nonuse value. In: Braden, J.B., and Kolstad, C.D. (eds.) Measuring the Demand for Environmental Quality. Elsevier Science Pulishers, B. V. North Holland, pp. 303–332.
Riera, A. (2000) Valuing natural goods. In: Georgantís, N., and Tarrazona, I.J.B. (eds.) Spatial Economics and Ecosystems: The Interaction between Economics and the Natural Environment. WIT Press, Southampton, UK, pp. 39–66.
Robinette, C. (1972) Plants, People and Environmental Quality. Department of the Interior, National Park Services, Washington, DC.
Rosenfeld, A.H., Romm, J.J., Akbari, H., and Pomerantz, M. (1998) Cool communities: strategies for heat islands mitigation and smog reduction. Energy and Buildings 28:51–62.
Rowntree, R.A. (1986) Ecology of the urban forest—introduction to part II. Urban Ecology 9:229–243.
Rowntree, R.A., and Nowak, D.J. (1991) Quantifying the role of urban forests in removing atmospheric carbon dioxide. Journal of Arboriculture 17:269–275.
Samaras, Z., and Sorensen, S.C. (1998) Mobile sources. In: Fenger, J., Hertel, O., and Palmgren, F. (eds.) Urban Air Pollution: European Aspects. Kluwer Academic, Dordrecht, pp. 63–92.
Sanders, R.A. (1986) Urban vegetation impacts on the urban hydrology of Dayton Ohio. Urban Ecology 9:361–376.

Savard, J.P.L., Clergeau, P., and Mennechez, G. (2000) Biodiversity concepts and urban ecosystems. Landscape and Urban Planning 48:131–142.

Schiller, A., and Horn, S.P. (1997) Wildlife conservation in urban greenways of the mid-southeastern United States. Urban Ecosystems 1:103–116.

Schneider, S.H. (1989) The changing climate. Scientific American 261(3):70–79.

Schroeder, H.W. (1986) Psychological value of urban trees: measurement, meaning and imagination. Proceedings of 3rd National Urban Forests Conference, American Forests, Washington, DC, pp. 55–60.

Schroeder, H.W., and Anderson, L.M. (1984) Perception of personal safety in urban recreation sites. Journal of Leisure Research 16:178–194.

Scott, K.I., McPherson, E.G., and Simpson, J.R. (1998) Air pollutant uptake by Sacramento's urban forest. Journal of Arboriculture 24:224–233.

Shechter, M., and Freeman, S. (1994) Nonuse value: reflections on the definition and measurement. In: Pethig, R. (ed.) Valuing the Environment: Methodological and Measurement Issues. Kluwer Academic, Dordrecht, pp. 171–194.

Sheppard, S. (1999) Hedonic analysis of housing markets. In: Cheshire, P., and Mills, E.S. (eds.) Handbook of Regional and Urban Economics, Volume 3: Applied Urban Economics. Elsevier, Amsterdam, The Netherlands, pp. 1595–1635.

Simpson, J.R. (1998) Urban forest impacts on regional cooling and heating energy use: Sacramento county case study. Journal of Arboriculture 24:201–214.

Simpson, J.R., and McPherson, E.G. (1996) Potential of tree shade for reducing residential energy use in California. Journal of Arboriculture 22:10–18.

Smardon, R.C. (1988) Perception and aesthetics of the urban environment: review of the role of vegetation. Landscape and Urban Planning 15:85–106.

Smith, W.H. (1990) Air Pollution and Forests. Springer-Verlag, New York.

Smith, V.K. (1991) Household production functions. In: Braden, J.B., and Kolstad, C.D. (eds.) Measuring the Demand for Environmental Quality. Elsevier Science, Amsterdam, pp. 41–76.

Soil Conservation Service (1986) Technical Release 55: Urban Hydrology for Small Watersheds. US Department of Agriculture, Soil Conservation Service, Washington, DC.

Spash, C. (1997) Environmental management without environmental valuation? In: Foster, J. (ed.) Valuing Nature: Economics, Ethics and Environment. Routledge, London, pp. 170–185.

Spitsyna, N.T., and Skripal'shchikova, L.N. (1991) Phytomass and dust accumulation of birch forests near open-pit mines. Soviet Journal of Ecology 22:354–359.

Starrett, D.A. (2000) Shadow pricing in economics. Ecosystems 3:16–20.

Stern, D.I. (1999) Use value, exchange value, and resource scarcity. Energy Policy 27:469–476.

Taha, H. (1996) Modeling the impacts of increased urban vegetation on the ozone air quality in the South Coast Air Basin. Atmospheric Environment 30:3423–3430.

Taha, H. (1997) Modeling the impacts of large-scale albedo changes on ozone air quality in the South Coast Air Basin. Atmospheric Environment 31:1667–1676.

Taha, H., Douglas, S., and Haney, J. (1997) Mesoscale meteorological and air quality impacts of increased urban albedo and vegetation. Energy and Buildings 25:169–177.

Taylor, A.F., Kuo, F.E., and Sullivan, W.C. (2001) Coping with ADD: The surprising connection to green play settings. Environment and Behavior 33:54–77.

Taylor, A.F., Kuo, F.E., and Sullivan, W.C. (2002) Views of nature and self-discipline: evidence from inner city children. Journal of Environmental Psychology 22:49–63.

Todorova, A., Asakawa, S., and Aikoh, T. (2004) Preferences for and attitudes towards streets flowers and trees in Sapporo, Japan. Landscape and Urban Planning 69:403–416.

Toman, M. (1998) Why not to calculate the value of the world's ecosystem services and the natural capital. Ecological Economics 25:57–60.

Tso, C.P. (1996) A survey of urban heat island studies in two tropical cities. Atmospheric Environment 30:507–519.

Tyrväinen, L. (1997) The amenity value of the urban forest: an application of the hedonic pricing method. Landscape and Urban Planning 37:211–222.

Tyrväinen, L. (2001) Economic valuation of urban forest benefits in Finland. Journal of Environmental Management 62:75–92.

Tyrväinen, L., and Miettinen, A. (2000) Property prices and urban forest amenities. Journal of Environmental Economics and Management 39:205–223.

Tyrväinen, L., Silvennoinen, H., and Kolehmainen, O. (2003) Ecological and aesthetic values in urban forest management. Urban Forestry and Urban Greening 1:135–149.

Tyrväinen, L., and Väänänen, H. (1998) The economic value of urban forest amenities: an application of the contingent valuation method. Landscape and Urban Planning 43:105–118.

Ulrich, R.S. (1984) View through a window may influence recovery from surgery. Science 224:420–421.

Ulrich, R.S. (1986) Human responses to vegetation and landscape. Landscape and Urban Planning 13:29–44.

Ulrich, R.S. (1999) Effects of Gardens on health outcomes: theory and research. In: Marcus, C.C., and Barnes, M. (eds.) Healing Gardens: Therapeutic Benefits and Design Recommendations. John Wiley and Sons, New York, pp. 27–86.

van den Berg, A., Vlek, C.A., and Coeterier, J.F. (1998) Group difference in the aesthetic evaluation of nature development plan: a multilevel approach. Journal of Environmental Psychology 18:141–157.

White, J.G., Antos, A. J., Fitzsimons, J.A., and Palmer, G.C. (2005) Non-uniform bird assemblages in urban environments: the influence of streetscape vegetation. Landscape and Urban Planning 71:123–135.

White, R.R. (2002) Building the Ecological City. CRC Press, Cambridge, MA.

Whitford, V., Ennos, A.R., and Handley, J.F. (2001) City form and natural process—indicators for the ecological performance of urban areas and their application to Merseyside, UK. Landscape and Urban Planning 57:91–103.

Wilson, E.O. (1999) The Diversity of Life. W.w. Norton, New York.

Xiao, Q., McPherson, E.G., Simpson, J.R., and Ustin, S.L. (1998) Rainfall interception by Sacramento's urban forest. Journal of Arboriculture 24:235–244.

Yokohari, M., Brown, R.D., Kato. Y., and Yamamoto, S. (2001) The cooling effect of paddy fields on summertime air temperature in residential Tokyo, Japan. Landscape and Urban Planning 53:17–27.

6
Benefits of Urban Green Space for Improving Urban Climate

Volker Heidt and Marco Neef

Urban settlements transform the natural environment so greatly that people tend to see the city only as an employment site, and economic and cultural center. Thus a growing number of people prefer to reside in greener suburbs or rural areas. This results in increased automobile commuter traffic, accompanied by traffic jams, accidents, stress, and ever more damage to the environment. Concepts of sustainable development or the ecological city represent strategies for changing these negative trends. The purpose for doing so is principally the well-being of a city's residents. Often this entails bringing more of the natural environment back into the city, because urban green space fulfills several critical functions in an urban context that benefit people's quality of life. There is a broad consensus about the importance, and therefore the value, of urban green space in cities as currently constructed, in addition to its value in planning ecological cities.

Steadily growing traffic and urban heat not only damage the environment, but also incur social and economic costs. As we explain further, we can save costs even by making small changes to existing situations. Furthermore, we maintain and show that an integrated approach is needed for designing and maintaining urban green space. The main thesis of this chapter, therefore, is as follows: To provide sufficient quality of life in high-density cities, it is important to maintain and restore an urban green space system; moreover, urban green space and a comfortable urban climate also produce social and economic benefits.

Urban Climate

One of the fundamental characteristics that set a city apart from its rural surroundings is the altered climate that prevails over urban environments. As compared to rural areas nearby, a distinctive urban climate occurs, involving differences in solar input, rainfall patterns, and temperature. Solar radiation, air temperature, wind speed, and hence relative humidity, cloud cover, and precipitation, can vary significantly due to the built environment in cities and according to a city's topography and local surroundings. There are complex and diverse factors within a city that also affect its climate, such as urban density, street orientation, shade caused by buildings of

M.M. Carreiro et al. (eds.), *Ecology, Planning, and Management of Urban Forests: International Perspectives.*

Table 6.1 Average difference in climatic parameters of built-up areas compared with surrounding rural areas

Climatic parameters	Characteristics	Compared to the surrounding area
Air pollution	Gaseous pollution	5–25 times more
Solar radiation	Global solar radiation	15–20% less
	Ultraviolet radiation	15–20% less
	Duration of bright sunshine	5–15% less
Air temperature	Annual mean average	0.5–1.5°C higher
	On clear days	2–6°C higher
Wind speed	Annual mean average	15–20% less
	Calm days	5–20% more
Relative humidity	Winter	2% less
	Summer	8–10% less
Clouds	Overcast	5–10% more
Precipitation	Total rainfall	5–10% more

Source: Gilbert (1991, p. 26).

varying height, and the type and amount of urban vegetation. Layout density, for example, can influence noise and atmospheric pollution (Sukopp and Wittig, 1998).

Compared to the countryside, urban climate is generally characterized by higher temperatures, weaker winds and solar radiation inputs that vary according to the degree of pollution (Gilbert, 1991; Table 6.1). All of these and other factors contribute, often in synergistic ways, to the urban climate differences measured in cities. Air pollution (e.g., carbon dioxide, sulfur dioxide, ozone, aerosols, cadmium, lead) in urban areas is high, often five to 25 times higher than nearby rural areas, due to pollutant emissions, especially from transportation and industry. Among other effects, high air pollution results in less solar input, but greater heat trapping, in cities (Fezer, 1995). Studies also show that clouds and rainfall can increase in cities due to higher atmospheric particulate concentrations that provide condensation nuclei for water (Bonan, 2002).

One aspect of urban climate that has received much study, however, is the urban heat-island effect. Almost every city in the world today is usually 1° to 4°C (2° to 8°F) warmer than its surrounding rural area; this clearly shows that cities behave as "heat islands" (Oke, 1973; Ammer and Bechet, 1978). Urban heat islands, which have been intensifying throughout this century, are isolated pockets of increased temperature located over cities and urban areas. The causes of this phenomenon are as follows:

- Heat absorption by building roofs and walls, as well as by pavement: Buildings and pavement absorb solar radiation instead of reflecting it, causing the temperature of the surfaces and their environment to rise 10° to 20°C (18° to 36°F) higher than ambient air temperatures (Taha et al., 1992).
- Greater percentage of impervious surfaces (buildings and pavement) and less area with vegetation or bare soil: This means that there are fewer trees, shrubs, and other plants to shade buildings and intercept solar radiation, and less "evapotranspiration" of moisture from vegetation and unpaved soil to cool urban surroundings (Bonan, 2002).

As a result of the urban heat island, the annual mean temperature of cities is several degrees warmer than their surrounding rural area. In some small open spaces in cities this difference can be as much as 10°C. During the day, wide streets, squares, and unplanted areas are the hottest parts of a town, while at night, narrow streets have higher temperatures than the rest of the city (Kuttler, 1998).

In summary, due to the urban heat island effect:

- The number of hot days above 25°C per year increases.
- The increased heat has negative effects on residents' well-being. During the hot months a heat island creates considerable discomfort and stress. In fact, extreme heat is held responsible for more deaths than violent weather events such as tornadoes, blizzards, or floods. In the summer of 1995, heat killed 700 elderly people in Chicago (Pomerantz et al., 1999). In August 2003 an extreme heat wave in western and southern Europe was responsible for more than 20,000 deaths, particularly among the aged population (Commission of the European Communities, 2005, p. 14). Urban areas were particularly affected. In Frankfurt, Germany, from August 3rd to 12th, maximum temperatures constantly exceeded 35°C, with minimum temperatures constantly above 21°C. In June, mortality was 14 per day with a maximum of 21 per day. From August 6th onward, daily mortality increased sharply to a maximum of 51 per day on August 13th (Heudorf and Meyer, 2005).
- There is increased demand for electricity for air-conditioning and, therefore, greater economic costs. For every degree Celsius rise in temperature, electricity generation rises by 4% to 8% (Pomerantz et al., 1999).
- Increased electricity generation by power plants leads to higher emissions of sulfur dioxide, carbon monoxide, nitrous oxides, and suspended particulates, as well as carbon dioxide, a greenhouse gas known to contribute to global warming and climate change.
- In summer when day length and solar intensity is greatest, the formation of harmful photochemical smog is accelerated, since ozone precursors, nitrous oxides (NO_x), and volatile organic compounds (VOCs) react more rapidly at warmer temperatures (Chameides and Cowling, 1995). For every degree Celsius rise in temperature, smog production increases by 7% to 18% (Pomerantz et al., 1999).

Therefore, as described below, modifications in urban planning and the enhancement of green space have the potential to mitigate the adverse effects of urbanization on urban climate in sustainable ways.

Functions and Objectives of Urban Green Space

Experience has shown that it is important to maintain and restore an urban environment that provides a healthy quality of life especially in high-density cities. Urban green spaces have important ecological benefits. However, they also play other roles in defining quality of life in a city, and even a city's identity (e.g., Central Park in New York City). Green spaces, by their areal extent, distribution, and other,

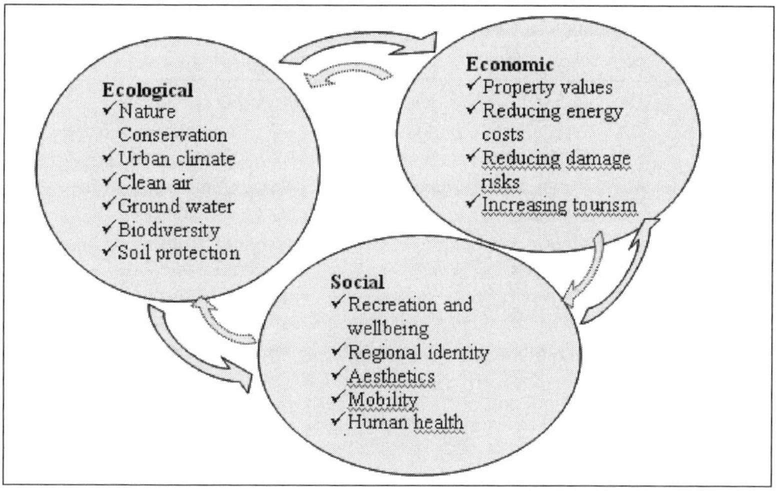

Fig. 6.1 Economic, social, and ecological functions and objectives of urban green space management and sustainable urban land use and their interactions

more qualitative criteria, may define a city's structure and identity through their social and aesthetic functions, and thereby affect the quality of life of its inhabitants. The combination of well-designed and maintained urban green space and urban planning can provide improvements in the ecological, economic, and social functioning of a city (Fig. 6.1).

Creating a Network System of Urban Green Space

Urban greening is a city planning instrument. Urban green space design consists of using different elements or types of green space, each of which fulfills special functions in the urban green concept or philosophy as a whole. *Punctiform elements* (nature parks/urban forests, neighborhood parks, cultural landscape parks) are solitary, often spatially isolated, green elements. Optimally, these solitary elements should be connected using linear elements that stitch the urban green system together to improve various environmental effects, such as biodiversity, nature conversation, or urban climate. *Linear elements* (like trails, greenways, waterways, highway verges, and green corridors) can serve to link urban parks together and also to connect the city center with areas at its outskirts. *Green corridors*, which can follow development axes, have several environmental functions. Corridors are more than parks habitats for different species of plants and animals. Depending on their structure, linear elements serve as conduits for organisms, as barriers or filters for pollutants, and can separate different urban areas to improve city structure.

Table 6.2 Selected elements of urban green space and examples of their functions

Setting or context	Park type	Park function
Urban parks	Urban "pocket" parks	Neighborhood use
	Urban nature parks/urban forests	Urban recreation/leisure
	Urban cultural-landscape park	Accentuate regional identity
Urban to suburban	Green corridors, greenways, trails	Linking urban parks for people, fauna, and flora
		Linking city center and outskirts for people, fauna, and flora
Country parks, regional parks	Suburban parks and green zone territories/suburban forests	Recreation for town inhabitants
		Sources of natural resources (water, air, etc.) for the city
	Linkage of open spaces	Protection of open space and landscape
		Linking suburban forests and suburban parks
		Instrument for landscape structuring

Green corridors may serve as dispersion corridors for flora and fauna and contribute to higher biodiversity in cities. To achieve a more unified and integrated urban ecological system, the spatial arrangement of urban green areas and elements is crucial. An urban green ecological system, therefore, is an urban green network system. Table 6.2 lists selected elements of urban green space and their functions in the urban environment (cf. Kaerkes, 1987, for the ecological relevance of different elements of urban green space).

The quality of various urban green elements as ecological compensation areas thereby depends on several factors:

- Size
- Location and distribution in the city
- Diversity in the composition and variation of vegetation structural types
- Combination of different green area types
- Linking and integration in green area systems
- Strain and strain resistance (temperature, air pollution)

Thus, if improvements in the general ecological and environmental quality of a city are an important goal for urban planning and decision makers, an integrated approach for strengthening a green network system should be on top of the planning agenda. For example, the Development Program for Urban Forests, as proposed by the East China Normal University in Shanghai, China, meets this requirement for increasing green space networks. The general goal of the Urban Forest Program for Greater Shanghai is to build up an urban open-space system to an area of 6340 km^2. Both urban and the suburban green areas are proposed to be developed synchronously. In the suburban area, two "rings," eight "longitudinal lines," five "large pieces," a

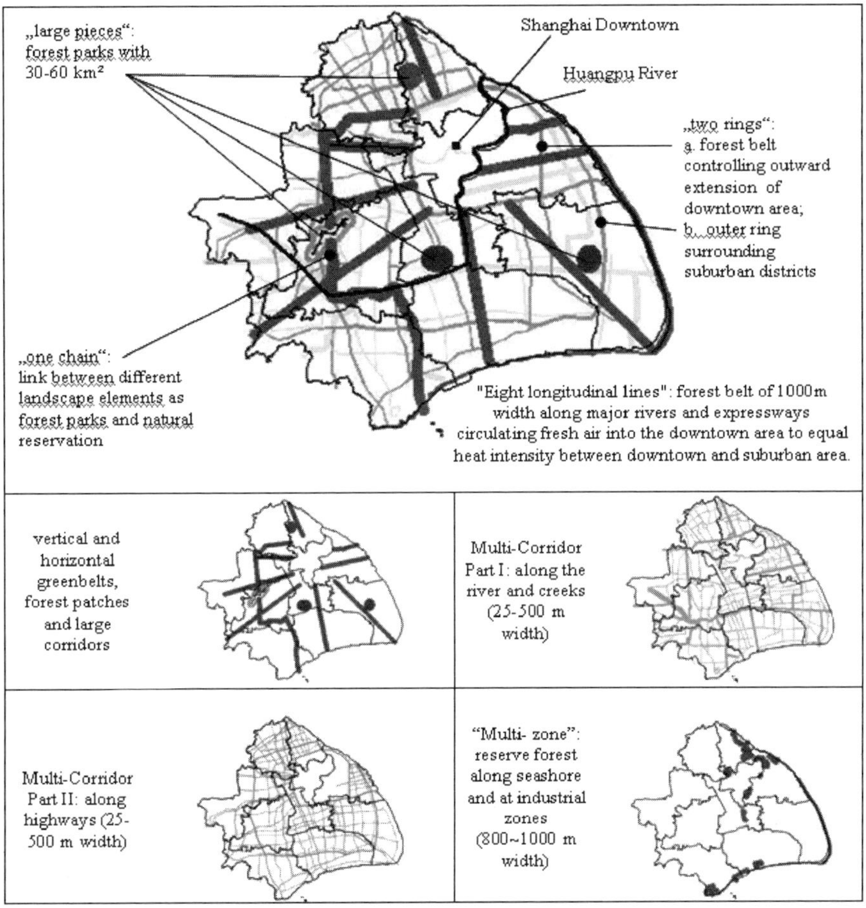

Fig. 6.2 Network system of urban green space in greater Shanghai (islands are not displayed). (Data from the Institute of Environmental Science, Shanghai East China Normal University, 2002.)

"multi-corridor," a "multi-zone," and one "chain" are connected to each other; by so doing a network is formed. This network is portrayed in Figure 6.2 (the map does not show Chongming, Changxing, and Hengsha Island). We also can relate the single elements and their functions to the terminology we defined earlier (Table 6.2). To translate the different terminologies, "eight longitudinal lines" and a "multi-corridor" refer to suburban green corridors, greenways, and trails. The "large pieces" refer to suburban parks and forests in the context of country or regional parks. A regional park composed of a chain of green spaces are the "two rings" and the "one chain." The function of the inner ring, for example, controls the outward extension of the downtown area by being highly valued recreational area for the town inhabitants.

The Ecological and Environmental Quality Benefits of Green Space

There are many reasons for enlarging a networked system of green space in cities. Urban green space ameliorates the climate; filters the air, water, and soil of many pollutants; and provides a habitat for fauna and flora (Kaerkes, 1987). It has also been shown that biodiversity can be higher in urban areas than in their rural surroundings, since with a sufficient supply of urban green spaces, cities can provide numerous ecological niches for many species—sometimes even endangered species (Sukopp and Wittig, 1998; Wittig, 2002). Green corridors linked to various vegetation patches are likely to play important roles in maintaining high biodiversity in urban areas, since they provide mobility axes for species.

Vegetated areas also provide locations where unsealed soils exist, thus simultaneously decreasing excessive surface run-off and combined sewer flows that damage local streams, and reducing the urban heat-island effect via greater evapotranspiration. These vegetated soils also may contain a greater diversity of microbes, such as mycorrhizal fungi that are beneficial to trees and other plants. Moreover, the effects of vegetation on the urban climate are important even in the case of small green spaces, like neighborhood parks. Inner-city green spaces are especially important for improving air quality via uptake of pollutant gases like ozone and via the high particulate dust-binding capacity of leaves. In the case of small parks, the amount, kind, and ratio of trees and shrubs are important. A "protection plantation" consisting of trees as tall walls with shorter bushes in between the trees is more efficient at filtering out air particulates than a forest of the same size consisting only of trees. A small park with both trees and shrubs can bind up to 68 metric tons of dust per hectare per year. A street with trees and small parks contain about 25% and 20% of the atmospheric dust load found in city centers without trees (Meyer, 1997). Even some trees in high-density neighborhoods decrease the amount of dust in the air. Trees in a street also produce small air circulations, which dilute pollutants and so reduce the risk of inversions and smog. Green spaces of 50 to 100 m depth improve air quality up to 300 m away in their neighborhoods (Meyer, 1997).

Economic Benefits of Urban Green Space

Quantifying the benefits and costs of urban green space can be important especially if a campaign has to be justified from an economic point of view (see Chapter 5). Although the qualitative effects of urban vegetation on urban thermal conditions are beyond controversy, there is a lack of quantitative information for assessing the benefits of greening campaigns. Still, indicators are very strong that green space and landscaping increases property values and the financial returns for land developers. Studies have found increased financial returns of 5% to 15% depending on the type of project (McMahon, 1996). Also, 70% to 80% of consumers rated natural open space as the feature they desired most in a new home development (McMahon, 1996).

Using vegetation to reduce the energy costs of cooling buildings has been increasingly recognized as a cost-effective reason for increasing green space and tree plantings in cities (Fezer, 1995). Even small green spaces decrease temperatures in urban environments in manifold ways:

1. They produce small-scale air circulations. These air movements, due to wind or thermal upswing, cause air exchange. Such effects of urban vegetation on local climate were studied by Bruse (1999) using a simple case study of a street canyon with homogeneous buildings on both sides (see Small Structural Changes, below).
2. Even though a park may be exposed to solar radiation throughout the morning, evaporation from the grass surface and trees create lower ground surface temperatures and consequently lower air temperatures in the park than in the surrounding urban area. With strong winds, a cooler park can significantly contribute to the reduction of the heat intensity in the town. For example, a park of 1.2 km × 1.0 km can produce an air temperature difference between the park and the surrounding city that is detectable up to 4 km away (Takashi and Vu Than, 1998).
3. Urban vegetation counteracts the urban heat-island effect by providing shade. Even single trees, strategically planted to shade homes, can reduce air conditioning bills significantly. Up to 227 kWh can be saved by each tree through cooling by evapotranspiration and 61 kWh through direct shading of a home (McPherson et al., 1999). Simulations of energy saving benefits for the cities of Sacramento and Phoenix found that three mature trees around homes cut annual air conditioning demand by 25% to 40% (McPherson et al., 1999). Figure 6.3 illustrates how trees and bushes provide shade and thus contribute to energy savings.

Social Benefits

Urban green space also plays a role in improving the social health of their inhabitants. It is not only the cleaning and cooling ability of plants that show direct positive effects on human health by providing shade, reducing heat strain, reducing risks of cancer, and cutting down on noise. In addition, urban green areas, particularly urban parks of all sizes, serve as a nearby resource for relaxation and recreation. Green areas in cities provide contact with nature, for example, marking the rhythm of the changing of the seasons: autumn when leaves fall, the flowering of plants and trees in spring, the presence of seasonal birds. Thus green spaces and trees provide an emotional warmth and softness to city life, as opposed to the hardness of concrete and pavement. They can also add a sense of privacy. Urban green spaces are also educational resources, providing locations for structured and informal lifelong learning about nature, and ecological and environmental processes.

Finally, urban green spaces are very useful in urban planning, because they are elements that bring order to the surrounding area. They imbue the area with aesthetic dignity and they often serve as a link between various neighborhoods.

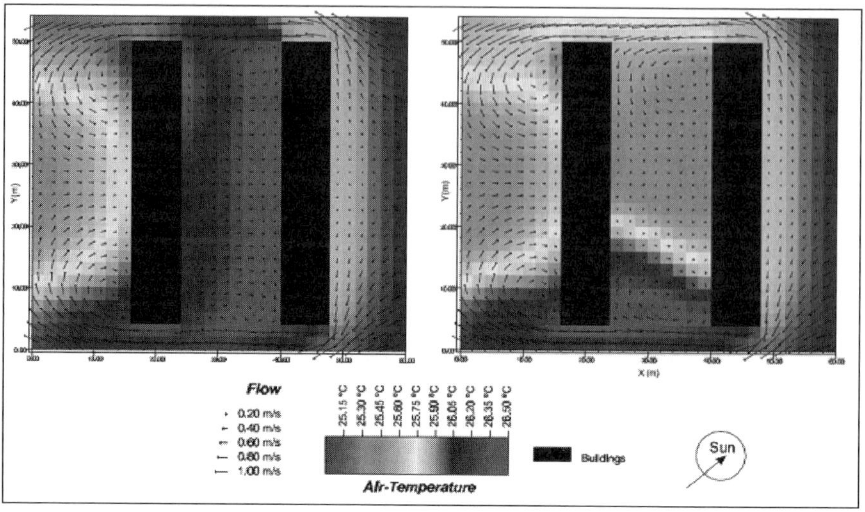

Fig. 6.3 Flow field and air temperature for the street canyon without trees (left) and with trees (right) in 1.8-m height at 14:00 Central European Time (CET). (From Bruse, 1999, with permission.)

Thus, they accentuate the identity of the neighborhood, often becoming an aesthetic or symbolic reference point, making streets and neighborhoods more alive. In addition, parks serve as links between neighborhoods, often becoming a place to socialize and bringing people together.

Opportunities for Changing Urban Conditions: Impacts of Structural Modification on Local Climate

Most research indicates that the interactions between different natural and artificial elements in the lower urban boundary layer produce patterns of varying local climate conditions that are very sensitive to structural changes. Intelligent planning of the urban environment leading to an improved local climate can have many benefits, such as energy savings and the reduction of health risks. Thereby, urban structural changes can be divided into two groups: small-scale changes within existing structures, and large-scale changes such as the complete redevelopment of urban areas.

Small Structural Changes

Decision makers often ask this question: Can we quantify the benefits of urban greening, and if so, how do we quantify the effects of small structural changes in the environment, like planting trees along a particular road? The multitude of different processes affected often makes it impossible to assess the impact of changes on local climate without the help of quantitative models. Such models are useful for communicating the relative benefits of alternative greening strategies to planners, decision makers, and the public.

To demonstrate the effects of urban vegetation on local microclimate, a computer simulation by Michael Bruse from the University of Bochum (Bochum, Germany) has been used (Bruse, 1999). The following example represents a simulation that was carried out for a typical street in Bochum, Germany (53° N 7.5° E). The street had a north-south orientation and was 16 m wide with homogeneous buildings (height 16 m) on both sides. The vegetation is represented by 20-m-high deciduous trees with a dense crown layer. In Figure 6.3 we see the temperature difference of the canyon without (left) and with trees (right). The different gray scales indicate the cooling effect of this small-scale structural change.

> Inside the green canyon, the air temperature is around 1.0 to 1.3 K lower than in the treeless case. Below the trees, wind speed is a little lower and vortex eddies at the end and at the beginning of the canyon can extend a bit further into the street because the overlying vertical vortex eddy between the buildings is suppressed by the tree crown layer. On the southern end of the street a new hot spot can be observed. Here the trees do not shade the ground surface but reduce wind speed so that air exchange is less effective than in the street without trees. [Bruse, 2000, p. 4]

Large Structural Changes

In addition to small-scale structural modifications involving green space, urban planning has the possibility of making large-scale structural changes. The goals of these structural changes are to increase the ecological, social, and economic value of the city and turn some old urban structures, characterized by high density and dark building materials into lower density multistory and open residential and commercial areas. The aerial photo in Figure 6.4 provides an example of general structural changes from old to new urban patterns. The old urban structure (right) shows high building density with dark roofs, which contributes to weak air circulation patterns and greater heat and pollutant trapping. On the left we see an example of a newer form of urban structure characterized by low density and light-colored roofs that consequently create an improved microclimate.

Building Materials: Reflectivity of Conventional Roofing and Pavement Materials as an Important Factor Contributing to the Urban Heat-Island Effect

When regarding urban climate and the urban heat-island effect, it is essential to understand the influence of the building materials on the degree to which they reflect sunlight, that is, the albedo value for the material. Black or dark structural materials absorb sunlight strongly (have low albedo), heat the air in their surroundings, and so create human discomfort, particularly in the summer. The way to fix the problem is to make surfaces brighter, so that they reflect more solar radiation

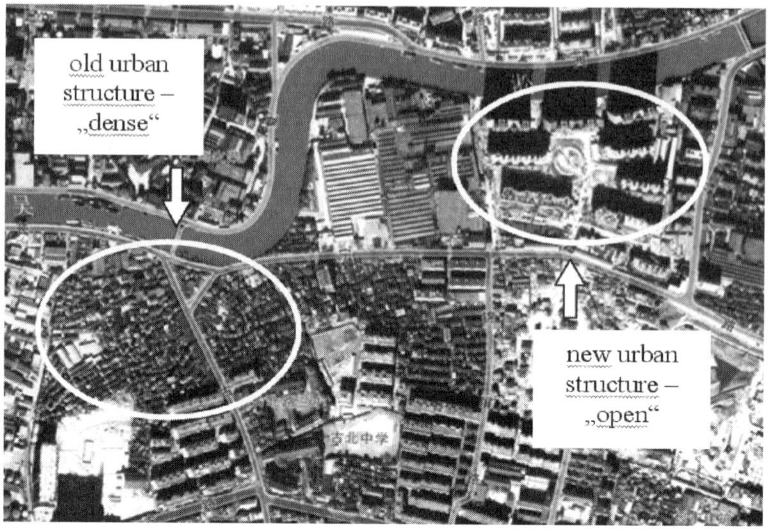

Fig. 6.4 Large-scale structural changes in building density that promote a cooler urban environment (Shanghai, China). (Photo courtesy of Institute of Geodesy of the City of Shanghai, 1991.)

(increase albedo) and stay cooler. Low albedo implies higher surface temperatures since larger amounts of solar energy are absorbed. Reflective materials or painting can be applied to pavements, walls, and roofs. A cooler roof on a building benefits it directly and immediately by contributing significantly to energy savings. For bright buildings of North American cities, numeric model simulations proved that in summer months the energy consumption could be lowered up to 15%, as opposed to dark building covers, which absorb solar radiation strongly (Akbari et al., 1999).

Conclusion: Approaches to a Sustainable Urban Development

As we have pointed out, there are three major elements of urban land use that influence the quality of urban climate. The first element is the concept of an integrated urban green system, which is determined by the presence of its single elements, their distribution, and its inner structure. To perform in an ecologically and socially optimal way, urban green space must be constructed as a networked system of open areas and recreation zones in an interrelated pattern. Understanding the importance of these green networks is a principle of integrated urban planning. Increasingly, high-density urban areas use nature trails to link urban green spaces and biotopes into a network. These nature trails are often located along rivers and creeks, which often form the main framework of a city's urban green space and forest. Greenways allow us to treat land and water as a system, as interlocking pieces in a puzzle, not as isolated entities. Second, we have seen the important potential of manipulating city structure, its building density and alignment, to improve urban climate and air quality. Third, the selection of building materials also plays an important role in reducing the urban heat island and its negative impacts on human health. Hence we have to develop and create demand for ecologically favorable building materials in urban construction. We conclude that the quality of an eco-city depends on these interrelated factors. To achieve the best results in urban planning, it is necessary to pursue such an integrated approach.

References

Akbari, H., Konopacki, S., and Pomerantz, M. (1999) Cooling energy savings potential of reflective roofs for residential and commercial buildings in the United States. Energy 24:391–407.

Ammer, U., and Bechet, G. (1978) Entscheidungshilfen für die Freiraumplanung. Naturwissenschafticher Teil. In: Materialien zur Landes- und Stadtentwicklungsforschung des Landes Nordrhein - Westfalen, Band 4,007, Institut für Lan-desentwicklungsforschung und Bauwesen des Landes Nordrhein-Westfalen (ILS NRW), Dortmund.

Bonan, G.B. (2002) Ecological Climatology: Concepts and Applications. Cambridge University Press, Cambridge, UK.

Bruse, M. (1999) Modelling and strategies for improved urban climates. Invited paper. In: Proceedings International Conference on Urban Climatology and International Congress of Biometeorology, Sydney, November 8–12. http://klima.geographie.ruhr-uni-bochum.de/bruse/paper63_00.pdf

Chameides, W.L., and Cowling, E.B. (1995) The State of the Southern Oxidants Study (SOS): Policy-Relevant Findings in Ozone Pollution Research, 1988–1994. Southern Oxidants Study, College of Forest Resources, North Carolina State University, Raleigh, NC.

Commission of the European Communities. (2005) Communication from the Commission to the Council, the European Parliament, the European Economic and Social Committee and the Committee of the Regions—Winning the Battle Against Global Climate Change. http://europa.eu/press_room/presspacks/climate/com2005_0035en01.pdf

Fezer, F. (1995) Das Klima der Städte. Justus Perthes Verlag Gotha, Gotha. Germany

Gilbert, O.L. (1991) The Ecology of Urban Habitats. Chapman & Hall, London, UK.

Heudorf, U., and Meyer, C. (2005) Heat Waves and Health—Analysis of the Mortality in Frankfurt, Germany, During the Heat Wave in August 2003. Das Gesundheitswesen 67(5):369–374.

Institute of Environmental Science, East China Normal University Shanghai. (2002) The Development Program of Urban Forest in Shanghai. Shanghai (unpublished).

Institute of Geodesy of the City of Shanghai. (2001) Shanghai Shi Yingxiang Ditu Ji: Zhongxin Cheng Qu (Photo and Imagery Atlas of Shanghai: City Centre). Shanghai.

Kaerkes, W.M. (1987): Zur ökologischen Bedeutung urbaner Freiflächen. Dargestellt an Beispielen aus d. mittleren Ruhrgebiet. Bochum. In: Materialien zur Raumordnung des Geographischen Instituts Bochum, Band 35. Geographisches Institut der Ruhr-Universität Bochum, Bochum, Germany.

Kuttler, W. (1998) Stadtklima. In: Sukopp, H., and Wittig, R. (eds.) Stadtökologie. Fischer-Verlag, Stuttgart, Germany, pp. 113–150.

McMahon, E.T. (1996) Green enhances growth. PlannersWeb: Planning Commissioners Journal 22:4. http://www.plannersweb.com/articles/look22.html

McPherson, G., Simpson, J.R., Peper, P.J., and Xiao, Q. (1999) Benefit-cost analysis of Modesto's municipal urban forest. Journal of Arboriculture 25:235–248.

Meyer, J. (1997) Die zukunftsfaehige Stadt. Nachhaltige Entwicklung in Stadt und Land. Werner Verlag, Düsseldorf, Germany.

Oke, T.R. (1973) City size and the urban heat island. Atmospheric Environment, Oxford 14:769–779.

Pomerantz, M., Akbari, H., Berdahl, P., and Taha, H. (1999) Physics and Public Policy for Urban Heat Island Mitigation. Summary of a presentation to the American Physical Society, Atlanta, GA, March. http://eetd.lbl.gov/HeatIsland/PUBS/APS-PressRelease/.

Sukopp, H., and Wittig, R. (eds.) (1998) Stadtökologie, 2nd ed. Gustav Fischer Verlag, Stuttgart.

Taha, H., Akbari, H., and Sailor, D. (1992) On the Simulation of Urban Climates: Sensitivity to Surface Parameters and Anthropogenic Heating. Technical Note, Lawrence Berkeley Laboratory, Berkeley, California.

Takashi, A., and Vu Than, C.A. (1998) Case study: effects of vegetation on the climate in the urban area: In: Breuste, J., Feldmann, H., and Uhlmann, O. (eds.) Urban Ecology. Springer, Berlin.

Wittig, R. (2002) Siedlungsvegetation. Ulmer, Stuttgart.

7
Applying Ecosystem Management to Urban Forestry

Wayne C. Zipperer

During the 1990s, the United States Department of Agriculture Forest Service shifted from commodity production management to ecosystem-based management (Overbay, 1992). Although definitions of ecosystem-based management vary by objectives, the principle had four primary elements: (1) maintaining viable populations of native species, (2) representing native ecosystems across their range of natural variability, (3) maintaining ecosystem processes, and (4) ensuring ecosystem goods and services for future human generations (Grumbine, 1994). In general, ecosystem management approach becomes a way of thinking more broadly about a system (Yaffee et al., 1996). For example, a forester must consider how management activities affect not only timber production but also ecosystem processes, biodiversity, and natural populations, all of which influence forest productivity. This way of thinking enables managers to look at the entire forest as a single entity and assess how management goals and objectives affect ecosystem integrity.

During the 1990s, urban forestry in the United States began to shift from single-tree to ecosystem-based management (Zipperer et al., 1995). This new approach recognizes the importance of urban vegetation (both public and private) as part of the urban ecosystem and as a source of many ecological services and benefits (Nowak and Dwyer, 2000). These benefits include cleaning air and water, enhancing human health, and providing wildlife habitat, recreational opportunities, and aesthetics. By taking an ecosystem approach to management, urban foresters can maximize benefits from the forest while minimizing the cost to maintain it.

Yet, an urban forester manages by altering the structure of only public trees through single-tree management. Does this mean that an ecosystem-based management is not a viable objective for urban forest management? Throughout the International Symposium on Urban Forestry and Eco-Cities held in 2002, speakers promoted the need to take a holistic approach to management and the need to better understand the social and ecological processes influencing the livability of a city. This chapter provides a succinct overview of ecosystem principles as they pertain to urban landscapes, and applies the theory of vegetation dynamics as a means of clarifying for managers how they may take a holistic approach through single-tree management.

M.M. Carreiro et al. (eds.), *Ecology, Planning, and Management of Urban Forests: International Perspectives.*

Ecosystems

An ecosystem is defined as a spatially and temporally explicit place that includes all the organisms, all abiotic factors in that environment, and their interactions (Likens, 1992). For an urban ecosystem, this includes the entire set of social, ecological, and physical components that define an urban area. One might ask, What is an urban ecosystem and how might it differ from other ecosystems? McIntyre et al. (1990) reviewed the concept of "urban" and concluded that no single definition exists because of the different perspectives of those who study or work in urban systems. I propose that rather than trying to define an urban area spatially, consider thinking of it as a system where ecological, physical, and social patterns and processes interact to create a unique environment. This environment represents both the green (e.g., vegetation) and gray (e.g., buildings and roads) infrastructure. In their paper on urban ecosystems, Pickett et al. (1997) presented a simple model to reveal the interconnectedness of social, ecological, and physical components. They asserted that by changing one component, the other components are directly or indirectly affected. So, from an urban forest management perspective, a manager, by altering some aspect of ecological structure (e.g., composition and diameter distribution of trees), can influence the social and physical components of the system, and all these factors (ecological, social, physical) must be taken into account when making management decisions, particularly since they will affect the extent of ecosystem services provided by the forest.

To achieve an ecosystem approach to management, the entire urban forest needs to be considered. A manager accomplishes this by looking beyond the particular management site and evaluating the effect of the site on adjacent land uses, and congruently, the effect of adjacent land uses on the site. In other words, the site should not be viewed independently of the context in which the site occurs, since context will affect the site and the site will affect its context. By viewing management activities from this broad perspective, the manager moves beyond simply planting a tree at a particular site or location, and asks how this activity affects ecosystem process and subsequent services to the site and adjacent areas. This perspective is important because an ecosystem is an open system, in which energy, materials, and organisms move into, through, and out of the system. By altering the urban forest structure or the physical environment of the site, the manager influences this movement. For example, by increasing the canopy cover by planting trees, a manager can influence the amount of particulate material and rain intercepted by the trees. A greater interception of material leads to cleaner air and less storm runoff. By taking a broad perspective, a manager can evaluate potential planting sites in the context of surrounding vegetation and ask if the proposed planting achieves the desired management goals and objectives, or if resources should be directed to other sites. So, a broad perspective enables managers to prioritize sites for planting, and this may maximize benefits while minimizing costs (also see Chapter 13).

To illustrate this point, I will use a figure representing the vacant lots and buildings in Baltimore, Maryland (Fig. 7.1). One objective for an urban forester might be to

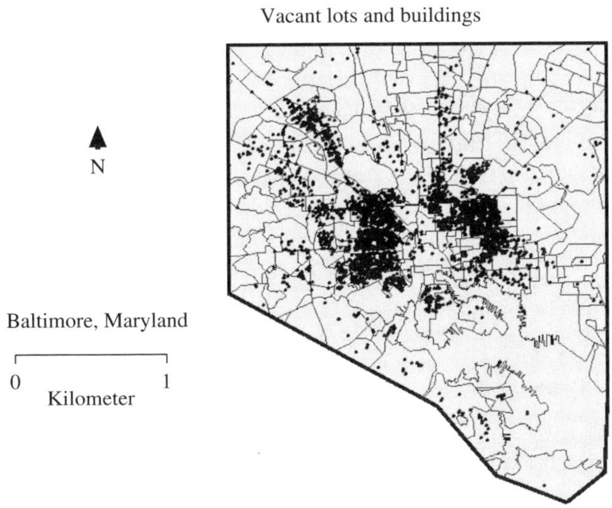

Fig. 7.1 A map showing locations of vacant lots and buildings in Baltimore, Maryland. (From Parker et al., 1999.)

afforest vacant lots, but which ones and which ones first? Which vacant lot has the greatest effect on water quality, on neighborhood well-being, and on city beautification? By asking these questions, the manager can determine which lots would most improve the quality of life in Baltimore. The link between site management and context could only be achieved by taking a broad perspective and asking what key ecosystem processes (social, ecological, and physical) influence the site and how these processes can be modified or enhanced by afforestation.

Managers should also keep in mind that ecosystems are dynamic. They are continually changing because of management activities, species natural history, natural succession, and natural and human disturbances. Throughout a city, public trees are being planted to maintain canopy cover and removed to reduce safety risks. These activities represent change. Furthermore, each city has its own disturbance regime. A disturbance regime defines the type, size, frequency, severity, and dispersion of disturbances influencing the city. For example, hurricanes can significantly alter the structure of an urban forest (Duryea et al., 1996). Although this disaster can be catastrophic to human well-being, it may provide the urban forester with a unique opportunity to restructure the forest by creating new planting opportunities, changing species diversity, and balancing its age structure (see Richards, 1983). By restructuring the public forest to meet an objective of sustaining or enhancing ecosystem goods and services, a manager may begin to take a long-term view of the forest and its benefits, and how to optimize those benefits.

An ecosystem approach enables managers to see how their activities of planting trees are interconnected with the entire urban forest and the ecosystem goods and services the forest provides. Similarly, an ecosystem management perspective plans for changes that may occur through natural and human disturbances. This holistic approach has been echoed throughout the International Symposium on Urban Forestry and Eco-Cities in 2002 and called by various names: ecoscape, ecoindustry, and ecoculture. No matter what it is called, a holistic or ecosystem approach to management creates a framework for improving the livability of our cities by maintaining or enhancing ecosystem services through influencing ecosystem structure and altering ecosystem processes. But a manager must still consider how to link ecosystem management to single-tree management. I propose that we adapt the concept of vegetation dynamics to urban forest management (Fig. 7.2 and 7.3).

Vegetation Dynamics

The concept of vegetation dynamics was proposed to account for successional changes on a site at a single species or individual level (Pickett et al., 1987a,b). The concept has three primary components: site availability, species availability, and species performance (Fig. 7.2). Succinctly, from a natural succession perspective, site availability refers to the creation of space for an individual to germinate, grow, and reproduce. Sites become available through the death of an individual or through a disturbance (Brand and Parker, 1995). Disturbance type dictates the frequency and size of site formation. Species available to colonize these sites currently exist in the seed bank or disperse there from adjacent areas. Once an individual species is planted on a site, its performance determines its survivability. Factors influencing survival include species autecology, environmental conditions and resources, and interactions with other site elements, such as other species. Autecological factors include life history and phenotypic plasticity. Examples of environmental conditions include climate, air pollution, heavy metal toxicity, and site history. Examples of resources include light, nutrients, and water. Examples of species interactions include competition, herbivory, disease organisms, mutualistic symbioses, and allelopathy. I will use this framework to discuss the application of ecosystem management to urban forest management in greater detail.

Site Availability

Within the urban landscape, site availability represents an array of sizes ranging from a single-tree pit, to a vacant lot, to an entire urban park (Zipperer et al., 1997) (Fig. 7.3). For example, in Chapter 8, Nerys Jones describes the reforestation of derelict industrial sites. To promote natural recruitment of species, industrial debris was removed and soils were prepared. As predicted by the vegetation dynamic model, an array of native and nonnative species from adjacent areas colonized these sites (also see Chapter 23). Local residents now use these "naturalized" areas for recreation.

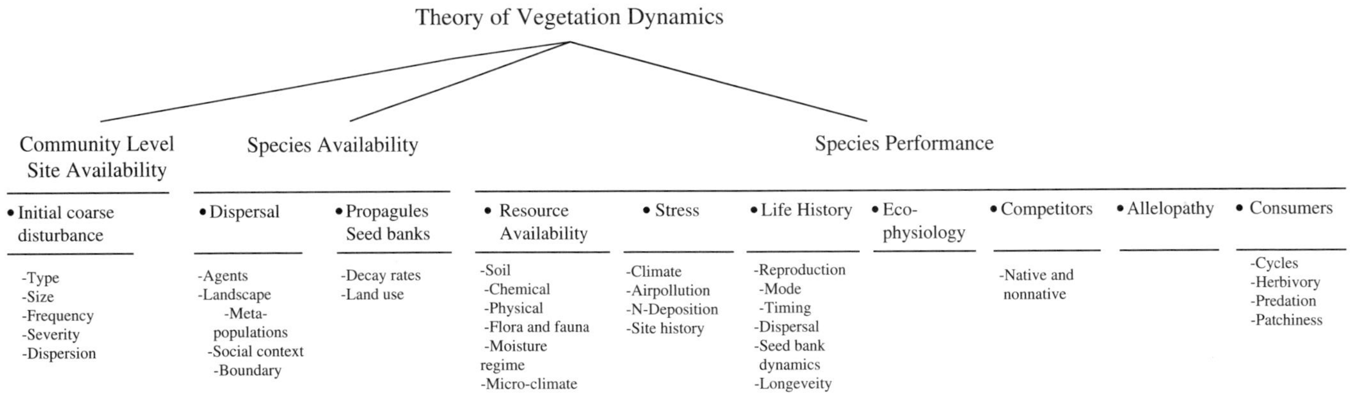

Fig. 7.2 Components of the theory of vegetation dynamics used to account for successional changes. N, nitrogen. (From Pickett et al. 1987a,b.)

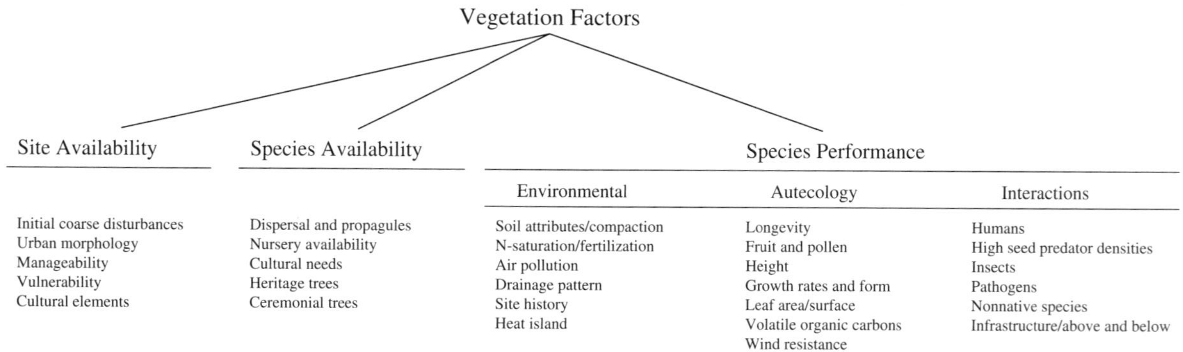

Fig. 7.3 Theory of vegetation dynamics modified for application of ecosystem management in urban landscapes by incorporating elements of the urban ecosystem used in the management-decision process

Site availability also is applicable at a citywide scale. The City of Shanghai demonstrated this by creating three new urban parks where none existed before. Site selection was based not only on the logistics of where to place a park but also on the social context of the site. These new parks occupy sites that offer an array of social and ecological benefits not previously enjoyed by residents (also see Chapter 20).

The selection of sites for these parks, as well as sites for single-tree management, is based on urban morphology. Urban morphology is the pattern of urban development, both vertically and horizontally (Sanders, 1984), and includes the buildings, streets, sidewalks, parking lots, and other human structures. Where human structures and surfaces already occur, the possibility of planting spaces is eliminated unless considerable effort and cost are expended to remove existing structures or surfaces. Therefore, the more densely packed a city is, the fewer the places for trees to grow. In Baltimore, for example, urban foresters use a geographic information system (GIS) to select vacant lots to rehabilitate (see Fig. 7.1). The selection process included not only biophysical factors but also social factors. Recognizing that community members were essential to the success of their projects, foresters worked with local community leaders to plant and maintain sites (Grove and Burch, 1997). Through this socioecological partnership, managers rehabilitated sites and community leaders revitalized their neighborhoods (also see Chapters 9 and 12).

Contextual elements and processes influence a site and its availability. For example, in Chapter 15 James Kielbaso discusses the importance of site manageability, and the benefit–cost ratio of managing a site. Shanghai created urban parks where there were none before. Only time will tell if the benefits of creating these parks will exceed their cost for development. Likewise, the selection of sites to plant trees must account not only for manageability but also other contextual influences such as vulnerability (damage by humans and natural events such as droughts, frost, and pollution) and cultural elements. In Chicago, forest managers work with local planners to maintain the connectivity of natural areas not only to maintain genetic flow among natural populations, but also to provide corridors for recreation (Gobster and Hull, 2000).

Planting sites also become available through catastrophic disturbances. Not only can these disturbances have devastating effects on the existing urban forest, but they also can create opportunities for the urban forest manager to replant, balance age and size structure, and enhance species diversity. Storms also provide insights into which species are capable of withstanding local disturbances. In their work, Duryea et al. (1996) assessed how different species survived a hurricane and used this information to make recommendations for future tree plantings in affected areas.

Species Availability

In a natural system, species availability depends on dispersal from adjacent areas and emergence from the soil seed bank. For the urban landscape, species availability is more complex and involves both ecological and social elements (Fig. 7.3). Species dispersal and seed banks play a critical role in reforesting abandoned or restoration sites (Robinson and Handel, 2000) and colonizing an existing remnant or regenerated forest patches. Because of the abundance of nonnative species growing in the urban landscape, many of the species colonizing remnant and regenerated forest patches are often nonnative (Moran, 1984; Guntenspergen and Levenson, 1997; Zipperer, 2002). This observation is of particular importance when considering new species for planting. One of the primary avenues for introduction of nonnative species into remnant vegetation is arboricultural and horticultural plantings (Reichard and White, 2001). As managers, we need to ask how our actions will affect not only the site but also the area around it. In other words, how does site content affect site context? Because ecosystems are open systems, propagules from plantings can be dispersed into remnant and regenerated forest patches of vegetation, potentially changing their species composition and structure and subsequent functions in the broader landscape (Rudnicky and McDonnell, 1989).

The debate over whether or not to use nonnative species in urban plantings can be acrimonious at times. The premise for using nonnative species is that the environmental conditions in urban landscapes have been altered, and native species can no longer survive or compete with nonnative species (MacDonald, 1993). However, the data documenting native species responses to urban conditions are limited. Realizing nonnative species may become invasive, selection protocols need to be implemented to eliminate introductions of invasive species when selecting nonnative species for plantings (Reichard and White, 2001).

In urban landscapes, social factors play a key role in species availability and selection. For example, nurseries may stock only a limited number of species, thus limiting species selection for plantings. Another presentation at the International Symposium on Urban Forestry and Eco-Cities in 2002 described new nurseries that are being created around various Chinese cities to meet projected demands of future tree plantings. Unfortunately, it seems that most of these nurseries contain a limited number of species and they were principally nonnative. From a holistic perspective, species diversity plays an important role in maintaining a system's resiliency and stability (Tilman et al., 1997). If the purpose of management is to enhance ecosystem services, then activities (e.g., greater species diversity for nursery stock) that achieve this goal are desirable and should be encouraged. Also, since many of the species planted in urban landscapes are cultivars, managers need to recognize cultivars' limited genetic diversity and account for it when selecting which species to plant.

If managers have a diverse selection of species to work with, they will be able to select appropriate species to meet site and contextual needs. However, plantings in our cities not only need to meet biological diversity criteria, but also need to balance management costs and capabilities (Richards, 1983, 1993; Nowak et al., 2001). This balance may reduce the number of species available to managers because of the cost of subsequent management. However, over time a manager can develop a list of species to meet diverse management needs once new species have been tested under different site and contextual conditions (see Chapters 24 and 25).

Other social considerations include conserving heritage and ceremonial trees (Jim, 2005a,b; also see Chapter 9). Heritage trees represent species that have local, regional, or national significance. For example, American Forests, a nonprofit organization in the United States, offers homeowners an opportunity to plant seeds and seedlings from historically important trees (http://americanforests.org/). In the United States, species may be selected to memorialize victims of homicides or accidents. Often these species may represent the favorite tree of an individual or an entire community. With time, these memorial plantings can become an important component of the social fabric of a neighborhood, town, city, or state.

Species Performance

Urban forest managers can influence site and species availability, but they have little influence on species performance (unless the species is genetically manipulated). However, the manager can increase the probability of tree survival by selecting the right species for site and contextual conditions. In the urban environment, examples of site content factors that affect species performance include soil compaction, poor nutrient availability, minimal planting space, and inadequate drainage (Fig. 7.3). Through best management practices, managers can minimize the negative impacts of these factors, thus decreasing mortality and increasing the effectiveness of plantings (Miller, 1988).

Contextual influences include not only air pollution, pathogens, and urban heat-island effects but also new development patterns. Air pollution assails the health of individual trees and the entire urban forest. By neglecting site condition or selecting the wrong species for those conditions, the manager may inadvertently increase its susceptibility to insect and pathogen outbreaks. As these outbreaks develop, they may move beyond the urban landscape into rural forests, hence increasing economic losses beyond a municipality's boundary. For example, a southern pine bark beetle infestation in Florida originated in Gainesville and progressed outward into neighboring counties. Although the beetle is native and was not considered a pest, environmental circumstances (4 years of drought), new development patterns, and stress from the urban environment created favorable conditions for a species outbreak. Similarly, a change in urban morphology (e.g., adding more buildings or developing vacant lots) may alter microclimatic conditions and increase heat-island effects (see Chapter 6). The additional heat load adds to the existing environmental stresses on individual trees.

A species' autecological traits not only are important for its survival in an urban environment but also have important contextual value. For example, a species' leaf area, emissions of volatile organic carbon compounds (VOCs), pollen production potential, and longevity are important elements when management objectives include reducing particulate matter and air pollution. A tree with high leaf area, and low VOC emission can improve air quality by intercepting more particulate material, cooling ambient temperatures through evapotranspiration and shading, and releasing lower VOCs than a tree without such traits. So, when selecting individuals to plant, the manager must consider not only species tolerant of high temperatures, but also those species that may contribute to ozone production from VOCs (Nowak et al., 2001) or high pollen loading to susceptible people in the vicinity. Likewise, longevity and growth rates are important traits influencing carbon sequestration. Slower growing species, such as those in the genus *Quercus*, may sequester carbon less quickly than a fast-growing species, such as those in the genus *Populus*, but because of their greater longevity, some *Quercus* species can sequester and store carbon for a longer time. Similarly, context will influence whether trees bearing fruits and nuts are to be planted. In one neighborhood, fruits and nuts may be viewed as a nuisance, whereas in a different neighborhood they may play an important role in supplementing local dietary needs, as occurs in agroforestry but in an urban landscape. As managers, we need to realize that matching species to the social context may be just as important as matching species to site conditions.

A manager also needs to acknowledge the interactions within and among ecosystem components that influences species performance. These interactions are both natural and anthropogenic. For example, a street tree needs to be large enough to minimize vandalism (e.g., breaking branches, bending, pulling the tree out of the ground). Natural interactions include increased seed predation and herbivory, which can significantly affect reforestation projects. With the planting of nonnative species in urban landscapes, competition may increase between native and nonnative species in colonizing available sites within forest remnants. Similarly, homeowners and managers may select nonnative rather than native species, thus reducing the likelihood of nurseries carrying more native species (a negative feedback loop reinforcing continued sale of nonnatives in nurseries). Also, due to international imports, urban landscapes are often exposed to new pests and pathogens (e.g., cities were Dutch Elm disease and chestnut blight infection foci in the 20th century). A recent example is the presence of Asian longhorned beetle in New York City, Chicago, and some cities in Connecticut (http://www.na.fs.fed.us/fhp/alb/index.shtm). This pest, which was unintentionally introduced on wooden pallets and boxes from China, has spread to the urban forests of several cities in the U.S. By not accounting for the variety of interactions that affect species performance, planting and restoration projects may fail.

In previous sections, I have modified Pickett et al.'s (1987b) and Pickett and McDonnell's (1989) theory of vegetation dynamics to include attributes associated with urban forest management (Fig. 7.3). This is not to say that the original vegetation dynamic model should be ignored, but rather, it should be complemented with additional ecosystem level attributes particular to urban areas that influence urban forest management. Similarly, this list of attributes is not meant to be exhaustive, but rather meant to increase a forest manager's awareness of factors influencing management actions and outcomes in urban areas. Managers will need to add to this list to account for the unique conditions and interactions created by the ecological, physical, and social components in their own urban landscapes that affect site availability, species availability, and species performance.

Conclusion

As urban forest managers, we need to think more broadly about the landscapes in which we work to identify the key ecological processes affecting a site, evaluate how they will affect our plantings, and assess how our plantings will affect these processes. Through our management, we can alter urban forest structure to improve ecological processes, thereby enhancing ecosystem goods and services. To meet these management goals, managers need to identify both site content and context factors when selecting species and sites. To be effective, an array of diverse species is needed to maintain urban forest stability and resilience. This diversity, however, will undoubtedly be tempered by management costs. Through proper education, managers and other individuals involved in urban forest management (e.g., nursery growers, politicians, and residents) can maintain a healthy urban forest to yield benefits for healthier lives.

References

Brand, T., and Parker, V.T. (1995) Scale and general laws of vegetation dynamics. Oikos 73:375–380.

Duryea, M.L., Blakeslee, G.M., Hubbard, W.G., and Vasquez, R.A. (1996) Wind and trees: a survey of homeowners after Hurricane Andrew. Journal of Arboriculture 22:44–49.

Gobster, P.H., and Hull, R.B. (eds.) (2000) Restoring Nature: Perspectives from the Social Science and Humanities. Island Press, Washington, DC.

Grove, J.M., and Burch, W.R. (1997) A social ecology approach and applications of urban ecosystem and landscape analyses: a case study of Baltimore, Maryland. Urban Ecosystems 1:259–275.

Grumbine, R.E. (1994) What is ecosystem management? Conservation Biology 8:27–38.

Guntenspergen, G.R., and Levenson, J.B. (1997) Understory plant species composition in remnant stands along an urban-to-rural land use gradient. Urban Ecosystems 1:155–169.

Jim, C.Y. (2005a) Monitoring the performance and decline of heritage trees in urban Hong Kong. Journal of Environmental Management 74(2):161–172.

Jim, C.Y. (2005b) Floristics, performance and prognosis of historical trees in the urban forest of Guangzhou city (China). Environmental Monitoring and Assessment 102:285–308.

Likens, G.E. (1992) The Ecosystem Approach: Its Use and Abuse. Ecology Institute, Oldendor/ Luhe, Germany.

MacDonald, L. (1993) Go native or exotic. Urban Forests 13:9–13.

McIntyre, B.M., Scholl, M.A., and Sigmon, J.T. (1990) A quantitative description of a deciduous forest canopy using a photographic technique. Forest Science 36:381–393.

Miller, R.W. (1988) Urban Forestry: Planning and Managing Urban Greenspaces. Prentice Hall, Englewood Cliffs, NJ.

Moran, M. (1984) Influence of adjacent land use on understory vegetation of New York forests. Urban Ecology 8:329–340.

Nowak, D.J., Crane, D.E., and Stevens, J.C. (2001) Syracuse's urban forestry resource. In: Nowak, D.J., and O'Connor, P.R. (eds.) Syracuse Urban Forest Master Plan: Guiding the City's Forest Resource into the 21st Century. USDA Forest Service, Newtown Square, PA, pp. 9–14.

Nowak, D.J., and Dwyer, J.F. (2000) Understanding the benefits and costs of urban forest ecosystems. In: Kuser, J.E. (ed.) Handbook of Urban and Community Forestry in the Northeast. Kluwer Academic/Plenum, New York, pp. 11–25.

Overbay, J.C. (1992) Ecosystem management. Speech delivered at the national workshop on taking an ecological approach to management. USDA Forest Service, Washington, DC, Salt Lake City, UT.

Parker, J.K., Sturtevant, V.E., Shannon, M.A., et al. (1999) Some contributions of social theory to ecosystem management. In: Sexton, W.T., Malk, A.J., Szaro, R.C., and Johnson, N.C. (eds.) Ecological Stewardship: A Common Reference for Ecosystem Management, vol 3. Elsevier, Amsterdam, pp. 245–277.

Pickett, S.T.A., Burch, W.R., Jr., Dalton, S.E., Foresman, T.W., Grove, J.M., and Rowntree, R.A. (1997) A conceptual framework for the study of human ecosystems in urban areas. Urban Ecosystems 1:185–201.

Pickett, S.T.A., Collins, S.C., and Armesto, J.J. (1987a) A hierarchical consideration of causes and mechanisms of succession. Vegetatio 69:109–114.

Pickett, S.T.A., Collins, S.C., and Armesto, J.J. (1987b) Models, mechanisms and pathways of succession. Botanical Review 53:335–371.

Pickett, S.T.A., and McDonnell, M.J. (1989) Changing perspectives in community dynamics: a theory of successional forces. Trends in Ecology and Evolution 4:241–245.

Reichard, S.H., and White, P.S. (2001) Horticulture as a pathway of invasive plant introductions in the United States. Bioscience 51:103–113.

Richards, N.A. (1983) Diversity and stability in a street tree population. Urban Ecology 7:159–271.

Richards, N.A. (1993) Reasonable guidelines for street tree diversity. Journal of Arboriculture 19:344–349.

Robinson, G.R., and Handel, S.N. (2000) Directing spatial patterns of recruitment during an experimental urban woodland reclamation. Ecological Applications 10:174–188.

Rudnicky, J.L., and McDonnell, M.J. (1989) Forty-eight years of canopy change in a hardwood-hemlock forest in New York City. Bulletin of the Torrey Botanical Club 116:52–64.

Sanders, R.A. (1984) Some determinants of urban forest structure. Urban Ecology 8:13–27.

Tilman, D., Knops, J., Wedin, D., Reich, P., Ritchie, M., and Siemann, E. (1997) The influence of functional diversity and composition on ecosystem processes. Science 277:1300–1302.

Yaffee, S.L., Phillips, A.F., Frentz, I.C., Hardy, P.W., Maleki, S.M., and Thorpe, B.E. (1996) Ecosystem Management in the United States: Assessment of Current Experience. Island Press, Washington, DC.

Zipperer, W.C. (2002) Species composition and structure of regenerated and remnant forest patches within an urban landscape. Urban Ecosystems 6:271–290.

Zipperer, W.C., Foresman, T.W., Sisinni, S.M., and Pouyat, R.V. (1997) Urban tree cover: an ecological perspective. Urban Ecosystems 1:229–246.

Zipperer, W.C., Grove, J.M., and Neville, L.R. (1995) Ecosystem management in urban environments. In: Kollin, C., and Barratt, M. (eds.) Proceedings of the Seventh National Urban Forest Conference. American Forests, Washington, DC, pp. 6–9.

8
Approaches to Urban Forestry in the United Kingdom

Nerys Jones

The term *urban forestry* is widely understood in North America and many European and Asian countries, but it has been introduced into the United Kingdom only relatively recently. However, the planting and care of urban trees and woodland—the practice of urban forestry—has long been established in the U.K. In recent years, the terms *urban forestry* and *community forestry* have tended to be used rather interchangeably. The term *urban forest* may be defined as all the trees, woods, and associated open spaces within an urban area, and the term *urban forestry* applies to the management of this resource. This chapter briefly reviews the history of urban forestry in the U.K., a very urbanized country with over 90% of its population living in towns, and examines the core principles for success in current practice.

History

The U.K. has been industrialized for over 200 years, and so it has a relatively long history of environmental degradation associated with traditional industrial processes.

Early Greening Initiatives

The greening of U.K. towns and cities dates back to the creation of the Royal Parks in London in the 17th century. From the early 1800s onward, there was occasional interest in greening the scars caused by industrial dereliction. For example, almost 200 years ago, the Earl of Dudley planted trees in the urban West Midlands region to reclaim worked-out limestone quarries, and the resulting woodland still survives today. Urban parks were created in many of the major industrial cities during the 19th century by industrialists who were keen to establish green areas and pleasure parks as a recreational facility for their work force. Between 1903 and 1924 the Midlands Reafforesting Association, a community-based voluntary organization, planted new woodlands on industrial spoil heaps in and around the city of Birmingham, a region that, at the time, was particularly badly disfigured by mining and metal smelting operations. Because of the atmospheric pollution from heavy industry, there was a

M.M. Carreiro et al. (eds.), *Ecology, Planning, and Management of Urban Forests: International Perspectives.*

widely held view that nothing green could be grown successfully. However, the association raised money by public subscription and bought land. It also planted on local authority and privately owned land (Bastin, 1914). The association's principles were very similar to those adopted in modern urban forestry:

- Concern about the scale of wasted land
- Recognition that industrial spoil can blight an area
- Recognition that trees can be grown on most types of despoiled land
- Recognition that more trees, as well as being attractive, would be good for people's health

Some of the association's woodland planting projects from the early 1900s have now become significant features in the modern landscape. These projects used a limited range of species, in particular, those species that could successfully tolerate difficult physical conditions such as wind exposure, air pollution, low soil fertility, and high acidity.

The Growth of Modern Urban Forestry

The more recent impetus for the development of urban forestry came with the need to treat the large amount of derelict land left by the decline of much of Britain's heavy industry in the 1970s and 1980s. In 1990, the first major modern urban forestry initiative, involving woodland creation on a strategic scale, was established in an area known as the Black Country, near Birmingham, about 150 km from the northwest of London. The Black Country refers to the formerly scarred appearance of the region, blighted by coal waste tips.

The Black Country Urban Forest program provided a focus for greening activity, involving a partnership of public, private, and voluntary sector organizations. Its original aim was to use urban forestry to improve the image of the region, through transformation of the landscape, while also improving the quality of life for those who were already living and working there. Over a 10-year period there was significant achievement in an extremely urban area of around 360 km^2; 800 hectares (ha) of new woodland were planted and 400 ha of existing woodland were brought under active management (1 ha = 2.47 acres). Strategic transport corridors were targeted to maximize the impact for those traveling through the region, and thousands of local people were actively involved in caring for the Black Country Urban Forest. This initiative has proved to be a very significant model for large-scale urban greening programs elsewhere across the U.K. (Johnston, 1999).

Since the early 1990s, there has been impressive growth in urban forestry practice in the U.K., and there are now over 40 major programs at regional or city levels (Fig. 8.1). These include the National Forest, the Community Forests program, and other significant initiatives in all four of the U.K. countries of England, Scotland, Wales, and Northern Ireland. Many of these initiatives have transformed substantial areas of derelict, despoiled, and underused land into woodland, for example, nearly 450 ha in the Central Scotland Forest between 1994 and 2001, but they are essentially long-term

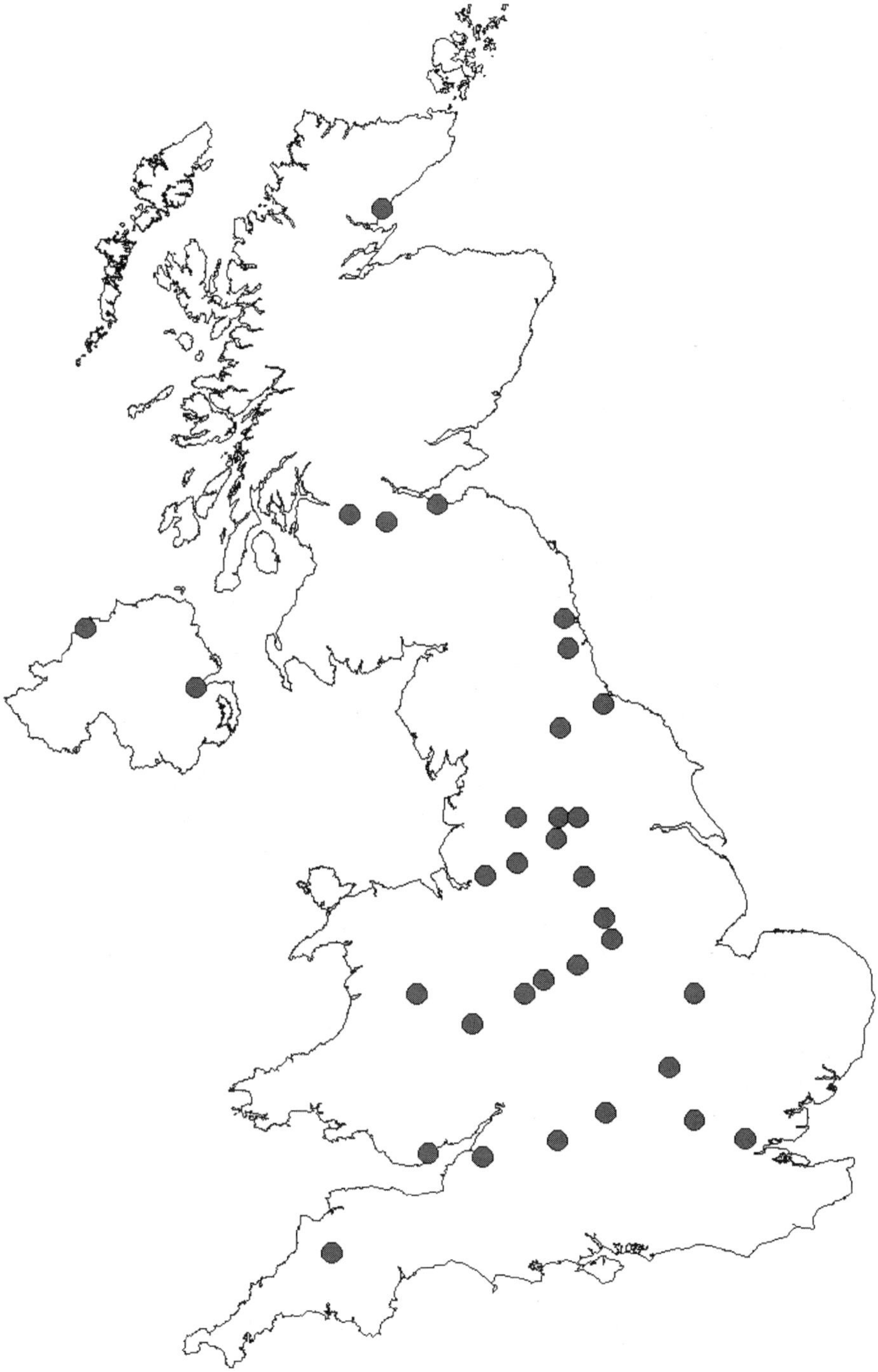

Fig. 8.1 Urban Forestry Programs in the United Kingdom (source NUFU, 2000)

projects (Countryside Agency, 1999). Since derelict and despoiled land is usually closely associated with areas of high population density in the U.K., its reclamation and the benefits that this can bring for local people generally form a major component of urban forestry initiatives. One of the most interesting current programs is in east London, an area known as the Thames Gateway, which is the focus of a major regeneration initiative. There is considerable potential for the new urban forest to provide a framework for new development and over 30 organizations have made a commitment to an urban forestry strategy for this region (National Urban Forestry Unit, 2002).

The following principles characterize the approach to successful urban forestry in the U.K.

Using Applied Ecology

Importance of Natural Colonization

Broad-leaved woodland is the natural form of vegetation in much of lowland Britain, and deciduous trees tend to recolonize vacant open spaces quite naturally. A comparative survey of land cover in the industrial West Midlands, using aerial photographs, showed that 49% of that region's tree cover was made up of young emergent woodland, formed through natural colonization. In one part of the area, the borough of Sandwell, a comparative study was made of the change in woodland cover over the 12-year period from 1977 to 1989. This showed an overall increase in woodland of 74%, with natural regeneration being by far the most significant reason (National Urban Forestry Unit, 1995). The potential for natural recovery had been recognized as long ago as the 1940s. As the survey of Birmingham and the Black Country by the West Midland Group (1948) noted in its publication, *Conurbation*:

> Land cannot be rendered permanently derelict: in the course of time a natural covering of soil and herbage will return. This process can be speeded up and used to advantage in the rehabilitation of the Black Country landscape.

The effectiveness of this ecological approach has been recognized by a number of researchers (University of Liverpool Environmental Advisory Unit, 1986). It is also argued that natural succession may well be preferable to planting new woodland because the trees will be better adapted to site conditions and genetic distinctiveness will be preserved (Rodwell and Patterson, 1994).

Pioneer Species

The principle of pioneer species has helped to inform the more ecological approach to urban woodland design adopted under the Black Country urban forestry initiative: bold, simple designs, using only two or three species, leading to the creation of robust, pioneer woodland (National Urban Forestry Unit, 2001)—a strong echo of the approach taken by the Midland Reafforesting Association almost a century earlier. Pioneer species such as *Betula pendula* and *Salix caprea* thrive on soil-less sites and

Quercus petrea and *Crataegus monogyna* are successful colonizers of open grassland sites across lowland Britain. Use of these and other colonizing species in the creation of new woodland increases the likelihood of successful establishment. At the outset, it is advisable to establish a simple plant community rather than a more complex mixture of species. Species diversity can be encouraged to develop naturally over time.

Combining Ecology and Forestry

The best of new urban woodland planting uses a combination of applied ecology, as described above, and orthodox forestry establishment techniques, such as ripping of the ground to relieve compaction (caused by previous industrial use) and effective chemical or mechanical weed control to reduce competition and improve initial establishment and early tree growth. Using this approach, it is possible to create a robust style of urban woodland which is largely self-sustaining in silvicultural terms. This releases scarce resources to be directed toward improved access and management for people. Clear signage, maintenance of paths and entrances, and litter collection all encourage use and enjoyment by urban people, who may be unfamiliar with woodland. By contrast, more complex, horticultural-type plantings require much higher silvicultural management, are much more expensive and demanding to maintain and often become neglected (National Urban Forestry Unit, 2001).

Nevertheless, the use of larger nursery stock for a more dramatic instant impact approach is, regrettably, still favored by many landscape designers. The frequent failure of larger trees is often blamed on vandalism, but this cause is generally exaggerated. Tree loss is much more likely to be due to poor planting and aftercare, drought stress, and the inherent loss of root systems, which usually occurs when large stock is transplanted (Bradshaw et al., 1995).

Involving People

The relevance of the ecological and social value of urban wild space was identified almost 30 years ago (Mabey, 1973), and in the late 1970s the government's nature conservation agency, the Nature Conservancy Council, commissioned an ecological survey of Birmingham and the Black Country (Teagle, 1978). This seminal work revealed extensive areas of wild land. By the mid-1980s, an urban nature conservation movement had become established in many of the U.K.'s towns and cities, and the importance of building popular and political recognition of the value of more natural green space and of involving people in urban landscape change began to be recognized (Baines, 1986).

Involvement of people helps to build "ownership" and generally reduce the problem of vandalism. This approach continues to the present day, and increasingly sophisticated techniques have been developed for understanding people's views of their local environment. A number of nongovernmental organizations have developed particular expertise in working closely with communities, and local government often works in partnership with such organizations to obtain the best results from community consultation (Groundwork U.K., 2001).

Techniques have been developed for helping people to survey and map their local neighborhood and to influence the way the information is used. People are also often involved in the growing, planting, and care of trees, the enhancement of new or existing woodland through the building and installation of bird and bat boxes to encourage wildlife, and the use of arts projects such as sculpture, theater, or storytelling.

There is strong evidence that people appreciate woodland close to where they live, but it can take time to build their confidence to use and enjoy it (Millward and Mostyn, 1989). The U.K. is a country with very low tree cover (8.4%, mainly rural woodland cover in England) and it has a very urbanized society. Many people, therefore, are unused to living close to woodland and need to feel comfortable with the scale of landscape change that may be involved when new woodlands are planted. Public participation in the urban forestry process holds the key to public acceptance and long-term support.

A Strategic Approach

Strategic Greening

The potential for a more strategic approach to urban wasteland was recognized over 30 years ago by the landscape architect and author Nan Fairbrother (1970). In her book, *New Lives, New Landscapes*, she argued that all the wasteland in and around towns should be identified and filled in with trees:

> If we can transform our present disturbed areas to good green-urban landscape, it will be more effective than any other single reform in upgrading our general outdoor environment.... In many places, this alone would frame our towns in green.... By planting areas of otherwise unused land we could travel into our cities through wooded landscape and unless there are definite reasons against it, tree planting could be the accepted and universal practice on all such land.

Land Use

There are numerous opportunities for woodland creation in towns. Even in the most heavily built-up urban areas, land suitability is rarely a seriously limiting factor. The National Land Use Database currently estimates the amount of so-called brownfield land in England and Wales to be around 60 000 ha. However, it is important to consider the scope for planting in the context of existing tree cover and to consider all types of available land (National Urban Forestry Unit, 1999). Derelict land, which cannot otherwise be easily developed, can be greatly improved by the addition of new woodland, and a significant proportion of such land in urban areas could readily be converted to woodland (Perry and Handley, 2000). However, technical factors are not the only consideration. Legal and risk management issues can often add significantly to the costs of apparently straightforward regeneration exercises.

Operational land around factories and other industrial complexes may offer scope for substantial areas of new woodland. It is usually difficult to offer public access on such sites, but, for this reason, they can develop as significant sanctuary areas for wildlife and still deliver many of the other benefits that come from urban woodland. They are also of direct value to the company employees.

Development sites themselves offer considerable potential for the creation of woodland around new housing and industry. There is particular merit in creating this green setting in advance of any new building, although developers and landowners are often reluctant to do this. Although such advance planting can be extremely cost effective, the landscape treatment of most developments in the U.K. is regarded as a concluding element, to be installed once building work is at or near completion. This unfortunately tends to discourage developers from investing in what they may perceive to be speculative planting before the detailed layout of the development has been implemented.

Land such as grassland in parks or schools, on hospital grounds, or in housing areas that is already designated as green space can also offer potential for new woodland. The simple, robust style of woodland described earlier can be considerably more cost-effective to manage in the longer term than mown grass. Converting even a small percentage of grassland to woodland can generate considerable potential savings in landscape management costs, as well as contributing to greater landscape diversity (National Urban Forestry Unit, 1998).

Transport corridors, such as motorways and other major roads, rail, and canal routes, are particularly important as the focus for new woodland creation. They often help to determine the image of a city and therefore affect investment decisions and economic regeneration potential. They are also important ecological corridors, allowing the movement of wildlife through urban areas (Baines, 1986).

Urban Forestry Strategies

The extent of the existing urban forest—trees in streets, private gardens, and school grounds—also needs to be recognized. The way in which this existing resource and any new trees and woodland relate to one another needs to be considered in a holistic way. It is particularly helpful if the whole picture can be published in an urban forestry strategy for a city or even a region. There are now a number of examples of such strategies, and they help to gain the support of a wide range of key organizations. They can also be particularly helpful in gaining access to certain kinds of national and European funds (National Urban Forestry Unit, 1999).

Partnership Working

There is now a very strong emphasis in the U.K. on partnership working between different sectors of society. It is recognized that it takes more than an understanding of trees to sustain a successful urban forest. A wide range of environmental initiatives have demonstrated that greater success can be achieved through the collaboration of different organizations, working in partnership toward a common ambition (see Chapter 12).

Recognizing the Wider Benefits of Urban Trees and Woodland

It is important to promote the wider relevance of urban trees and woodland to policy makers, since it is not always well understood (Jones, 1999). Some of the principal environmental, social, and economic benefits are as follows:

Environmental

Increased shelter and energy conservation
Improved biodiversity
Cleaner air
Increased shade and reduction in urban heat island effect

Social

Increased community cohesion and social inclusion
Increased potential for recreation and exercise
Improved public health

Economic

Reduced maintenance costs for public parks and open spaces
Improved porosity, storm water management, and flood alleviation
Increased land and property values

Interestingly, many of these wider benefits were known at the beginning of the 20th century, but it has taken a long time for them to become embedded in official policy. The government's England Forestry Strategy (Forestry Commission, 1998) now redefines forestry in terms of its wider benefits to society. In contrast to earlier emphasis on timber production, the strategy has four themes:

- Forestry for rural development
- Forestry for economic regeneration
- Forestry for recreation and tourism
- Forestry for environment and conservation

The publication of this document has proved to be a turning point for forestry in the U.K., with the more traditional, mainstream forestry players now aligning themselves much more directly with the more social agenda that characterizes urban forestry.

Conclusion

The fact that the wide-ranging benefits of urban trees and woodland have environmental, social, and economic relevance means that urban forestry provides an excellent opportunity to put sustainable development into practice. The U.K.

government's brochure for the 2002 World Summit on Sustainable Development featured urban forestry is just such an example (Department for Environment, Food, and Rural Affairs, 2002). The challenge now is to ensure that an increasing proportion of policy makers and practitioners view urban trees and woodland as *functional green infrastructure*. The trees and woods in towns are much more than a leafy green backdrop to development: they make a very significant contribution to improving people's quality of life.

References

Baines, C. (1986) The Wild Side of Town. Elm Tree Books/Hamish Hamilton, London.

Bastin, S.L. (1914) Tree planting in the Black Country. Journal of the Royal Agricultural Society of England 14:70–75.

Bradshaw, A.D., Hunt, B., and Walmsley, T.J. (1995) Trees in the Urban Landscape. E. and F.N. Spon, London.

Countryside Agency. (1999) Regeneration Around Cities. Countryside Agency, Cheltenham.

Department for Environment, Food, and Rural Affairs. (2002) Reaching the Summit: Johannesburg, the U.K. and Sustainable Development, DEFRA, London.

Fairbrother, N. (1970) New Lives, New Landscapes. The Architectural Press, London.

Forestry Commission. (1998) A New Focus for England's Woodlands: England Forestry Strategy. Forestry Commission, Cambridge.

Groundwork U.K. (2001) Breaking the Mould. Groundwork U.K., Birmingham.

Johnston, M. (1999) The Development of Urban Forestry in Britain and Ireland. PhD dissertation, University of Ulster.

Jones, N. (1999) The future for the urban forest. In: Community Forestry—A Change for the Better. Proceedings of the Community Forestry Conference, London, December 7–8, 1998.

Mabey, R. (1973) The Unofficial Countryside. William Collins, Glasgow.

Millward, A.M., and Mostyn, B.J. (1989) People and nature in cities the social aspects of planning and managing natural parks in urban parks. Urban Wildlife Now 2 Nature Conservancy Council, Shrewsbury.

National Urban Forestry Unit. (1995) Black Country Open Space Survey. National Urban Forestry Unit, Sandwell.

National Urban Forestry Unit. (1998) Trees or Turf?: Best Value in Managing Urban Greenspace. NUFU, Wolverhampton.

National Urban Forestry Unit. (1999) Trees and Woods in Towns and Cities: How to Develop Local Strategies for Urban Forestry. NUFU, Wolverhampton.

National Urban Forestry Unit. (2001) A Guide to Designing Urban Woodland. NUFU, Wolverhampton.

National Urban Forestry Unit. (2002) Green Gateway: Urban Regeneration in the Thames Gateway London, Using Trees and Woods. NUFU, Wolverhampton.

Perry, D., and Handley, J. (2000) The Potential for Woodland on Urban and Industrial Wasteland in England and Wales, Forestry Commission Technical paper No. 29. Forestry Commission, Edinburgh.

Rodwell, J., and Patterson, G. (1994) Creating New Native Woodlands, Forestry Commission Bulletin 112. HMSO, London.

Teagle, W.G. (1978) The Endless Village. Nature Conservancy Council, Shrewsbury.

University of Liverpool Environmental Advisory Unit. (1986) Transforming our Wasteland: the Way Forward. HMSO, London.

West Midland Group. (1948) Conurbation. The Architectural Press, London.

9
Opportunities and Alternatives for Enhancing Urban Forests in Compact Cities in Developing Countries

C.Y. Jim

That cities need to be greened is almost a foregone conclusion, if not *de rigueur*, for any plans for urban development or redevelopment. A green city is an ideal with a universal appeal that traverses temporal, spatial, and cultural divides (Hestmark, 2000). For many people, the greening of urbanized areas conjures up a deep innate desire to connect with the natural world and its diversified assemblages of organisms. It is natural for people to harbor a psychological and emotional attachment to beautiful natural objects, such as admirable amenity vegetation (Kaplan, 1984; Ulrich, 1986). Different socioeconomic strata develop similar levels of appreciation and preference for urban nature (Kuo et al., 1998).

Urban greening entails introducing natural elements into the largely cultural fabric of cities. The fundamental requirement is the provision of planting spaces by design, or leaving such spaces unpaved by default. Greening is realized to different degrees in cities, and the quality and amount of green space is dictated by fashion, and so subject to changing contemporaneous societal attitudes and political will (Mumford, 1961; Attorre et al., 2000). Nature and culture make an enigmatic pair in relation to the history of urbanization. It is from nature that humans obtained sustenance and inspiration to develop our culture. Yet upon acquiring culture, in characteristic human fashion, we unthinkingly began to damage and reject nature (Jim, 2002a). Most cities customarily are dominated by cultural artifacts that overshadow nature, and in some places nature is thoroughly eradicated.

Subconsciously and subliminally, humans need nature for a balanced physical and mental development (see the biophilia concept of Wilson, 1984). Yet consciously or unconsciously, we create conditions in cities that are often inhospitable to plant growth. Different cities, due to inherent natural biota and topography, and their development and redevelopment history, have engendered urban forms that can either accommodate or constrain vegetation growth. The most intense human-nature interactions and conflicts occur in cities, and densely populated, compact cities are particularly deprived of greenery. In recent years, especially in some developing cities, past excesses and paradoxical attitudes were moderated. We have renewed our partnership with nature and relearned to embrace the notable emblems of nature, such as amenity vegetation, in our attempts to reestablish our tenuous psychological link with nature.

M.M. Carreiro et al. (eds.), *Ecology, Planning, and Management of Urban Forests: International Perspectives.*

The inborn desire to green cities has been widely echoed by government policies and practices. A city with high-quality and generous green spaces epitomizes good planning and management, has a conscientious vision for taking care of present and future generations, and promotes a healthy environment for humans and their companion plants and valued wildlife species (Adams and Leedy, 1987; Johnston, 1990; Godefroid, 2001). Meritorious green cities bestow a strong sense of pride and ownership on the part of their citizenry and government. However, the intangible and indirect nature of the benefits of vegetation has until recently hindered our understanding of their societal roles beyond those involving common ornamental and amenity functions. Now the multiple functions and benefits of urban vegetation have been increasingly recognized (Mole and Young, 1992; Petit et al., 1995) and their ecological services have been translated into monetary units (McPherson et al., 1997; Nowak and Dwyer, 2000; see also Chapter 5) to facilitate their understanding and appreciation, and to inform policies, decisions, and budgets on urban forest programs.

Such refined and necessary natural goods, however, could be sidetracked, if not suppressed, by political expediency, bureaucracy, and apathy (Foster, 1977; Duvernoy, 1995). Fortunately, many enlightened and informed politicians, administrators, planners, and citizens aspire to create outstanding green cities and eco-cities (Hough, 1994; Bradley, 1995) that are reminiscent, if not emulative, of the garden-city ideal (Howard, 1902). Cities have to compete in a global marketplace for investment and talent, not only by providing conventional infrastructure, but also by providing a greener and cleaner environment. This motivates many developing cities to supply greenery for new developments and preserve existing greenery during redevelopment and expansion projects (Gordon, 1990; Beatley, 2000). Urban sustainability requires abating pollution plus adding positive features, notably trees, to ameliorate unhealthy conditions in rapidly growing and industrializing cities (Finco and Nijkamp, 2003). Maintaining and increasing a city's green infrastructure is one important means for increasing human quality of life in cities.

At present about half of the world population dwells in cities, with developed countries reaching 76% and less developed ones 40%. By 2030, 60% of the world population is expected to live in cities, with the bulk of the new urban increase happening in developing countries (United Nations, 2005). Such rapid urbanization, involving many Asian cities, would drive more people into the realm of stressful urban existence. The toll of city life on the physical and mental state of the Asian population could be ameliorated by urban development that cares for environmental quality and human health. As most Asian cities are compact with the world's highest population densities (Wendell Cox Consultancy, 2006), and will continue to develop in this mode, it is essential to plan for the coexistence of greenery and the built-up fabric. Greening Asian cities could contribute to sustainable and healthy city objectives that will benefit many millions of urban inhabitants. Their growth offers opportunities to plan and develop in environmental friendly ways.

However, competition for space in compact cities is intense and may impose short-term socioeconomic constraints on urban green-space retention and development. This chapter surveys the pertinent limitations to greening in compact cities especially in developing countries and identifies practical opportunities, alternatives, and solutions for inserting new greenery despite development pressure. Recent research findings in a cluster of related disciplines, including arboriculture, urban forestry, urban ecology, landscape ecology, landscape architecture, urban planning, and urban geography, have nurtured an interdisciplinary convergence that could be translated into effective greening practices. Extensive field studies in different compact city environments also provide first-hand information for this review. Since they have a longer life span, larger biomass, and more notable environmental functions, trees are used as a surrogate for urban vegetation in this discussion.

The Need to Green Compact Cities

Overcoming Barriers to Greening Compact Cities

The main features of compact urban areas are close juxtaposition of buildings and roads, limited interstitial space to insert greenery, mixed land use, and a union of form and function (Jenks et al., 1996). A compact city exhibits a high-density built form (Burton, 2002), with a large proportion of the land surface covered by buildings and other artificial structures and surfaces. The high value of land and property is both the cause and the effect of high density, resulting in maximum utilization of the land resource for commercial, residential, or industrial purposes. The ratio of impervious to pervious land is typically high, often exceeding 90% (Arnold and Gibbons, 1996), thus the degree of naturalness and opportunity for nature to exist in such environments is severely constrained.

Cities in developing countries often inherit an old compact form, and many are expanding rapidly but remaining in this high-density pattern. With more of the world's population choosing to live in cities, especially in developing nations, the type of urban growth (Thomas et al., 1999) will affect urban environmental quality. Growth that includes strong consideration of the roles that vegetation plays in ameliorating many negative impacts of development on the environment must be fostered. For example, the urban heat island effect, which is more intensified in compact cities (Wong and Yu, 2005), could be reduced by greening. Some developed countries have adopted plans for compact city development as well (Burton, 2002), but still embrace enhancing greenery as a means of stimulating environmental and economic revival (Hughes, 1991). Such urban renaissance projects call for a different approach to greening. Urban growth and greening experiences in developed and developing worlds need to inform each other more often.

Compactness brings both advantages and disadvantages for urban environmental quality, provision of infrastructure and services, and transfers of people and goods. Cities containing a high-density commercial-business district (CBD) at its center, or an old high-density urban core as a legacy of the past, are not uncommon. But some cities deliberately develop new compact areas or infill existing ones to a higher density (Williams, 1999). Whether old, modified, or new, such compactness needs proportionately more attention devoted to green-space provision and environmental well-being, since it can be overlooked or sacrificed (Jim, 1989a). With an inherently tight urban fabric, continual attempts to raise development density by infilling and renewal of old lots have eliminated locations for nature to survive in hastily developed cities. The exigencies of meeting basic needs and development aspirations may overshadow greenery and other environmental concerns (Marcotullio, 2001). The urge to take the myopic path of developing first and making amends later (Olembo and de Rham, 1987) reflects a common failure to learn from other cities' experiences.

The urge to have rapid urbanization and intensification in developing cities could compromise environmental planning. Whereas individual cities have unique problems and limitations in implementing a greening imperative, most physical and physiological constraints that beset vegetation growth are common across many cities (Grey and Deneke, 1986; Bradshaw et al., 1995). Condensed development renders these constraints more acute and pervasive. The keen competition for space means less is available for greening and the greening sites are more stressful to plant growth (Morell, 1992). Existing green spaces and greenery are more likely to be harmed, intruded upon, or eliminated (Jim and Liu, 2001a,b). Green spaces in cities are increasingly recognized as pertinent elements for urban environmental quality, a healthy city and citizens, and attractiveness to investments and talents in an increasingly globalized world (Jackson, 2003; Perdue et al., 2003; Ellaway et al., 2005).

Compact cities, beset by deficits in green spaces, could find innovative ways to capitalize on the extensive benefits of urban vegetation (Tyrväinen et al., 2005). Understanding the tangible and intangible restrictions (Grey, 1996; Miller, 1997) could provide hints for maximizing greening opportunities that make cities more sustainable. The social and psychological barriers to increasing green spaces in compact cities are deeply rooted, due to the common resigned attitude that little could be done to improve the overwhelming, if not oppressive, artificial excesses of a compact cities. Economic barriers, due to unwillingness to invest in the greenery infrastructure, also need to be overcome. With planning and vision, policies and practices could be molded to equip compact cities with a reasonable amount of amenity vegetation, certainly more than these cities contain currently.

Emphasizing the Urban Greening Imperative

Generous and high-quality greenery is a necessary ingredient for maintaining environmental quality and quality of life. The amenity, environmental, and socioeconomic benefits of urban vegetation are well recognized (McPherson et al., 1997; Nowak and Dwyer, 2000; Stone and Rodgers, 2001). Lack of amenity vegetation in compact cities is increasingly recognized, generating earnest calls for improvement (Duvernoy, 1995) and stimulating a search for solutions. The fundamental constraint to greening dense cities is the dearth of suitable habitats. Any attempt to plant trees will soon exhaust the number of potential planting sites. Even if difficult, it is essential to continually provide new planting sites in developing or redeveloped areas. To move in this direction, legal and administrative policies could be modified so that more greenery can penetrate from public into private lands. This is most feasible in locations where development extends into natural habitats, or in farmland where more green spaces could be set aside. Various incentives could be instituted to persuade land users or developers to allocate green spaces on a voluntary basis. A survey of potential development sites could identify portions with high nature content for preservation and incorporation into the future development. Incentives could be offered to developers to respect existing natural endowment, which are often of high ecological and amenity values that could hardly be replaced by artificially created green spaces.

However, enthusiastic greening policy can be obstructed at the implementation stage. Such constraints should be thoroughly evaluated to find practical solutions and alternatives. A steady provision of high-caliber planting sites at strategic locations and times is needed to sustain the greening program. Proposals are needed to overcome encumbrances and institute new measures for make greening an integral component of the development process (explained below). The ultimate objective is to modify the mindset within and outside the administration so that green space is recognized as indispensable urban infrastructure. Public and private development projects could incorporate green features that dovetail with a citywide greening master plan.

Finding and Allocating New Green Spaces

Overcoming Hurdles Presented by Tight City Plans

A city plan and its constituent plantable space are the result of many interacting factors over a city's development history. Developments that fill most of the land surfaces with buildings, roads, and other artificial structures and surfaces could create a tight city plan. Planning standards and policies that permit inordinately high development intensity by way of site coverage and plot ratio will generate tight city plans. Some city areas are densely packed from the outset, whereas others are gradually infilled as they are redeveloped to a higher density. Such an excessively high-density development mode precludes green spaces. The city plan, once made, is often close to immutable and can present a social and psychological hurdle for effecting change, since many people believe at the outset that a high-density city has little hope of provisioning adequate green space. A great deal of determination and effort is needed to open up the city plan in the course of urban renewal in order to bring relief. Partnerships among government, the private sector, and citizens are needed to overcome these difficulties (see Chapter 12).

The smart growth strategy (Stone and Rodgers, 2001) should recognize a city's greenery as economic and ecological assets (Hughes, 1991; Platt et al., 1994; Atkinson et al., 1999). Any attempt to develop or redevelop a compact city should adopt the enlightened strategy to reap benefits from green spaces. In old neighborhoods, city-plan restructuring is needed to increase green spaces. The size and shape of lots, and alignment and width of roads, could be reorganized, reconfigured, and reconstituted. The visionary urban renewal of Paris in 1850s under the direction of Baron Haussmann and Emperor Napoleon III represented a pioneering large-scale attempt to introduce tree-lined boulevards, parks, and gardens into the cramped and dilapidated old city quarters. Such a fine example of restructuring the city plan in conjunction with urban redevelopment has since been imitated at different scales in many Western and Asian cities. However, related legal and administrative procedures, especially those pertaining to land ownership and rights, could become intimidating and intractable. The multiple ownership of land in high-rise tenement blocks makes urban renewal vexing and sluggish. These formidable and persistent obstacles could extinguish initial optimism and enthusiasm. That is why the ills of excessive urban intensification should be recognized (Williams, 1999) and prevented in new developments.

Most compact cities have tight city plans that lack planting sites. To increase a compact city's green spaces, plantable spaces have to be found or more fully utilized. A city's plantability index can be described as the ratio of plantable space to impervious cover (mainly roads and buildings). City precincts with a porous layout that includes a generous provision of open spaces at ground level would have more niches for plants in places not taken up by buildings and roads. Amenity vegetation would then have a greater opportunity to penetrate the built-up matrix to furnish a pleasant sylvan ambience. Some city precincts, unfortunately, have a tight city plan with hardly any residual space for greenery. Fortunately, most urban areas have a plantable space level lying between the extremes of tight and porous forms. In redeveloping a compact urban area, finding innovative ways to convert it into a porous plan could bring sorely needed rejuvenation. For instance, building frontage could be set back, intra-lot green passages could be inserted, lanes between buildings could be converted into linear garden strips, and underused road could be partly or wholly converted into pedestrianized green corridors that traverse the neighborhood.

Providing Abundant and High-Quality Sites for Greening

Many city plans promote an excessive development density that is tree-unfriendly. One short-term solution would be searching more intentionally for residual plantable sites in a city. Long-term solutions require loosening the tight city plan in urban renewal areas, setting buildings back from lot boundaries, demarcating roadside tree strips and amenity plots, mandating that trees be considered an essential element of urban infrastructure, instituting a city-wide long-term landscape plan, and ascertaining that all new development, including infilling of vacant and porous lots, or when redeveloping older buildings, will switch to a tree-friendly geometry. The traditional concept of infilling with buildings could be reversed, to become infilling with green spaces.

The crux of this green urban reform is to find and keep plantable spaces (PSs), which are urban lands that have not been built upon or covered up by impervious road or other surfaces, or that have not been zoned for such purposes in future. Six types of PSs can be identified based on their location and land ownership: (1) roadside PS as a constituent part of a road with land in the public domain; (2) lot-frontage PS associated with private or public land; (3) intra-lot PS situated within private or public land; (4) open-space PS on public land as urban parks and other public amenity spaces; (5) remnant PS usually located where development has been constrained, such as steep or unstable slopes; and (6) largely vertical PS found on roofs and vertical faces on buildings, boundary walls, and noise barriers. Plantable spaces in densely developed areas tend to be small, isolated, and unevenly distributed, and are precious due to scarcity. Each of these PS categories has plantable spaces of differing quality for growing vegetation of different types.

Even where PS is available, planting sites in compact cities are often beset by stressful site conditions both above and below the ground surface. The main above-ground problems are cramped sites, intrusions into tree-growth space, poor air quality, vandalism, and accidental damage. Short-term solutions include trimming overgrown trees and removal of intrusions. Lasting solutions include reserving amenity corridors and spaces for trees, selecting species with final sizes that match site dimensions, and selecting species that are tolerant of polluted air and cramped soil space.

New developments and redevelopments should assign green spaces following spatial and conservation planning guidelines (Dramstad et al., 1996). The preoccupation with green-space acreage and tree counts could be directed toward the geometry of the green network and quality of the greenery. New sites should nurture high-quality vegetation, especially large trees for more substantial visual and environmental benefits. Areas with existing high-quality vegetation, notably mature woodlands, should be preserved to blend sympathetically with future buildings and roads (Löfvenhaft et al., 2002). This was recommended for Stockholm, Sweden. This can be done for cities like Hong Kong, Guangzhou, Xiamen, Nanjing, Taipei, or Kuala Lumpur, which have wooded areas situated adjacent to the contiguously built-up urban core. Any attempt to expand the city into such areas should be accompanied by an ecological assessment to identify natural areas within the new urban areas, preferably to be blended sympathetically into a green-space network.

To create interesting and diverse urban vegetation, both green coverage and content are important. The anachronistic 19th-century idea of containing, controlling, and conquering urban green spaces (Jorgensen et al., 2002) could be overhauled to meet modern aspirations. Instead, informal and somewhat wild green sites would complement manicured ones (Thompson, 2002). For an urban park of a certain size, a portion of it could be devoted to a semi-wild type with native plants. Such a naturalistic approach has been incorporated into the design of some renowned urban parks, such as the Bois de Boulogne in Paris, Holland Park in London, and Central Park in New York. For a city such as Hong Kong, which has extended into periurban hill-slope areas with natural vegetation, some hillside pockets offer ready-made naturalistic urban parks.

Creating Roadside Tree Corridors by Setback Zoning

Setting back of buildings at the road frontage is a principal way to gradually relax the stifling city plan and introduce greenery. A minimum setback of 3 m from lot boundaries would provide a reasonable roadside tree corridor. The corridor strip should have a soil depth of at least 1.2 m that is free from underground utilities to facilitate tree growth. Achieving notable greening effects through frontage setbacks will take decades, however, since the exercise is tantamount to restructuring or retrofitting the city plan to make it tree friendly. Its implementation demands a long-term vision and lots of sustained determination to persuade and to enforce.

Setbacks could be encouraged by incentives, such as transference of development rights to the remainder of the plot, or better still, the reward of a bonus plot ratio. The plot ratio, a planning and development control measure, is the area of building floor space (summed for all floor levels) per unit land area. For example, for 1000 m^2 of land, a plot ratio of 5 means that 5000 m^2 of floor space could be constructed. Voluntary provision of wider setback strips should be encouraged to accommodate large trees for more detectable landscape improvements, especially for >0.5-ha lots with >100 m of aggregate road frontage. For sites that can provide generous setbacks wider than 3 m, consideration could be given to creating a road-median planting strip in addition to the lot-frontage strip to enhance the extent and hence the visual and environmental benefits of vegetation.

These measures are particularly worthy of promoting in compact cites because roadside trees are the most cost-effective and conspicuous way to upgrade the cityscape. They occupy little space, sharing the above-ground spaces with vehicles and pedestrians, and yet could impart notable scenic and environmental benefits. A coordinated landscape plan could identify roads or road segments for uniform setbacks and tree planting, and to avoid discontinuous setback along a given road section (Jim, 1998e). An integrated approach could realize the potential of this method to introduce greenery into old city areas. Planting site design, especially in confined roadside strips, needs innovative approaches to overcome severe physical constraints (Kuhns et al., 1985; Evans et al., 1990) and protect roadside trees against acute stresses (Chevallerie, 1986; Hauer et al., 1994).

Using Rooftops and Facades for Greening

The lack of plantable spaces at the street level of compact cities could be partly compensated by enlisting the vertical dimension of the city fabric. Cities have plenty of surfaces on buildings that could be planted with amenity vegetation, yet they are grossly underutilized. Many roof tops remain barren in Asian cities, and they tend to absorb solar radiation, impose heat loads on indoor spaces below them, and increase air-conditioning energy costs (Saaroni et al., 2000). These gray roofs afford chances to convert them into the extensive or intensive types of green roofs (ZinCo GmbH, 2000; Earth Pledge Foundation, 2004). New buildings could be required by building regulations or by law to installed green roofs, an approach that has been successfully implemented in Tokyo. Many German cities stipulated this requirement several decades ago (Köhler, 2005). Existing buildings could be retrofitted with green roofs based on the load-bearing capacity of the roof. Incentives could be given to home-owners and developers to install vegetation on the roof surfaces. A clear government policy with the support of developers and home-owners could take the green roof idea forward.

In many Asian cities, where the green roof idea is not widely accepted, it is necessary to provide publicity and public education to advocate the multiple environmental, economic, and social benefits of green roofs (Osmundson, 1999; Brenneisen, 2004) to promote their adoption. Barriers to green roof installation are largely due to misconceptions about their establishment and maintenance costs, lack of technical information and know-how, and unfounded worries about leakage and damage to the waterproofing and insulation layers. The government in conjunction with the landscape industry could establish a demonstration green roof site in association with a resource center to provide technical and price information on various commercial products and methods (Chicago Department of Environment, 2001). Research on the design, material, and species selection for Asian cities in different climatic zones would help develop the capability of the roof greening industry (Chicago Department of Environment, 2003).

In addition to green roofs, building facades could also be used where appropriate for greening with suitable climbers (Dunnett and Kingsbury, 2004) to increase the total green surfaces of a compact city. The concept of green plot ratio (Ong, 2003) for a given development site recognizes the contribution of vegetation at the street, roof and facade levels. The government could adopt this enlightened idea to encourage vertical greening in compact cities, and to provide an objective and quantifiable way to reckon the greening contributions of a development project.

Increasing Trees in Private Sites

Away from roadside and public green spaces, amenity vegetation is seldom nurtured in compact cities. Some porous public lands may have ground-level spaces for trees, but they are increasingly threatened by redevelopment at a higher intensity and infilling. In compact Asian cities, private lots for residential, commercial, and industrial buildings usually occupy the bulk of the city's areas, yet they harbor few trees. A citywide landscape plan could be developed to collate and encourage private-sector contributions to the greening endeavor. A building setback requirement could bring coordinated improvement of the streetscape in the long run.

The government must enlist the partnership of developers to form a concerted public–private greening endeavor. The willingness and ability of developers in Asian cities to plant trees vary greatly. Small lots and small developers have less latitude and are less inclined to earmark land for tree planting. Large lots and big developers have more flexibility and means to do so. The prevalent high-density development style has commonly resulted in 100% site coverage of the ground, leaving no room for street-level greening. Planting at a raised elevation on building terraces or roofs in the form of "hanging gardens," while softening the hardscape, do not improve environmental conditions at the street level. Besides, these green sites are accessible often only to residents or tenants, and often have poor landscape design, poor maintenance, and low green cover.

To mobilize developers, a high-level of coordination for their greening efforts would help (Ames and Dewald, 2003). Guidelines for open space and tree planting for different statutory land use zones could give developers an unambiguous message (Singapore Government, 2004; Taipei Government, 2004), and allow them to discern their role in the integrated landscape plan. Technical support and information supplied by government professional officers will help to maintain standards of private tree work. The government could take the lead by providing demonstration greening in public development projects. A package of clearly written and enforced specifications will furnish the necessary groundwork.

Concrete incentives to save and plant more trees could be introduced into the planning approval system. For instance, a more flexible award of bonus plot ratio (see Creating Roadside Tree Corridors by Setback Zoning, above) could be introduced for voluntary building of setbacks for roadside planting strips and amenity areas. Developers could be given direction to contribute toward the overall greening of the city. Small sites and small developers, in particular, could be assured of the importance of their collective contribution to the greening endeavor, and their civic-minded efforts could be given due identity and recognition.

In compact cities, few residents dwell in houses with private gardens. But where they exist, these low-density sites afford an unusual opportunity for citizens, rather than public agencies or commercial enterprises, to contribute toward greening and improving environmental quality in their city (Thompson et al., 2003; Gaston et al., 2005). Such neighborhoods could add a different dimension to the green stock of compact cities. Often, the independent decisions of many individual owners bring surprisingly diversified assemblages of amenity vegetation and landscape styles. They reflect the earnest desire of citizens ordinarily trapped in the cramped city milieu to nurture their own greenery (Jim, 1987a,b). Some old suburban enclaves with dense tree canopies, often containing outstanding specimen trees, have been isolated by urban sprawl. With a high land value, they are susceptible to being redeveloped at higher density. Measures are needed to identify such land uses with mature sylvan components with a view to preserving them as tree conservation area zones.

Modifying Conventional Practices

Reducing Above-Ground Roadside Constraints

Roadside space in compact cities is commonly inadequate or unsuitable for trees situated above and below the ground (Jim, 1997a,b). Buildings are often constructed at or near 100% site coverage at the street level without setback from the lot frontage. Busy commercial areas have advertisement signs and overhead electrical cables as physical obstructions. The pavement is usually too narrow or heavily used to accommodate tree planting. Building awnings above narrow sidewalks block vertical plant growth for even shrubs and small trees. In medium- and low-density precincts, the setback areas are usually cordoned off by a wall and paved to serve as access roads, car parking, or other non-green uses. A similar pattern is found in public-use lands (including schools, police and fire stations, hospitals, etc.), where the grounds are also commonly paved. A citywide program to replace unnecessarily concreted areas with soil and vegetation could noticeably increase the amount of pervious and environmentally friendly covers of a city.

For developing Asian cities, the recent installation of mass transit systems could be accompanied by measures to contain the growth of private cars. A reliance mainly on public transport would demand fewer roads and flyovers, so that fewer existing trees will be hampered by vehicular transport, and more roadside sites could be planted.

Side streets with low vehicle use could be pedestrianized and planted, and underused road spaces could be similarly converted. However, emergency vehicle access (EVA) often severely restricts planting opportunities in narrow pedestrianized streets. To overcome this constraint, trees of small final size can be planted on the EVA, and they could more easily be sacrificed under pressing circumstances, and then replanted.

Road alignment is usually quite persistent, and hence roadside tree corridors are normally protected. However, the need to widen roads could convert street tree corridors to carriageways. In fact, planners often consider roadside tree corridors to be ready land reserves for road widening. Roads with fine mature trees could be designated as landscape conservation areas or heritage roads. Discouraging car ownership could reduce the pressure on roads and hence on roadside planting areas.

Tackling Underground Utility Restrictions

In compact cities, competition for underground spaces is also rather keen. Burying utility lines underground, mainly below the narrow pavement at a shallow depth often <1 m, would directly conflict with tree roots. Many potential planting sites along paved areas cannot be used for greening, because the underground space has been usurped. Trenching to repair cables or pipes or to install new ones commonly injures roadside trees, and is an important cause for their poor performance, premature decline, and collapse during storms (Jim, 1997a,b, 2003). Routine trenching practice can weaken trees, predisposed to other stresses and diseases, and induce premature tree decline. Such practices also increase tree maintenance costs.

A spatial database on buried utilities underneath footpaths could identify sites where soil volumes for tree roots are available. Tree location and site design could reduce conflicts between buried utilities and tree roots, and permit future growing space for roots (Urban et al., 1988). Trench routing and trenching work would benefit from improved technical guidelines for avoiding root injury, especially for outstanding, heritage trees (National Joint Utilities Group, 1995; Jim, 2005a–c).

Trenching that goes near trees with >15-cm trunk diameter should not use a conventional open-and-cover method within the tree protection cordon (TPC) defined by the extent of the tree crown. The trench shall either be diverted away from the TPC, or directed under the tree at a depth below 1 m using the microtunneling or trenchless technique. Tree roots that are >3 cm in diameter encountered during trench excavation shall not be cut or dislocated, and shall be protected from damage and desiccation.

In road overhaul projects where sufficient underground room is available, the buried utilities could be shifted to one side to release space for planting. Otherwise, fixed or movable planters (containers) could be more widely installed. There is a lack of experience in using planters to grow trees especially in tropical regions. Comprehensive research could be initiated on planter design and the selection and care of plants that can perform reasonably well in such sequestered and stressful environments (Jim and Ho, 2000).

In the long term, municipal authorities could accommodate utilities in dedicated tunnels, especially along the busiest roads in the core areas (Gong et al., 2005). Underground spaces along sidewalks could then be released for tree planting, and roadside trees could escape frequent root damage from trenching. Additional benefits include the reduction of nuisance and hazards from trenching works, and improvement of landscape quality. New development areas could adopt the cost-effective and environmentally friendly utility tunnels. The savings from road opening, accumulated over some years, could repay the installation cost. The massive social and economic cost of causing inconvenience, delays, and disruptions to pedestrian and vehicular traffic, not to mention the risks of accidents, injuries, and mortality, should be factored into the costing package. Some major compact cities such as Shanghai, Singapore, and Taipei are moving in this direction.

Improving Poor Soil Conditions

Poor-quality soil is rather prevalent in compact cities. Soil as an important green-site attribute has been widely neglected or misunderstood in urban greening projects (Bullock and Gregory, 1991; Craul, 1992). Site soil conditions are often unsuitable for plants. Small tree pits without ameliorating the site soil are particularly unfriendly to trees. Field and laboratory soil assessment could judge soil suitability for plant growth and identify improvement methods (Jim, 1998a,b; Craul, 1999).

The main physical problems are excessively stony and sandy soil, and limited rooting volume in terms of depth and lateral spread (Perry, 1994; Jim and Ng, 2000). Planting sites are commonly beset by the presence of rocks, building foundations, utility junction and control boxes, and sealing of the soil surface by concrete or asphalt (Jim, 1993). These inert and sequestering materials seriously limit the usable soil rooting volume and block root expansion. The roadside soil underneath the paving is usually densely compacted according to engineering requirement to support load. The paucity of soil porosity would restrict air and water passage, and water storage (Jim, 1998c). The soil suffers from poor aeration, low water-holding capacity, inadequate moisture supply, and sometimes sluggish drainage (Jim and Ng, 2000).

Some chemical soil properties are unfavorable to plant growth, notably the lack of available nutrients, limited nutrient-holding capacity, and alkalinity leading to micronutrient deficiency and nutrient imbalance (Craul, 1980; Jim, 1998d). Urban soils suffer from widespread contamination by construction rubble that contains calcareous concrete, cement, and mortar fragments, and raises the soil reaction into the harmful alkaline range. The solubility and supply of micronutrients such as iron, manganese, copper, and zinc are suppressed under the high pH environment. Urban soils are the sink for various pollutants brought by run-off water, rainfall, gravity settlement of particulates from the atmosphere, and canopy drip washing of dust particles deposited on leaves. Common shortages of essential nutrients, especially nitrogen and phosphorus, could curtail tree growth (Jim, 1998c,d).

Poor soils dumped or trapped in the urban landscape, after buildings and roads have been completed, are difficult to ameliorate or replace, and intractable deficiencies will linger. Trees in such a poor growing medium cannot perform well and may incur heavy management liabilities. Short-term solutions could improve soil conditions by amendments and physical manipulation. Long-term solutions could adopt improved urban soil specifications for trees, construct soil corridors, and install dedicated tree strips that are separate from the utility zone. The improvement or replacement of poor site soils with an imported soil mix prepared according to a specification could be implemented before the buildings or roads are installed.

Selecting Species and Planting Materials

Long-term planting success, especially for trees, depends on properly assessing the match between the desired plant species and site conditions. For rapidly developing cities in tropical and subtropical regions, the lack of scientific information on tropical tree species and cultivars suitable for urban planting hinders this process. In compact cities, more intense conflicts between trees and the city matrix make species selection all the more important. A common problem is the planting of the trees in spaces that cannot accommodate final mature tree dimensions (Jim, 1998e). Short-term solutions could include more thoughtful matching between trees and sites. Long-term solutions involve systematic research on the suitability of tree species and cultivars for cramped and suboptimal urban habitats, particularly for tropical and subtropical cities. Greater scientific dialogue among arboricultural researchers and practitioners needs to occur, so that a shared database of recommended trees for different types of urban sites can be created.

The quality of plant materials should be ascertained before following through with planting. This cardinal factor must be vigilantly scrutinized at different stages in the planting process, and especially at delivery. Many vexing and chronic arboricultural problems could be traced to poor-quality trees, often in the sapling stage of growth. Common weaknesses include trees lacking vigor, crossed branches, V-shaped crotches, unbalanced crowns, crooked or curved trunks, multiple stems, poor scaffold and branching habits, lack of corrective pruning and branch training, wounded or decayed trunks, and sparse foliage. Such inherited maladies at seedling and sapling stages would develop in time into long-term liabilities and potential hazards. Trees that may in due course become unstable, unsightly, poorly structured, weak, diseased or hazardous must not be planted.

A trained arborist could identify the telltale signs of weakness and reject such plants. The root cause of most tree problems lies with the nursery, particularly in the production method. Quality control must begin in the nurseries, from the selection of seeds, seedlings, and saplings, to the continual culling of weaklings. Public-sector nurseries should compete with private ones in supplying planting materials to maintain the quality of planting material raised by both sectors. Otherwise, domination by public nurseries could lead to lax quality control and conservative attitude in species selection. Detailed and enforceable specifications dealing with plant quality will fill a major gap in the quality assurance process that should be stringently observed by suppliers and demanded by users. The development of a strong and professional landscape industry could promote higher standards in planting materials and methods. A rigorous scheme of inspection and rejection of delivered planting materials would be a necessary adjunct.

To assess the long-term landscape consequences of newly planted trees, a scientifically refined species-selection strategy needs to be achieved. Research on the suitability of species and cultivars for stressful urban habitats could inform species choice. Since developing countries have limited capabilities for conducting long-term planting trials in experimental plots, an alternative strategy of systematically evaluating the growth and survival of planted trees would inform future selection choices. A national or global database of amenity trees suitable for urban sites varying in harshness and for different climatic zones would expand our knowledge more rapidly by including experiences in many different cities (Jim, 1990).

Planting sites in compact cities are commonly small and scanty, and there is the urge to utilize them fully. Landscape design should match site characteristics and use. Site geometry, including size and shape in the three-dimensional sense both below and above ground, should be considered in matchmaking. In particular, large sites that are precious should be filled with trees with sizable final dimensions. Otherwise, compact cities will have few large trees to serve as living landmarks and cityscape anchors for people.

Applying Landscape and Urban Ecology Principles

Designing Shape and Connectivity of Green Spaces

The size, shape, location, environs, and interface with adjacent land uses are key determinants of green-space quality. Green-space design could employ relevant landscape ecology principles (Dramstad et al., 1996; Chen and Jim, 2003) to maximize their ecological and environmental functions. In land-use plans, the location and configuration of green spaces could be demarcated by well-established spatial planning concepts. The basic landscape components to be considered are patches (the green areas), corridors (the linear belts), and matrix (the surrounding built-up areas), and the ancillary issue of edges or boundaries (the interface between patch or corridor with the matrix). Some geometrical properties of green spaces could enhance their ecological and social functions, such as size, shape, orientation, and distance from and connectivity with other green spaces.

A good design would include some large green areas, wide corridors, long linear belts (offers greater exchange and contact for people and organisms with the surrounding built-up matrix), orientation parallel to natural or artificial linear features (streams, coastlines, or roads) to serve as environmental buffers, proximity to other green spaces, and connection between green spaces to form a green network. High-quality boundaries between green and nongreen areas are wide, curvilinear, gradual, and dominated by vegetation rather than artificial materials. Such a configuration, emulating the natural transitions between vegetation types, permits the interpenetration of the two components, a longer interface between them, and more beneficial natural influence on developed areas. Landscape ecology concepts could be applied with imagination to green-space planning (Cook, 2002; Leitão and Ahern, 2002). Vegetation could serve as a buffer between noncomplementary land uses. As far as possible, green spaces could be configured to form a landscape structure linking patches with green corridors or greenways (Flink and Searns, 1993) to form an integrated green-space system to enmesh built-up areas (Flores et al., 1998; Jim and Chen, 2003).

To bring spatial permeation and connectivity of green spaces, amenity strips could be planned along new roads at roadsides, medians, roundabouts, and incidental amenity parcels. Within lots, green spaces should be allocated in the grounds of residential, office, government, institutional, and community land uses with a view to linking the otherwise disconnected green enclaves. Remnant natural areas within new developments should especially be salvaged. Planting opportunities could be maximized at linear greenway sites (Flink and Searns, 1993), such as promenades in a coastal area, at a lakeside, and along the banks of rivers and canals.

A comprehensive green plan could knit together disparate greening endeavors, with specific recommendations on locations, dimensions, ingredients, and functions of green spaces, to be tailor-made for different land uses and urban habitats (Jim, 1999). Combining high-density and high-rise residential development with adequate provision of fine green spaces is feasible, as exemplified by the Tampines new town in Singapore (Foo, 2001). Residual plantable sites, which are often omitted in formal but piecemeal greening projects, could be systematically enlisted into the green network. Amenity corridors and wedges are especially valuable in a green-space web (Schabel, 1983; Valk, 2002).

Inserting Natural Pockets in Conventional Green Spaces

Nongeometric characteristics also need attention, notably the degree of naturalness of the green site, to be effected through substrate preparation and species composition. Green spaces larger than a certain size (say 1000 m^2) should earmark about 20% of the area for naturalistic planting by creating natural habitats (e.g., lowland and high-altitude forests, shrub land and grassland, sandy and rocky substrates, riverine and coastal habitats), using native and nonhorticultural species. This approach been successfully implemented in some urban green spaces, such as the Russell Square in London, the Botanic Gardens in Singapore, and the Kasai Rinkai Park and the Jindai Botanic Gardens in Tokyo.

Different sites could be given different functions to be fulfilled by dedicated designs that should involve more ecological elements in addition to visual-scenic-ornamental ones. To meet modern aspirations, the neat, tidy, manicured, and horticultural type of urban green space design could be complemented by the naturalistic-ecological approach (Henke and Sukopp, 1986). In fact, the nature-oriented designs are often less expensive to build and are largely self-sustaining with minimum maintenance needs (Bos and Mol, 1979; Manning, 1979). Hence they reduce recurrent upkeep cost of green spaces. The strong demand for natural areas within and near cities (Johnston, 1990) could be satisfied with features that many overdesigned and expensive (capital and recurrent costs) urban parks fail to deliver (Thompson, 2002).

It is necessary to tackle the political pressure to manicure green sites that could unfortunately frustrate the realization of this approach. The public could be convinced through education and other publicity measures to realize the values and benefits of having natural pockets in the city. The importance of contacts with and exposure to the multifarious stimuli offered by natural vegetation in children's intellectual development (Stearns, 1972; Taylor et al., 1998, 2001) could be highlighted to win citizen acceptance.

Creating Green Fingers, Wedges, and Pockets

Some elements of naturalness or wildness are often welcome as pleasing diversions to regularity, linearity, and formality. They contribute significantly to urban ecological diversity and interests. The most valuable configuration is an intimate mingling of green spaces and the built-up matrix, so that green areas are situated close to people to create a nature-in-city ambience. The most welcome and used venues are situated within about 400 m or a 10-minute walk from work or home (Müller-Perband, 1979; Burgess et al., 1988). The Urban Environmental Accords recommends that by 2015 every urban resident should have access within 500 m to a public park or recreational open space (United Nations Environment Program, 2005).

At the urban edge where it interfaces with the countryside, tongues or wedges of periurban woodlands should be preserved to extend into the built-up areas in an interfingering pattern (Frey, 2000). Small pockets of remnant nature embedded in developed areas, such as hillocks and remnant slopes, should be kept in the wild state (Jim, 1989a,b). With peninsulas of nature extending from the countryside into the city, and islands of nature punctuating the city, landscape, amenity, and air-quality benefits can be improved. The remnant natural enclaves are particularly valuable, and such gap sites could be guarded against conversion to built land to preserve the high degree of naturalness and wildlife habitats, and to enhance their contribution to urban environmental and scenic qualities (Parsons and Daniel, 2002) and outdoor recreation (Tartaglia-Kershaw, 1982). These natural sites can fulfill the increasingly popular ecocentric environmental value, preferring informal and wild sites that provide solitude and escape from city existence at convenient locations (Kaltenborn and Bjerke, 2002; Thompson, 2002). More people using natural areas will need an unconventional type of management to maintain the ecological integrity and health of natural areas, such as reduction of exotic species invasions, soil erosion, and soil compaction.

In land-use planning, opportunities for interpenetration between city and nature should be assiduously preserved. Such city wilds could be comprehensively surveyed to ascertain their status and conservation worth, as exemplified by the treatises on London (Fitter, 1945) and Portland, Oregon (Houck and Cody, 2000). To conserve these precious areas, they could be designated on zoning plans as a new land use category labeled natural enclaves. The green spokes or fingers of Stuttgart, Germany (Schabel, 1983), the proposed green plan for Nanjing (Jim and Chen, 2003) and Beijing (Li et al., 2005) in China provide cases in point. Management inputs should be commensurate with the cardinal objectives of sustaining their ecological and environmental functions. There is no need to tame them and to dilute their naturalness with exotic and horticultural plants and concrete footpaths.

Whereas formal green spaces are seldom threatened by development, natural pockets embedded within or on the fringe of the city are often subject to intrusion and damage (Jim, 2002b). As a city intensifies its land use and expands, such natural sites are often sacrificed (Swenson and Franklin, 2000). Sometimes the "residual" enclaves are erroneously considered as wasted resources or impediments to development. Due to property rights issues, protection of private land with high conservation value needs special policies (Bowers, 1999). Otherwise, in the contest for limited land resources, private economic gains will continue to prevail over public conservation objectives.

In terms of conservation priority, it is pertinent to protect natural habitats situated near homes and within easy access from built-up areas, as they could be frequently used by nearby residents and workers for passive recreation (Müller-Perband, 1979; Burgess et al., 1988). Where appropriate, existing green sites with convenient access could be upgraded based on ecological principles and methods to augment their natural contents and to imitate natural habitats. Existing soil and vegetation should be preserved, and native species that are naturally associated with the existing species could be added. Such enrichment planting programs could aim at assisting or accelerating the successional progression toward high biodiversity, complex biomass structure, and ecosystem maturity that are commensurate with site conditions and context. Nature can best be preserved; if the best is not available, emulated nature could be created as a substitute.

Providing Natural and Artificial Green Spaces

Green-space development sometimes adopts an ingrained and perhaps distorted urbanized mentality that everything has to begin from scratch. Existing natural habitats and vegetation, even of high quality, could be eliminated by design or damaged by default due to inadequate protection against construction impacts (Williams et al., 2005). The natural land form is often drastically altered, resulting in equally drastic disturbance of soil and hydrological and microclimatic conditions. Nature's remnants are sometimes regarded as inferior or out of place in the built environment, and are uncommonly incorporated into the development framework (Johnston, 1990; Mazzotti and Morgenstern, 1997). The attitude of excluding nature in the humanized milieu, and replacing it with a poor imitation, has kept nature at bay in many urban areas.

Even though development may intrude into natural and mature woodlands, there are few incentives to preserve woodland enclaves in landscape design (Goldsmith, 1988). This common neglect calls for the infusion of ecological knowledge into the landscape design curriculum. Thus tongues and pockets of natural habitat have been commonly eradicated in the process of urbanization, and such destruction occurred both in the old city core and in new towns. Only occasionally were isolated trees of outstanding character and performance preserved in development sites. These remnants of nature are often trapped in incongruous sites that are unsuitable or even harmful for their continued existence. Accommodating nature-in-city (Henke and Sukopp, 1986) is an idea that could be more enthusiastically embraced by landscape architects and policy makers, and in due course by more citizens with the help of formal and public education programs.

Some sites have inherited good tree cover, especially natural woodland pockets tucked away within the built-up areas, often left by design due to religious or superstitious reasons, or by default due to geotechnical constraints associated with the difficult terrain. They could be left alone for nature to take care of itself and run its own course. Sites with plantable space gaps could benefit from enrichment planting. Degraded forest areas could be subject to ecological restoration measures (Borgmann and Rodewald, 2005). Prepared sites usually have lower ecological value than natural sites, due to synthetic design with simple composition and biomass structure, limited vegetation coverage, isolated configuration, low habitat and species diversities, and lack of attraction to wildlife (Fernández-Juricic, 2000; Hess and King, 2002). Parts of such urban parks could be converted into natural areas based on ecological design (Henke and Sukopp, 1986).

The creation of a diversity of wild habitats in a naturalistic setting to be filled with native species would be welcomed by residents influenced by modern ecological thinking (Johnston, 1990). With active planting and nurturing of suitable native species at the start, thus preempting the spaces and resources, the likelihood of invasion by nonnative species into such sites could be much reduced. Even small pockets could create interesting ecological diversity and attract both wildlife and human visitors. Inputs of expert ecologists could be enlisted into the design team.

Embracing Naturalistic Green-Space Design

Nature-in-the-city could be adopted as an urban design principle (Cole, 1986; Henke and Sukopp, 1986) for more earnest translation into practices. For new developments that extend into well-vegetated natural areas, portions with high ecological value should be demarcated for sympathetic incorporation into the future built environment. Whereas countryside fringing a city is precious, countryside occluded within a built-up area is a gem. The new development areas could inherit a ready-made and high-quality green space with plenty of natural ingredients. Such natural sites tend to sustain themselves with little management input. The concern of urban planners and managers about the escalating cost of developing and maintaining formal green spaces could be partly resolved. The increasingly tight municipal budget would not be strained by the need to maintain more formal open spaces. By providing a continuum of urban green spaces from the highly formal to the entirely natural, different tastes and demands of citizens could be catered to.

Understanding the natural assets of development sites provides the prelude to keeping nature in cities. Assessing the naturalness of areas designated for development (Mazzotti and Morgenstern, 1997) could be conducted early in the development stream, so that important sites will not be inadvertently damaged. Periurban woodlands with high diversity of habitats, communities of flora and fauna, soil-water conservation functions, fresh air sources (Schabel, 1983), and passive recreational and nature-educational potentials constitute a natural heritage in the vicinity of beneficiaries. That these natural areas in cities can provide key ecosystem services, which cannot be emulated by manicured urban parks, should be emphasized (Jim, 2004; Jim and Chen, 2006). Natural areas situated close at hand, particularly intra-urban woodland enclaves, as islands of nature, is a prized possession. As much of the original organic structure, associations and constituents should be preserved intact. Future activities and management should respect the integrity and continuity of natural features and processes.

A spatially oriented landscape planning strategy could be developed to provide an optimal green-space configuration, aiming at a green network linking patches by greenway corridors or stepping-stone sites to maximize connectivity (Langevelde et al., 2002; Vuilleumier and Prelaz-Droux, 2002). Conversely, habitat fragmentation and associated landscape degradation should be minimized (Cook, 2002; Valk, 2002), to forestall biodiversity depauperization and invasion by weedy and alien species (Smale and Gardner, 1999; Godefroid and Koedam, 2003). The massive green belts around cities such as Berlin and Seoul could serve their ecological and recreational functions better by creating links to intra-urban greenways and green spaces. The size and shape of patches and their edge structure at the city-nature (matrix-patch and matrix-corridor) interfaces, should foster ecological functions and services (Dramstad et al., 1996; Jim and Chen, 2003). The group value of tree clusters and woodlands should take precedence over the narrow focus on species rarity as a conservation yardstick. Meritorious natural areas could be designated as future parks and passive recreation venues (Johnston, 1990), to be incorporated into an urban green-space system linked to urban-fringe and extra-urban natural areas.

Establishing a well-connected and pervasive green-space system would require both passive preservation and active provision. Where suitable sites are unavailable, new urban woodlands could be created with innovative afforestation techniques using a diversified assemblage of native species and sensitive site preparation (Baines and Smart, 1991; Harmer, 1999). Research into urban habitat creation and conservation techniques could be conducted to suit different geographical and ecological circumstances (Carr and Lane, 1993; Wheater, 1999; Lee and Thompson, 2005). Brown fields and derelict sites could be transformed into green areas (Sousa, 2003) in a reverse land conversion process. A pertinent measure of success is the attraction of indigenous wildlife into the wooded enclaves (Fernández-Juricic, 2000; Livingston et al., 2003). The city-countryside synergy could be fully tapped by designing for their juxtaposition. The cardinal principles of nature-reserve design based on island biogeography theory, namely large size, contiguity, proximity, and connectivity, could enhance green-site quality (Davey, 1998).

Improving Institutional Support

Removing Institutional Constraints and Enhancing Community Involvement

The main institutional constraints to providing appropriate green cover in many Asian compact cities involve the lack of resources and long-term commitment, poor coordination among government departments, little involvement of private-sector initiatives, shortage of trained staff, inadequate leadership, and absence of an overall greening strategy and plan (Jim and Liu, 2000; Jim, 2002a). Solutions could be sought from mobilizing wide support within and without the government, establishment of clear and visionary high-level policies for a green city, appointment of a dedicated government unit equipped with adequate budget and expertise for trees, motivation and coordination of private-sector participation, and an overarching public–private partnership organization to oversee and coordinate all urban tree efforts.

A cramped city milieu generates heavy demands for roads. Rightly or wrongly high priority is often accorded to transport requirements, which in some instances have overridden other landscape needs. The safety clearance for vehicles, especially for double-decker buses in some cities, imposes a rather unyielding constraint on tree planting along many roads. Tree crowns are not supposed to block the carriageway and have to retreat behind the curb line, and narrow pavement aggravates this situation. Traffic signs have to remain visible to motorists, and lay-by bays (shoulders of roadways) and bus stops have to be kept clear of obstructions. The extensive visibility clearance at road junctions and in front of traffic signs, stringently demanded by road safety, could quash many roadside planting opportunities. In compact cities, small city blocks and high road and traffic sign densities have aggravated the restrictions on tree planting. The design and positioning of traffic signs in compact cities could be modified to accommodate the needs of roadside greening.

The limited pavement spaces are often usurped by a host of street paraphernalia, such as traffic signs, lampposts, railings, control boxes, parking meters, and fire hydrants. The inflexible enforcement of such "rules" has excluded taking advantage of many marginal plantable sites. In a densely packed roadside environment, the reality is that not too many sites are unambiguously free of obstructions or can fully abide by the traffic requirements. There are too many claims on limited air space and too many forbidden locations for urban trees. The unclear demarcation of tree responsibilities and authority in the city administration does not help urban foresters to navigate this regulatory mine field. A meeting of all the stakeholders could find compromises and solutions.

The use of trees for environmental and ecological education could foster public discourse on our living plant companions. Urban tree knowledge could enter school curricula to strengthen awareness and nurture informed citizens who could be rallied to the cause. Tree walks with cultural, historical, and popular science themes could be designed to encourage more contact between people and trees. A citizenry concerned with and knowledgeable about trees will develop a love for trees and a desire for more and better trees, form a closer partnership with the government, and lend earnest support to planting and preservation activities. Involving people in tree planting and care, and nurturing them as tree advocates could cultivate a partnership and a sense of ownership to induce contribution and support (see Chapter 12). An umbrella organization such as a tree council could be formed to oversee, coordinate, and mobilize urban-tree efforts from different quarters.

Developing a Citywide Greening Strategy and Plan

Planning is the key to successful urban greening. Greening that is planned well in advance of development or redevelopment projects is more likely to be successful. Trees should be mandated as an essential urban infrastructure, and a statutory green-space zone would enhance provision at the land-use zoning stage. A green code for development sites that stipulates intra-lot and lot-frontage green-space standards could trigger widespread and coordinated private-sector participation in urban greening. Requiring a proportion of a lot to be designated as green space could open up the tight city plan in new developments and renewal areas.

Similarly, a green code for roads could be developed, such as the one adopted in Singapore (Singapore National Parks Board, 2003) to ensure that adequate and high-quality plantable spaces are reserved for trees when new roads are built and old ones overhauled. To position green spaces at strategic locations, development rights could be transferred from a preserved green site to a development site elsewhere. Where justifiable on the grounds of ecological value, amenity, and landscape contribution, and in general in the interest of the community, land could be purchased by a land trust or public funds for conversion to green spaces. The collective contributions of individual lots, small or large, will in time bring significant city-level improvements and add to the value and prestige of properties.

Developers are commonly required by planning laws to prepare a master landscape plan (MLP) for individual sites. This piecemeal approach, with plans for preparing individual sites and approved in isolation, may not be coordinated with the city's greening vision. It is pertinent to develop an overarching greening blueprint (Nowak and O'Connor, 2001) in the form of a citywide greening strategy (CGS). This high-order plan could focus on the following cardinal objectives: (1) to increase planting sites in developed urban areas and along existing roads; (2) to reserve planting sites in new development areas and along new roads; (3) to ensure that planting sites can support high-quality tree growth and will not be degraded by other impacts and activities; and (4) to develop planting themes in different neighborhoods or districts, so as to build unique landscape characters and identities in different parts of the city (Jim, 1999). The CGS can be reviewed once every 5 years to keep it up to date. Individual MLPs are expected to align with the CGS to form parts of the whole and to collectively contribute to the city's overall landscape improvement.

The CGS can include the following specific measures to improve the quantity and quality of urban greenery: (1) to assign tree management responsibilities and related powers to an urban tree authority; (2) to maximize the use of existing planting sites for tree planting; (3) to provide incentives to developers to introduce trees within lots at the street level especially at the lot frontage; (4) to upgrade the performance of existing and new urban trees; (5) to minimize damage and destruction of existing urban trees due to development and other causes, with particular attention to champion caliber trees; (6) to identify streets or street segments to institute setback of building footprints from property lines for conversion to roadside tree strips; (7) to identify trees and urban fringe woodlands and natural areas for conservation; (8) to ascertain that adequate resources are made available to fulfill greening endeavors; and (9) to ensure that all tree works are conducted at a high professional standard in keeping with international best practices and norms.

The spirit and purpose of the CGS could be translated into action plans by developing public as well as private planting plans on an annual basis with a 5-year rolling horizon (Jim, 1999). Such planting plans can be prepared for each district, to cover both developed areas and the urban fringe including green belts. Each public tree planting plan could include the following: (1) the boundaries of planting sites marked on maps not smaller than 1:1000 scale; (2) the locations, species, and size of trees to be planted in the planting sites; (3) the time frame for the completion of the planting work; (4) the parties responsible for planting and maintaining the trees; (5) anticipated problems and proposed solutions; and (6) the budget and funding sources for the proposed planting plan. For private tree planting plans, the developers are expected to follow the principles embodied in the CGS to provide planting spaces.

Official open space standards stipulated in planning statues are sometimes set at a low level in compact cities. For instance, Hong Kong designates a mere $2\,m^2$/person, of which $1\,m^2$ is district open space and $1\,m^2$ local (Planning Department, 2003). Moreover, the standards set an active-to-passive green-space ratio of 3:2, in terms of open space area devoted to these two forms of recreation. Since greenery is considered passive recreational land, in practice the green-space provision is only $0.8\,m^2$/person, which is inadequate by any measure. Raising such low standards could bring improvement in the long term. Green space could be designated as a separate category in the zoning plan, so that it will not be usurped by active recreational uses. To ensure that a good proportion of green space is devoted to tree planting, a minimum tree-stocking rate (as aggregate canopy cover area of trees) could be stipulated at 50% of green space, to be averaged over a given neighborhood.

Some cities do not have the benefits of a dedicated urban tree ordinance. They rely instead on a confusing jumble of administrative measures and indirect applications of other laws, which are often ineffective in promoting and improving urban greening. It will be worthwhile to enact a tree law to encompass the spirit and stipulation of the overall urban greening strategy, and to usher more assured compliance.

Conclusion

Compact cities in developing countries present many inherent physical and institutional constraints to greening, many of which are unique to them or are more intensively expressed than in cities with a less dense growth form. Many municipal authorities would want to enhance urban greenery, but have been frustrated by inertia and apparently insurmountable difficulties. The tight city plan, in particular, seems to be ossified and immutable. There is a tendency to adhere persistently to environmental determinism in a resigned manner, creating a barrier to change. The formation of developable land in compact cities is often an expensive and time-consuming enterprise. There is thus the urge to maximize land use intensity often to the exclusion of vegetation, sacrificing environmental quality and the quality of life.

Research findings and field surveys of the situation in a number of compact city areas in different parts of the world point to several opportunities and alternatives to improve both the quantity and quality of urban greenery. However, inadequate and low-quality plantable spaces gravely limit greening. By modifying existing approaches or adopting innovative ones, cities could gradually introduce more plantable spaces and amenity vegetation into the built-up matrix. Rather than subscribing to the belief that there is little space to maneuver, city authorities could proactively seek and create space. Opportunities abound and they should not be allowed to slip away through inaction.

Different parts of a city, even compact ones, tend to have varied coverage of natural and semi-natural areas with potential to support higher quality vegetation. In the spirit of precision planning, different site conditions call for specific approaches to realize these potentials. Urban greening is by nature a multivariate venture that demands the union of knowledge and expertise from disparate disciplines. The experience of one city can often be shared with others. Many fine models of good practices, however, remain obscure or fail to propagate and be applied outside their home range. The gap between research and application and between science and practice, in the field of urban greening, could be narrowed by more effective interactions, integration, exchanges, and communication.

Of all constraints, inertia is probably the most restrictive. The principal obstacles are administrative and political, involving policy (Duvernoy, 1995; Bowers, 1999) and the exigent urge to cater to the wishes of the ignorant, conservative, and apathetic. The multiple benefits of urban greenery with long-term synergistic effects for improving environmental quality (Dochinger, 1980; Nowak and Dwyer, 2000) could be more effectively communicated to stakeholders to gain their support. The emphasis on physical planning at the expense of landscape quality could be shifted toward ecological planning (Henke and Sukopp, 1986; Gordon, 1990; Cook and Lier, 1994). The bias toward physical infrastructure at the expense of natural infrastructure could be fine-tuned by integrating development with greenery (Herz et al., 2003). Generous and meritorious greening can coexist comfortably with the compact urban form. If trees were to be incorporated into a city plan (Petit et al., 1995), that is, to plant wherever and whenever we build, we would have achieved the goal of greening difficult compact city sites.

Greening cities, especially upgrading compact urban areas with greenery, is widely advocated as a key feature of a livable (Lennard and Lennard, 1987) and sustainable city (Roseland, 1998; Newman, 1999; Marcotullio, 2001). Greening could serve as a necessary but not sufficient condition to attain urban sustainability, presenting a partial answer to this quest. It provides promises as humanity tries to find an alternative urban-growth paradigm that departs from conventional and conservative ideas. The idea of eco-cities calls for flexibility toward major environmental and economic trends and changes, and greening is a constituent part of the greater improvement process. Whereas socioeconomic benefits carry environmental costs, it has to be recognized that environmental benefits also incur socioeconomic costs. The latter conversion, as conducted by enlightened municipal authorities, is costly, because of the dual task of repairing past ills as well as augmenting future benefits. The inaction of previous generations has left a legacy of greenery deficit in many compact cities, and present and future generations would have to compensate for these to settle this transgenerational debt.

Acknowledgments The author is grateful to the research grants provided by the Hui Oi Chow Trust Fund, the University of Hong Kong Committee on Research and Conference Grants, and the Green Fun Committee.

References

Adams, L.W., and Leedy, D.L. (1987) (eds.) Integrating Man and Nature in the Metropolitan Environment. National Institute for Urban Wildlife, Columbia, MD.
Ames, B., and Dewald, S. (2003) Working proactively with developers to preserve urban trees. Cities 20:95–100.
Arnold, C.L. Jr., and Gibbons, C.J. (1996) Impervious surface coverage, the emergence of a key environmental indicator. Journal of the American Planning Association 62:243–258.
Atkinson, D., Smart, R.A., Fairhurst, J., Oldfield, P., and Lageard, J.G.A. (1999) A history of woodland in the Mersey Basin. In: Greenwood, E.F. (ed.) Ecology and Landscape Development: A History of the Mersey Basin, Liverpool. Liverpool University Press and National Museums and Galleries on Merseyside, pp. 91–107.
Attorre, F., Bruno, M., Francesconi, F., Valenti, R., and Bruno, F. (2000) Landscape changes of Rome through tree-lined roads. Landscape and Urban Planning 49:115–128.
Baines, C., and Smart, J. (1991) A Guide to Habitat Creation. London Ecology Unit, London.
Beatley, T. (2000) Green Urbanism: Learning from European Cities. Island Press, Washington, DC.
Borgmann, K.L., and Rodewald, A.D. (2005) Forest restoration in urbanizing landscapes: interactions between land uses and exotic shrubs. Restoration Ecology 13:334–340.
Bos, H.J., and Mol, J.L. (1979) The Dutch example: native planting in Holland. In: Laurie, I.C. (ed.) Nature in Cities. Wiley, Chichester, pp. 393–416.
Bowers, J. (1999) Policy instruments for the conservation of remnant vegetation on private land. Biological Conservation 87:327–339.
Bradley, G.A. (1995) (eds.) Urban Forest Landscapes: Integrating Multidisciplinary Perspectives. University of Washington Press, Seattle, WA.
Bradshaw, A.D., Hunt, B., and Walmsley, T. (1995) Trees in the Urban Landscape: Principles and Practice. Spon, London.
Brenneisen, S. (2004) Green roofs: how nature returns to the city. Acta Horticulturae 643:389–393.
Bullock, P., and Gregory, P.J. (1991) (eds.) Soils in the Urban Environment. Blackwell, Oxford.
Burgess, J., Harrison, C.M., and Limb, M. (1988) People, parks and the urban green: a study of popular meanings and value for open spaces in the city. Urban Studies 25:455–473.
Burton, E. (2002) Measuring urban compactness in UK towns and cities. Environment and Planning B Planning and Design 29:219–250.
Carr, S., and Lane, A. (1993) Practical Conservation Urban Habitats. Open University in association with Nature Conservancy Council. Hodder & Stoughton, London.
Chen Sophia, S., and Jim, C.Y. (2003) Quantitative assessment of the treescape and cityscape of Nanjing, China. Landscape Ecology 18:395–412.

Chevallerie, H. de la. (1986) The ecology and preservation of street trees. In: Bradshaw, A.D., Goode, D.A., and Thorp, E.H.P. (eds.) Ecology and Design in Landscape. Blackwell, Oxford, pp. 383–397.

Chicago Department of Environment. (2001) A Guide to Rooftop Gardening. Chicago Department of Environment, Chicago.

Chicago Department of Environment. (2003) Green Roof Test Plot: 2003 End of Year Project Summary Report, Chicago.

Cole, L. (1986) Urban opportunities for a more natural approach. In: Bradshaw, A.D., Goode, D.A., and Thorp, E.H.P. (eds.) Ecology and Design in Landscape. Blackwell, Oxford, pp. 417–431.

Cook, E. (2002) Landscape structure indices for assessing urban ecological networks. Landscape and Urban Planning 58:269–280.

Cook, E., and Lier, H.V. (1994) Landscape planning and ecological networks: an introduction. In: Cook, E., and Lier, H.V. (eds.) Landscape Planning and Ecological Network. Elsevier, Amsterdam, pp. 1–4.

Craul, P.J. (1980) Characterization of streetside soils in Syracuse, New York. Metropolitan Tree Improvement Alliance (METRIA) Proceedings 3:88–101.

Craul, P.J. (1992) Urban Soil in Landscape Design. John Wiley and Sons, New York.

Craul, P.J. (1999) Urban Soils: Applications and Practices. John Wiley and Sons, New York.

Davey, A.G. (1998) National System Planning for Protected Areas. World Conservation Union, Gland, Switzerland.

Dochinger, L.S. (1980) Interception of airborne particles by tree plantings. Journal of Environmental Quality 9:265–268.

Dramstad, W.E., Olsen, J.D., and Forman, R.T.T. (1996) Landscape Ecology Principles in Landscape Architecture and Land-use Planning. Island Press, Washington, DC.

Dunnett, N., and Kingsbury, N. (2004) Planting Green Roofs and Living Walls. Timber Press, Portland, OR.

Duvernoy, G. (1995) Keeping it green, political and administrative issues in the preservation of the urban forest. In: Bradley, G.A. (ed.) Urban Forest Landscapes: Integrating Multidisciplinary Perspectives. University of Washington Press, Seattle, WA, pp. 78–87.

Earth Pledge Foundation (2004) Green Roofs: Ecological Design and Construction. Schiffer Publishing, Atglen, PA.

Ellaway, A., Macintyre, S., and Bonnefoy, X. (2005) Graffiti, greenery, and obesity in adults: secondary analysis of European cross sectional survey. British Medical Journal 331: 611–612.

Evans, M., Bassuk, N., and Trowbridge, P. (1990) Sidewalk design. Landscape Architecture 80:102–103.

Fernández-Juricic, E. (2000) Avifaunal use of wooded streets in an urban landscape. Conservation Biology 14:512–521.

Finco, A., and Nijkamp, P. (2003) Pathways to urban sustainability. Journal of Environmental Policy and Planning 3:289–302.

Fitter, R.S.R. (1945) London's Natural History. Collins, London.

Flink, C.A., and Searns, R.M. (1993) Greenways: A Guide to Planning, Design, and Development. Island Press, Washington, DC.

Flores, A., Pickett, S.T.A., Zipperer, W.C., Pouyat, R.V., and Pirani, R. (1998) Adopting a modern ecological view of the metropolitan landscape: the case of a greenspace system for the New York City region. Landscape and Urban Planning 39:295–308.

Foo, T.S. (2001) Planning and design of Tampines, an award-winning high-rise, high-density township in Singapore. Cities 18:33–42.

Foster, R.S. (1977) Roots: caring for city trees. Technology Review July/August:29–34.

Frey, H.W. (2000) Not green belts but green wedges: the precarious relationship between city and country. Urban Design International 5:13–25.

Gaston, K.J., Smith, R.M., Thompson, S.K. and Warren, P.H. (2005) Urban domestic gardens II: experimental tests of methods for increasing diversity. Biodiversity and Conservation 14:395–413.

Godefroid, S. (2001) Temporal analysis of the Brussels flora as indicator for changing environmental quality. Landscape and Urban Planning 52:203–224.

Godefroid, S., and Koedam, N. (2003) How important are large vs. small forest remnants for the conservation of the woodland flora in an urban context? Global Ecology and Biodiversity 12:287–298.

Goldsmith, F.B. (1988) Threats to woodland in an urban landscape: a case study in Greater London. Landscape and Urban Planning 16:221–228.

Gong, X.H., Zhang, J.S., and Shu, Y. (2005) Research on the development and utilization strategy of Shanghai urban underground space. Urban Management 84(14):33–38 (in Chinese).

Gordon, D. (ed.) (1990) Green Cities: Ecologically Sound Approaches to Urban Space. Black Rose, Montreal.

Grey, G.W. (1996) The Urban Forest: Comprehensive Management. John Wiley and Sons, New York.

Grey, G.W., and Deneke, F.J. (1986) Urban Forestry, 2nd ed. John Wiley and Sons, New York.

Harmer, R. (1999) Creating New Native Woodlands: Turning Ideas into Reality. Forestry Commission Information Note 15, Edinburgh, UK.

Hauer, R.J., Miller, R.W., and Ouimet, D.M. (1994) Street tree decline and construction damage. Arboricultural Journal 20:94–97.

Henke, H., and Sukopp, H. (1986) A natural approach in cities In: Bradshaw, A.D., Goode, D.A., and Thorp, E.H.P. (eds.) Ecology and Design in Landscape. Blackwell, Oxford, pp. 307–324.

Herz, R., Schroeder, H., Stahre, P., Stal, Ö., and Woodward, S. (2003) Towards the Integrated Management of Urban Vegetation and Infrastructure. Final Report of COST C3 Working Group Vegetation and Urban Civil Engineering.

Hess, G.R., and King, T.J. (2002) Planning open spaces for wildlife. I. Selecting focal species using a Delphi survey approach. Landscape and Urban Planning 58:25–40.

Hestmark, G. (2000) Temptations of the tree. Nature 408:911.

Houck, M.C., and Cody, M.J. (2000) (eds.) Wild in the City: A Guide to Portland's Natural Areas. Oregon Historical Society Press, Portland, OR.

Hough, M. (1994) Design with city nature, an overview of some issues. In: Rutherford, H.P., Rowntree, R.A., and Muick, P.C. (eds.) The Ecological City: Preserving and Restoring Urban Biodiversity. University of Massachusetts Press, Amherst, MA, pp. 40–48.

Howard, E. (1902) Garden Cities of Tomorrow. Faber and Faber, London.

Hughes, R. (1991) The role of urban forestry in environmental strategies and the economic regeneration of the post industrial town. European Environment 1:1–5.

Jackson, L.E. (2003) The relationship of urban design to human health and condition. Landscape and Urban Planning 64:191–200.

Jenks, M., Burton, E., and Williams, K. (1996) (eds.) The Compact City: A Sustainable Urban Form. Spon, London.

Jim, C.Y. (1987a) Land use and amenity trees in urban Hong Kong. Land Use Policy 4:281–293.

Jim, C.Y. (1987b) The status and prospects of urban trees in Hong Kong. Landscape and Urban Planning 14:1–20.

Jim, C.Y. (1989a) The distribution and configuration of tree cover in urban Hong Kong. GeoJournal 18:175–188.

Jim, C.Y. (1989b) Tree canopy characteristics and urban development in Hong Kong. Geographical Review 79:210–225.

Jim, C.Y. (1990) Trees in Hong Kong: Species for Landscape Planting. Hong Kong University Press, Hong Kong.

Jim, C.Y. (1993) Soil compaction as a constraint to tree growth in tropical and subtropical urban habitats. Environmental Conservation 20:35–49.

Jim, C.Y. (1997a) Roadside trees in urban Hong Kong: tree size and growth space. Arboricultural Journal 21:73–88.

Jim, C.Y. (1997b) Roadside trees in urban Hong Kong: tree growth and environmental condition. Arboricultural Journal 21:89–106.

Jim, C.Y. (1998a) Physical and chemical properties of a Hong Kong roadside soils in relation to urban tree growth. Urban Ecosystems 2:171–181.

Jim, C.Y. (1998b) Soil characteristics and management in an urban park in Hong Kong. Environmental Management 22:683–695.

Jim, C.Y. (1998c) Soil compaction at tree planting sites in urban Hong Kong. In: Watson, G.W., and Neely, D. (eds.) The Landscape Below Ground II. International Society of Arboriculture, Champaign, IL, pp. 166–178.

Jim, C.Y. (1998d) Urban soil characteristics and limitations for landscape planting in Hong Kong. Landscape and Urban Planning 40:235–249.

Jim, C.Y. (1998e) Impacts of intensive urbanization on amenity trees in Hong Kong. Environmental Conservation 25:146–159.

Jim, C.Y. (1999) A planning strategy to augment the diversity and biomass of roadside trees in urban Hong Kong. Landscape and Urban Planning 44:13–32.

Jim, C.Y. (2002a) Planning strategies to overcome constraints on greenspace provision in urban Hong Kong. Town Planning Review 73:127–152.

Jim, C.Y. (2002b) Heterogeneity and differentiation of the tree flora in three major land uses in Guangzhou City, China. Annals of Forest Science 59:107–118.

Jim, C.Y. (2003) Protection of urban trees from trenching damage in compact city environments. Cities 20:87–94.

Jim, C.Y. (2004) Characteristics of urban park trees in Hong Kong in relation to greenspace planning and development. Acta Horticulturae 643:123–128.

Jim, C.Y. (2005a) Monitoring the performance and decline of heritage trees in urban Hong Kong. Journal of Environmental Management 74(2):161–172.

Jim, C.Y. (2005b) Floristics, performance and prognosis of historical trees in the urban forest of Guangzhou city (China). Environmental Monitoring and Assessment 102:285–308.

Jim, C.Y. (2005c) Outstanding remnants of nature in compact cities: Patterns and preservation of heritage trees in Guangzhou city (China). Geoforum 36:371–385.

Jim, C.Y., and Chen, S.S. (2003) Comprehensive greenspace planning based on landscape ecology principles in compact Nanjing city, China. Landscape and Urban Planning 65:95–116.

Jim, C.Y., and Chen, W.Y. (2006) Recreation-amenity use and contingent valuation of urban green spaces in Guangzhou, China. Landscape and Urban Planning 75(1–2):81–96.

Jim, C.Y., and Ho, S.S.M. (2000) Soil moisture regime of landscape tree planters in urban Hong Kong. In: Burghardt, W., and Dornauf, C. (eds.) Soils of Urban, Industrial, Traffic and Mining Areas I: The Unknown Urban Soil, Detection, Resources and Faces. University of Essen, Essen, Germany, pp. 51–56.

Jim, C.Y., and Liu, H.T. (2000) Statutory measures for the protection and enhancement of the urban forest in Guangzhou City, China. Forestry 73:311–329.

Jim, C.Y., and Liu, H.T. (2001a) Species diversity of three major urban forest types in Guangzhou City, China. Forest Ecology and Management 146:99–114.

Jim, C.Y., and Liu, H.T. (2001b) Patterns and dynamics of urban forests in relation to land use and development history in Guangzhou City, China. Geographical Journal 167:358–375.

Jim, C.Y., and Ng, J.Y.Y. (2000). Soil porosity and associated properties at roadside tree pits in urban Hong Kong. In: Burghardt, W., and Dornauf, C. (eds.) Soils of Urban, Industrial, Traffic and Mining Areas III: The Soil Quality and Problems: What Shall We Do? University of Essen, Essen, Germany, pp. 629–634.

Johnston, J. (1990) Nature Areas for City People. Ecology Handbook 14, London Ecology Unit, London.

Jorgensen, A., Hitchmough, J., and Calvert, T. (2002) Woodland spaces and edges, their impact on perception of safety and perception. Landscape and Urban Planning 60:135–150.

Kaltenborn, B.P., and Bjerke, T. (2002) Associations between environmental value orientations and landscape preferences. Landscape and Urban Planning 59:1–11.

Kaplan, R. (1984) Impact of urban nature: a theoretical analysis. Urban Ecology 8:189–197.

Köhler, M. (2005) Green facades and green roofs with a long tradition in Berlin, Germany. Journal of British Columbia Society of Landscape Architects 7(1):22–25.

Kuhns, L.J., Meyer, P.W., and Patterson, J. (1985) Creative site preparation. Agora 5:7–10.

Kuo, F.E., Bacaicoa, M., and Sullivan, W.C. (1998) Transforming inner-city landscapes, trees, sense of safety, and preference. Environment and Behavior 30:28–59.

Langevelde, F. van, Claassen, F., and Schotman, A. (2002) Two strategies for conservation planning in human-dominated landscapes. Landscape and Urban Planning 58:281–295.

Lee, J.T., and Thompson, S. (2005) Targeting sites for habitat creation: an investigation into alternative scenarios. Landscape and Urban Planning 71:17–28.

Leitão, A.B., and Ahern, J. (2002) Applying landscape ecological concepts and metrics in sustainable landscape planning. Landscape and Urban Planning 59:65–93.

Lennard, S.H.C., and Lennard, H.L. (1987) Livable Cities: People and Places, Social and Design Principles for the Future of the City. Gondolier Press, Southampton, NY.

Li, F., Wang, R.S., Paulussen, J., and Liu, X.S. (2005) Comprehensive concept planning of urban greening based on ecological principles: a case study in Beijing, China. Landscape and Urban Planning 72:435–336.

Livingston, M., Shaw, W.W., and Harris, L.K. (2003) A model for assessing wildlife habitats in urban landscapes of eastern Pima County, Arizona (USA). Landscape and Urban Planning 64:131–144.

Löfvenhaft, K., Björn, C., and Ihse, M. (2002) Biotope patterns in urban areas: a conceptual model integrating biodiversity issues in spatial planning. Landscape and Urban Planning 58:223–240.

Manning, O. (1979) Designing for nature in cities. In: Laurie, I.C. (ed.) Nature in Cities. Wiley, Chichester, pp. 3–36.

Marcotullio, P.J. (2001) Asian urban sustainability in the era of globalization. Habitat International 25:577–598.

Mazzotti, F.J., and Morgenstern, C.S. (1997) A scientific framework for managing urban natural areas. Landscape and Urban Planning 38:171–181.

McPherson, E.G., Nowak, D., Heisler, G., et al. (1997) Quantifying urban forest structure, function, and value, the Chicago Urban Forest Climate Project. Urban Ecosystems 1:49–61.

Miller, R.W. (1997) Urban Forestry: Planning and Managing Urban Green Spaces, 2nd ed. Prentice Hall, Englewood Cliffs, NJ.

Mole, G., and Young, S. (1992) Growing Greener Cities: A Tree Planting Handbook. Living Planet Press, Los Angeles.

Morell, J.D. (1992) Competition for space in the urban infrastructure. Arboricultural Journal 18:73–75.

Müller-Perband, E. (1979) The modern town park in Germany. In: Laurie, I.C. (ed.) Nature in Cities. John Wiley and Sons, New York, pp. 297–326.

Mumford, L. (1961) The City in History: Its Origins, its Transformations, and its Prospects. Secker and Warburg, London.

National Joint Utilities Group. (1995) Guidelines for Planning, Installation and Maintenance of Utility Services in Proximity to Trees. Publication Number 10. NJUG, London.

Newman, P.G.W. (1999) Sustainability and cities, extending the metabolism model. Landscape and Urban Planning 44:219–226.

Nowak, D.J., and Dwyer, J.F. (2000) Understanding the benefits and costs of urban forest ecosystems. In: Kuser, J.E. (ed.) Handbook of Urban and Community Forestry in the Northeast. Kluwer Academic, New York, pp. 11–25.

Nowak, D.J., and O'Connor, P.R. (2001) (eds.) Syracuse Urban Forest Master Plan: Guiding the City's Forest Resource into the 21st Century. U.S. Department of Agriculture Forest Service, Northeastern Research Station, General Technical Report NE-287. Newtown Square, PA.

Olembo, R.J., and de Rham, P. (1987) Urban forestry in two different worlds. Unasylva 39:26–35.

Ong, B.L. (2003) Green plot ratio: an ecological measure for architecture and urban planning. Landscape and Urban Planning 63:197–211.

Parsons, R., and Daniel, T.C. (2002) Good looking, in defense of scenic landscape aesthetics. Landscape and Urban Planning 60:43–56.

Perdue, W.C., Stone, L.A., and Gostin, L.O. (2003) The built environment and its relationship to the public's health: the legal framework. American Journal of Public Health 93:1390–1394.

Perry, T.O. (1994) Size, management and design of tree planting sites. In: Watson, G.W., and Neely, D. (eds.) The Landscape Below Ground. International Society of Arboriculture, Savoy, IL, pp. 3–15.

Petit, J., Bassert, D.L., and Kollin, C. (1995) Building Greener Neighborhoods: Trees as Part of the Plan. American Forestry Association, Washington, DC.

Planning Department (2003) Hong Kong Planning Standards and Guidelines. Hong Kong SAR Government, Hong Kong.

Platt, R.H., Rowntree, R.A., and Muick, P.C. (1994) (eds.) The Ecological City: Preserving and Restoring Urban Biodiversity. University of Massachusetts Press, Amherst, MA.

Roseland, M. (1998) Toward Sustainable Communities: Resources for Citizens and their Governments. New Society, Gabriola Island, British Columbia.

Saaroni, H., Ben-Dor, E., Bitan, A., and Potchter, O. (2000) Spatial distribution and microscale characteristics of the urban heat island in Tel-Aviv, Israel. Landscape and Urban Planning 48:1–18.

Schabel, H.G. (1983) Urban forestry, some lessons from Germany. In: Proceedings 1982 Annual Meeting of Society of American Foresters. American Forestry Association, Washington, DC, pp. 340–345.

Singapore Government (2004) The Parks and Trees Act 2004. Singapore.

Singapore National Parks Board (2003) Guidelines on consultation on types of development submission to parks and trees regulation section. http://www.nparks.gov.sg/development/dev_sub-pro.shtml

Smale, M.C., and Gardner, R.O. (1999) Survival of Mount Eden Bush, an urban forest remnant in Auckland, New Zealand. Pacific Conservation Biology 5:83–95.

Sousa, C.A. de (2003) Turning brownfields into green space in the City of Toronto. Landscape and Urban Planning 62:181–198.

Stearns, F. (1972) The city as a habitat for wildlife and man. In: Detwyler, T.R., and Marcus, M.G. (eds.) Urbanization and Environment: The Physical Geography of the City. Duxbury Press, Belmont, CA, pp. 261–277.

Stone, B. Jr., and Rodgers, M.O. (2001) Urban form and thermal efficiency: How the design of cities influences the urban heat island effect. American Planning Association Journal 67:186–198.

Swenson, J.J., and Franklin, J. (2000) The effects of future urban development on habitat fragmentation in the Santa Monica Mountains. Landscape Ecology 15:713–730.

Tartaglia-Kershaw, M. (1982) The recreational and aesthetic significance of urban woodland. Landscape Research 7:22–25.

Taylor, A.F., Kuo, F.E., and Sullivan, W.C. (2001) Views of nature and self-discipline: evidence from inner city children. Journal of Environmental Psychology 22:49–63.

Taylor, A.F., Wiley, A., Kuo, F.E., and Sullivan, W.C. (1998) Growing up in the inner city: green spaces as places to grow. Environment and Behavior 30:3–27.

Thomas, V., Dailami, M., Dhareshwar, A., et al. (1999) The quality of growth. Ekistics 394:13–20.

Thompson, C.W. (2002) Urban open space in the 21st century. Landscape and Urban Planning 59:59–72.

Thompson, K., Austin, K.C., Smith, R.H., Warren, P.H., and Gaston, K.J. (2003) Urban domestic gardens I: putting local garden diversity in context. Journal of Vegetation Science 14:71–81.

Tyrväinen, L., Pauleit, S., Seeland, K., and de Vries, S. (2005) Benefits and uses of urban forests and trees. In: Konijnendijk, C.C., Nilsson, K., Randrup, T.B., and Schipperijn, J. (eds.) Urban Forests and Trees: A Reference Book. Springer, Berlin, pp. 81–114.

Ulrich, R.S. (1986) Human response to vegetation and landscapes. Landscape and Urban Planning 13:29–44.

United Nations. (2005) World Urbanization Prospects, the 2005 Revision. http://www.prb.org

United Nations Environment Program. (2005) Urban Environmental Accords, United Nations World Environment Day 2005. http://www.wed2005.org

Urban, J., Sieveert, R., and Patterson, J. (1988) Trees and space: a blueprint for tomorrow. American Forests 94(7/8):58–74.

Valk, A. van der (2002) The Dutch planning experience. Landscape and Urban Planning 58:201–210.

Vuilleumier S., and Prelaz-Droux, R. (2002) Map of ecological networks for landscape planning. Landscape and Urban Planning 58:157–170.

Wendell Cox Consultancy (2006) Demographia and the Public Purpose. Belleville, IL. http://www.demographia.com/db-intlua-area2000.htm

Wheater, C.P. (1999) Urban Habitats. Routledge, London.

Williams, K. (1999) Urban intensification policies in England, problems and contradictions. Land Use Policy 16:167–178.

Williams, N.S.G., McDonnell, M.J., and Seager, E.J. (2005) Factors influencing the loss of an endangered ecosystem in an urbanising landscape: a case study of native grasslands from Melbourne, Australia. Landscape and Urban Planning 71:35–49.

Wilson, E.O. (1984) Biophilia. Harvard University Press, Cambridge, MA.

Wong, N.H., and Yu, C. (2005) Study of green areas and urban heat island in a tropical city. Habitat International 29:547–558.

ZinCo GmbH (2000) Green roofs: Recommended Standards for Designing and Installation on Roofs: Planning Guide, 6th Ed. Zinco GmbH, Unterensingen, Germany.

10
Urban Ecology Studies in China, with an Emphasis on Shanghai

Yong-Chang Song and Jun Gao

Urban ecology as a new scientific discipline was derived from the convergence of ecology and urban science, and began to develop in the early 1970s. In 1971, the United Nations Educational, Scientific, and Cultural Organization (UNESCO) launched an international cooperative research program, Man and the Biosphere (MAB), which addresses the impact of increased human activity on the whole biosphere, current environmental pressures and resource shortages, and conducts a search for rational approaches and methodologies for managing the biosphere. Among the 14 research projects of the program, was the Ecological Prospects for Energy Utilization in Urban and Industrial Systems. This project brought about a great advance in the study of urban ecology worldwide. Around this period a number of studies were conducted in Brussels, Tokyo, Hong Kong, Frankfurt, Rome, Moscow, Berlin, and elsewhere, and the results were published broadly (Duvigneaud, 1974; Kunick, 1974; Numata, 1977; Vester and Hesler, 1980; Boyden et al., 1981; Giacomini, 1981; Bornkamm et al., 1982; Yanitsky, 1982; Bonnes, 1984). When the concept of urban ecology was introduced into China in the early 1980s, Chinese ecologists, economists, geography specialists, and scientists in urban planning were attracted by the new discipline and started studying this field in China from their different specialized perspectives. This chapter briefly reviews general trends as this discipline grew in China over the last 20 years, and details the current emphasis and new developments in urban ecological study in China, especially in Shanghai.

Study of Urban Ecology in Retrospect

The development of urban ecology studies in China over the last 20 years can be divided into three phases:

Starting Phase (1982–1990)

As soon as the idea of the urban ecology was introduced into China, the Shanghai Ecology Association started to deliberate on how to approach studies in this field.

M.M. Carreiro et al. (eds.), *Ecology, Planning, and Management of Urban Forests: International Perspectives.*

Later, the first national conference of urban ecology was held in Shanghai in December, 1984, and in 1986 a second national conference was held in Tianjing. During this period a number of studies emphasized clarifying and prioritizing objectives, goals, tasks, and methods in urban ecology. As a result, many key papers and theses were published (Song, 1983; Chen, 1987, 1988; Wang, 1988; Zhou, 1989; Zhou et al., 1990). In October 1987 an international symposium, Urban-Periurban Ecosystems Research and Its Application to Planning and Development, was organized by the MAB committee of UNESCO in Beijing, for the promotion of international cooperation and exchange in urban ecology studies in China. Even at the beginning of urban ecology studies in China the concept of a city as a social-economic-natural complex ecosystem (Ma and Wang, 1984) strongly influenced its development.

During this early period urban ecology projects in some cities were completed, including, for instance, an international cooperative project in Tianjin (Tianjin Municipal Bureau of Environmental Protection, 1988; Cooperative Ecological Research Project (CERP), 1995), Beijing, and a few cities in southern Jiangsu Province. In Shanghai, studies were completed on urban climate (Zhou and Zhang, 1985), urban soils (Wang, 1992), urban rodents (Zu and Zhou, 1990, 1991), urban birds (Shanghai Municipal Bureau of Environmental Protection, 1986), and biomonitoring of the urban environment (Song and Gu, 1988; Steubing and Song, 1991, 1993). Other integrated studies were also in progress on the complex ecosystem of an animal farm in Nanhui County, Shanghai (Zhou, et al., 1986), the complex ecosystem of the Changxing Islands (Song and Wang, 1991), and the aquatic ecosystem of Dianshanhu Lake and control of its eutrophication (Song and Wang, 1992; Song et al., 1992).

Growth Phase (1991–2000)

The development of urbanization in China reached full-speed in the early 1990s. The sustainable development principle was adopted in the declaration of the Rio 1992 convention on the earth' s development in the 21st century. As a result, the study of urban ecology in China advanced to a new stage. In 1992 an international symposium, Metropolitan Development and Ecology, was hosted in Shanghai with an emphasis on the exploration of the new Shanghai–Pudong district as one of the topics. Shortly thereafter, the soon-to-be mayor of Shanghai clearly advocated at the 1993 International Conference on the Water-Metropolis the goal of transforming Shanghai into an eco-city. At the same time a number of other cities also committed themselves to a similar objective.

Political commitments of adopting a path toward urban ecological sustainability are important. However, defining the fundamental characteristics of an eco-city can lead to much debate. What is an eco-city? There are different understandings domestically and abroad (Yanitsky, 1982, 1984; Wang and Lu, 1994; Wu et al., 2000; Register, 2002). From an ecological viewpoint, the city is an artificial terrestrial ecosystem that is dominated by human beings and influenced by human activities over long periods of time. Such a high and sustained degree of human impact over time results in changes in city structure, alterations in the circulation of material, and changes in pathways and efficiency of energy transfer. With this understanding in mind, we regard the eco-city to be an ecosystem of sound structure, efficient function, and harmonious relationships between people and nature. Sound structure refers to a moderate population density, proper use of land, high environmental quality, sufficient green space, functioning infrastructure, and effective protection of biodiversity. Efficient function refers to unblocked material flows, sufficiently recycled substances, greater energy efficiency, smooth and rapid transmission of information flows, and reasonable movement of people throughout the city. A harmonious relationship refers to the dynamic interaction between people and nature that leads to an ecologically, economically, and socially sustainable city, where resource use matches supply rate, where the environment is capable of dealing with stresses, and where good cooperation exists between urban and rural areas (Song, 1994; Song et al., 2000). In other words, the eco-city should be a settlement where inhabitants have ample opportunities to realize their individual potentials, where the physical and mental health of citizens and health of the environment are maximally protected, where resources are used efficiently, where technology is environmentally friendly, where increasing material recycling and natural cycles in the city are a major goal, and where the benefits of a city's geographic location can be optimized.

An eco-city evaluation system (Fig. 10.1) that includes three hierarchical levels of social, environmental, and economic data for calculating the Urban Quality Index (UQI) was established based on the above concepts and goals. The first level of factors consists of information dealing with a city's structure, function, and harmonious relationships. The second level of factors nested within the above three factors consist of 10 elements, with human population structure, city infrastructure, environmental quality, and green space making up the Structural Index; material cycling, resource supply, and production efficiency making up the Functional Index; and social guarantee, civilization, and sustainability making up the Harmonious Relationship Index (similar to the Quality of Life Index). The third level consists of 30 elementary indices that are nested hierarchically as shown in Figure 10.1. All of these factors are used to calculate the Urban Quality Index (UQI). The criteria for each elementary index in Shanghai have been based on data from cities that have been developing these criteria (Table 10.1), and from a corresponding method (Box 10.1) for assessing the characteristics of an eco-city (Song et al., 1999).

The evaluation system and assessment method were used to evaluate the situation in 1996 in Shanghai as well as the target for the program in 2010. In 1996 Shanghai achieved a UQI value of 0.371 (the highest value for the UQI is 1.0), while the program for the year 2010 received a UQI of 0.71. Where five assessment levels were used instead of the three described here (see Box 10.1), Shanghai in 1996 was given a grade of III, indicating that it was functioning at just an acceptable level ecologically. If the targets for 2010 are met, Shanghai would receive a grade of II, indicating that its ecological functioning and quality of life would improve (Song et al., 1999, 2000).

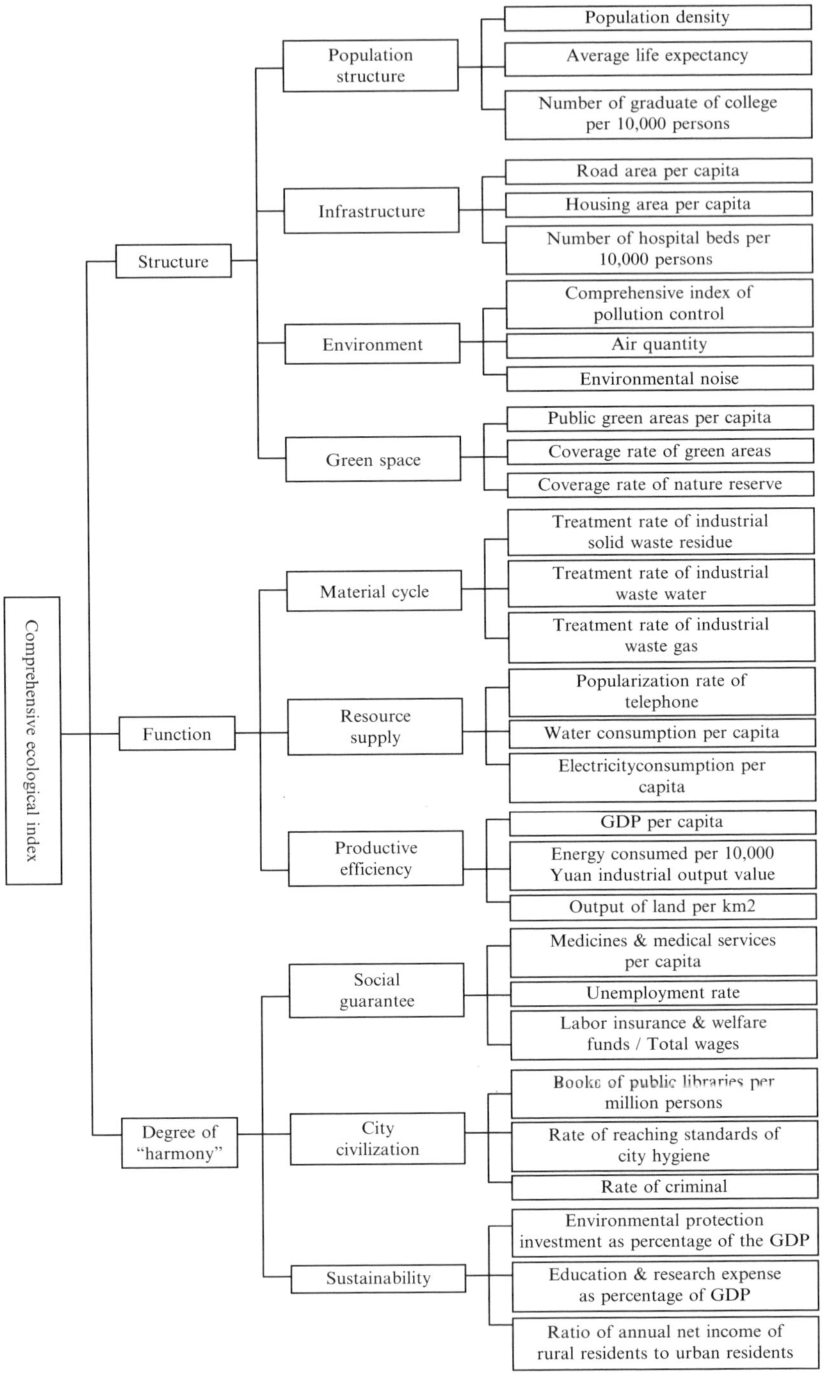

Fig. 10.1 Evaluation system for eco-city performance. GDP, gross domestic product

Table 10.1 Standard values for elementary indices for eco-city performance for Shanghai

Elementary indices	Standard value	Reference location	Shanghai in 1996 Actual value	Shanghai in 1996 Index value
Population density (persons /km^2)	3500	Average of West Berlin, Warsaw, and Vienna	4672	0.460
Average life expectancy (year)	78	Present value for Tokyo	76	0.734
Number of college graduates per 10,000 people	1180	Present value for Seoul	904	0.527
Road area per capita (m^2)	28	Present value for London	2.9	0.005
Housing area per capita (m^2)	16	Present value for Tokyo, Seoul	8.7	0.322
Number of hospital beds per 10,000 people	90	Present value for an advanced city in China (e.g., Taiyuan)	51.6	0.458
Comprehensive index of pollution control (ranges from 0 to 50 with 50 being the best)	Full mark is 50	Standard of National Environmental Protection Bureau, China	44.7	0.569
Air quality (SO^2 μg/L)	15	Present value for Shenzhen	53	0.466
Environmental noise [dB(A)]	<50	First class of national standard	67	0.165
Public green area per capita (m^2)	16	Maximal value for Chinese cities at present	1.9	0.006
% green area coverage in a city	45	Present value of Shenzhen	17	0.028
Nature reserve coverage (%)	12	Mid-target for eco-construction in China	0.15	0.001
Amount of industrial solid waste residue treated (%)	100	International standard	90.3	0.458
Amount of industrial waste water treated (%)	100	International standard	87.4	0.248
Amount of industrial waste gas treated (%)	100	International standard	94.6	0.207
Telephone ownership (sets/100 people)	76	Present value for Tokyo	30.1	0.202
Water consumption per capita (L/d)	455	Average for Tokyo, Hong Kong, Seoul, Taipei, Paris, New York	308	0.559
Electricity consumption per capita (kWh/d)	8	Average for Tokyo, Osaka, Hong Kong, Seoul, Taipei, Paris, Singapore	0.73	0.031
GDP per capita (¥*)	400 000	Present value for Tokyo	22275	0.027

(continued)

Table 10.1 (continued)

Elementary indices	Standard value	Reference location	Shanghai in 1996 Actual value	Index value
Energy consumed per 10,000 ¥ of industrial output value (in metric tons of coal)	0.5	Present value for Hong Kong	1.24	0.468
Value of land per km^2 (10,000 ¥)	70 000	Present value for Hong Kong	4577	0.053
Medicines and medical services per capita (¥)	21 00	Present value for Hong Kong and Guangzhou	401	0.114
Unemployment rate (%)	1.2	Lowest values in large cities in the world	2.7	0.083
Labor insurance and welfare funds/total wages (%)	50	Maximum value	49.8	0.996
Books in public libraries per 10,000 persons (volumes)	34 000	Present value of Tokyo, Seoul, Moscow	12672	0.227
Rate of reaching standards of environmental sanitation of city (%)	100	National standard	88	0.64
Crime rate (cases /10,000 persons)	0.05	Determined by the demand for societal security	0.07	0.984
Environmental protection investment as percentage of the GDP (%)	2.5	Determined by the existing value of a well-balanced city	2	0.742
Education and research expense as percentage of GDP (%)	2.5	Determined by the existing value of well-balanced city	1.93	0.718
Annual net income of rural residents/annual net income of urban residents (0–1)	1	Determined by the demand for reducing differences between rural and urban areas	0.67	0.049

*¥(RMB Yuan) is the Chinese monetary unit, 1 ¥≈ 0.125 U.S.$.
Source: Based on data in Song et al. (1999).

During this period, a number of urban ecology studies addressed the establishment of a sustainable urban ecosystem. This involved the ecological planning and design of new settlements, clean production technologies, sustainable development of industry, and environmental and ecological risk assessments during the construction of a city (Wang and Lu, 1994). In December 1997, the third national conference of urban ecology, Ecology of Sustainable Development of Cities and Towns, was held in Shenzhen and Hong Kong.

Box 10.1 Calculating the Urban Quality Index

The formula for calculating the elementary urban quality index (Q_i) is as follows:

$$Q_i = 1 - [(S_i - C_i)/(S_i - S_{min})]$$

or

$$Q_i = 1 - [(C_i - S_i)/(S_{max} - S_i)]$$

where Q_i denotes the elementary urban quality index, S_i is the baseline value of the elementary index corresponding to an established environmental standard, C_i is the current value of the elementary index for the selected city, S_{min} is the minimum value of C_i divided by 1.05, and S_{max} is the maximum value of C_i multiplied by 1.05. The first formula is used when greater values of C_i mean better urban quality (e.g., average life expectancy or per capita public green area), whereas the second formula is used when smaller values of C_i correspond to a better quality of urban environment (e.g., environmental noise or unemployment rate).

Q_i can be calculated in terms of a number of environmental, economic, and social variables (Table 10.1), and thus an overall measure of urban quality for a given city can be obtained by taking the average value of all individual Q_i:

$$V_i = \frac{1}{m}\left(\sum_{i=1}^{m} Q_i\right)$$

where V_i is the average urban quality index, and m is the total number of elementary urban quality indices considered.

At this conference progress and new findings in urban ecology research both domestically and abroad were summarized and exchanged, and future challenges and strategies discussed.

Urban ecology research in Shanghai during this period mainly focused on environmental planning, green-space development, and re-vegetation of waste disposal sites. Studies included environmental planning for the Waigaoqiao Free Trade Zone in Pudong (Wang, 1995, 1996), re-vegetation of a waste disposal site in Laogang (Hou, 1994; Song and Shi, 1995), and green space construction in the downtown area (Yan, 1998; Che, 2000). Aerial photography was used in the study of city greening to estimate coverage by green areas and their spatiotemporal dynamics (Zhou and Sun, 1995). The so-called "3S" technique (Remote Sensing, global positioning system (GPS), and geographical information system (GIS)) was used to analyze green areas using a landscape ecology approach (Gao and Wang, 2002). Additionally, we have been cooperating with the International Center of Ecology in Japan on restoration experiments involving natural vegetation in urban areas using the Miyawaki method (Miyawaki, 1993; Wang and Chen, 1999; also see Chapter 12). Within 3 years, two forest stands with near-natural vegetation have been built in the outer-ring forest belt and a nearby business site (Da et al., 2004).

Development Phase in the New Millennium (2001–Present)

At the beginning of the new century, studies in urban ecology have developed a full vitality from the south to the north of China. The 4th national conference of urban ecology, The New Challenge of Ecological City Construction in the 21st Century, was held in Zhuhai and Macao in December 2000. Two years later the 5th International Eco-City Conference was held in Shenzhen. These conferences provided a platform for direct discussion of common problems among colleagues at home and abroad, strengthened international academic exchanges, and advanced the development of urban ecology in China. The Shenzhen Declaration on Eco-City Development was adopted by the attendees of the 5th International Eco-City Conference. This declaration aimed to promote the construction of ecological cities, not only in China but also around the world.

In the meantime, several forums on urban ecological planning convened in Changsha, Beijing, Yangzhou, and other cities. In these forums, the scientists discussed the ecological challenges of rapid and dense urban construction at a large national scale. A series of papers and books about eco-city planning, regulation, and sustainable development of urban ecosystems, and construction of eco-cities were published. Based on the principles of the social-economic-natural complex, as well as ecosystem and traditional Chinese ecological concepts, Wang and his colleagues set forth a holistic approach to urban planning that integrates a highly efficient eco-industry, harmonious eco-culture, and eco-landscapes (Wang, 2003; Wang and Xu, 2005; Wang et al., 2000, 2001, 2004). Studies in urban ecology have now been extended to include applied research in developing ecological industries and a circular economy, where waste stream outputs from some industries are used as resources by others.

Current Emphases of Urban Ecological Studies in Shanghai

A new development plan for Shanghai city has been approved. The ambitious goal is for Shanghai not only to develop into an international center of economy, finance, trade, and shipping by 2020, but also to develop into an eco-city. In addition, an international fair will be held in Shanghai in the year 2010 with the theme of Better City, Better Quality of Life! All of these trends place more urgent demands urban ecological research programs.

Developing Ecological Regionalization and Implementing Urban Ecological Planning in Shanghai

Urbanization is developing very rapidly in Shanghai, whose urban area and boundary is continuously expanding. The level of urbanization in the Shanghai metropolitan area is estimated to reach 80% by the year 2010. This means 80% of the population of the Shanghai administrative area (6340.5 km^2) will live in a city and town. To avoid uncontrolled and unsustainable development, implementing an ecologically based urban plan is an important mission. As Sukopp and Wittig (1993) pointed out, ecological planning must connect and integrate the individual plans of special interest groups. Showing such interconnections is necessary in order for ecological requirements to be incorporated as a high priority in urban planning. Consequently the urban eco-planning of Shanghai will synthetically analyze and assess a number of individual plans that deal with distinct issues, such as plans for its population, environmental quality, green space, housing, and transportation, as well as larger scale social and economic development plans. By pointing out negative and positive ecological interrelationships and interactions that could result from implementation of individual plans, eco-planning can make recommendations that integrate overall urban structure, land use, and landscape development for meeting sustainable development targets for the city.

To make such an ecological urban plan a high-quality one, a comprehensive study of the city's eco-region must be conducted. To carry this out, the urban area will be divided into sectors on the basis of the structure, function, and development rate and type, and the relationships among these factors. For each sector, its natural conditions, eco-environmental sensitivity, eco-services, socioeconomic status, and development direction will be analyzed, and a strategy for protecting ecosystems, eco-environmental protection, and reasonable use of resources will be formulated.

Biotope Mapping and Monitoring Landscape Change During the Process of Urbanization

Biotope mapping (also called landscape mapping) is not only a cornerstone of urban ecological planning, but also a means of monitoring landscape dynamics. As an example of biotope mapping, we describe an urban-rural ecotone in southwest Shanghai. Aerial photos from 1984, 1989, and 1994 were used and combined with fieldwork to produce biotope maps of three stages in the city's development. These maps showed that the diversity and evenness of biotope types before major urban development in 1984 were higher than those in 1994 when urban development had advanced (Fig. 10.2). Due to urbanization, biotypes such as farmland, green space, and water-bodies had decreased greatly

Fig. 10.2 Biotope map of the urban-rural ecotone of southwest Shanghai (1984 and 1994) (revised from Gao, 2000, 2003)

and shifted further from the urban core. At the same time, biotope types such as factories, storage facilities, and residences had increased rapidly, and the development of large cultural and educational installations had stalled. As urban area expanded, biotope diversity and evenness decreased as agricultural biotopes were transformed into urban land. The rapid rate of urbanization over these 10 years indicates that urban-rural development in southwest Shanghai is rapid and not well planned (Gao, 2000; Gao and Song, 2003). At present Shanghai is facing a new round of rapid development. The area of factories, residences, and public facilities rose from 11.2% of Shanghai's total area (6340 km^2) in 1994 to 22% in 2000 (Li et al., 2004). To avoid unnecessary loss in quality of life, there is a pressing need to determine the objectives of such continuous development in this metropolis, to plan the position of each type of landscape, especially the distribution pattern of industry and residential settlement in the country, to pay more attention to water protection, to increase the construction of green space, and to make land-use development more ecologically sensitive. To meet these needs, a comprehensive biotope mapping of the entire Shanghai metropolis is urgently needed.

Constructing the Urban Forest, and Establishing a Greenbelt Network Connecting Urban and Rural Areas

One of the glaring problems that urbanization causes worldwide is the destruction of natural ecosystems and the segregation people from nature. Urban greening will be an important solution to this problem, owing to its immense value in restoring and preventing the destruction of natural ecosystems, decreasing the size of sealed ground surfaces, protecting biodiversity, improving urban eco-quality, and promoting eco-equilibrium. Finally, urban greening will permit the city to emerge with nature, and raise the quality of urban life.

In recent years Shanghai has made considerable gains in green-space expansion. The public green area per capita in the central districts of Shanghai rose from 4.6 m^2 in 2000 to 9.2 m^2 in 2003. The coverage rate of green space increased from 22.2% to 35.2%. A forest belt 99 km long and 100 m wide surrounding the city was completed in 2003. Landscaping in the city has been significantly improved, and the heat island in the central district was reduced (Li and Song, 2003). At present there are still some deficiencies in the greening activity of Shanghai. For example, greening has been limited to the central city (inside the outer-ring road); green space design gave preference to aesthetic plantings, while ignoring native species and zonal plant communities, and gave insufficient recognition to the growth needs of plants and to rules of plant community formation.

To raise greening levels and improve Shanghai's ecological efficiency, as well as to attain the goal of becoming an eco-city, a pilot project for urban forest construction in Shanghai was carried out in 2001 under the leadership of the Shanghai Municipal Agricultural Commission. Based on such factors as heat-island mitigation, improving the balance between carbon dioxide and oxygen, environmental protection, and the recreational needs of residents, we calculated that the urban forest area for Shanghai should be 2243 km^2. This extent of coverage means that 35% of the total area of Shanghai would be devoted to its urban forest. A forest network with the framework of two rings, eight lines, five zones, multi-corridors, multi-grids, and one chain was proposed (Fig. 10.3; also see Chapter 6). This means planting two ring-shaped forests, an inner ring 500 m wide by 97 km in length surrounding the central district, and an outer ring 180 km long in suburban land. In addition, the plan includes eight longitudinal forest belts 1000 m wide along expressways and major rivers, five large forest parks about 30 km^2 each in area scattered in the suburbs, multiple green corridors 25 to 500 m in width along smaller rivers and roadways, grids of protective shelter-belt forests along the seashore and in industrial areas, and one chain linking various stepping-stone habitats stretching from Hangzhou Gulf in the south to the Changjiang river bank in the north (Song et al., 2002). A year after this plan was proposed, a program entitled "Study of the Development of the Modern Urban Forest in Shanghai" has been approved by the Shanghai Municipal Government and Chinese Academy of Forestry. As a result, volumes of reports, including concepts, plans, techniques, management, and evaluation of urban forests, has been published (Peng, 2003). As this program was being enacted, an international symposium, Urban Forests and Construction of an

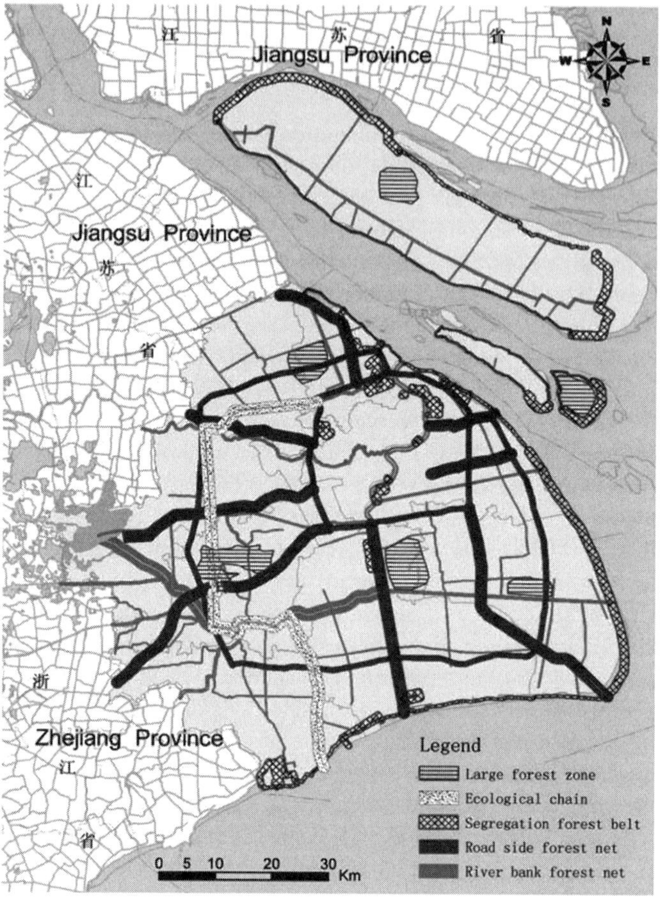

Fig. 10.3 Framework for the urban forest plan in Shanghai

Eco-City, was held in September 2002 in Shanghai. A national and international group of scientists and practitioners made many suggestions and comments on the plans for urban forest construction in Shanghai. At present the construction of a suburban forest park is being carried out by the Shanghai Landscape Administration Bureau.

Strengthening Water Environmental Protection, Rational Use of Water Resources, and Restoration of the Aquatic Ecosystem of Shanghai

Shanghai is located along the ocean at the mouth of two major rivers. The Yangtze River, Huangpu River, and Suzhou Creek flow through Shanghai, with crisscrossing streams forming an extremely dense river network. The river network density is 3.4 km/km^2, and the water surface area makes up 7.75% of the total area of Shanghai. More than 88% of the rivers and creeks are heavily polluted due to dumping of huge amounts of wastewater from industrial, human, and livestock sources. Eutrophication of Dianshan Lake, Shanghai's drinking water source, is still increasing. The Huangpu River from Dianfeng to Wusongkou has a length of 113.4 km, and while the water quality of its upper reaches can meet potability criteria, the middle and lower reaches are seriously polluted. Suzhou Creek, a major branch of the Huangpu River, runs through Shanghai and has 53.1 km of channels, 23.8 km of which occur in the central city. Due to industrial development and the rapid population increase along the creek since the 1950s, the pollutant inputs have risen several times, so that by the 1970s it became a very heavily polluted, open sewer.

Engineering projects to restore the Huangpu River have been carried out since the 1980s. An artificial conduit for wastewater disposal was completed in 1993. It intercepted 1.4 million km^3 of wastewater per day originally flowing from the center of the city to the Suzhou Creek. Consequently water quality has improved. In 1996 progress was made on an integrated engineering plan. Because of the interception of wastewater, construction of floodgates, regulation of the water flow, aeration, and sediment dredging in upper reaches, the concentration of DO, BOD_5, and NH_4 in the water were reduced, water transparency has risen, and the "black and stink" in stretches of urban areas has disappeared for the most part. However, organic pollutants remain high, and water quality in stretches running through urban areas was generally worse than the fifth grade of the surface-water criterion (according to National Surface Water Criterion GHZB 1-1999). The community composition of aquatic species was also depauperate with both biodiversity and biomass being very low (Department of Environmental Science, East China Normal University, 2002).

The planktonic community was dominated by small and pollutant-tolerant species. The benthos also had low biomass and was dominated by one species of pollutant-tolerant Oligochaete worm. Some macro-aquatic plants were limited only to the upstream reaches. There were very few fish species with only some pollutant-tolerant species in the surface waters. The decomposition function of the microorganisms in the water and the sediment was quite insufficient. There was a large amount of polluted sediment with an average thickness of approximately 1.5 m with a heavy metal content was rather high and toxic to organisms. Branches of Suzhou Creek were heavy polluted too (Department of Environmental Science, East China Normal University, 2002). Therefore, the restoration of Suzhou Creek has become the most important and difficult environmental protection issue for Shanghai.

The natural river system and channel network has also been destroyed by urban expansion; the percentage of surface waters in Shanghai decreased from 11.1% in 1980 to 7.75% in 2001. According to the plan of the Shanghai Municipal Bureau of Hydrology, the percentage of water surface will increase to 9% in 2010, and will reach 10% in 2020. If this plan is to be practical, it is necessary to study the whole water system ecologically, and combine the construction of both the water and forest networks together. Additionally, because surface water is seriously polluted, some factories and enterprises excessively extracted ground water, which has led to subsidence problems. Therefore, the conservation of the aquatic environment, reasonable use of water resources, and optimization of the aquatic ecosystem are among the main tasks for transforming Shanghai into an eco-city.

Developing Eco-Industries and Stimulating the Construction of Ecological Industrial Parks

Wastes discharged during production processes are the main sources of urban pollutants. Therefore, if an eco-city is to be constructed, waste streams must be reduced at the point of production. Clean production technologies should be encouraged, and at the same time the development of ecological industrial parks should be promoted. Ecological industrial parks would contain factories of different businesses that can be networked into an integrated production system. The material flows among and within factories should be organized to reduce the waste stream. Waste needs to be assimilated as soon as possible, and recycling processes implemented. Construction of industrial parks demands the participation of entrepreneurs and relevant institutions to identify and optimize the different types of factories that can co-occur in these parks to promote such a mutual symbiosis among businesses that reduces waste. (For example, the famous eco-industrial park in Kalundborg, Denmark, is a good example (http://www.symbiosis.dk); also see models in China (http://www.chinacp.com/eng/cpcasestudies/ce_cases.html). The government has to create information systems, promote renovation plans, and provide technological support for developing more industrial parks throughout the city and country.

Stressing the Construction of Eco-Communities, Developing Eco-Buildings, and Calling for an Eco-Morality

The residential community is a fundamental element of a city, a place where people live and socialize. The construction of an eco-community is directly concerned with the improvement of the living environment and quality of life of its residents. This task has two aspects. One involves the establishment of material and spiritual welfare, encompassing areas such as housing, green space, environmental hygiene, and service facilities. Another covers matters such as ecological education, promoting an ecological lifestyle, inheriting and developing traditional culture, personal participation and cooperation in social activities, and the greater social welfare. The major purpose of developing eco-communities is to strengthen the roles of natural and social features, thus combining development and management from an ecosystem point of view. The ultimate goal is to create human communities that are healthy and safe and ecologically aware.

The Prospects for Urban Ecology Studies in China

The importance of urban ecology lies in its practice. Its study should be closely related to the establishment of an eco-city, based both on theoretical guidelines and practical procedures. Besides the above-mentioned projects, urban ecological studies in China need to enhance research in the following areas.

Exploration of the Optimum Models of Urban Development

Continued urbanization is an inevitable trend as human societies grow and develop. At present, approximately 50% of the population of the world lives in cities; in developed countries this percentage can exceed 75%. Undoubtedly, the rate of urbanization will continue to accelerate, and therefore the well-being of most of the world's people will depend on conditions in our cities. Urban ecology studies must strengthen research on the influence of urban size and structure on resource use and the state of the environment so as to establish a model of their relationships, and to determine optimal urban size. Such information will establish the boundary conditions for the possible realization of an eco-city in different geographic and climatic contexts and stages of socioeconomic development.

Searching for Harmonious Development Models Between a City and Its Region

A city is closely connected to its surroundings economically, socially, and ecologically. The biological productivity and environmental conditions of its surroundings are the foundation for existence and development of a city. The lure and attraction of the city to the surrounding people is inevitable, so a metropolis will doubtlessly affect its surroundings. For instance, Shanghai is the head of the Yangtze economic zone and also the center of the city group of the Yangtze Delta. Its development is closely related to that of seven provinces, especially Jiangsu and Zhejiang. It will be affected by all the construction in its outer reaches, especially by a number of important engineering projects now taking place. At the same time environmental change in Shanghai will probably have a reciprocal effect on these surrounding areas. Therefore, studies of the interaction between a city and its surrounding region, including how resources and the environment change during urbanization, are absolutely necessary to investigate pathways toward harmonious development between the city and surrounding regions that will safeguard regional eco-security.

Study of the Metabolism of Energy and Material in Urban Ecosystem

Energy flows and material cycles are essential aspects of all urban ecosystems. Their strengths and patterns are the main indices for measuring the development level of a city. Studies involving energy and resource use serve as the basis for establishing a model of energy flow and material cycling and assist in finding ways to save energy, water, and other resources. Through such studies one can better understand the basic characteristics and dynamics of urban ecosystems, promote the efficiency of the city, and raise the level of services it provides.

Studying the Influence of Urbanization on Global Change and Its Responses

The city is one of the main factors causing global warming and climate change. Approximately 80% of anthropogenic CO_2 comes from cities (Intergovernmental Panel on Climate Change, 1995). Vehicle exhaust leads to heat islands in big cities, causes smog, initiates thunderstorms, and reduces productivity of terrestrial ecosystems (Herring, 1996). Additionally, climate change will threaten the development of most cities, particularly those located on coasts and deltas, such as Shanghai and New Orleans. Research on the influences of and responses to urbanization during global change is related not only to a city's security, but also to the orientation and the manner of city development in the future.

Studying the Influence of Urbanization on Biodiversity

Urbanization is one of the main reasons for the loss of biodiversity. Urbanization brings about the elimination of habitats, and makes some species disappear completely. Furthermore, some exotic species introduced by urbanization may lead to biological invasions into rural areas, causing local species loss. According to the floristic statistics of Shanghai, in 1959 there were 1719 plant species (including infraspecific taxa), among them 590 species of wild plants (Xu, 1959). Forty years later the number of species has increased to 2296 (including infraspecific taxa), but the wild plants have decreased to 500 species (Science and Technology Academy of Shanghai, 1999). Additionally, environmental pollution and other changes in living conditions in the city may also cause genetic changes in living species populations (Prus-Glowacki, 1999; Cook, 2000; Chen et al., 2003). Therefore, research on the influence of urbanization on biodiversity and potential ramifications for native biodiversity loss are important tasks for both conservation and urban ecology.

References

Bonnes, M. (1984) Mobilizing scientists, planners and the local community in a large-scale urban situation: the Rome case study. In: Di Castri, F., Baker, F.W.G., and Hadley, M. (eds.) Ecology in Practice, Dublin, UNESCO. Tycooly International Publishing, Paris, pp. 53–61.

Bornkamm, R., Lee, J.A., and Seaward, N.R.D. (1982) Urban Ecology. The Second European Ecological Symposium. Berlin, September 1980. Blackwell Scientific Publications, Oxford/London/Edinburgh/Boston/Melbourne.

Boyden, S., Millar, S., Newcombe, K., and O'Neill, B. (1981) The Ecology of a City and its People: the Case of Hong Kong. Australian National University Press, Canberra.

Cooperative Ecological Research Project (CERP). (1995) Towards A Sustainable City—Methods of Urban Ecological Planning and Its Application in Tianjin, China. Berlin: Urban System Consult GmbH.

Che, S.-Q. (2000) The Study on the Green Space Landscape Ecology of Shanghai City. Doctoral dissertation, the Department of Environmental Science, East China Normal University (in Chinese with English summary).

Chen, Y.-Q. (1987) Urban economics. In: Xu, T.X. (ed.) Eco-Economics. Hangzhou: Zhejiang People's Press, pp. 314–327 (in Chinese).

Chen, Y.-Q. (1988) Theory and Practice of Urban Economy. Shanghai: Shanghai Social Academy Press (in Chinese).

Chen, X. Y., Li, N., and Shen, L. et al. (2003) Genetic structure along a gaseous organic pollution gradient: a case study with *Poa annua* L. Environmental Pollution 124:449–455.

Cook, L.M. (2000) Changing views on melanic moths. Biological Journal of the Linnean Society 69:431–441.

Da, L.-J., Yong, Y.-C., and Chen, M. (2004) The method of ecological greening and its application in the construction of the approaching nature plant community in Shanghai. Chinese Landscape Architecture 3:38–40 (in Chinese with English summary).

Department of Environmental Science, East China Normal University. (2002) Ecological Study on Sediment Dredge of Suzhou Creek. The Final Report, unpublished (in Chinese).

Duvigneaud, P. (1974) L'ecosysteme "Urbs." Mem. Society of the Royal Bot. Belg. Memoire. Societe. Royale. Botanique. Belgique. 6:5–35.

Gao, J. (2000) Studies on landscape ecology in the urban-rural ecotone in the southwest area of Shangai. Doctoral dissertation, the Department of Environmental Science, East China Normal University (in Chinese with English summary).

Gao, J., and Song, Y.-C. (2003) On landscape dynamics of the urban-rural ecotone based on remote sensing and GIS: a case study of southwest Shanghai. Acta Ecologica Sinica 3(4):805–813 (in Chinese with English abstract).

Gao, J., and Wang, Y.-Q. (2002) Remote sensing and GIS based urban forest ecological analysis and assessment. In: He X.-Y., and Ning, Z.-H. (eds.) Advances in Urban Forest Ecology. Beijing: China Forestry Publishing House, pp. 269–276 (in Chinese with English abstract).

Giacomini, V. (1981) Rome considered as an ecological system. Nature and Resources 17(1):13–19.

Herring, D. (1996) Interaction of smoke, clouds, and radiation over Brazil (SCAR-B). Earth Observer. 18(2). http://eospso.gsfc.nasa.gov/eos_observ/3_4_96/p46.html.

Hou, C.-R. (1994) Distribution and translocation of heavy metal in soil-plant system in Laogan garbage disposal site of Shanghai. Master's thesis, Department of Environmental Science, East China Normal University (in Chinese with English summary).

Intergovernmental Panel on Climate Change. (1995) IPCC working group I summary for policy-makers. Cambridge University Press, Cambridge, UK.

Kunick, W. (1974) Veränderungen von Flora und Vegetation einer Groβstadt. Dargestellt am Beispiel von Berlin. West. Diss. Technishe Universtät, Berlin.

Li, J.X., and Song, Y.C. (2003) Correlation between land surface temperature and rate of green space coverage in central area of Shanghai. Shanghai Environmental Science 22(9):599–601.

Li, J.X., Wang, Y.-J., Shen, X.H., and Song, Y.C. (2004) Landscape pattern analysis along an urban-rural gradient in the Shanghai metropolitan region. Acta Ecologica Sinica 24(9):1993–1980 (in Chinese with English abstract).

Ma, S.-J., and Wang, R.-S. (1984) Social-Economic- Natural Complex Ecosystem. Acta Ecological Sinica 4(1):1–9 (in Chinese with English abstract).

Miyawaki, A. (1993) Restoration of native forests from Japan to Malaysia. In: Lieth, H., and Lohmann, M. (eds.) Restoration of Tropical Forest Ecosystems. Kluwer Academic Publishers, Dordrecht, pp. 5–24.

Numata, M. (ed.) (1977) Tokyo Project: Interdisciplinary Studies of Urban Ecosystems in the Metropolis of Tokyo. Chiba University.

Peng, Z.H. (2003) Development of Modern Urban Forest in Shanghai. Chinese Forestry Press, Beijing (in Chinese).

Prus-Glowacki, W., Oleksyn, J., and Reich, P.B. (1999) Industrial pollutants tend to increase genetic diversity: evidence from field-grown European Scots pine populations. Water Air and Soil Pollution 116:395–402.

Register, R. (2002) Eco-cites: Building Cities in Balance with Nature (in Chinese, translated by Wang and Hu). Social Science Literature Press, Beijing.

Science and Technology Academy of Shanghai. (1999) The Plants of Shanghai,Vol.1 and 2. Shanghai Scientific and Technological Literature Publishing House, Shanghai.

Shanghai Municipal Bureau of Environmental Protection. (1986) Protected Birds in Shanghai. Xuelin Press, Shanghai (in Chinese).

Song, Y.-C. (1983) A brief description of urban ecology. The Studies of Environmental Science 1:1–13 (in Chinese).

Song, Y.-C. (1994) Construction eco-city toward to 21 century. Science and Technology of Shanghai Construction 13-14 (in Chinese).

Song, Y.-C., and Gu, Y.-J. (1988) Monitoring air pollution with heavy metal contents in trees. Urban Environment and Urban Ecology 1:34–38 (in Chinese with English abstract).

Song, Y.-C., and Shi. G. (1995) Study on ecological planning and re-vegetation for Laogan garbage disposal site of Shanghai. The final report, 43 pp. (in Chinese).

Song, Y.-C., and Wang, Y. (eds.) (1991) Study on the complex ecosystem of Changxing Island. East China Normal University Press, Shanghai (in Chinese with English abstract).

Song, Y.-C., Qi, R.-H., You, W.-H., Wang, X.-R., and Zhu, L.-B. (1999) A study on indices system and assessment criterion of eco-city. Urban Environment and Urban Ecology 12(5):16–19 (in Chinese with English abstract).

Song, Y.-C., You, W.-H., and Wang, X.-R. (eds.) (2000) Urban Ecology. East China Normal University Press, Shanghai (in Chinese).

Song, Y.-C., Li, J.-X., Da, L.-J., and Qi, R.-H. (2002) Preliminary study on the establishment of the urban forests in Shanghai. In: He, X.-Y., and Ning, Z.-H. (eds.) Advances in Urban Forest Ecology. China Forestry Publishing House, Beijing, pp. 277–282 (in Chinese with English abstract).

Song, Y.-C., Wang, Y., and Qi, R.-H. (eds.) (1992) Study on the ecosystem and eutrophication of the Dianshan Lake. East China Normal University Press, Shanghai (in Chinese with English abstract).

Steubing, L., and Song, Y.-C. (1991) Umweltbelastungen in der Stadt Shanghai unter besonderer Berucksichtigung der lufthygienischen Situation. Forum Stadte-Hygiene 42:311–314.

Steubing, L., and Song, Y.-C. (1993) Schwermetalle in Nahrungsnetz in der Region Pudong (China). Phytocoenologie 23:269–276.

Sukopp, H., and Wittig, R. (1993) Stadtökologie. 2. Auflage. Gustav Fischer, Stuttgart.

Tianjin Municipal Bureau of Environmental Protection. (1988) Urban Ecosystem and Comprehensive Prevention of the Pollutions. Chinese Environmental Science Press, Beijing (in Chinese).

Vester, F., and von Hesler, A. (1980) Sensitivitäsmodell. Regionale Planungsgemeinschaft Untermain, Frankfurt am Main.

Wang, R.-S. (1988) High Efficiency and Harmonious Relationship: The Principles and Methodology of Urban Ecological Regulation. Hunan Educational Press, Changsha (in Chinese).

Wang, R.-S. (ed.) (2003) Complex Ecosystem and Circular Economy. Meteorology Press, Beijing (in Chinese).

Wang, R.-S., Chi, J., and Ouyang, Z.-Y. (2001) Eco-Integration for Sustainable Development of Middle and Small Sized Towns. Meteorology Press, Beijing (in Chinese).

Wang, R.-S., Hu, D., Wang, X.-R., and Tang, L.-J. (2004) Urban Eco-Service. Meteorology Press, Beijing (in Chinese).

Wang, R.-S., and Lu, Y.-L. (1994) Urban Ecological Development: Research and Application. China Environmental Science Press, Beijing.

Wang, R.-S., and Xu, H.-X. (2005) A Comprehensive Approach for Yangzhou Eco-City Development. China Science and Technology Press, Beijing (in Chinese with English summary).

Wang, R.-S., Zhou, Q.-X., and Hu, D. (2000) Methods of Regulation and Control of Urban Ecosystem. Meteorology Press, Beijing (in Chinese).

Wang, X.-H., and Chen, X.-Y. (1999) Application of Miyawaki method to the construction of Shanghai Ecological Environment. Shanghai Environmental Sciences 18(2):100–104 (in Chinese with English abstract).

Wang, Y. (ed.) (1992) Background value of soil of Shanghai. Chinese Environmental Science Press, Beijing (in Chinese).

Wu, R.-J., et al. (2000) Principles and Approaches to Eco-City Construction—Analysis Concurrently on the Status and Future of Shanghai. Fudan University Press, Shanghai (in Chinese).

Xu, B.-S. (1959) Plants Name List of Shanghai. Shanghai Science and Technology Press, Shanghai.

Yan, L-.Z. (1998) On the sustainable development of greening in Shanghai of 21th Century. Chinese landscape Architecture 14(56):44–46 (in Chinese).

Yanitsky, O.N. (1982) Towards an eco-city: problem of integrating knowledge with practice. International Social Science Journal 34(3):469–480.

Yanitsky, O. (1984) Integration of social and natural sciences for urban planning. In: Di Castri, F., Baker, F.W.G., and Hadley, M. (eds.) Ecology in Practice, Dublin, UNESCO. Tycooly International Publishing, Paris, pp. 30–46.

Zhou, J.-H., and Sun, T.-Z. 1995. Study on remote sensing model of three dimensional green biomass and the estimation of environmental benefits of Greenery. Remote Sensing of Environment 10(3):162–174.

Zhou, J.-L. (ed). (1989) Eco-Economic System of Urban and Rurality. Chinese Environmental Science Press, Beijing (in Chinese).

Zhou, J.-L., Wang, R.-S., and Zheng, S.-Z. (eds.) (1990) Study Method and Case Studies of Urban Eco-Economy. Fudan University Press, Shanghai (in Chinese).

Zhou, J.-L., Wu, R.-J., and Li, S.-F. (1986) The study on the comprehensive management and sustainable development mode of agro-stock-fishery—development of aquatics of Nanhui, Shanghai. Rural Eco-Environment 1:1–5 (in Chinese).

Zhou, S.-Z., and Zhang, C. (1985) An Introduction of Urban Climatology. East China Normal University Press, Shanghai (in Chinese).

Zu, L.-B., and Zhou, Y.-L. (1990) Ecological characteristics of the remnant rodents in urban area, I. Study on community characteristics of urban remnant rodents. Journal of Biology and Control of Chinese Intermediary Agent 1(6):350–353 (in Chinese with English abstract).

Zu, L.-B., and Zhou, Y.-L. (1991) Ecological characteristics of the remnant rodents in urban area, II. Study on community niche of urban remnant rodents. Journal of Biology and Control of Chinese Intermediary Agent 2:(1)26–28 (in Chinese with English abstract).

11 Using the Urban–Rural Gradient Approach to Determine the Effects of Land Use on Forest Remnants

Margaret M. Carreiro

The Creation and Ecological Importance of Urban and Suburban Forest Remnants in the Eastern United States

During the past 50 years, urban and suburban areas in the United States have been expanding rapidly at the expense of agricultural land and natural ecosystems (Richards, 1990; Douglas, 1994). Between 1960 and 1990, 12.6 million hectares (ha) of cropland, forest, and pasture in the United States were converted to urban and suburban land (Frey, 1984; Dougherty, 1992). Urban development rates have been accelerating, as evidenced by the fact that 4.5 million ha of rural lands were developed in the 5 years between 1992 and 1997 (United States Department of Agriculture National Resources Inventory, 2001). While cities and towns now cover 3.5% of the lower 48 states in the U.S., their associated sprawl into adjacent counties, designated as metropolitan areas, has resulted in 24.5% of the area of the U.S. being categorized as urban (Dwyer et al., 2000). Currently, 80.3% of the U.S. population of 300 million lives in cities and their surrounding suburbs (U.S. Census Bureau, 2001).

Urban growth provides many benefits to people by bringing economic and cultural vitality to a region. However, the landscape modification and pollution that follow urbanization profoundly alter the organisms and ecological functions occurring on land and in water bodies, like streams, that drain the land. As stress on our natural environment increases, we lose the ecosystem services they once provided free of charge to society (air and water purification, soil stabilization, flood control, meso- and microclimate modification, pest and disease control) (Daily et al., 1997; Millennium Ecosystem Assessment, 2005; Farber et al., 2006; see also Chapter 5). Often these ecological services must then be replaced with expensive engineered solutions to reduce negative impacts on people's physical and psychological well-being (Farber, 2005). To decrease the unintended negative consequences of city growth, there needs to be increased recognition among planners and policy makers that natural habitats and other green spaces in metropolitan areas are vital infrastructural components of a functioning urban ecosystem, as vital to our quality of life as our engineered infrastructure (Alberti et al., 1999; see also Chapter 2). There is an obvious and pressing need to find

M.M. Carreiro et al. (eds.), *Ecology, Planning, and Management of Urban Forests: International Perspectives.*

integrated ways to work "with" instead of "against" nature to create cities that are more resource efficient and more resilient to fluctuations in economic and environmental forces, and that provide healthier conditions so people can not only exist but also thrive (Grimm et al., 2000). While some solutions are useful for many cities, others must be tailored to the geographic location, history, culture, and resources of each individual city. Indeed, many chapters in this book offer examples of such general as well as locally tailored solutions and plans.

Where possible, one component of these holistic strategies should be preventing the destruction of natural habitat remnants in and around cities, so that these remnants can continue to provide ecological benefits to the human community. This may be difficult in many of the world's cities, where occupation of the land by dense human settlements and associated agricultural systems has been long and continuous. In such cases, natural habitat remnants within an urban–suburban matrix may be few to nonexistent. However, in many regions of the U.S. where cities are relatively young compared to those in Europe and Asia, this preservation strategy is still tenable within an urbanizing context. For instance, in the eastern third of the country, human settlements are expanding mostly into forested land (Dwyer et al., 2000). Between 1990 and 1996 the states that experienced the greatest population growth per unit land area have occurred in the east (Dwyer et al., 2000), demonstrating that urban intensification is also part of the expansion process. In the eastern third of the U.S., temperate deciduous forests cover much of the northern and northeastern states, while deciduous and coniferous pine forests dominate in the southeast. These eastern forests have actually been expanding over the last 100 to 150 years, particularly in the northeast, due to initiation of secondary succession processes after farmland abandonment in the 19th and early 20th centuries (Foster, 1993). Subsequent urban expansion and a complicated system of public and private land ownership has resulted in many forest fragments of different sizes being created in urbanizing landscapes in the eastern U.S. (Medley et al., 1995; Vogelmann, 1995; Heilman et al., 2002). Since many of these forests are not intensely managed, the suburban and urban conditions that surround them will play a large role in determining their long-term condition and species composition, and consequently their level of ecosystem service support to society.

Therefore, posting signs declaring these areas "saved" from development will not guarantee the long-term existence of these habitats in a reasonably healthy state. These habitats consist of living organisms, not static museum specimens, and each species population reacts to the physical, chemical, and biotic conditions changing around it as urban sprawl approaches and intensifies and if climate changes rapidly. In addition, if these areas are parks, the quality of these habitats will also respond to changing patterns and intensities of public use over time. Early signs of undesirable change and degradation must be detected before damage becomes large and more expensive to repair and restore. To maintain high ecological quality, natural areas in cities will need to be actively and intelligently managed, using adaptive strategies that are based on scientific information collected from the local environments themselves. This requires greater inter- and transdisciplinary interactions among urban planners, environmental managers, and social and ecological scientists than has generally occurred in the U.S. until now (Daly and Klemens, 2005; Wilkinson et al., 2005). The reasons for this are many, involving separate educational training and the evolution of different professional cultures that are not normally motivated to seek each other out for information and advice during the land-use planning process or for conducting research. Therefore, until the last decade, research opportunities in these urban habitats have not often been sought by academics, while urban practitioners in the U.S. more often seek specific kinds of scientific expertise from consulting firms, rather than from academia. Greater dialogue among these groups would reap benefits from both an applied management and basic research perspective.

In addition to serving the pragmatic need for managing natural remnants sustainably, research in natural remnants surrounded by cities and suburbs offer the academic and scientific community places to explore how varying land-use contexts might affect the biotic structure and ecosystem functions of native remnants such as forests (McDonnell et al., 1997; Carreiro and Tripler, 2005). As forests become fragmented and their edge-to-interior ratios increase relative to their original condition, they may become exposed to greater inputs of energy, matter, and species from their altered surroundings (Saunders et al., 1991; Matlack, 1993a). External influences on these ecological systems may become as important as, if not more important than, internal regulation in defining their ecological communities. This may be especially true for forest remnants that become surrounded by cities, since urban land cover properties (e.g., impervious surfaces) and activities (e.g., fossil fuel combustion) contrast greatly with those of forests. Compared with rural forests, these urban forest remnants would likely be subjected to greater external inputs of thermal energy (urban heat-island effect; Oke, 1995), greater inputs of injurious as well as beneficial pollutants (e.g., O_3 vs. NO_3^-) (Turner et al., 1990; Lovett et al., 2000), increased colonization by exotic species (Kowarik, 1990; Rebele, 1994; McKinney, 2002), and increased human visitation (sociological phenomena such as recreation and suburban residential activities at forest boundaries; Matlack, 1993b). If society sets management goals of maintaining urban and suburban forests for their value in providing (1) ecosystem services for the public, (2) sites for ecological education and experiences with nature, (3) recreational opportunities, and (4) reinforcement of a cultural identity with the original native landscape, then it becomes important to evaluate the extent to which an urban land matrix affects ecosystem processes and species communities in these remnant forests.

The rest of this chapter describes one approach for increasing our understanding of how landscape context may affect native remnants "stranded" within urban and suburban areas. While the focus here is on native forest remnants, the approach can be extended to understanding the impact of land-use context on any patch type, be it a natural remnant of any kind (e.g., grassland), a restored near-natural habitat (see Chapter 26), or a human-created one, like a lawn. Therefore, unlike most of the other chapters in this book, which involve structural descriptions and past or planned modifications of entire urban–suburban landscapes and their vegetation, this chapter focuses specifically on describing one approach for determining the effects of varying intensities of human settlement on one type of patch (native forest remnants) within the entire landscape. The focal variable described in this chapter is the availability of nitrogen (N) to such forest remnants, because it is an important nutrient for forests, affecting plant species composition and ecosystem productivity (Chapin et al., 2002), and because sources of N are both internal and external to a particular ecological patch. Patterns of atmospheric deposition and soil N cycling in forest remnants along urban–rural gradients in two metropolitan areas in the U.S. are reported, allowing comparisons between two cities. In addition, results from one of these gradient studies is used to demonstrate the need to go beyond descriptive correlations with land use if we are to identify the proximate factors, affected by land use, that directly control nutrient availability or other ecological response variables in forests surrounded by varying degrees of urbanized land.

The Urban-Rural Gradient Approach

A great deal of the credit for the recent revived interest in urban–suburban areas among ecologists in the U.S. can be traced to the ideas laid out by Mark McDonnell and Steward Pickett (1990) in a conceptual paper in the journal *Ecology*, and in a later book, *Humans as Components of Ecosystems* (McDonnell and Pickett, 1993). In these and subsequent publications (McDonnell et al., 1997; Pickett et al., 1997), they and their colleagues introduced and developed the concept of the urban-to-rural land-use gradient and suggested that researchers could adopt the broader, ecological gradient paradigm (Bray and Curtis, 1957; Whittaker, 1967; Greig-Smith, 1983; Ter Braak and Prentice, 1988) to structure their studies of urban effects on nature (McDonnell et al., 1993). During the conceptual development of gradient analysis, distinctions have been made between simple and complex and between direct and indirect gradients. McDonnell et al. (1993) provide detailed examples and distinctions among these types of gradients that are especially important to consider when conducting studies within spatially and temporally complex urban environments.

To summarize their ideas briefly, a simple gradient involves detecting and identifying continuous change in a single factor across the environment. If the environmental gradient is judged to be large, such as soil moisture content from the top to the bottom of a hill slope, then it could be hypothesized as being an important determinant of an ecological response variable, such as abundance of a particular plant species along that hill slope. In this case, univariate statistics are used to relate variation in species abundance to variation in soil moisture and determine that factor's power to explain or predict the response variation. However, most ecological gradients are complex rather than simple. For example, while moisture may vary with placement along a hill slope, soil type (toposequence), soil organic matter content, and herbivore abundance are likely to co-vary as well. These multiple variables then constitute a complex gradient, and their interactions might then improve our ability to predict the abundance of a particular plant species along that hill slope. Of course, strong correlations of either single or multiple independent factors with a response variable may or may not indicate true causation by any or all of the factors. Experiments, where single or multiple factors are varied, and the degree of response by the dependent variable is measured, are used to invoke mechanistic causation.

Direct gradients, either simple or complex, occur contiguously in real space. The soil moisture gradient on this hypothetical hill slope, therefore, is both simple and direct. Complex gradients can also occur along continuous, uninterrupted spatial transects. Elevation gradients on mountains are often used as examples of complex, direct gradients. In this case, elevation can also serve as a surrogate or "dummy" variable that integrates multiple factors that change along with elevation, such as soil type, temperature, precipitation, radiation, and pressure. Ecological responses like plant community composition can be related in a simple way to elevation gradients, because they respond not to elevation in and of itself but rather to the multiple factors that co-vary with elevation up a mountain. In contrast, indirect gradients do not occur in a linear, continuous fashion in real space. Often multivariate ordination techniques are used to order response variables, like species in a community, graphically in ordination space. The axes of these graphs imply the existence of either simple or complex gradients in factors that may explain the differences among these responses. However, those gradients are often not obvious and do not exist in an unbroken line across real space. The "ordering" of sites according to changes in level or intensity of single or multiple independent variables is manipulated during the analysis and graphic display of results. Again, measurement of suspected causal factors and manipulative experimentation is needed to reveal the underlying reasons for the abstract order of the ecological responses.

McDonnell et al. (1993) made the case that using the concepts and tools of gradient analysis could make the ecological study of cities more tractable. Since many older cities in the United States have grown more or less concentrically from an urban center, factors like human density, extent of impervious land cover, road density, traffic volume, fossil fuel use, and industrial complexes tend to increase as one moves closer to the urban core. As a consequence, natural habitat remnants in such a landscape become surrounded by increasing degrees of urban influence as their distance from the city decreases. Since the urban to rural morphology of New York City was strongly concentric in form, early studies along the New York City urban–rural gradient (McDonnell et al., 1997) often successfully used distance (kilometers) from the urban core as a surrogate explanatory variable for many ecological responses. Of course, it is not distance per se that causes ecological response variation, but rather the fact that distance happened to integrate many strong but initially unknown urban effects on forest remnants. This allowed distance to be used successfully in these early urban–rural gradient studies (similar to the use of elevation as a surrogate variable that usefully explained ecological variation up a mountain).

Of course, the environmental factors and conditions caused by urban-to-rural land uses do not necessarily vary linearly or in spatial synchrony with distance from the city. Also because multiple factors changing across these landscapes sometimes interact to intensify or counteract each other, both the magnitude of the land-use variation and of the habitat responses are not always ordered linearly with distance from a city center. These landscape-level gradients, therefore, are complex and, especially in cities that are not monocentric in form, often indirectly ordered in real space. In addition, distance from a city center is not a useful integrative or standardized index of urban intensity for comparing urban impacts across cities of different extent and form. Therefore, improving our understanding of urban impacts necessitates the use of remote sensing and Geographical Information Systems (GIS) tools to quantify variation in land use, land cover, and landscape ecology metrics (Luck and Wu, 2002) to make these gradients more structurally explicit and useful for cross-city comparisons. Since many rapidly growing cities today have morphologies that are multicentric or sectored, urban–rural gradient studies in many cities are probably more usefully explored by assuming that gradients are often complex and indirect. Ecological responses can then be ordinated according to the structural aspects of cities, many of which, like impervious surface coverage, are common across cities. Explicitness and standardization of landscape metrics, as well as social metrics, are needed to move the study of urban effects on ecological systems forward. Luck and Wu (2002), Wu et al. (2003), and Hahs and McDonnell (2006) are examples of studies that develop explicit spatial indices that can link landscape structural characteristics to ecological states and processes in urban areas. Use of such indices and approaches can also advance cross-system comparisons among different cities to seek commonalities and differences in land transformation characteristics and their ecological responses along urban–rural gradients.

However, it must be remembered that these types of studies are still comparative and correlative, taking advantage of the environmental variation caused by urbanization to characterize how specific organisms, processes, and natural communities might vary along particular land-use and land-cover gradients. To go beyond correlative studies using aggregate land-cover, land-use, and social attributes, factors that respond to land-use variation, such as temperature or pollutant inputs, should be identified and measured along with landscape metrics and land-use types. These are the more proximate factors that affect organisms and processes within a patch like a forest remnant. Then direct causative relationships with ecological responses inside the remnant patch can be more strongly linked with land-use and land-cover differences outside the patch. Follow-up studies must be performed involving on-the-ground measurement of both the responses and their suspected direct causes, and if possible their manipulation in lab or field experiments, to make the strongest inferential case for causation between land-use differences and ecological processes within a patch.

To summarize, urban–rural gradient studies can increase our scientific understanding of how a particular city might affect the species composition and functional behavior of natural areas presently within the city's boundaries by comparing urban forests (or other habitat types) with similar reference areas further from urban influence. By developing standard protocols for describing land-use characteristics (Luck and Wu, 2002; Wu et al., 2003; Hahs and McDonnell, 2006) and for measuring ecological responses (Niemala et al., 2002), comparative studies among several cities could be initiated to learn which habitat types are more sensitive than others to particular factors derived from the surrounding land, which biotic or ecosystem responses are idiosyncratic to specific cities, and which responses are shared by natural communities in and near cities in different regions of the world. In addition, managers and planners would find the comparative study of urban, suburban, and rural natural areas particularly useful for predicting how parks and preserves currently far from the city may change once the sprawl front advances to surround them in the future. This may stimulate more proactive rather than reactive societal responses, such as reevaluation of land purchasing priorities and development of appropriate and timely adaptive management strategies for that region.

Using the Urban–Rural Gradient Approach: Two Case Studies Involving Nitrogen Availability in Native Forest Remnants

To illustrate the usefulness of this approach, the results of two urban–rural gradient studies (one in New York City and one in Louisville, Kentucky) are presented. Both studies focused on determining whether relationships existed between land use and atmospheric inputs to and soil nitrogen (N) cycling in forests within and near these two cities. The availability of N in both its organic and inorganic forms (typically inorganic ammonium, NH_4^+, and nitrate, NO_3^-) is important to quantify, since it has been identified as the nutrient that most often limits plant growth in terrestrial ecosystems (Vitousek and Howarth, 1991). Also N in excess of plant or microbial demand (N saturation) may also become microbially transformed to chemical species that can be readily exported in soil leachate (NO_3^-) to contaminate aquatic systems, and exported as volatile gases (NO_x and N_2O) that pollute the air or contribute to global warming (Aber et al., 1998). Nitrogen sources to a forest are both external (atmospheric N deposition) and internal (N cycled among plants, animals, microbes, and soils within the forest) (Chapin et al., 2002). Large forests far from high-density human settlements typically receive from 60% to 80% of their annual nitrogen needs from internally recycled (mostly microbially mineralized) nitrogen (Likens and Bormann, 1995) with the remainder derived from atmospheric deposition. Proximity to cities may increase the amount of N available for forest plant uptake due to (1) high rates of fossil fuel combustion in cities (urban atmospheres often contain higher concentrations of NO_x gases and nitrate than atmospheres in more remote areas [Gatz, 1991]); (2) the urban heat-island and other factors that may accelerate microbial production of inorganic N (NH_4^+ and NO_3^-) from organic N through the process of soil N mineralization, and (3) changes in species composition of plant and soil faunal communities. Plants access N through direct foliar uptake of NO_x, NH_4^+, and NO_3^- from wet and dry precipitation (Garten and Hanson, 1990; Latus et al., 1990) and by root uptake from soil NH_4^+, and NO_3^- pools. Although several of these forms of N (NO_x, NO_3^-, NH_3) are measured throughout the year as criteria air pollutants in cities throughout the U.S. (Environmental Protection Agency Web site: http://www.epa.gov/air/oaqps/montring.html), there is surprisingly little information on deposition rates of N as wet and dry precipitation to vegetated areas in and near metropolitan areas in the U.S. There are also few studies of soil N-mineralization in urban forest remnants. Therefore, the objectives of the two studies below were to quantify and compare fluxes of atmospheric N deposition and rates of soil N-mineralization in remnant oak forests along their respective urban–rural land-use gradients.

The urban–rural gradient transect of sites established in the New York City area extends northeastward 130 km from the Bronx, through Westchester County, to Litchfield County in Connecticut. New York City (40°47′ N 73°58′ W) was settled by Europeans in the early 17th century and attained a population of 8,008,278 with a mean density of 10,324 individuals/km^2 in the year 2000 (U.S. Census Bureau, 2001). The New York City Metropolitan Statistical Area (MSA) with a population of approximately 20 million is the largest among the 280 MSAs in the U.S. (U.S. Census Bureau, 2000). During these studies, the population density at the urban end of this gradient around the forest remnants was 10,000/km^2 and declined to 10 people/km^2 at the rural end (Medley et al., 1995).

In Louisville, Kentucky, the transect of forest remnants extended from the Iroquois Park forest in Louisville, southward 45 km through suburban Jefferson Memorial Forest to the Bernheim Forest in rural Bullitt County. Louisville (38° 15′N, 085° 46′W) was first settled in the late 18th century and by 2000 had an inner city population of 259,000 with a mean density of 1600/km^2. After a city-county merger in 2003, the city's population became 700,000 with a mean density of 704/km^2, and now ranks 17th in population size in the nation. The Louisville MSA, with a population of 1,200,000, ranks 50th in the nation (U.S. Census Bureau, 2000). Population density within a 1.5-km radius of the forest remnants varied from 1665 individuals/km^2 at the urban end, 183/km^2 in the suburban section, to 20/km^2 at the rural end. Both Louisville and New York City have similar climate and are located within the eastern deciduous forest biome where the oak-hickory (*Quercus-Carya*) forest type is common. The climate in the New York City region consists of warm humid summers and cold winters with a mean annual air temperature of 12.5°C. Precipitation is evenly distributed throughout the year and averages 1260 mm annually (New York State Climate Office, 2006). Louisville's climate is of the mid-latitude type with a mean annual air temperature of 13°C, and mean annual precipitation of 1143 mm, also distributed evenly throughout the year (Ulack et al., 1998).

To maximize the ability to detect the possible impact of urban, suburban, and rural land uses on these forests, one to three 30 × 30 m plots were chosen within a remnant to be as similar as possible, using the following criteria: (1) dominance (≥50% of plot basal area) by the same tree species (*Quercus rubra* in New York, *Q. prinus* in Louisville); (2) location on the same or closely related soil series (Charlton-Hollis soils in New York, Tilsit-Carpenter soils in Louisville); (3) in the case of Louisville where forests are on Knob hills, the same aspect; (4) no signs of recent natural or human disturbance like canopy gaps, fire, severe insect infestation, or selective logging; and (5) location at least 30 m from a heavily trafficked road. Again, the reason for such stringent plot selection criteria is that we wanted to detect whether *external* land use may be affecting atmospheric input chemistry and internal N cycling in soil. By minimizing *internal* variation in factors known to affect throughfall chemistry and soil N cycling, such as tree species, soil type, and disturbance, our ability to detect whether land use may be affecting these processes in forest remnants is improved. If differences in the forest plots are found, then our confidence that ecological response variables are related to variation in land use surrounding the stands is enhanced.

Urban–Rural Patterns in Atmospheric Nitrogen Deposition and Soil Nitrogen Cycling in the New York City Metropolitan Area

Over a 77-day period in the summer of 1996, Lovett et al. (2000) measured atmospheric bulk wet deposition and throughfall fluxes of NO_3^-, NH_4^+, SO_4^{2-}, Ca^{2+}, and Mg^{2+} to oak forests located along a 130-km-long urban-to-rural transect in the New York City metropolitan area. Bulk deposition contains ions in wet deposition (i.e., rain) as well as particulates that collect with rainfall. Throughfall is the wet deposition that percolates through the canopy and has an ionic composition that reflects chemical exchanges by leaves (canopy processing) and dry deposited particulates that coat leaf surfaces. Lovett et al. (2000) found that the combined flux of NO_3^--N and NH_4^+-N in *net* throughfall (net throughfall = throughfall – bulk deposition, and provides an estimate of dry deposited particulates) to the forest floor during that period was 17 times greater in New York City oak forests than in outlying suburban and rural forests (15.5 mmol N m^{-2} in city forests vs 0.9 mmol N m^{-2} in rural forests). In addition, particulate dry deposition of NO_3^- was found to be 7.3 times greater in these urban forests than in the rural stands. The large NO_3^- deposition fluxes to the city's forests and its virtual absence in suburban forests only 45 km to the north could be explained by the reaction of acidic anions like NO_3^- with alkaline dust particles (Ca and Mg) thought to originate mostly from construction and demolition activity within the city (i.e., concrete dust). Since the particles were large (>2 µm), most sedimented in the city, and were not distributed by wind to outlying areas. Moreover, these particulate N inputs would have underestimated total atmospheric deposition to these forests, since inputs from gaseous N, which is more concentrated near cities (Baumbach et al., 1989) and can be taken up by leaves and incorporated into organic N (Latus et al., 1990), were not measured.

If the trends observed for N inputs in New York City can be generalized to other cities, then relative to forests in outlying areas, urban forests receive a large N subsidy in dry deposition during the growing season. Forest response to this added N would depend on where these forests lie along a continuum of N saturation. If the forests grow on nutrient poor soils and are undersaturated with N, then the added N may actually stimulate primary production, as long as other injurious factors like high atmospheric O_3 do not constrain a plant growth response to N. If the forests are already at or approaching N saturation, then the additional atmospheric N inputs could cause the forest's condition to deteriorate (Aber et al., 1998).

Nitrogen needed to support primary production does not only enter a forest from outside the system. Most of a forest's annual need for nitrogen is derived from the internal recycling of a portion of the nitrogen capital stored in its biomass and soil. Organic forms of nitrogen in soil are eventually converted by microbes and invertebrates into NH_4^+, and in some forests varying amounts of that NH_4^+ pool can be converted to NO_3^- by the process of nitrification, which is most often performed by autotrophic nitrifying bacteria (Paul and Clark, 1996). From 1996 to 1997, Zhu and Carreiro (2004) conducted a field study to determine if the net production rate of inorganic N (combined NH_4^+ and NO_3^-) and nitrification (only NO_3^-) from soil organic matter varied in a predictable fashion with land use in eight oak forest remnants across the New York City urban–rural gradient. The gradient sites consisted of three oak forests in the Bronx, two in suburban Westchester County, north of the Bronx, and three in rural Litchfield County, Connecticut. Predicting the direction of the trend for the process of N mineralization was not straightforward because rates of N mineralization could be faster at the urban end of the gradient due to warmer temperatures, but they could also be slower near the city if microbes and decomposers found the chemical and physical quality of the urban litter materials (e.g., dead leaves) not as easy to decompose as that of rural litter (Carreiro et al., 1999; Pouyat and Carreiro, 2003). Rates could also differ depending on the composition of different functional groups of decomposer organisms along this gradient. Zhu and Carreiro (2004) found that over an entire year more net NH_4^+ and NO_3^- were produced from organic matter in the top 7.5-cm soil horizon in the urban and suburban forests than in the rural forests. But even more interestingly, they learned that the amount and proportion of NO_3^- produced was greater in the urban (48% NH_4^+ converted to NO_3^-) and suburban (44% NH_4^+ converted) forest plots than in the rural (2.8% NH_4^+ converted) forest plots (also see Pouyat et al., 1997). While the N-mineralization rates (total NH_4^+ and NO_3^-) increased linearly with decreasing forest distance from the city, the pattern of increase in nitrification rates was decidedly nonlinear since virtually no *net* NO_3^- production was measured in the rural forests, but was very high in urban and some suburban forests. This suggested that the factors controlling these two related processes were not tightly coupled in space along this gradient. The temperature differences between the urban and rural forests (about 2.5°C throughout the year; McDonnell et al., 1997) may explain part of the N-mineralization pattern, but not the entire difference in nitrification pattern. Which factor(s) might then explain this spatially disjunctive pattern of N mineralization and nitrification?

To answer this question, we examined other factors like soil organisms. More or less simultaneously with these N-mineralization experiments, we were quantifying the distribution of earthworms in forests along the urban–rural gradient, since we had learned that most of the earthworms in these northeastern forests were exotic species (most were in the genus *Amynthas*, native to Asia; Patrick Bohlen, personal communication). In the summer of 1998, we found that the biomass of earthworms was 10 times greater in the urban forests than in the rural forests (urban oak forests

8.88 ± 0.876 g ash-free dry mass [AFDM] worms m^{-2}; suburban, 2.87 ± 0.145 g AFDM m^{-2}; rural, 0.866 ± 0.774 g AFDM m^{-2}). Therefore, earthworm distribution appeared to correlate well with the disjunct nitrification patterns observed in these forests; but were the worms indeed responsible for increased nitrification rates in the urban and some suburban forests?

We used microcosm experiments conducted in the laboratory to address this question. These large Asian worms produce a surface cast layer in the soil of the urban forests that in wet years can be as deep as 7 cm. These casts consist of partially digested organic material and soil, and when first deposited sustain active microbial growth. Carreiro and Zhu hypothesized that the worms may indirectly stimulate nitrification by providing a soil microhabitat (the casts) that preferentially stimulates the growth of nitrifying bacteria. To determine this, earthworm casts and mineral soil (down to 10 cm depth) directly below each cast sampling point were collected separately from 10 randomly selected locations within each of two forest plots exhibiting the highest nitrification rates (urban: Van Cortland Park, Bronx, NY; suburban: Mianus River Gorge, Bedford, NY). Casts and soils were brought to water holding capacity in 20 microcosms and incubated in the lab at room temperature for 14 days. At the end of this incubation period we extracted and measured NH_4^+ and NO_3^- produced in these microcosm samples. We found that for both forests, nitrification rates were much greater in the earthworm cast samples than in the mineral soil directly below the cast layer. For example, in the suburban forest we found that microbial nitrification was 10 times greater in earthworm casts than in the bulk soil below (respectively, 9.13 ± 1.94 vs. 0.920 ± 0.92 μg NO_3-N per g dry mass soil per day, $p = .0005$). These results agreed with those of Steinberg et al. (1997), who manipulated earthworm densities in microcosm experiments to determine their role in nitrification in these urban forest soils.

In an earlier microcosm experiment, Zhu and Carreiro (1999) used the acetylene block technique to determine that chemoautotrophic nitrifying bacteria rather than heterotrophic microbes were entirely responsible for conducting nitrification in these urban and suburban forest soils. So not only were these exotic earthworms more abundant in urban and suburban forests, but evidence from these field and lab studies strongly suggest these worms had redirected more N into a different nitrogen transformation pathway in these forests. In contrast with the rural forests in which the dominant inorganic soil nitrogen form was NH_4^+, in these urban and suburban forests NO_3^- was more prevalent and in a subset of these forests was the dominant inorganic form of N (Zhu and Carreiro, 2004).

The greater inputs and production of N in urban rather than rural forests in the New York City gradient, especially the greater inputs of N as $NO_3^-$and the higher nitrification rates, could have several important implications for forests close to New York City. Plant species respond to N availability differently and some are able to take up N as NO_3^-more competitively than other species (Templer and Dawson, 2004). Over the long term, then, the total amount and ratios of $NO_3^-$to NH_4^+could affect plant species composition in these forests (Tamm, 1991). In addition, if a greater proportion of the N capital of a forest is in the form of NO_3^-, such a forest could lose more N particularly during periods when plant uptake is low and microbes still active (late autumn through early spring). This N loss could be due to the greater leachability of NO_3^- compared with NH_4^+ in most soils, and to the increased potential for greater N loss to the atmosphere as N_2O, a greenhouse gas, during nitrification and denitrification processes (Paul and Clark, 1996). In these cases, the landscape level behavior of these urban forests with respect to N retention would differ from rural forests having the same soil types and dominant tree species composition. If N-saturated, urban forests would then become sources of N to their surroundings rather than net sinks for N, as forests normally behave in many rural areas of the U.S. (Likens and Bormann, 1995). Assumptions about forest remnant behavior and their source-sink landscape level roles in urban contexts cannot therefore be based on models derived from rural forests.

To summarize, from the study of forests along the New York City urban–rural gradient we have learned that urban forests may receive larger inputs of exogenous N from the atmosphere than neighboring suburban and rural forests, and that an invasion by an exotic species has likely contributed to altering the biogeochemical cycling rate and pathway of an important element (N) in urban and suburban forests compared with their rural counterparts. We now know that the forests in and near New York City have greater active pools of N than forests of similar tree composition and soils further from the city. The exotic earthworm study has also provided us with an example of how correlative patterns emerging from the comparative gradient study (the nitrification pattern along the urban–rural gradient) can stimulate the generation of hypotheses that can be tested through experimentation in the lab or in the field to provide stronger mechanistic explanations for those spatial trends.

Are Factor Gradients and Ecological Response Variables Similar in Other Cities?

Since New York City is the U.S.'s largest MSA, with a population of 20 million, it may not provide a typical example of urban effects on forests. How applicable are the N cycling results discovered in New York City forests to forests in other urban areas in the eastern U.S.? For example, due to its size New York City's atmospheric N concentrations may be anomalously high when compared with other, more numerous smaller cities. We might expect that the amount of externally derived N deposited onto forests in smaller cities might be less in both absolute and proportional terms when compared to the total amount of N internally cycling within these same forests. Inter-city comparisons would help us separate ecological patterns that are idiosyncratic to particular cities from those that may be similar across cities. Since atmospheric N deposition is related to fossil fuel emissions, one can compare New York City's N emissions profile to that of Louisville, Kentucky, an MSA one-twentieth its size, using the EPA Air Data National Emissions Trends database (http://www.epa.gov/air/data/geosel.html). Emissions data for 1999 show that Louisville, with a city-county population 8.7% the size (693,000 vs. 8,000,000) and

6.8% the density (704 vs. 10,324 people/km^2 in 2000) of the five counties making up New York City proper (http://www.demographia.com/db–2000city50kdens.htm), produces a surprisingly higher proportion of NO_x gases on an areal basis than would be predicted from population density alone (22% that of New York City's 245 metric tons/km^2 yr^{-1}). These emission rates and Louisville's annual mean atmospheric NO_2 concentration (24 ppb, 67% that of the Bronx site) suggest that forests in moderately sized cities of approximately 500,000 to 1 million may receive N inputs that are almost as great as those in cities with an order of magnitude larger population (New York State Department of Environmental Conservation, 1998; Kentucky Division of Air Quality, 2000).

To determine whether differences exist in the flux rates of N to urban, suburban, and rural forests in the Louisville metropolitan area, Tripler and I began collecting bulk precipitation and throughfall in oak forests along an urban–rural gradient in and near Louisville (unpublished data). From May to October 2002, rainfall (bulk precipitation) was collected weekly from a total of three stations in open areas near the urban, suburban, and rural forest sites. Throughfall was also collected simultaneously from beneath the canopies of 27 *Q. prinus* trees, nine trees each in the urban, suburban, and rural forests. We found that urban–rural gradient trends in atmospheric N deposition were similar in both the Louisville and New York City area forests. The amount of total inorganic N (combined NH_4^+- N and NO_3^--N) entering the urban forests in throughfall was 31% and 53% greater than that entering suburban and rural forests, respectively. As found in New York City, the dry particulate component in throughfall was responsible for most of the difference along the Louisville gradient (7.01, 4.66, 1.43 mmol N m^{-2}, urban, suburban, rural, respectively) rather than the amount entering via bulk precipitation. Along the Louisville gradient the proportion of combined NH_4^+- N and NO_3^--N that entered as NO_3^- ranged from 65% to 72%, as was found in the New York study. Bulk deposition fluxes of Ca^{2+} and Mg^{2+} to the urban stands in Louisville were also two to three times greater than those to rural forests. Since greater inputs of these basic cations was also found in New York City, urban forests may generally receive greater inputs of these nutrients than rural forests nearby.

In summary, the trends in deposition fluxes of N, Ca^{2+}, and Mg^{2+} to forests in the Louisville area are very similar to those in New York City, which were also collected over the growing season and over a similar number of weeks. In addition, the absolute amounts of N, Ca^{2+}, and Mg^{2+} entering the urban forests in both cities were very similar. Louisville had approximately half the N inputs of New York, but slightly greater inputs of Ca^{2+} and Mg^{2+}, despite the fact that the Louisville metropolitan area (1.2 million inhabitants) has a population just 5% that of New York's, and a mean population density 6.8% that of New York's. City size and density alone, therefore, are unlikely to explain most of the variation in atmospheric deposition trends. This is not surprising since other geographic, sociopolitical, and economic factors can influence air quality in a particular city. For example, the Louisville area has a number of large, coal-burning power plants nearby along the Ohio River, and depending on dominant wind directions, they can contribute to atmospheric deposition in the local area. The fact that N deposition to the rural plots in the Louisville gradient was greater than the amount entering the rural plots in the New York gradient is perhaps indicative of greater contribution of emissions from these large point sources to the region surrounding Louisville. In addition, automobile traffic may also be greater on a per capita basis in Louisville than in New York, since the fraction of New Yorkers who own or drive cars within their city limits is likely less than in Louisville. Local and state air quality regulations on emissions from both stationary and mobile sources (automobiles) may differ between the two cities as well.

Nitrogen mineralization studies were conducted from July 2001 to December 2002 in the Louisville area forests using cores of the upper 10-cm soil horizon. Unlike the urban–rural trends observed in New York, N-mineralization rates from December 2001 to December 2002 were greatest in the rural plots, followed by the urban and then suburban plots. On a dry mass soil basis, the upper 10-cm soil horizon of the rural stands mineralized 26% and 69% more N than the urban and suburban stands, respectively. This pattern was unexpected since we had predicted that the higher inputs of NH_4^+ and the warmer temperatures at the urban plots would stimulate soil N production. The urban forests mineralized 2.2, suburban 1.96, and rural 3.23 mg N/kg Soil Organic Matter (SOM)/day over that 1-year period. Compared to forests along the New York City urban–rural gradient, annual N mineralization on an SOM basis in the Louisville urban and suburban plots was 50% and 56% that of their New York forest counterparts. However, the rural plots in Louisville mineralized 130% more N than the rural forests in the New York City gradient. The nitrification pattern across the sites in Louisville differed greatly from that in New York as well. In the urban and suburban stands in New York as much as 70% of mineralized NH_4^+ was transformed to NO_3^-, with net nitrification being negligible in the rural forests. However, in Louisville nitrification in the rural stands was 10 times that in the urban plots, and negligible in the suburban forests. On average 65% of the total N mineralized was converted to nitrate in rural stands in Louisville. These results cannot be fully explained at this time. However, potential explanations may include the fact that exotic earthworms are not as obviously abundant in the urban, suburban, or rural forests in Louisville, as they were in New York City.

Conclusion

Human activities profoundly alter distributions of organisms and ecosystem functions throughout the world, but nowhere do they modify the earth's surface more directly and continuously than in cities. Cities share a number of similar attributes (high impermeability, dense human populations, road traffic, and pollutant loading of air, soil, and water), regardless of the wide variety of climatic zones, biome types, or physiographic provinces on which they are overlain. These attributes provide ecologists with an opportunity to explore the impact that cities of different sizes, ages, and growth rates have on a variety of ecosystem patches within them, and to compare similarities and differences in the responses of contrasting biotic communities and ecosystem processes to the urban "template" superimposed on them. If humanity is to create more livable cities surrounded by resilient natural and managed habitats that can provide dependable ecosystem services, there is a pressing need to learn which ecosystem types are more sensitive to urban disturbances, which biotic or ecosystem responses are particular to specific cities, and which responses are shared by natural communities in and near cities in different regions and biomes. This goal should stimulate communication and integration of knowledge and approaches to the study and planning of urban ecosystems by people in various fields both academic and practical (e.g., ecology, geography, sociology climatology, urban planning, engineering).

This chapter demonstrated that the urban–rural gradient approach has been used successfully to increase our understanding of how the urban land-use matrix differs from nearby suburban and rural areas as a source of energy, matter, and species, and how ecosystem processes in the more natural components of an urban ecosystem (in this case native forest remnants) may vary sharply from those in a less urban land-use context. Use of this approach enhanced our ability to detect differences in factors originating in the urban land-use matrix that can strongly affect forest ecosystem functions (e.g., atmospheric N deposition) and potentially the amount and kind of ecological services they provide. For instance, it must be remembered that while forests and trees in other landscape contexts contribute to human health by filtering pollutants from the air, these same pollutants are either taken up by the trees or collect in the soil where they can affect decomposer organisms as well as plant roots. Over the long term these pollutants can either stimulate or harm the trees or alter ecological processes in forest remnants (or restored forest patches) such as tree species successions, primary production, and soil nutrient cycling.

In the case of N deposition, forests in both New York City and Louisville exhibited similar trends in that urban forest remnants received greater amounts of N input in net throughfall (an estimate of particulate dry deposition) than their respective suburban and rural forests. On the other hand, this approach also allowed us to hypothesize that urban forest remnants in New York City can potentially perform differently at the landscape level than their rural counterparts. Urban forest remnants in New York City may already be or more quickly become sources of N to their surroundings rather than serving as N sinks, as would be assumed from ecosystem studies of rural forests in the region. Such potential shifts in landscape-level roles could affect the quality of the ecological services these remnants provide to society, since high levels of exported N in soil leachate contribute to eutrophication of waterways. In addition, comparisons across these two cities demonstrated that the direction of an ecosystem response (soil N mineralization) may differ along gradients in two different cities. Soil N mineralization rates were highest at the urban end of the gradient in New York City, but were highest in the rural end of the gradient in Louisville. In New York City the gradient approach also revealed the importance of land-use legacies in affecting the present rate of N cycling in forest soils, because unlike the rural forests, urban and suburban forests contained high populations of Asian earthworm species that affected the N cycle. It appears that this biotic land-use legacy not only could explain the high N mineralization rates encountered in New York City's urban and suburban forests, but also even more strongly explain the nitrification variation in forests along this gradient. The extent to which the atmospheric N deposition patterns are consistent for other cities and the explanation for differences in direction for the N mineralization trends across the two gradients await further comparative and manipulative experiments to improve our knowledge of the ecology of urban areas and the functioning of natural habitats embedded within them.

References

Aber, J.D., McDowell, W., Nadelhoffer, K., et al. (1998) Nitrogen saturation in temperate forest ecosystems: hypotheses revisited. Bioscience 48:921–934.

Alberti, M. (1999) Modeling the urban ecosystem: a conceptual framework. Environment and Planning B: Planning and Design 26:605–630.

Baumbach, G., Baumann, K., and Droscher, F. (1989) O_3 and NO_x measurements in forest and urban areas. In: Bojkov, R.D., and Fabian, P. (eds.) Ozone in the Atmosphere. Deepak, Hampton, VA, pp. 502–507.

Bray J.R., and Curtis, J.T. (1957) An ordination of the upland forest communities of southern Wisconsin. Ecological Monographs 27:325–349.

Carreiro, M.M., Howe, K., Parkhurst, D.F., and Pouyat, R.V. (1999) Variations in quality and decomposability of oak leaf litter along an urban-rural gradient. Biology and Fertility of Soils 30: 258–268.

Carreiro, M.M., and Tripler, C.E. (2005) Forest remnants along urban-rural gradients: examining their potential for global change research. Ecosystems 8:568–582.

Chapin, F.S. III, Matson, P.A., and Mooney, H.A. (2002) Principles of Terrestrial Ecosystem Ecology. Springer Science+Business Media, New York.

Daily, G.C., Alexander, S., Ehrlich, et al. (1997) Ecosystem services: benefits supplied to human societies by natural ecosystems. Issues in Ecology, Number 2. Ecological Society of America.

Daly, J., and Klemens, M.W. (2005) Integrating conservation of biodiversity into local planning. In: Johnson, E.A., and Klemens, M.W. (eds.) Nature in Fragments. Columbia University Press, New York, pp. 313–334.

Dougherty, A.B. (1992) Major uses of land in the United States. Agricultural Economics report no. 723. Natural Resources and Environment Division, Economic research Service, US Department of Agriculture, Washington, DC.

Douglas, I. (1994) Human settlements. In: Meyer, W.B., and Turner, B.L. (eds.) Changes in Land Use and Land Cover: A Global Perspective. Cambridge University Press, Cambridge, U.K.,pp. 149–169.

Dwyer, J.F., Nowak, D.J., Noble M.H., and Sisinni, S.M. (2000) Connecting people with ecosystems in the 21st century: an assessment of our nation's urban forests. USDA Forest Service Technical Report PNW-GTR-490.

Farber, S. (2005) The economics of biodiversity in urbanizing ecosystems. In: Johnson, E.A., and Klemens, M.W. (eds.) Nature in Fragments. Columbia University Press, New York.

Farber, S.R., Costanza, D.L., Childers, J., et al. (2006) Linking ecology and economics for ecosystem management. BioScience 56:121–133.

Foster, D.R. (1993) Land-use history and forest transformations in central New England. In: McDonnell, M.J., and Pickett, S.T.A. (eds.) Humans as Components of Ecosystems. Springer-Verlag, New York, pp. 91–110.

Frey, H.T. (1984) Expansion of urban area in the United States: 1960–1980. USDA Economic Research Service Staff Report No. AGES830615. Washington, DC.

Garten, C.T. Jr., and Hanson, P.J. (1990) Foliar retention of ^{15}N-nitrate and ^{15}N-ammonium by red maple (Acer rubrum) and white oak (Quercus alba) leaves from simulated rain. Environmental and Experimental Botany 30:333–342.

Gatz, D.F. (1991) Urban precipitation chemistry: a review and synthesis. Atmospheric Environment 25B:1–15.

Greig-Smith, P. (1983) Quantitative Plant Ecology. University of California Press. Berkeley, CA.

Grimm, N.B., Grove, J.M., Pickett, S.T.A., and Redman, C.L. (2000) Integrated approaches to long-term studies of urban ecological systems. Bioscience 50:571–584.

Hahs, A.K., and McDonnell, M.J. (2006) Selecting independent measures to quantify Melbourne's urban–rural gradient. Landscape and Urban Planning 78:435–448.

Heilman, G.E. Jr., Strittholt, J.R., Slosser, N.C., Dellasala, D.A. (2002) Forest fragmentation of the conterminous United States: assessing forest intactness through road density and spatial characteristics. Bioscience 52:411–422.

Kentucky Division of Air Quality. (2000) Annual Report 2000. Department for Environmental Protection, Natural Resources & Environmental Protection Cabinet, Frankfort, KY. http://www.air.ky.gov/programs/airmonitoring/.

Kowarik, I. (1990) Some responses of flora and vegetation to urbanization in Central Europe. In: Sukopp, H., et al. (eds.) Urban Ecology. SPB Academic Publishing, The Hague, Netherlands, pp. 45–74.

Latus, C., Forstel, H., and Fuhr, F. (1990) Quantitative measurement of NO_2 uptake and metabolism by sunflower plants. Naturwissenschaften 77:283–285.

Likens, G.E., and Bormann, F.H. (1995) Biogeochemistry of a Forested Ecosystem. 2nd ed. Springer-Verlag, New York.

Lovett, G.M., Traynor, M.M., Pouyat, R.V., Carreiro, M.M., Zhu, W.-X., and Baxter, J.W. (2000) Environmental Science and Technology 34:4294–4300.

Luck, M., and Wu, J. (2002) A gradient analysis of urban landscape pattern: a case study from the Phoenix metropolitan region, Arizona, USA. Landscape Ecology 17:327–339.

Matlack, G.R. (1993a) Microenvironment variation within and among forest edge sites in the eastern United States. Biological Conservation 66:185–194.

Matlack, G.R. (1993b) Sociological edge effects: spatial distribution of human impact in suburban forest fragments. Environmental Management 17:829–835.

McDonnell, M.J., and Pickett, S.T.A. (1990) Ecosystem structure and function along urban-rural gradients: an unexploited opportunity for ecology. Ecology 71:1231–1237.

McDonnell, M.J., and Pickett, S.T.A. (eds.) (1993) Humans as Components of Ecosystems: The Ecology of Subtle Human Effects and Populated Areas. Springer-Verlag, New York.

McDonnell, M.J., Pickett, S.T.A., and Pouyat, R.V. (1993) The application of the ecological gradient paradigm to the study of urban effects. In: McDonnell, M.J., Pickett, S.T.A. (eds.) Humans as Components of Ecosystems: The Ecology of Subtle Human Effects and Populated Areas. Springer-Verlag, New York, pp. 175–189.

McDonnell, M.J., Pickett, S.T.A., and Pouyat, R.V., et al. (1997) Ecosystem processes along urban-to-rural gradients. Urban Ecosystems 1:21–36.

McKinney, M.L. (2002) Urbanization, biodiversity and conservation. Bioscience 52:883–890.

Medley, K., McDonnell, M.J., and Pickett, S.T.A. (1995) Forest-landscape structure along an urban-to-rural gradient. Professional Geographer 47:159–168.

Millennium Ecosystem Assessment. (2005) Ecosystems and Human Well-Being: Synthesis. Island Press, Washington, DC. http://www.millenniumassessment.org/en/Products.aspx.

Niemala, J., Kotze, D.J., Venn, S., et al. (2002) Carabid beetle assemblages (Coleoptera, Carabidae) across urban-rural gradients: an international comparison. Landscape Ecology 17:387–401.

New York State Climate Office. (2006) Department of Earth and Atmospheric Sciences, Cornell, University, Ithaca, NY. http://nysc.eas.cornell.edu/.

New York State Department of Environmental Conservation. (1998) Air quality report.

Oke, T.R. (1995) The heat island characteristics of the urban boundary layer: characteristics, causes and effects. In: Cermak, J.E., Davenport, A.G., Plate, E.J., and Viegas, D.X. (eds.) Wind Climate in Cities. Kluwer Academic Publishers, Netherlands, pp. 81–107.

Paul, E.A., and Clark, F.E. (1996) Soil Microbiology and Biochemistry, 2nd ed. Academic Press, Inc., New York.

Pickett, S.T.A., Burch, W.R. Jr., Dalton, S.E., Foresman, T.W., Grove, J.M., and Rowntree, R. (1997) A conceptual framework for the study of human ecosystems in urban areas. Urban Ecosystems 1:185–199.

Pouyat, R.V., and Carreiro, M.M. (2003) Controls on mass loss and nitrogen dynamics of oak leaf litter along an urban-rural land-use gradient. Oecologia 135:288–298.

Pouyat, R.V., McDonnell, M.J., and Pickett, S.T.A. (1997) Litter decomposition and nitrogen mineralization in oak stands along an urban-rural land use gradient. Urban Ecosystems 1:117–131.

Rebele, F. (1994) Urban ecology and special features of urban ecosystems. Global Ecology and Biogeography Letters 4:173–187.

Richards, J.F. (1990) Land Transformation. In: Turner, B.L., Clark, W.C., Kates, R.W., Richards, J.F., Matthew, K.T., Meyer, W.V. (eds.) The Earth as Transformed by Human Action. Cambridge University Press with Clark University, Cambridge, pp. 163–178.

Saunders, D.A., Hobbs, R.J., Margules, C.R. (1991) Biological consequences of ecosystem fragmentation: a review. Conservation Biology 5:18–32.

Steinberg, D.A., Pouyat, R.V., Parmelee R.W., and Groffman, P.M. (1997) Earthworm abundance and nitrogen mineralization rates along an urban-rural land use gradient. Soil Biology and Biochemistry 29:427–430.

Tamm, C.O. (1991). Nitrogen in Terrestrial Ecosystems. Springer-Verlag, New York.

Templer, P.H., and Dawson, T.E. (2004) Nitrogen uptake by four tree species of the Catskills Mountains, New York: impications for forest N dynamics. Plant and Soil 262:251–261.

Ter Braak, C.J.F., and Prentice, I.C. (1988) A theory of gradient analysis. Advances in Ecological Research 18:272–327.

Turner, B.L. II, Clark, W.C., Kates, R.W., Richards, J.F., Mathews, J.T., and Meyer, W.B. (eds.) (1990) The Earth as Transformed by Human Action: Global and Regional Changes in the Biosphere Over the Past 300 Years. Cambridge University Press and Clark University, New York.

Ulack, R., Raitz, K., and Pauer, G. (1998) Atlas of Kentucky. University Press of Kentucky, Lexington, KY.

United States Department of Agriculture. (2001) National Resources Inventory Summary Report 1997 (Revised December 2000). United States Department of Agriculture, Natural Resources Conservation Service. http://www.nhq.nrcs.usda.gov.

U.S. Census Bureau. (2000) http://www.census.gov/population/cen2000/phc-t3/tab03.pdf

U.S. Census Bureau. (2001) Population change and distribution 1990–2000 in brief. http://www.census.gov/prod/2001pubs/c2kbr01–2.pdf.

Vitousek, P.M., and Howarth, R.W. (1991) Nitrogen limitation on land and sea: how can it occur? Biogeochemistry 13:87–115.

Vogelmann, J.E. (1995) Assessment of forest fragmentation in southern New England using remote sensing and geographic information systems technology. Conservation Biology 9:439–449.

Whittaker, R.H. (1967) Gradient analysis of vegetation. Biological Review 49: 207–264.

Wilkinson, J., Vickerman, S., and Lerner, J. (2005) Conserving biodiversity through state and regional planning. In: Johnson, E.A., and Klemens, M.W. (eds.) Nature in Fragments. Columbia University Press, New York, pp. 284–312.

Wu, J., David, J.L., and Jenerette, G.D. (2003) Linking land-use change with ecosystem processes: a hierarchical patch dynamic model. In: Guhathakurta, S. (ed.) Integrated Land Use and Environmental Models. Springer, New York, pp. 99–119.

Zhu, W.-X., and Carreiro, M.M. (1999) Chemoautotrophic nitrification in acidic soils in forests along an urban-rural transect. Soil Biology and Biochemistry 31:1091–1100.

Zhu, W.-X., and Carreiro, M.M. (2004) Temporal and spatial variations in nitrogen cycling in deciduous forest ecosystems along an urban-rural gradient. Soil Biology and Biochemistry 36:267–278.

12
A Philosophical Basis for Restoring Ecologically Functioning Urban Forests: Current Methods and Results

Akira Miyawaki

Introduction: Purposes of Tree Planting in Japan

There are several purposes of planting trees, but the main purpose in many countries has been lumber production. For example, in Japan needle-leaved trees such as Japanese cypresses (*Cryptomeria japonica, Chamaecyparis obtuse)*, pine trees (*Pinus thunbergii, P. densiflora*), and larches (*Larix kaempferi*) have long been monocultured in plantation rows on mountains for timber. At one time lumber was one of Japan's main industries. However, recently Japan's timber industry has been overtaken by much cheaper lumber imported from other countries. Many plantations were abandoned with little subsequent management, and as a result these forests have degraded greatly.

Today, trees in urban areas are being planted for many different reasons. Aesthetic beautification is one main purpose in urban environments, especially those around industrial sites and transportation corridors. Contemporary landscape architectural designs in Japan often consist of hardscapes with little vegetation as "softening" design elements (see Chapter 9). Buildings of metal, concrete, and other nonliving materials occupy a major part of the limited urban space, with adult trees of rare, exotic species and fast-growing pioneer species scattered in openings. Along roadsides adult trees are planted in rows. In many urban parks exotic trees are chosen and scattered on lawns. These are the typical tree planting palettes and arrangements in urban areas of Japan and many other countries.

Philosophy and Significance of Urban Forests

Residents' need for green space in cities varies considerably, and so, for example, we recognize the positive roles that exotic trees can sometimes play in harsh urban environments. However, forests, especially native forests of indigenous trees, also have quite significant roles to play in urban environments. They serve as green oases that help relieve stress and renew our spirits by providing calm and comfortable surroundings for physical activities as well as contemplation, providing

M.M. Carreiro et al. (eds.), *Ecology, Planning, and Management of Urban Forests: International Perspectives.*

a respite from today's physically confining world of computer screens and information technology (Miyawaki, 2002b, 2004).

These native forests also provide many ecological and social service functions, such as disaster prevention and mitigation. Native forest trees in temperate and subtropical zones of Asia, especially in Japan, have deep taproots and will not fall over easily (Miyawaki, 1992). Their leaves are evergreen and will not burn easily. When earthquakes, typhoons, and their attendant large-scale fires and tsunamis occur, native forests mitigate disaster by providing buffer zones that reduce the ability of these disasters to percolate through the densely settled landscape. The potential disaster mitigation function of native forests was demonstrated during the Great Hanshin Earthquake in January 1995 (Miyawaki, 1998; Miyawaki and Box, 2006). Not a tree from natural vegetation areas fell, and native trees saved people by preventing the spread of fire and by stopping roofs and pillars from falling. On the other hand, some sections of elevated highways and railways, which involved the latest science and techniques, were destroyed in an instant (Miyawaki, 1999). Because native forests are multilayered and have green surface areas nearly 30 times larger than those of unilayered vegetation like lawns, they are also much more effective at providing ecological services such as air and water purification, and the blocking of sound, wind, and dust. At the local scale these forests reduce the urban heat-island effect, and at the global scale, they contribute to reducing global warming (Miyawaki and Meguro, 2000). Therefore, from a philosophical perspective, the planting of urban forests should be planned and justified not only for beautification purposes, but also for their ecological service functions. For this latter reason, we strongly recommend that urban forests consist of "natural forests with native trees"(Miyawaki, 2001).

Methods and Proposals for Constructing Ecologically Functional Urban Forests and Their Expected Effects

When planting trees in a large area to form an urban forest patch, we have to consider appropriation of property for these plantings, sapling production, and the hiring of a labor force for performing the planting. After planting, maintenance costs, including weeding and pruning of offshoots, are entailed for the first few years. These financial considerations are not small. A solution to these difficulties can be found by forming public–private cooperative partnerships among governmental agencies, private companies, and local residents (Miyawaki and Golley, 1993; Miyawaki, 2002a,b; Miyawaki et al., 2004).

The plants seen outdoors at present differ greatly from the original indigenous vegetation. If we are to make greater use of native species in an urban context, we need to obtain information on natural vegetation in the surrounding region by conducting phytosociological field surveys. Finding natural vegetation remnants can often be difficult in densely populated Asian cities, such as those in Japan. But remnants of the potential natural vegetation of the area are seen in the local Chinju-no-mori (shrine or temple groves) and forests abutting older houses. Land use, topography, and soil profiles are also investigated. Geographic information systems can expand our ability to locate areas that potentially have natural vegetation by examining maps and map overlays of factors like soils, topography, elevation, and land use.

Such procedures help us determine the "potential natural vegetation" (Tüxen, 1956) of an area. This is the theoretical vegetation that the natural environment of a local area could finally support if human influence were completely stopped. The trees from the potential natural vegetation fit the climate and the soil of the area, and grow to form a quasi-natural forest. Therefore, for the purposes of reforestation, we use tree species from what we determine to be the potential natural vegetation of the area (Miyawaki, 1992). We identify the local potential natural vegetation through exhaustive field surveys, and choose the dominant and companion tree species that are found (Miyawaki, 2004). These tree species have deep or tap roots, and are so difficult to transplant that even landscape gardeners dislike dealing with them. But once rooted, they survive and grow well. We collect seeds of tree species from the potential natural vegetation, germinate them, and pot seedlings so that their root systems fully develop in the containers (Miyawaki et al., 1993). These potted seedlings can be transplanted without damaging root systems, especially the delicate root hairs, so their survival rate is quite high. They are also easy for local residents to handle, even children. Planting of seedlings in a linear fashion is not normally done. Instead, seedlings are mixed and planted densely, as observed in actual natural forests. Because they are densely planted, the seedlings resist strong winds, changes in temperature, and low humidity. They have the potential to grow tall, and after natural selection has occurred, they develop into naturalized forest stands.

It is critical that the potted seedlings have fully developed root systems, because plants live or die on the strength of their roots. We prepare rich topsoil, because roots live or die on the strength of their soil. If the planting site is bare land, topsoil rich in soil fauna should be added before planting to a depth as deep as 20 cm. Topsoil is very important for the success of the planting, since seedlings absorb water and nutrition only from topsoil. Topsoil also contains most of the microorganisms needed to make the soil fertile. In Germany, topsoil is called *Mutter Borden*, that is, "mother soil," which reflects its importance in nurturing plant growth. Right after planting, the site should be mulched with rice straw or other organic materials at an application rate of about 4 kg/m^2. Mulching protects seedlings from too many freeze-thaw episodes, helps prevent drying out of the soil surface, and reduces weed growth and soil erosion after heavy rains. Within a few years, the mulch also adds to the organic matter content of the soil via decomposition.

The timing of such plantings is also important and is not driven by suitability of season alone. To stimulate public involvement, we often take advantage of events like planting festivals. Reforestation should be conducted with the help of various organizations, as well as individuals. Reforestation can be viewed as analogous to dramas: vegetation ecologists write play scenarios, government and private companies work as producers and directors, and residents, including schoolchildren, play the part of leading characters on the stage. They all have the opportunity to play a role in reforesting their region (Fig. 12.1).

Fig. 12.1 Planting festival on a slope of a park in Yokohama where 650 residents planted 15,000 seedlings in March 2003 under the leadership of Mayor Hiroshi Nakada (standing with children) and guidance of Professor Akira Miyawaki (kneeling on right). Government-private partnerships and public involvement greatly enhance the success rate of such plantings, and create greater public demand for urban forest plantings and their maintenance

Weeding is needed for the first few years after planting. Weeded grass should not be burned, but rather should be placed on the forest floor as mulch. As trees grow, they spread their branches and reduce the sunlight. Thus weeds seldom return, because in urban areas most are cosmopolitan species that are not shade tolerant. After 3 years of human management, nature manages itself through natural selection, and there is no need for much maintenance any longer. In our experience, these planted trees eventually maintain themselves as ecosystems consisting of a multilayered forest with tall canopy-level trees, subcanopy trees, shrubs, herbs, and a soil rich in fauna (Fig. 12.2).

For construction of an urban forest, a large space where many trees can be planted is desirable. But since we define a forest to be a collection of trees with multilayers, a miniature urban forest can be planted in even 1-m-wide strips. An eco-city should have small urban forests in belts along streets and rivers, around schools and public facilities, and link them with each other or with hedges around residential houses to create a network of green corridors. A large forest patch, like Central Park in New York City, could be constructed at the center or on the outskirts of the city. These larger forests provide spaces where residents can relax and enjoy nature near their home. In case of emergency, they can also use them as pedestrian escape routes. So these forests not only heal the tired hearts of city dwellers, but may simultaneously provide disaster mitigation and environmental protection benefits.

When a new town is designed, the layout of forest patches, their scale, and construction methods should be considered from the start, along with the blueprint

Fig. 12.2 Location where trees were planted during a planting festival in Pudong, Shanghai. Some 15,000 seedlings were planted by 1200 residents including students from China and Japan in June 2000. This photo was taken 6 years later. Large forest patches like this one provide many ecosystem services for urban populations, like air pollution filtration, heat-island mitigation, and rainfall absorption

for roads and buildings. For reforestation in an existing town, it is important to construct forests wherever it is possible. Sometimes large trees are transplanted into a development to achieve a mature appearance in the landscape from the start. Such an approach for urban revegetation is far from ideal. When only mature trees are planted, costs for the trees themselves and their continued maintenance, like the need for structural propping, are very high. Moreover if root systems are not well developed or are damaged during planting, trees will stop growing well for several years, thus nullifying the benefits of starting out with large trees. Planting one or two adult trees per 100 to 200 m^2 for aesthetic landscaping purposes is acceptable. However, seedlings of dominant tree species selected from the potential natural vegetation should be planted in between larger individuals. Within 3 years these seedlings grow 2.5 to 3 m high, and they form a quasi-natural urban forest after 5 to 10 years (Fig. 12.3).

Tree mortality should also be expected, and over several decades to centuries, individual trees will die. Some die sooner through natural forces. Such dead trees and withered branches in a forest should be left on site, for they become decomposed, help increase biodiversity, and promote forest reproduction. In cases where dead trees in urban forests interfere with the aesthetics of particular landscapes, they should be buried in the earth for decomposition or used as railings along paths. They should not be burned. Finally, to provide a more naturalized structure and function to the forest boundaries or along pathways through the forest, flowering shrubs can

(A)

(B)

Fig. 12.3 **(A)** Close-up of forest planted in Pudong, Shanghai, in June 2000. This photo was taken 3 years after seedlings were planted. **(B)** The same site after 6 years (April 2006). Tree growth in dense and mixed plantation of young seedlings with well-developed root systems is steady and rapid by light demanding effect, making transplantation of adult trees costly, and inadvisable

be planted as mantle communities that keep fallen leaves inside the forest. This will save on maintenance costs, allow nutrients to remain and be cycled in the forest, stimulate soil decomposer communities (Carreiro, 2005; also see Chapters 1 and 11), and give residents the pleasure of seeing flowers.

Results Over the Last 30 Years

Over the past 30 years we have constructed and restored quasi-native forests around industrial institutions (e.g., ironworks, power stations, car factories), traffic facilities (e.g., alongside highways and railroads), public institutions including schools, and in new towns from northern Hokkaido to southern Okinawa in the 3000 km-long Japanese Archipelago. We have also conducted reforestation experiments in other countries, including those in Southeast Asia, the Brazilian Amazon (Miyawaki, 1993; Miyawaki and Abe, 2004), China, and Inner Mongolia. In total we estimate that we have planted over 30 million trees in over 1500 sites.

For example, we planted trees to regenerate a forest serving primarily an environmental protection function. In 1984, we were asked to preserve a rocky hillside of a 40° slope that was excavated during the construction of school buildings of Kanagawa Prefectural Kurihama High School in central Japan. The conventional method for slope protection was to spray seeds of an exotic grass on slopes, or to pour cement on rocky hillsides. However, our intention was to restore a native forest on the slope and thereby stabilize the slope more effectively than by these other methods. To accomplish this, we cut horizontal, narrow V-shaped ditches running along the slope contours on the hillside, and filled them with rich topsoil. We planted potted seedlings from the potential natural vegetation identified through vegetation field surveys. Three years later, we found that the root systems of the seedlings had grown 4 m long through rifts, and that the average tree had attained a height of 3 m. At present, a 10-m-high quasi-natural forest has formed and protects the hillside and the human communities downslope (Miyawaki, 2002c).

We have also restored tropical rainforests on Borneo Island in Malaysia. Tropical rainforests are among the most productive of ecosystems on earth, but also among the most sensitive to human disturbance. When a road is constructed through such a forest, the heavy rainfalls wash away a great deal of soil, limiting the ability of this system to reestablish itself. Overgrazing, rampant tree harvesting, shifting cultivation, and establishment of oil palm and rubber tree plantations are primary causes of deforestation in the island of Borneo. Tropical rainforests in Malaysia are now nearly extinct, and are found only in limited areas such as the National Parks of Niah, Lambir Hills, and Similajau in Sarawak.

The site of our tropical reforestation experiment was barren land that had once been under shifting cultivation and was located on the Bintulu campus of the University of Agriculture, Malaysia (Universiti Putra Malaysia, UPM) on the island of Borneo. The project was funded by a far-sighted Japanese company and in cooperation with Yokohama National University and University of Agriculture,

Malaysia (UPM). Following the method of ecological reforestation we used in Japan, we conducted vegetation field surveys and identified the local potential natural vegetation. After collecting seeds, germinating them, and nursing potted seedlings with well-developed root systems, we began planting seedlings in 1991. Every year thereafter, we have continued planting potted seedlings to regenerate rainforests at Bintulu. We did not plant seedlings of exotic species, like *Eucalyputus, Acacia mangium*, and longleaf pine trees, but instead planted species of Dipterocarpaceae and other species selected from the potential natural vegetation of the region. Japanese volunteers even went to Borneo to join the planting teams. In total, we have already planted 390,000 potted seedlings of 91 species (Miyawaki, 1993; Meguro and Miyawaki, 1997). Some have grown 15 to 18 m high within 15 years of being planted (as of August 2005).

Conclusion

Unlike monocultured forests of needle-leaved trees, native forests of the potential natural vegetation help conserve and perhaps augment biodiversity not only in the multilayered plant communities above ground but also in soil communities below ground. Seeds that fall or are carried by small animals and birds into the network of green corridors in urban areas may germinate in the forest floor, and raise the biodiversity of the forest. Individual trees and species communities may change over time, but the forest system and the benefits they provide to urban citizens can be sustained for long periods.

In the new millennium, the concept of urban forests should not simply encompass older ideas of tree planting for lumber production or beautification. Instead, urban forests ought to be conceived as native forests containing the potential natural vegetation of the region that function as buffers for environmental protection and disaster mitigation. However, they should also be a source of intellectual excitement, and aesthetic and spiritual inspiration for residents of modern, standardized, "cement-desert" cities. These forests should also be constructed or restored to provide resources that enhance human existence and maintain plant gene pools for the future.

We have established basic principles for the restoration of vertically structured, naturalized forests and their ecological functions throughout Japan and Southeast Asia. We have conducted these restoration experiments since the 1970s at more than 1500 sites throughout the 3000-km long Japanese Archipelago, and, since the 1980s, in Southeast Asia, China, and South America. In each project the local residents were leading participants, while governments and private companies provided funding and other strategic support. These reforestation projects are an important part of the eco-city movement, which has the potential to improve the quality of life of urban residents throughout the world. Far-sighted people with the power to execute bold changes in Shanghai and Yokohama have already started constructing ecological urban forests on a large scale. Perhaps these large-scale urban reforestation experiments can serve as models to inspire the spread of the eco-city movement throughout the world (Fig. 12.4).

Fig. 12.4 **(A)** Dense and mixed planting of potted seedlings with fully developed root system on a 45° slope near the main entrance of Yokohama National University in 1978. Planted tree species were evergreen *Quercus (Q. myrsinaefolia, Q. glauca) Castanopsis sieboldii, Persea thunbergii*, etc., which are the main and companion species from the potential natural vegetation in the region. Fifteen tree species were planted here in June 1978. **(B)** Same site in June 2005

References

Carreiro, M.M. (2005) Effects of urbanization on decomposer communities and soil processes in forest remnants. In: Johnson, E.A., and Klemens, M.W. (eds.) Nature in Fragments. Columbia University Press, New York.

Meguro, S., and Miyawaki, A. (1997) A study of initial growth behavior of planted Diptrocarpaceae trees for restoration of tropical rainforests in Borneo/Malaysia. Tropical Ecology 38(2): 237–245. International Society for Tropical Ecology.

Miyawaki, A. (1992) Restoration of evergreen broad-leaved forests in the Pacific region. In: Wali, M.K.(ed.) Ecosystem Rehabilitation. 2. Ecosystem Analysis and synthesis. SPB Academic Publishing, The Hague, pp. 233–245.

Miyawaki, A. (1993) Restoration of native forests from Japan to Malaysia. In: Lieth, H., and Lohmann, M.(eds.) Restoration of Tropical Forest Eco-systems. Kluwer Academic Publishing, the Netherlands.

Miyawaki, A. (1998) Restoration of urban green environments based on the theories of vegetation ecology. Ecological Engineering 11:157–165.

Miyawaki, A. (1999) Creative ecology: restoration of native forests by native trees. Plant Biotechnology 16(1):15–25.

Miyawaki, A. (2001) Philosophy and practice of ecology in Japan in special regards to forest restoration. 86th annual meeting of Ecological Society of America. Abstracts. p. 27. Wisconsin, August 5–10.

Miyawaki, A. (2002a) Construction of anti-disaster, environment protection forest based on the vegetation ecology. Proceedings of the Symposium on Forest Conservation and Sustainable Development. Guizhou Provincial Peoples Publishing House, China, pp. 28–29.

Miyawaki, A. (2002b) Ecological restoration and creation of living environments—principles and applications. Plenary lecture, VIII INTECOL Ecology in a Changing World. p. 26. Seoul, Korea, August 11–18.

Miyawaki, A. (2002c) Restoring forests on slopes. Landscape Greening 9:6–7. Study Group of Landscape Greening, Tokyo (in Japanese).

Miyawaki, A. (2004) Restoration of living environment based on vegetation ecology. Theory and practice. Ecological Research 19:83–90. Blackwell Publishing Asia, Australia.

Miyawaki, A., and Abe, S. (2004) Public awareness generation for the reforestation in Amazon tropical lowland region. Tropical Ecology 45(1):59–65. International Society for Tropical Ecology.

Miyawaki, A., and Box, E.O. (1996) The Healing Power of Forests: The Philosophy behind Restoring Earth's Balance with Native Trees. 286 pp. Kosei Publishing, Tokyo.

Miyawaki, A., Fujiwara, K., and Ozawa, M. (1993) Native forests by native trees—restoration of indigenous forest ecosystem (reconstruction of environmental protection forest by Prof. Miyawaki's Method). Bulletin of the Institute of Environmental Science and Technology, Yokahama National University 19:73–107. Yokohama (in English and Japanese).

Miyawaki, A., and Golley, F.B. (1993) Forests reconstruction as ecological engineering. Ecological Engineering 2:333–345. Elsevier, Amsterdam.

Miyawaki, A., and Meguro, S. (2000) Planting experiments for the restoration tropical rainforest in Southeast Asia and a comparison with laurel forest at Tokyo Bay. Proceedings IAVS Symposium. Opulus Press, Uppsala, Sweden, pp. 249–250.

Miyawaki, A., and the reporting group of Mainichi Shinbun. (2004) Planting Tomorrow. Mainichi Shinbun-sha (in Japanese).

Tüxen, R. (1956) Die huetige potentielle natürliche Vegetation als Gegenstand der Vegetationskartierung. Angewandte Pflanzensozioligie 13:5–42. Stolzenau/Weser.

II
Planning, Managing, and Restoring Urban Forests

13 Strategic Planning for Urban Woodlands in North West England

Keith Jones

The North West was arguably the first region to pollute the environment on a structured, grand, even imperial scale in the desire for economic growth. This new millennium will be an age when we can set our sights on reversing that process, based on the principles of sustainable development.

—*Lord Thomas of Macclesfield*
Past Chair, North West Regional Development Agency

Overview of North West England

North West England (Fig. 13.1) covers the counties of Cumbria, Lancashire, Merseyside, Greater Manchester, and Cheshire, an area of 14,110 square kilometers (5448 square miles). The North West contains 11.3% (6.7 million) of England's total population. Population densities, especially in parts of Greater Manchester and Merseyside, are among the highest in Europe. The region has the fourth largest (out of 12) United Kingdom gross domestic product (GDP) at £77,652 billion, but the ninth lowest GDP per person at £11,273. Environmental quality is especially high in Cumbria and the Lake District, and exceptionally poor, by Western standards, in parts of Lancashire, Greater Manchester, and Merseyside. The region also has one of the lowest levels of woodland cover at 6.5% (96,000 hectares [ha]) in England. In urban parts of the Region woodland cover is so limited that there is only 1.8 ha per 1000 population (Table 13.1).

Having a large work force, natural harbors (such as Liverpool), large coal reserves, and ample water supplies, large areas of the North West developed as the hub of Britain's industrial revolution. Industries such as cotton, mining, chemicals, and munitions stamped large industrial footprints across the North West and generated vast wealth. However, during the late 20th century many of these heavy industries declined, leaving a legacy of unemployment and dereliction. Past industries' footprints became industrial scars on the environment and landscape, particularly in West Cumbria, South East Lancashire, Greater Manchester, Merseyside, and North Cheshire. As a result the North West now has around 25% of England's derelict land, perhaps as much as 30,000 ha. The 2002 Derelict,

M.M. Carreiro et al. (eds.), *Ecology, Planning, and Management of Urban Forests: International Perspectives.*

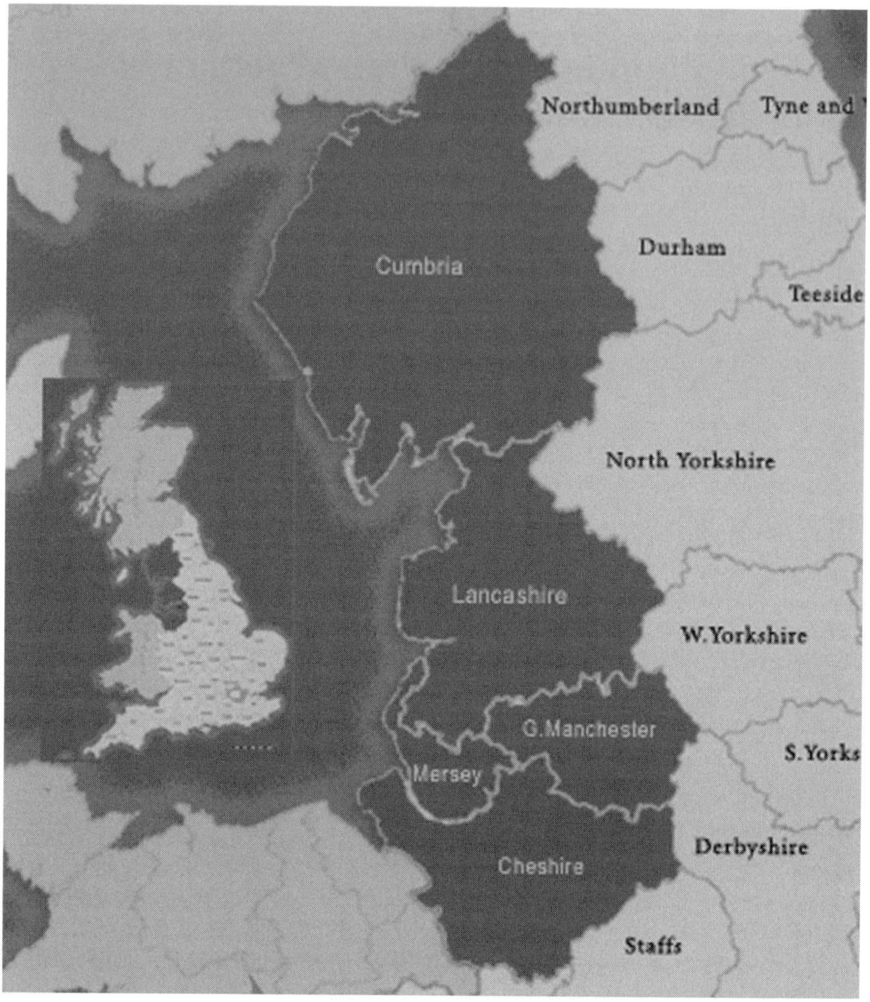

Fig. 13.1 Map of Great Britain with region of study

Table 13.1 North West England population and woodland statistics

County	Population	Total land area (ha)	Woodland area (ha)	Current woodland cover (%)	Woodland ha per 1000 population
Cumbria	495,000	680,400	64,582	9.5	130
Lancashire	1,434,000	298,900	14,078	4.6	9.8
Merseyside	1,386,000	66,000	2,478	3.7	1.8
Gtr Manchester	2,560,000	126,900	4,695	3.4	1.8
Cheshire	995,000	238,000	10,337	4.4	10.4
North West Totals	6,871,000	1,411,000	96,171	6.8	14

Source: Forestry Commission Inventory of Woodlands, 2001.

Underused, and Neglected (DUN) Land Survey recorded 3893 sites in North West England covering 26,385 ha with 1627 sites (14,915 ha) in previously developed land (North West Development Agency [NWDA], 2006; see also http://www.englandsNorth_West.com/englandsNorth_West_news/facts_and_figures/).

The Derelict Land Legacy and the Opportunity Provided by Woodlands

The huge mass of derelict land places an immense drag on the region's social, economic, and environmental well-being. It contributes to social deprivation and to a downbeat image that inhibits economic growth. People have to live and work surrounded by a very poor environment. This has led to a number of government departments and agencies as well as multidisciplinary partnerships assessing the problems and seeking solutions. Of particular note, the NWDA instigated and published the "Land Reclamation Review—Reclaim the North West." This report stated that commercial "hard-end development" (i.e., buildings and roads) would not reclaim sufficient derelict land. The report recommended that a new and imaginative "soft-end use" approach (development of woodlands and other green infrastructure) would need to be developed. At the same time, the U.K. government's first England Forestry Strategy (1998) advocated the multiple public benefit potential of woodlands to contribute to social, economic, and environmental growth (Table 13.2), as well as highlighting a woodland's role in economic regeneration. Thus in those urban areas where derelict land is concentrated and woodland cover is low, there is a huge opportunity to deliver sustainable public benefits via increasing woodland cover on derelict land. This opportunity was the foundation of a new partnership (entitled Newlands) between the NWDA and the Forestry Commission (FC) and other partners including the Red Rose and Mersey Community Forests. The FC is the government department responsible for forestry policy throughout Great Britain. Its mission is to protect and expand Britain's forests and woodlands and increase their value to society and the environment. The NWDA is responsible for the sustainable economic development and regeneration of England's North West and has five key priorities: business development, regeneration, skills and employment, infrastructure, and image.

Strategic Integrated Planning: The Public Benefit Recording System

It is vital to provide the best sustainable solution from the inevitable mixture of competing sectoral (social, economic, and environmental) interests. Competing priorities, strategies, needs, and opportunities need to be discussed and resolved so as to maximize the added value from the limited resources available.

Table 13.2 Example benefits of woodlands

Urban environment	Trees can save up to 10% of energy consumption through their moderation of the local climate. They also stabilize the soil, prevent erosion, and reduce the affects of air pollution and storm-water run-off.
Contaminated land	Woodland can assist in the remediation of contaminated land by reducing soil erosion and off-site particulate migration.
Healthier lives	Trees reduce the incidence of asthma, skin cancer, and stress-related illness by filtering polluted air, reducing smog formation, shading out solar radiation and by providing an attractive and calming setting for recreation.
Community development	Involving communities in the development of their environment assists community capacity building; reducing deprivation.
Education	Woodland makes an excellent outdoor classroom close to schools and homes and is capable of contributing to a wide range of the curriculum.
Wildlife	Trees play a vital role in the urban ecosystem by helping to attract and support a wide variety of wildlife, which people can enjoy close to home.
Landscape	Trees soften the landscape of towns and cities. making them greener and more attractive to live in. They are also particularly successful in tackling the scars left by industrial decline, mineral extraction, and landfill sites.
Local economy	Woodland can help to transform a local economy by making an area attractive to inward investment. It can also help to increase property values and provide jobs, particularly in the intermediate labour market.
Useful products	Even in towns, trees can yield useful products like timber, renewable fuel, wood chip mulch, charcoal, etc. These all help to provide a focus for small businesses and community life while generating some income to contribute to long-term management costs.
Efficient use of land	For an equivalent area, woodland is able to absorb greater numbers of people enjoying more diverse recreational pursuits than would be possible in open grassland.
Cost-effective land-use option	In the long-term, woodlands are low cost in relation to recovery for hard-end use and generally cheaper to sustain than mown grass. They are also capable of delivering a much greater range of sustainable public benefits.

To address this conundrum, a group of sectoral experts developed an integrated (social, economic, and environmental) assessment tool: the Public Benefit Recording System (PBRS). The tool was designed to do the following:

- Create a way of assessing and balancing competing agendas and priorities.
- Maximize the synergy and sustainability of actions so as to achieve maximum social, economic, and environmental outputs and outcomes via woodlands development.
- Enable the creation of a sustainable "joined-up strategy" and a shared vision.
- Facilitate ownership, partnership working, and delivery.

Overview of the Public Benefit Recording System

The PBRS was developed to highlight the potential public benefits arising from the regeneration of derelict land in the Mersey Belt via the creation of community and urban woodland. It was designed to be an objective tool that can be used to jointly evaluate potential economic, social, and environmental benefits. It was developed alongside an aerial survey of a DUN land survey of the North West of England. The PBRS creates a cohesive view as to how to target the woodland regeneration of derelict land and maximize the benefits arising from the creation of new public open space.

The PBRS uses four categories of public benefit: social benefit indicators, public access indicators, economic benefit indicators, and environmental benefit indictors. Within every category a range of relevant objective attributes has been created, and data for each site and its locality are recorded.

Social Benefit Attributes

This section of the PBRS uses different measures of social deprivation to assess how much benefit there would be to local communities from the establishment of urban or community woodland.

1. The Index of Multiple Deprivation (IMD) was developed by researchers at Oxford University and derived from scores and ranks given to domains of income, employment, health, education, housing, access, and child poverty. The overall IMD score measures the level of social deprivation within a particular ward, while the rank value indicates the level of deprivation within a ward in comparison to other wards within the U.K. For the purpose of allocating PBRS, only the IMD score has been used, as other wards within the U.K. influence the ranks and we are only concerned with wards in Greater Manchester, Merseyside, and North Cheshire.

    a. *IMD score of most deprived ward within 500 m of site perimeter*: The IMD score awarded to a site is taken as the most deprived ward (highest IMD score) within 500 m of the site boundary (acceptable walking distance). This technique ensures that if the site boundary covers two wards or a site is surrounded by several different wards within 500 m of its boundary, then the one that is most deprived within the 500m gets priority and is used to measure expected social benefit.
    b. *Index of Multiple Deprivation (National Context Score)*: The recorded IMD score is then used in both a national and local context within the PBRS. The national IMD social benefit score compares the level of social deprivation for the ward within which the site is located against other wards in the U.K. The thresholds are based on percentiles (50%, 62.5%, 75%, and 87.5%). If a ward has an IMD score that is at a level within the top (87.5%) percentile, then it will be within the top 12.5% highest scoring wards in the country.

c. *Index of Multiple Deprivation (Local Context)*. In the local context the IMD score is measured against the district average and the county average. Higher social benefit scores are awarded where the IMD score is greater than both the district and county average. Lower scores are given where the IMD score is less than both the district and county average.
d. Figure 13.2 illustrates DUN sites in relation to six bandings or levels of multiple deprivation. These are the national scores for the IMD. The lowest band (lightest blue) applies to wards whose IMD score is both lower than the national mean and the national median IMD score. The second band applies to wards whose IMD score is between the national mean and the national median. Subsequent bands apply in 12.5 percentile bandings above the national median IMD score, up to 100% (e.g., band 3 is for wards that rank between 50% and 62.5% in the national ranking; band 4 is for wards that are between 62.5% and 75% of national ranking of IMD). Again, high percentiles (darker blue) indicate greater social benefit from creation of public woodlands.
e. It is clear that there is a significant location of DUN sites in wards that are above 62.5% ranking of national IMD and so on.

2. *Proportion of 500 m site perimeter buffer occupied by housing*: There will be a greater need for good quality open space where the proportion of housing around a site is high, and there will be a larger number of people who will benefit from an enhanced environment. The proportion of land occupied by

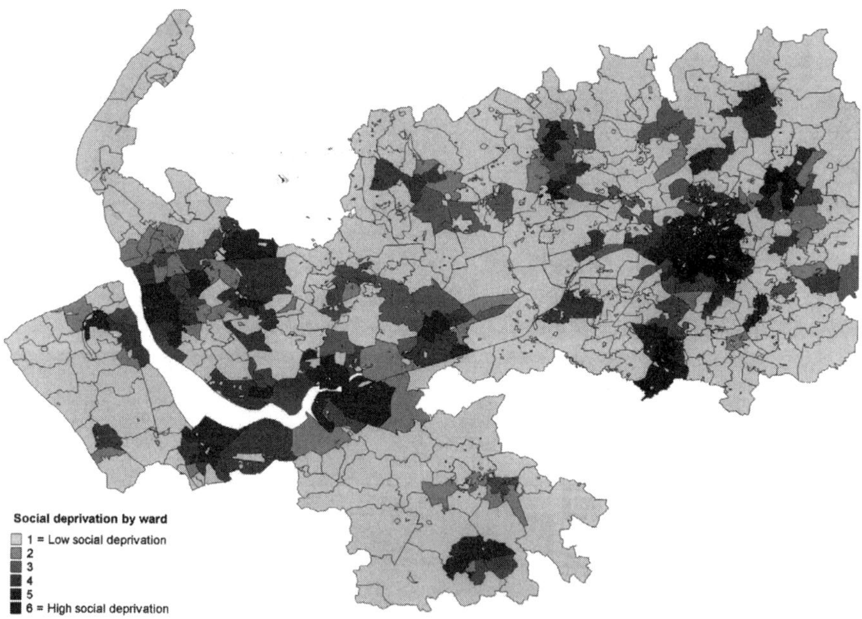

Fig. 13.2 Deprivation and derelict, underused, and neglected (DUN) land. The darker the area, the greater the social benefit derived from creation of public woodlands

housing is only a proxy for actual population within 500 m of the site. However, budgetary restrictions meant that population estimates from the 1991 census were not used.

3. *Site size (hectares)*: Higher scores are given to the larger sites. These are strategically more feasible to develop, but may score poorly with respect to other social benefit indicators, given that they are predominantly located in urban fringe and rural areas where social deprivation may be lower, or where there are fewer people to receive any social benefits.
4. *Designated Health Action Zones, New Deal for Communities, Education Action Zones, Employment Zones*: These are allocations given by the government to deprived areas in an attempt to improve the health of individuals and health care services provided, improve performance in schools, and encourage individuals to get involved in training programs and employment. These zones are used as indicators of social benefit because Community Woodlands contribute to each of these objectives, by encouraging people to take outdoor exercise in a safer, healthier environment, by providing a valued educational resource, and encouraging people to take part in training programs and job-seeking activities.
5. *Number of schools within 1 km of site perimeter*: In addition to the Education Action Zones, the social benefit scores include the number of primary and secondary schools within a 1-km radius of the site perimeter. The larger the number of schools, the greater will be the education benefit of increased access to nature and local ecological diversity.

Public Access Attributes

Public access indicators assess existing access within the site and transport links to the site from neighboring communities. High scores indicate that the DUN site can significantly increase the quantity and quality of local access to public open space.

The following indicators are used to score the DUN site in terms of public access.

1. *Is the site on a footpath (public rights of way only)*? Footpaths are the most important type of access, as adequate footpaths encourage people to walk to and around a site (encouraging exercise) and help to reduce pollution by discouraging the use of fuel-driven vehicles. They are most beneficial to individuals/families who are without the use of a car and who may find accessing similar woodlands difficult. The scores for this indicator are based only on established public rights of way mapped on ordnance survey plans, as this gives a better impression of the extent to which people legally have access to a community woodland, should one be established.
2. *Existing public use within the site*: Existing public use of the site indicates the popularity of the site. The appearance of well-used pathways (public rights of way) and unmarked and clearly marked areas of activity suggests a site is well used. Therefore, improved management of the site will have a guaranteed benefit

on the current users and encourage additional people to visit the site. Conversely, improvements to a poorly used site may go unnoticed or require additional input to encourage people to even visit the site.

3. *Is the site within 500 m of a train station? Is the site on a bus route*? Indicators relating to the proximity of a site to bus routes, railway stations, and metro stations allocate scores where public transport facilities are available. There are both environmental and social benefits to be gained from having access to the site via public transport through reductions in air pollution and accessibility to nature for households without a car.
4. *Does the site have direct cycleway/bridleway access*? Cycleway and bridleway access will also encourage people to take more exercise and further reduce reliance on cars.
5. *Is the site on a primary road*? Sites that are accessible from or visible from primary roads also score highly for public accessibility, as there is potential for these sites to benefit a wider catchment of people than sites that are located "off the beaten track" or hidden in the center of small housing estates.
6. *Is parking available*? The availability of parking adjacent to a site will make the site accessible to people who do not live within walking distance of the site or do not have access to public transport. It will also encourage the site to be used by passersby or people wishing to break up journeys or visit different areas.
7. *Proximity to other public open space (POS) greater than 1 ha*. This indicator gives higher scores that correspond to increasing distances between the site and other POSs. There will be increased benefits from new community woodland in areas that presently have little access to public open space.

Economic Benefit Attributes

One of the objectives of urban and community woodland establishment is to encourage investment in areas of economic blight and high unemployment by making business parks and industrial sites more aesthetically attractive to investors. Woodlands can also provide screening to housing areas located near large industrial activities or unsightly business parks. The economic benefit indicators measure the current economic climate of the area surrounding the site and the proximity of the site to existing and proposed business parks and industrial areas. The economic benefit section also takes into account the proximity of the site to retail developments. Environmental improvements may also help to encourage the development of new shops or associated leisure facilities by introducing more people to an area either as visitors or new residents.

The economic benefit section also scores the local economic climate by reference to prevailing house prices in the district. Proximity to woodland can increase house prices slightly, a trend that can encourage local regeneration of the housing sector. Finally, enhancing and improving sites adjacent to main transport corridors can benefit the local economy by creating a good impression of an area at locations that will receive the most attention from visitors and potential investors. In contrast, areas of derelict and neglected land are a strong disincentive to inward investment of high-value industry or residential development.

Environmental Benefit Attributes

This section takes into account the benefits that new community woodland may have on the environment surrounding a DUN site. The indicators used are based on information that can be obtained at a desktop level, as the project does not allow for site investigations. Detailed landscape and ecological surveys of the site would be needed as part of the community woodland design. Although biodiversity and landscape quality are not assessed in detail at this point in the project, it does not mean that they are unimportant. Careful establishment of community woodland will in many cases improve the environmental quality of a site by protecting and enhancing existing features of biodiversity. However, where DUN land is already designated as being ecologically important, then there will be a reduction in the score for environmental benefit as the designation will probably mean that existing open habitats are important and thus may not be appropriate for community woodland establishment.

The following indicators are used to score environmental benefit:

1. *Proximity to ancient woodland and proximity to other woodland (scored separately)*: In terms of biodiversity and in terms of wildlife corridors, larger woodland areas are significantly more valuable than small fragmented woodlands. A DUN site that could extend or link existing woodlands scores highly.
2. *Proximity to areas of ecological interest*: Where sites are adjacent to or within 500 m of existing areas of designated ecological interest, there is the potential to improve or protect the quality of the designated site through the establishment of community woodland. Additionally, community woodlands can serve to increase ecological diversity in locations where areas of ecological interest are scarce. If the DUN site is already designated for ecological value, then the score is reduced, as there would be no benefit from new woodland establishment.
3. *Proximity to waterbody*: Where sites are derelict and potentially contaminated, establishment of community woodland would be of great benefit to water bodies in and around the site through adsorption of contaminants by trees. Woodland can also increase diversity of riparian zones by providing protection from soil erosion and trampling, thus allowing new vegetation to establish.
4. *Air Quality Management Areas (AQMAs)*: AQMAs are designations given by local authorities to areas within a district that have poor air quality as a consequence of existing developments, or where new developments could detrimentally affect air quality due to the cumulative effects likely to occur

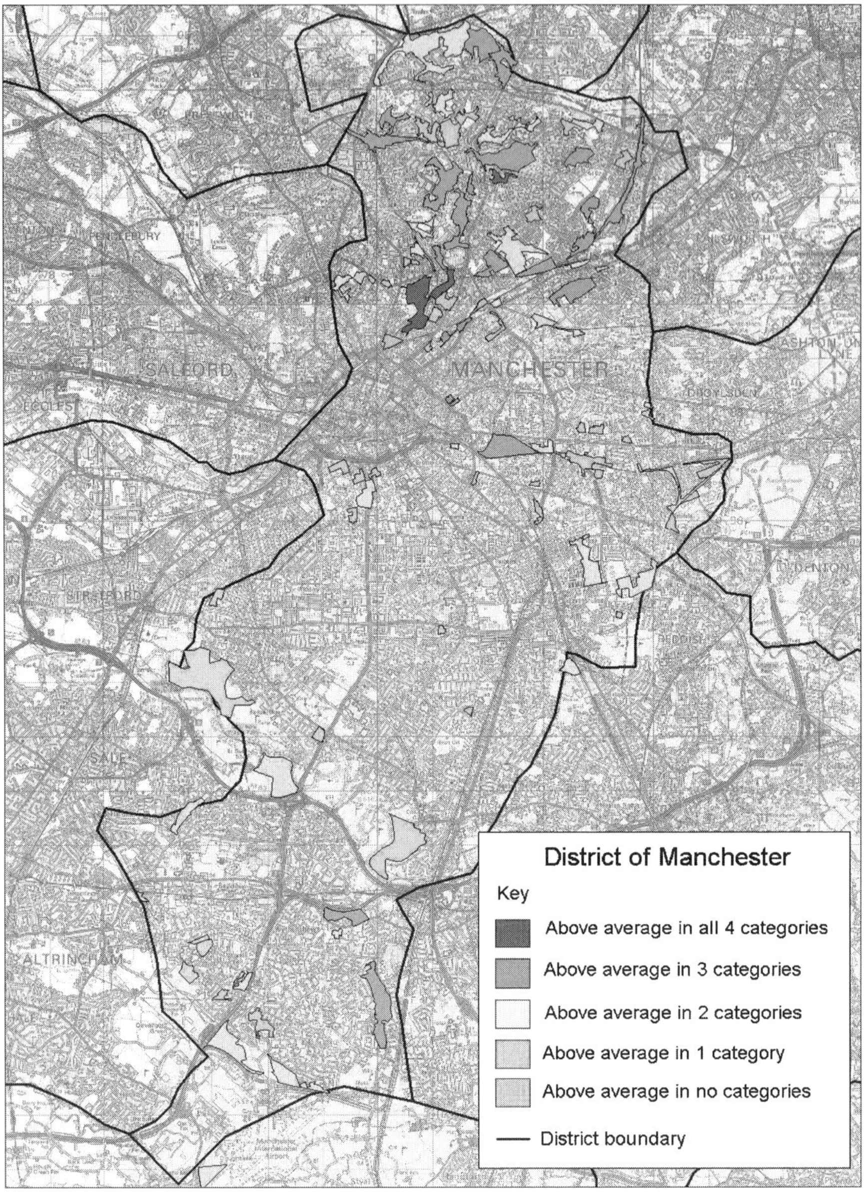

Fig. 13.3 Example of a district's (Manchester) derelict land sites scores using the Public Benefit Recording System (PBRS)

when added to existing developments. Therefore, AQMAs indicate that local air quality is poor. Introduction of community woodlands within an AQMA will benefit the environment by improving air quality through the uptake of air pollutants.

5. *Proximity to transport corridors*: Trees reduce air pollution from vehicles and provide a screen from visual and noise intrusion. Transport corridors are also used as an economic benefit indicator. Corridors are defined as motorways, primary roads, railways, major rivers, and roads near airports.

For every site, a separate PBRS score is derived for each attribute. These are then totaled to provide an overall category score for each site. To ensure that each category is given equal importance, category scores are not added together. Instead, scores for each category are assessed against district averages enabling the potential public benefit of a site, or groups of sites, to be assessed according to local and subregional needs and priorities (see Appendix). This assessment is made by color coding the scores and presenting them in map form using a Geographical Information System (GIS) system (Fig. 13.3).

Using the PBRS and presenting results in this way enables decisions to be made against a holistic overview of the potential social, economic, and environmental outcomes. Despite its original focus on urban woodland establishment, there is increasing interest from other agencies in developing and applying the PBRS to other land-use and development decisions as an aid to strategic planning and investment, and as a means of combining policies and priorities for promoting "joined-up" thinking. (For a detailed overview of the PBRS, go to http://www.pbrs.org.uk).

Application of the Public Benefit Recording System

One of the first applications of the PBRS was with the Newlands (New Economic Environments via Woodlands) program. Newlands, which is targeted at the regeneration of the North West's derelict land, is being developed by the Forestry Commission and the NWDA, and began operation in 2002. The program will eventually operate throughout North West England, but initially will build on the achievements of the Mersey and Red Rose Community Forests, Groundwork, and the Forestry Commissions Land Regeneration Unit, in the Mersey Belt. The program, funded by the NWDA, will be managed by the Forestry Commission and will contribute directly to the regional economic strategy and the economic regeneration theme of the England Forestry Strategy.

The program illustrates the strategic application of the PBRS, as it will be utilized to target activity and create holistic programs of activity at the district and county level. Indeed, a site's PBRS score will be an essential step in identifying suitable sites for Newlands activity and investment, where sites will be chosen according to the process shown in Figure 13.4.

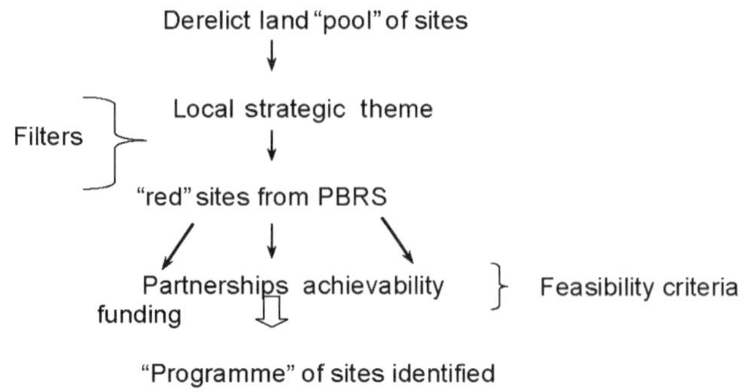

Where:

Local Strategic Theme – subregions, districts, or funding programs may have an overarching strategic theme. For example, the Mersey Belt has a thematic priority to improve the economic image of the area via improvements to its transport corridors.

PBRS 'red' sites – those sites identified in the DUN survey and highlighted as important via above average PBRS 'score' in all categories.

Fig. 13.4 Process for choosing sites for Newlands activity and investment

Conclusion

The Forestry Commission, working with partners, including the North West Regional Development Agency, has created a GIS-based holistic (social, economic, and environmental) aid to strategic planning and investment for woodlands called the Public Benefit Recording System. The PBRS is proving to be a valuable aid to strategic planning and for targeting investment. Across North West England, the PBRS approach has been adopted to create fresh social, economic, and environmental partnerships and joint added-value action plans. This chapter has explained how this tool has been used to target and prioritize new sustainable urban woodlands dealing with the regions legacy of derelict land. Up-to-date information on the PBRS can be found at www.pbrs.org.uk.

References

England Forestry Strategy. (1998) www.forestry.gov.uk. North West Development Agency. (2006) Reclaim the North West. www.nwda.co.uk.

Glossary of Terms

Attribute Feature of the site or its locality that is recorded and scored (e.g., site size, local social deprivation, proximity to schools).
Category Type of public benefit that is being assessed, divided into four categories of social, access, economic, and environmental benefits.

District An administrative subregion of an area (e.g., Liverpool is a district within Merseyside).
Record Factual information including all the attributes of an individual site (e.g., number of schools within 500 m of site, number of schools within 1 km of site, whether site is within a health action zone, etc.).
Score An arbitrary number given in respect of a record (e.g., sites over 30 hectares score higher than sites over 5 hectares).
Threshold/Criterion The point at which a score is given, or is increased (e.g., percentile rankings of Index of Multiple Deprivation).

Appendix: Public Benefit Scoring Form—List of Options and Threshold Values for Scoring

Section 1: Site Details (Information Only; Not Scored)

Location (drop-down table options)
Rural
Urban fringe
Inner City
Town
Village

Data source: aerial/map interpretation.

Section 2: Land Use and Planning Context (Information Only; Not Scored)

Current land status (drop-down table options)
Derelict
Underused
Neglected

Data source: aerial interpretation/desktop information sources.

Ownership (drop-down table options)
Private
Local authority
Other public sector (e.g., ministry of defense)
Public utilities (e.g., rail track)
Not known

Data source: consultation with local authority/community forest team.

Section 3: Social Benefit Section—Scoring System

Social deprivation in a national context (drop-down table options)

Indicator	Thresholds	Score
Index of Multiple Deprivation scores for worst ward within 500 m of site (derived from DETR Index)	<16.93, better-off than national average and national median ward	1
	16.93 to 21.69, between national average and national median (50%) ward	2
	21.69 to 22.10, in 50–62.5% percentile range for deprivation	3
	22.10 to 29.14, in 62.5–75% percentile range for deprivation	4
	29.14 to 41.01, in 75–87.5% percentile range for deprivation	5
	>41.01, in most deprived 12.5% of wards in England	6

Data source: DETR Index of Multiple Deprivation.

Social deprivation in a local context (drop-down table options)

Indicator	Thresholds		Score
Local Index of Multiple Deprivation (IMD) for worst ward within 500 m of site	Bolton	<34.26	1
		34.26 to 36.50	2
		>36.50	3
Score of 1 indicates that the highest IMD score (most deprived ward) within 500 m of the site is lower than both the district and county average; therefore, the ward is better off than average	Bury	<25.05	1
		25.05 to 36.50	2
		>36.50	3
	Manchester	<36.50	1
		36.50 to 57.24	2
		>57.24	3
	Oldham	<36.50	1
		36.50 to 38.71	2
		>38.71	3
Score of 2 indicates the ward has a level of deprivation that lies between the district average and county average	Rochdale	<36.50	1
		36.50 to 40.96	2
		>40.96	3
	Salford	<36.50	1
		36.50 to 43.71	2
		>43.71	3
Score of 3 indicates the ward has a level of deprivation that is higher than both the district and county average	Stockport	<20.27	1
		20.27 to 36.50	2
		>36.50	3
	Tameside	<35.33	1
		35.33 to 36.50	2
		>36.50	3
	Trafford	<21	1
		21 to 36.50	2
		>36.50	3
	Wigan	<34.65	1
		34.65 to 36.50	2
		>36.50	3
	Knowsley	<47.50	1
		47.50 to 59.57	2
		>59.57	3

(continued)

(continued)

Indicator	Thresholds		Score
	Liverpool	<47.50	1
		47.50 to 60.44	2
		>60.44	3
	St. Helens	<39.51	1
		39.51 to 47.50	2
		>47.50	3
	Sefton	<33.53	1
		33.53 to 47.50	2
		>47.50	3
	Wirral	<37.17	1
		37.57 to 47.50	2
		>47.50	3
	Ellesmere Port and Neston	<19.77	1
		19.77 to 26.25	2
		>26.25	3
	Halton	<19.77	1
		19.77 to 40.25	2
		>40.25	3
	Vale Royal	<18.49	1
		18.49 to 19.25	2
		>19.25	3
	Warrington	<19.25	1
		19.25 to 21.67	2
		>21.67	3

Data source: DETR Index of Multiple Deprivation.

Area of 500-m perimeter buffer occupied by housing (drop-down table options)

Threshold	Score
<10%	0
10% to 30%	1
30% to 50%	2
50% to 70%	3
70% to 90%	4
>90%	5

Data source: aerial and map interpretation.

Site size (hectares) (drop-down table options)

Threshold	Score
<1 ha	1
1 to 5 ha	2
5 to 10 ha	3
10 to 30 ha	4
>30 ha	5

Data source: digitization of OS landline maps.

Designated health action zones (drop-down table options)

Threshold	Score
Yes	1
No	0

Data source: local authority.

Designated new deals for the communities (drop-down table options)

Threshold	Score
Yes	1
No	0

Data source: local authority.

Designated education action zone (drop-down table options)

Threshold	Score
Yes	1
No	0

Data source: local authority.

Number of primary and secondary schools within 1-km radius of site perimeter (drop-down table options)

Threshold	Score
0 schools	0
<3 schools	1
4 to 6 schools	2
7 to 9 schools	3
>10 schools	4

Data source: aerial and map interpretation.

Section 4: Public Access Benefit Section—Scoring System

On a direct footpath link (drop-down table options)

Threshold	Score
Yes	1
No	0

Data source: aerial and map interpretation.

Current public use of the site (drop-down table options)

Threshold	Score
Poor	1
Good	2
Excellent	3

Data source: aerial and map interpretation.

Within 500 m of train station/metro station (drop-down table options)

Threshold	Score
Yes	1
No	0

Data source: aerial and map interpretation.

Within 25 m of cycleway/bridleway access (drop-down table options)

Threshold	Score
Yes	1
No	0

Data source: aerial and map interpretation.

Availability of adjacent car parking (drop-down table options)

Threshold	Score
Yes	1
No	0

Data source: aerial and map interpretation.

On a primary road (drop-down table options)

Threshold	Score
Yes	1
No	0

Data source: aerial and map interpretation.

On a bus route (drop-down table options)

Threshold	Score
Yes	1
No	0

Data source: aerial and map interpretation, local authority.

Proximity of nearest public open space of >1 ha (drop-down table options)

Threshold	Score
Adjacent	1
<500 m	1
500 m to 1 km	2
1 km to 2 km	3
>2 km	4

Data source: aerial and map interpretation.

Section 5: Economic Benefit Section—Scoring System

Proximity to business park (drop-down table options)

Threshold	Score
Within	3
Adjacent	3
<500 m	2
500 m to 1 km	1
>1 km	0

Data source: map and UDP interpretation.

Proximity to areas of major industrial activity (drop-down table options)

Threshold	Score
Within	4
Adjacent	4
<500 m	3
500 m to 1 km	2
1 km to 2 km	1
>2 km	0

Data source: map and UDP interpretation.

Proximity to UDP proposed employment area (drop-down table options)

Threshold	Score
Within	3
Adjacent	3
<500 m	2
500 m to 1 km	1
>1 km	0

Data source: map and UDP interpretation.

Designated employment zone (drop-down table options)

Threshold	Score
Yes	1
No	0

Data source: map and UDP interpretation.

Proximity to local centers of commerce and larger retail outlets (drop-down table options)

Threshold	Score
Within	3
Adjacent	3
<500 m	2
500 m to 1 km	1
>1 km	0

Data source: map and UDP interpretation.

House prices (drop-down table options)

Threshold		Score
Detached	<108,300	2
Detached	≥108,300	1
Semi-Detached	<57,700	2
Semi-Detached	≥57,700	1
Terraced	<38,000	2
Terraced	≥38,000	1

Data source: government housing relocation allowance figures for differing local authorities.

Proximity to a primary transport corridor (drop-down table options)

Threshold	Score
Adjacent	3
<500 m	2
500 m to 1 km	1
>1 km	0

Data source: map and UDP interpretation.

Section 6: Environmental Benefit Section—Scoring System

Proximity to areas of ancient woodland (drop-down table options)

Threshold	Score
Adjacent	3
<500 m	2
500 m to 1 km	1
>1 km	0

Data source: English nature inventory of ancient woodland.

Proximity to areas of woodland (drop-down table options)

Threshold	Score
Adjacent	3
<500 m	2
500m to 1 km	1
>1 km	0

Data source: aerial and map interpretation. Woodland Resource Surveys commissioned by community forests.

Proximity to areas of ecological importance (drop-down table options)

Threshold	Score
Within	0
Adjacent	3
<500 m	2
500 m to 1 km	1
>1 km	0

Data source: map and UDP interpretation.

Proximity to nearest water body (drop-down table options)

Threshold	Score
Adjacent	3
<500 m	2
500 m to 1 km	1
>1 km	0

Data source: aerial and map interpretation.

Within air quality management zone (drop-down table options)

Threshold	Score
Yes	1
No	0

Data source: local authority.

Proximity to a primary transport corridor (drop-down table options)

Threshold	Score
Adjacent	3
<500 m	2
500 m to 1 km	1
>1 km	0

Data source: map and UDP interpretation.

14 Landscape Corridors in Shanghai and Their Importance in Urban Forest Planning

Junxiang Li, Yujie Wang, and Yong-Chang Song

In the past century people have witnessed rapid rates of urbanization throughout the world. By 2030 it is projected that more than 60% of the world's population will live in cities (United Nations, 1999). Since the environment of many urban areas is becoming increasingly deteriorated, more attention is being given to making cities healthier, safer, and more sustainable. The role of green space, and urban forests in particular, in achieving these goals is acknowledged by urban planners, managers, and policy makers. Urban forests provide recreational and wildlife habitat and are appreciated for their aesthetic and architectural value. However, they also provide many ecological services to society by reducing the urban heat island, air pollution, noise, energy costs for buildings, and soil erosion. They also store and sequester carbon and perform hydrological functions such as flood control (Miller, 1997).

In recent decades, concern about the environmental quality and long-term livability of urban areas has been a driving paradigm for planning professionals (Flores et al., 1998). Recently, the science of green-space planning has conceptually adopted an ecological framework, one that promotes a biologically rich urban environment and interactions among sites across multiple spatial and temporal scales. Ecological principles such as content, context, temporal dynamics, heterogeneity, and hierarchy have been suggested as factors that should be considered in urban green-space planning (Flores et al., 1998). Urban forests are increasingly viewed as living, integral components of urban infrastructure and not simply aesthetic "window dressing" (Miller, 1997; Jim, 1999). This change in appreciation is due to the recognition that trees and other vegetation play important roles in improving conditions in urban environments. While urban forests have many social, ecological, and economic effects, how best to optimize these benefits are among the critical questions that ecologists, planners, decision makers, and practitioners encounter. For instance, when developing new cities, where and how is it best to construct urban forests? To answer these questions is a great challenge and involves integrated study of applied and basic scientific disciplines. The science of landscape ecology has much to offer applied fields such as land-use management, urban planning, and biodiversity conservation, since it deals explicitly with questions of how landscape pattern affects environmental and social processes (Wu, 2001; see also Chapter 2).

M.M. Carreiro et al. (eds.), *Ecology, Planning, and Management of Urban Forests: International Perspectives.*

Greenways are designed as linear landscape elements (Forman and Godron, 1986; Forman, 1995) that are commonly used as a landscape planning tool (Carlson et al., 1989; Little, 1990; Smith and Hellmund, 1993; Fábos and Ahern, 1995) aimed explicitly at conserving, enhancing, or restoring biological diversity. In such projects, the major goal is to propose design solutions or land conversion scenarios that produce predictable and desirable ecological consequences (Forman and Collinge, 1996). For example, the design of riparian greenways, in particular, usually considers important ecological functions, such as protecting water quality, providing animal habitat, and facilitating movement of organisms among remnant patches of native habitat (Fábos and Ahern, 1995).

In the 1980s and 1990s, the greenway movement resulted in thousands of plans and projects in the United States. Among them are two current greenway plans in the U.S.: the vision plan for the New England region and the national plan on "Greenways and Greenspaces for the United States" (Fábos, 2004). The intention of these plans are (1) to protect all nationally significant and environmentally sensitive corridors and other green spaces, (2) to provide the U.S. population with increased recreational opportunities, and (3) to restore all nationally significant historical and cultural greenway corridors.

Shanghai has now experienced more than 20 years of innovation and open experiments in promoting economic growth. However, rapid urbanization has been accompanied by increasingly serious environment pollution that already affects Shanghai's sustainable development goals. The Shanghai municipal government has realized that it is very important to develop urban forests to improve environment quality and provide livable places for its urban residents. In 2002, the pilot research program, Urban Forest Planning and Development in the Modern Shanghai Metropolitan Region, was launched by local government to initiate eco-city construction in Shanghai. Since we have been involved in this program (author Y.-C.S. was one of the two principal investigators [PIs], and author J.L. was one of the co-PIs, who was responsible for the urban forest planning), we present here the proposed framework of urban forest greenways and corridor network plans for the Shanghai Metropolitan Region. In addition, we summarize the types of landscape corridors that occur in Shanghai, and address the importance of urban forest planning in water protection, air pollution prevention, urban hydroclimate regulation, catastrophe prevention, biological conservation, and outdoor recreation.

Urban Forest Status in Shanghai

The Status of Green Space in Shanghai

Shanghai City lies in the Yangtze River delta. The area of the entire territory is 6340.5 km^2 and the current amount of land used for urban construction in its central area is 446 km^2, with the urban construction area per capita being 53.9 m^2. Land is a scarce resource in Shanghai, but Shanghai is also deficient in urban green space. In 1999 its total green area had increased to 3492 hectares [ha], with a public green space per capita of 3.5 m^2, compared to 88.3 ha and 0.132 m^2 per capita in 1949. Percent coverage of the city by green areas reached 20.34% in the central urban area by the end of 1999. The area devoted to forestry is about 371.9 km^2, with suburban forest stands occupying 309.33 km^2. Forest coverage is 4.88%. If we include forests in rural land use, and trees lining roads and urban streets, the total forest coverage in the Shanghai region is estimated to reach 10.4%[1].

Issues and Problems for Shanghai's Urban Forest

If Shanghai is to develop in a more ecologically sound fashion, it must address several issues concerning its urban forests: (1) its low forest coverage (10.4%) and per capita public green area (3.5 m^2) compared to the average for China's other cities, 25.5% and 5.54m^2, respectively; (2) the large gap between the forest coverage in Shanghai (10.4%) and the average (16%) for the whole country; (3) the imbalance between urban and suburban forest development; (4) the irrationality of the current spatial distribution of the urban forest with its uneven layout of parks and green patches, its excessive service radius of green space, and the isolation and lack of connectivity between the parks and large green spaces in the city; (5) the excessive emphasis on visual effects of the forest landscape and horticultural approaches in urban forest construction, and neglect of the integrated economic, sociological, and ecological roles played by urban forests; and (6) disregard for biodiversity conservation in current urban forest development by excessive planting of monocultural plantation forests.

The Study Area

The Shanghai metropolitan region is located on the eastern coast of China (between 30° 40′ to 31° 53′ N and 120° 51′ to 122° 12′ E), with the Yangtze River estuary lying to the north, the East Sea to the east, and Hangzhou Bay to the south. Shanghai has a northern subtropical monsoon climate, with an average annual temperature of 15.9°C, with the summer temperatures averaging 28°C and winters averaging 4°C. Average annual precipitation is approximately 1200 mm, with 60% of rainfall occurring during May to September. While climatic conditions generally promote plant productivity, vegetation growth can be limited by extremes in temperature (down to −12.1°C in winter and with summer highs that exacerbate drought), as well as frequent typhoons in summer and autumn. Shanghai lies on an alluvial delta plain, so its physiognomy is low and broad with an elevation that is generally 2.0 to 4.5 m. There are 13 hills in Shanghai, such as Sheshan and Tianma,

[1] Shanghai Agriculture and Forestry Bureau, 2000. Inventory of Forest resources in Shanghai

with elevations of less than 100 m scattered in Songjiang county, Qingpu county, and Jingshan county in the southwest. The coastline is about 471 km long with abundant beach resources. The native vegetation is characterized by subtropical evergreen broad-leaved forests dominated by *Castanopsis sclerophylla* (Fagaceae), *Cyclobalanopsis glauca* (Fagaceae), *Machilus thunbergii* (Lauraceae), *Schima superba* (Theaceae), and *Cinnamomum japonicum* (Lauraceae). Along the coast, wetlands are dominated by sedges and grasses, chiefly *Phragmites australis, Scirpus mariqueter*, and *Spartina alterniflora*.

Environmental Conditions

The air pollution in Shanghai is derived primarily from petroleum and coal combustion with particulates being a major component. Waste gas emissions totalled 790 billion m^3, mostly from industrial sources (744 billion m^3) and contains SO_2, soot, and small particulates (446,600 metric tons, 107,400 metric tons, and 15,000 metric tons, respectively; Shanghai Environmental Bulletin, 2002). The average atmospheric concentration of SO_2 is 0.051 mg/m^3 for the entire city with an average of 0.058 mg/m^3 in the central urban district, dropping to 0.039 mg/m^3 in the suburbs. The average concentration of NO_2 is 0.035 mg/m^3 in the central district, 0.017 mg/m^3 in the suburbs, and averaging 0.030 mg/m^3 for the entire city. The concentration of breathable small particulates increased from 0.100 to 0.108 mg/m^3 from 2001 to 2002, and the dust fallout in the city averages 9.02 tons/km^2. As expected, the spatial distribution of major pollutants is typically characterized by higher concentrations in the central urban districts. However, there are other localized high concentrations of pollutants in the Jiading and Baoshan districts in the northwestern suburbs, in Minhang District in the southeast, and in sections of Pudong New District in the outer beltway. Water pollution in Shanghai is another critical environmental issue, with eutrophication and heavy metals being particularly serious problems. More than 88% of the waterways in Shanghai were below the standard limits of category V (Shanghai Environment Protection Bureau, 2000; Wang, 2001).

Landscape Characteristics of the Shanghai Metropolitan Area

Synoptic Characteristics of the Shanghai Landscape

In 2000, the single largest land-use type in the Shanghai metropolitan area consisted of agricultural land, accounting for is the 60.15% of Shanghai's total area. Therefore, agricultural land could be regarded as the landscape matrix, while residential land use, industrial land use, green land, public facilities, and other miscellaneous land uses account for 14.26%, 6.09%, 2.35%, 1.74% and 1.82%, respectively,

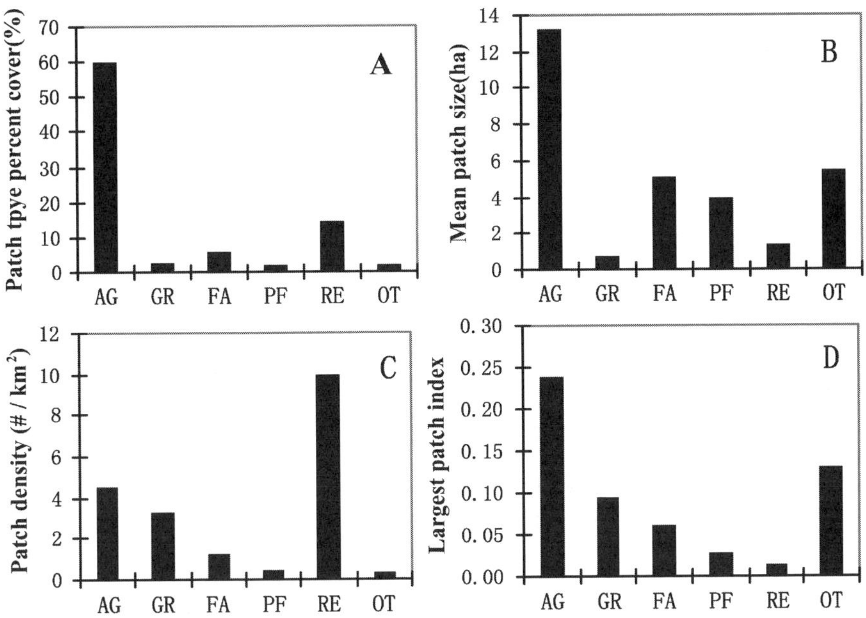

Fig. 14.1 Synoptic landscape characteristics for the Shanghai Metropolitan region (Li et al., 2004). AG, agriculture; GR, green land; FA, industry; PF, public facility; RE, residential; OT, other

of the city's area (Fig. 14.1A). Mean patch size of agricultural land is the largest of these land-use types (13.25 ha), while the categories of miscellaneous other, industrial, public facilities, residential, and green space have mean patch sizes of 5.52 ha, 5.07 ha, 3.91 ha, 1.43 ha, and 0.72 ha, respectively (Fig. 14.1B). The highest patch density was that in residential land use (9.99 patches/km^2), followed by agricultural (4.54/km^2), green space (3.25/km^2), industrial (1.2/km^2), public facilities (0.45/km^2), and construction (0.33/km^2) land (Fig. 14.1C). As for the largest patch index, the largest were found in agricultural land use with residential land use being the smallest in this category (Fig. 14.1D). All above landscape metrics were calculated with Fragstat's version 3.3 (McGarigal et al., 2002; Li et al., 2004a, b).

Waterways and roads, comprising 13.6% of the total area, form important corridor networks and are the most distinctive landscape elements in the Shanghai region. Shanghai contains 23,787 streams and rivers with a total length of approximately 21,646 km, and waterway density of 3.41 km/km^2. Road length in the central urban area is 2574 km, having a road density of 3.9 km/km^2. In the suburbs total road length is 4101 km, with a road density of 0.65 km/km^2. Because this dense network passes through other types of landscape patches, especially farmland, it is largely responsible for creating a highly fragmented landscape in the region. This can be demonstrated by comparing the patch density values, which are relatively low, with the other landscape metrics shown (Fig. 14.1).

Types of Landscape Corridors in Shanghai

A corridor is a narrow strip of land that differs from the land cover matrix on either side of it (Forman and Godron, 1986). However, this is only a structural definition. The word *corridor* is now used to describe both the structural and functional aspects of linear landscape features in the literature of many disciplines (Hess and Fischer, 2001). In general, landscape corridors can be categorized by the types of processes that created them. Therefore, one can distinguish between disturbance, remnant, environmental resource, planted/introduced, or regenerated corridors (Forman and Godron, 1986). In the Shanghai metropolitan region, some of the corridors, such as rivers and streams, are natural or seminatural. However, most are introduced, planted, or entirely man-made. For example, along some river banks and coastal zones, forest strips were planted in the 1970s to early 1980s.

In Shanghai, the availability of land imposes important limitations on urban forest development. With this constraint in mind, researchers and planning professionals should take advantage of space already available for forest construction. From the perspective of landscape ecology, existing landscape corridors, such as rivers, streams, and roads, and potential sites, such as coastlines and lake shores, are all suitable places for laying out urban forest stands. Therefore, it is necessary to obtain information on the types and respective lengths of landscape corridors that already exist to inform further planning designs for forest corridors in a city. We have initiated such a survey in Shanghai. Landscape corridors in the Shanghai metropolitan region can be separated into seven categories according to their origin and structure (Table 14.1).

The Roles of Landscape Corridors in Urban Forest Planning in Shanghai

Corridors perform at least six ecological functions in a larger landscape. They may serve as habitats, conduits, filters, barriers, sources, and sinks for organisms, matter, and energy (Forman, 1995). These ecological functions have been widely recognized and adopted by a number of disciplines, including conservation biology, wildlife management, landscape ecology, and landscape planning. From the perspective of landscape planning professionals, greenways or corridors are designed to provide multiple ecological services. Therefore, conservationists and planners should consider and document explicitly all of the possible functions of particular corridors when designing them. Addressing these functions explicitly during the design stage of the planning process should eliminate much of the confusion surrounding their roles, and focus attention on establishing design criteria for corridors that function as intended (Hess and Fischer, 2001). The roles that landscape corridors play in urban forest planning in Shanghai are based on specific functions described in detail below.

Table 14.1 Types of landscape corridors in Shanghai

Corridor types	Origins	Structural attributes	Current condition
Road	Disturbance	Curvilinearity, nodes, connectivity, width, length, edge	Most roadsides have no trees or forest stands
River and stream	Natural or seminatural	Curvilinearity, connectivity, width, length, edge	Riparian trees or woodlands along small rivers and streams. Urban rivers or streams lined with concrete, not trees
Seashore and lake shore	Natural	Curvilinearity, width, length, edge, porosity	Some sections of seashore have concrete seawalls, but no forest windbreaks. Most of the seashore covered with reed *(Phragmites australis)* or *Scirpus mariqueter*
Pollution isolation greenbelt	Planted	Width, length, height, population density,	Most factories in Shanghai have not built forest greenbelts for pollution reduction
Hedges or agroforestry greenbelt	Seminatural or planted	Width, height, porosity	Some fields have planted trees on edges, but most not afforested on edges
Railway or light railway	Planted	Width, porosity	No planted trees or shrubs on either side of railways
Power lines	Artificial		No trees or forest strips along power lines

Using Corridors to Protect Shanghai's Water Resources

The fact that riparian forests or greenways function to protect water quality has been well documented (Binford and Buchenau, 1993; Paul and Meyer, 2001). Riparian forest corridors in Shanghai are planned mainly to protect water resources, especially in the upper reaches of the Huangpu River, the source of Shanghai's drinking water. The principal function of the riparian forest corridor is water pollution prevention from nonpoint pollutants from nearby agricultural land. In addition, riparian forest corridors function to connect different habitats, filter out sediments due to soil erosion, stabilize the stream banks, and improve stream habitat for both fish and invertebrates (Schultz et al., 1995; Vought et al., 1995).

Using Forest Corridors for Air Pollution Isolation and Absorption

These kinds of forest corridors are planned and designed to reduce the amount of urban air pollution from nonpoint and industrial point sources moving into surrounding areas. These isolation greenbelts would function to impede air pollutant diffusion, reduce noise, and absorb waste gases like SO_2, NO_x, and particulate dust. These functions are optimized by constructing specific vegetation structures and plant species configurations.

Road networks with high traffic density also produce pollution such as exhaust gases, road dust, heavy metals, as well as noise, all of which are harmful to animals and people living near roads. Vegetation strips of a specified structure and width on both sides of the road have been proven to reduce these pollutants (Barrett et al., 2005). Moreover, forest corridors along the roadsides can function as conduits for animal movement, habitats (Beier and Noss, 1998), and barriers or filters for sediment that would otherwise run into nearby streams (Han et al., 2005).

Using Landscape Corridors to Mitigate Negative Urban Impact on Local Hydrology and Climate

Urban forests have large impact on local hydrology and climate. The benefits of urban forests on these environmental factors can be maximized by configuring vegetation in patterns that are unique to each landscape's purpose, such as aesthetics, greenbelts, wildlife, energy and water conservation, and fire-hazard reduction (Bradley, 1995). Several studies have established relationships between different urban forest structures and specific functions such as visual quality (Schroeder, 1986), energy savings (McPherson, 1993), removal of atmospheric carbon dioxide (Rowntree and Nowak, 1991), urban heat island mitigation (Huang et al., 1987; Oke, 1989; McPherson, 1994), sound reduction (Cook and Van Haverbeke, 1977), urban hydroclimate, air quality, and residential energy use (McPherson et al., 1997).

Urban forest corridors situated along main roads, rivers, and streams are well placed to optimize functions of urban heat-island mitigation, storm runoff reduction, and microclimate and hydroclimate regulation. Such locations may also create channels that allow cooler, cleaner air to be funneled from ex-urban and suburban areas into the center of Shanghai.

Promoting Urban Biological Conservation Using Landscape Corridors

The roles that landscape corridors play in biodiversity conservation have been well documented. Andrews (1993) described five functions of wildlife corridors, and Forman (1995) identified six societal goals that corridors could help achieve. Hess and Fischer (2001) reviewed the history of the word *corridor* in the context of conservation, and summarized its functional roles. Moreover, greenway design proposals, aimed at preserving or enhancing riparian habitat, may be especially important, given the widespread loss and fragmentation of this biologically rich habitat type (Naiman et al., 1993). Even though evidence linking the presence of greenways to enhanced movement of organisms through complex landscapes has not been statistically verified, it is clear that animals do move through vegetated corridors (Bennett, 1990; Merriam and Lanoue, 1990; Bennett et al., 1994). Resolution of this complex issue of greenways as wildlife corridors will depend on ecological research aimed at describing and quantifying animal movements in response to particular landscape elements. Hence, recent and ongoing landscape ecological studies are particularly relevant to current theory and practice of landscape architecture and planning. The goal of such ecological studies is to understand the implications of particular landscape spatial patterns, including the configuration of corridors, for various ecological processes (Forman and Godron, 1986; Turner, 1989; Forman, 1995; Hansson et al., 1995; Collinge, 1996). These kinds of ecological data, which are currently limited, could be usefully incorporated into future greenway or corridor design.

There are many organisms that might make use of these urban greenway corridors in Shanghai. There are 14 species of amphibians, 32 species of reptiles, 386 bird species, and 40 mammal species that have ever been observed in Shanghai's metropolitan region (Huang et al., 1991). Some are IUCN[2] Red-listed, threatened species, such as the Chinese bullfrog (*Rana tigrina rugulosa*), the hawksbill turtle (*Eretmochelys imbricata*), leatherback turtle (*Dermochelys coriacea*), the Pacific Ridley turtle (*Lepidochelys olivacea*), and the loggerhead (*Caretta caretta*), which is now in state-protected wildlife level II status and IUCN critically endangered. Over the last 100 years, some wildlife species have disappeared due to human activities. These include the small Indian civet (*Viverricula indica*) and the pintail duck (*Anas acuta*). So it is urgent that we act to protect the wildlife species still living in Shanghai. One of the most effective measures for affording such protection is to restore habitat and construct wildlife corridors, particularly forested ones.

Catastrophe Prevention

Two kinds of urban forest corridors are currently planned: coastal forests to resist storm tides and typhoons, and an agroforesty network to serve as suburban greenbelts or windbreaks for protecting farmland from wind erosion. Of course, these vegetation corridors can also function as buffer zones that can reduce nutrient leaching and pesticide drift from arable fields, and support more diverse wildlife in agricultural areas.

---

[2] IUCN is the acronym of The Internationl Union for the Conservation of Nature and Natural Resources. Now use the name "World Conservation Union" since 1990, but the full name and the acronym are often used together as many people still know the Union as IUCN. The World Conservation Union is the world's largest and most important conservation network. It is also a multicultural, multilingual organization with 1100 staff located in 40 countries. Its headquarters are in Gland, Switzerland.

The IUCN Red List is the world's most comprehensive inventory of the global conservation status of plant and animal species. It uses a set of criteria to evaluate the extinction risk of thousands of species and subspecies. These criteria are relevant to all species and all regions of the world. With its strong scientific base, the IUCN Red List is recognized as the most authoritative guide to the status of biological diversity. The overall aim of the Red List is to convey the urgency and scale of conservation problems to the public and policy makers, and to motivate the global community to try to reduce species extinctions. So if some species enter the red-list means that it is being threatened and needed to be protected.

For more details, please visit the website of IUCN at http://www.iucn.org/

Social Functions of Corridors

Greenway corridors not only improve environmental quality but also may provide opportunities for outdoor recreation (Little, 1990). Using corridors for this purpose has steadily grown in popularity, and the corridors have been used in the planning and design professions as an efficient and socially desirable goal for open space planning (Linehan et al., 1995). Greenways have been systematically considered integral to the protection of ecological structure and function, and central to the open space planning process (Ahern, 1991a). By linking ecological structure and function, a regional greenway system may be able to protect biodiversity, provide present and future open space needs, and allow for economic growth and development (Ahern, 1991b). Moreover, greenways together with an urban park system can provide open space for outdoor entertainment and recreation of urban residents (Zube, 1995).

According to the Shanghai Municipal Statistic Bureau (2002), total urban green space is now 18,758 ha, and includes 7810 ha of public green area, and 1411 ha of urban parks. However, Shanghai's population reached 13,342,300 in 2002. This means that there is now only 5.85 m^2 of public green space per person, and only 1.06 m^2 of public park space per person. This low amount of green space limits the ability of urban residents in using public space for outdoor recreation. Planning and constructing forested corridors along streams, roads, streets, and other community areas would increase the open space needs of Shanghai's citizens for walking, hiking, and biking.

Shanghai's Plans for Constructing Urban Forest Corridors

Coastal Windbreak Forest Planning

A coastal windbreak forest can reduce wind speed and protect coastal farmland and houses from catastrophic weather events, such as typhoons and windstorms, as well as tsunamis (Danielsen et al., 2005). The location of these forest corridors, therefore, is important and should be determined according to expected storm landing routes, and the frequencies and maximum wind speeds of typhoons over recent years. The forest corridor should be planted in at least a 1500-m-wide belt along the coast where landings are anticipated, oriented perpendicular to the prevailing wind direction. Corridor width should be 500 m in the secondary prevailing wind direction, while along the remaining coastal line not in the prevailing wind direction the width should be set to 300 m. In general, to serve as an effective windbreak it is recommended that the width of the forested strip should be approximately five to ten times tree height, and the forest structure should be designed to be sparse and penetrable to reduce the wind speed to the greatest degree (Wang et al., 1985). So the distance between tree rows should be 50 to 100 m, given an average tree height of 10 m. In our planning design, we adopted the smaller threshold value of five times tree height, meaning that every 50 m there is a forest strip. The width of each forest strip will be reduced to 10 m with four to five rows of trees.

According to the statistics of the landing routings, frequencies, and maximum wind speed of the typhoons over the past 50 years in Shanghai (Atlas of Shanghai, 1997), there are three sections along the mainland coast of Shanghai that are most vulnerable. The first section is 50 km long along the coast from Nanhui County to Chuansha County, where the highest frequency of typhoon landings have occurred over the past 50 years. The forest corridor here has been designed to be 1500 m wide. The second forested sections will be placed along the 68-km coast along northern Hangzhou Bay and the 27-km coast from Chuansha County to the Yangtze estuary. Here the forest corridor is designed to be 500 m wide. The third segment will be along the 22.8-km coast along the southern Yangtze River bank and will be 300 m wide.

In Chongming Island, the southeast part of the 22.8-km coastal forest corridor will be constructed in the prevailing wind direction, and will be 1500 m wide. The northern part of the island coast, which is 39 km long and in the second most common prevailing wind direction, will have a windbreak forest corridor 500 m wide. In the northeast and southwest coasts of the island, the forest corridor is designed to be 300 m wide by 47.5 km long. In Changxing Island, which lies in the prevailing wind direction, the forest corridor is designed to be 5.8 km long by 1500 m wide. In the remaining part of the island's coast, the forest corridor will be 47.5 km long by 300 m wide (Fig. 14.2).

Roadside Forest Corridor Planning

The term *road corridor* refers to the road surface, its maintained roadsides, and any vegetated strips parallel to the road (Forman and Alexander, 1998). The ecological effects of roads have been reviewed by Forman and Alexander (1998), and examples given of the ecological roles of roads as conduits, barriers, filters, habitats, sources, and sinks in the larger landscape (Bennett, 1991; Forman, 1995). The key corridor variables affecting these processes are width, connectivity, and intensity of usage. For landscape planning and design purposes, the multiple functions of the roadside forest corridor should be considered in an integrated fashion, so that as many as possible of the potential beneficial ecological services of the road corridor can be optimally realized.

Highly trafficked roads and adjacent roadside forest strips are important landscape corridors in the Shanghai metropolitan region. Shanghai's roads can be sorted into four classes: expressways, state highways, county roads, and rural roads. Since all these roads are connected to form a network across the entire landscape, they can provide a framework for a forest corridor system. Shanghai has a 756-km expressway, including two loop expressways, one forming an inner and the other an outer suburban ring road. Forest corridors for each of these beltways should be designed separately. The 97-km inner loop encircles the urban area and is designed to prevent urban sprawl. Therefore, the ex-urban side of this expressway is now planted with a 500-m-wide forest. The 180-km outer suburban loop, which is now being constructed, was originally planned to connect Shanghai with its satellite counties and towns. Here the 250-m-wide forest corridor is designed for each side of the

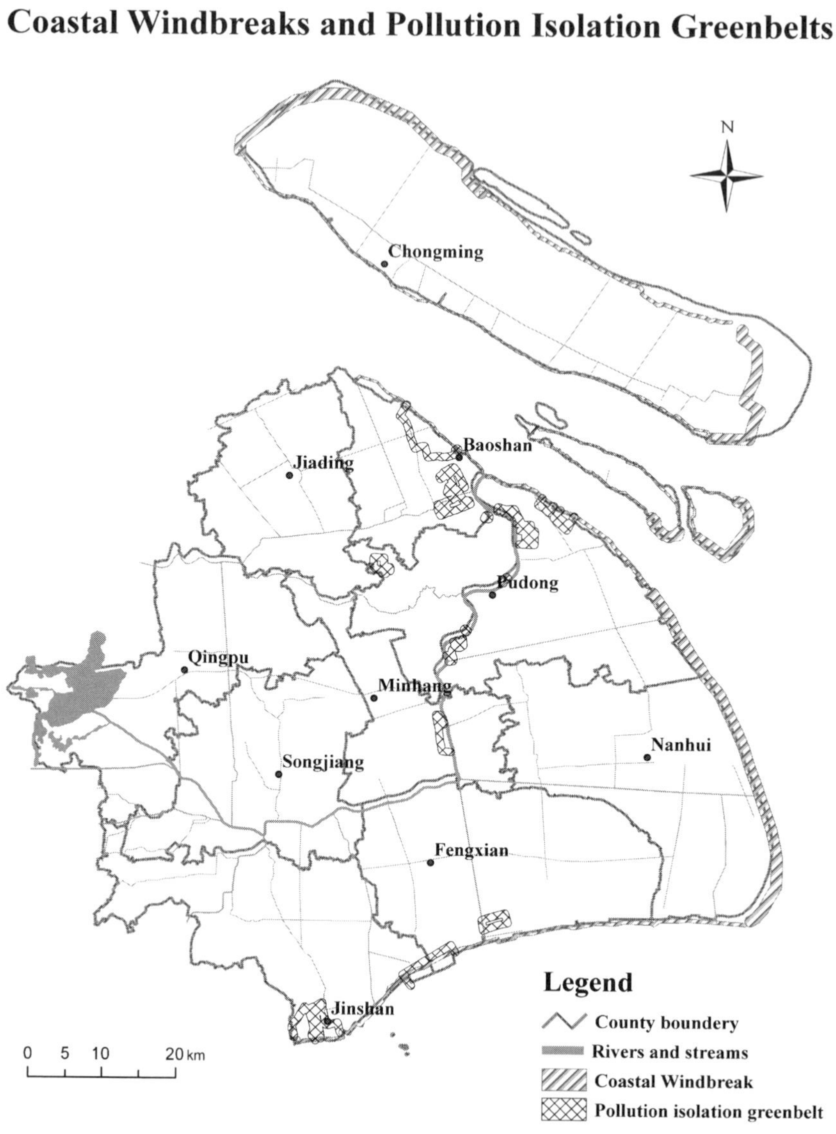

Fig. 14.2 Coastal windbreaks and pollution isolation greenbelts planned in the Shanghai Metropolitan region

expressway. These two forest corridors are multifunctional, providing wildlife habitat, a windbreak, a conduit, and greenbelt buffer, particularly where some sections of the corridor were enlarged and designed as urban–rural gardens for the recreation and entertainment of urban residents. For the remaining 479-km expressway, the forest corridors designed on both sides are 50 m wide.

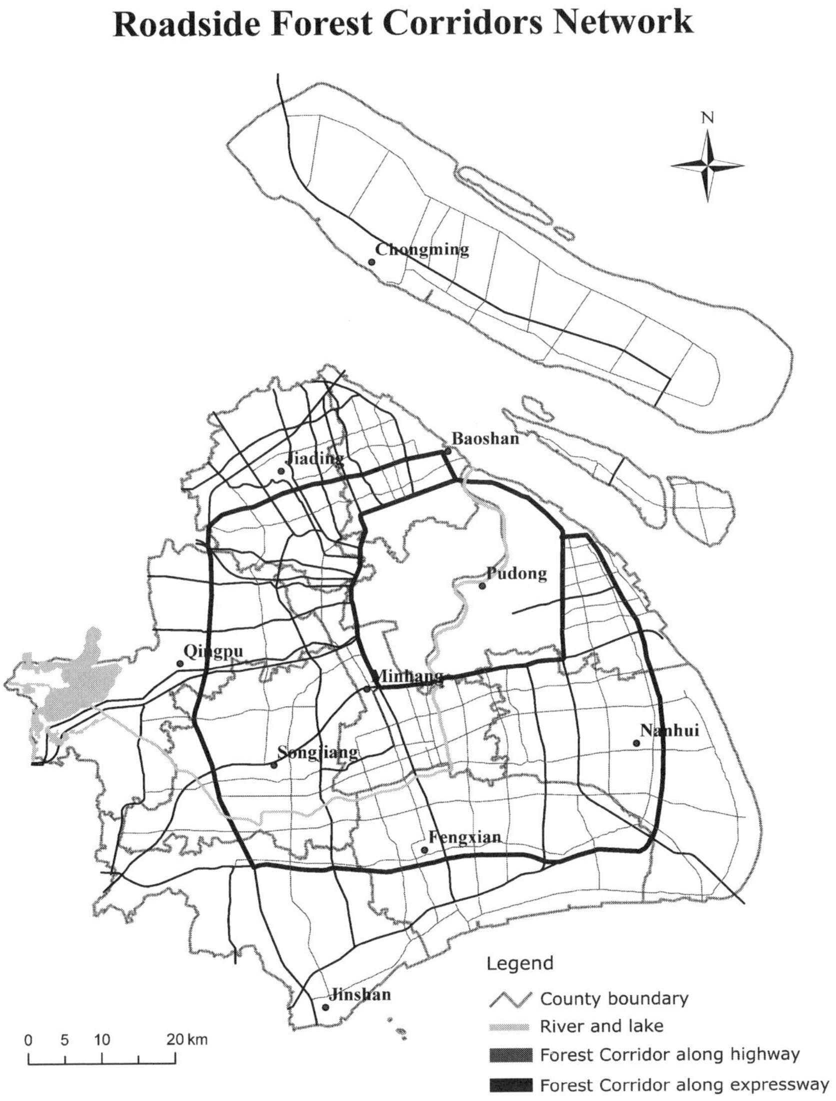

Fig. 14.3 Roadside forest corridors network planned in the Shanghai Metropolitan region

State highways that start in or pass through Shanghai are very important traffic arteries connecting Shanghai with other cities. Forest corridors for these roads are designed to be 20 m wide on both sides. For county and rural roads, forest corridors were set to 6 or 10 m wide on each side depending on the space that can be used. Railroads and light railways from Shanghai to the suburbs have 20-m-wide forest corridors on each side (Fig. 14.3).

Riverside or Stream-Side Forest Planning

The rivers and streams in Shanghai have been straightened, had their riverbanks lined with concrete, have few riparian woodlands, and are therefore much polluted. In the urbanized areas of Shanghai, river and stream coverage has decreased from 11.1% in 1980 to 8.4% in 2000 (Wang, 2001). Since riparian forest corridors have many ecological functions (addressed earlier in this chapter), constructing as many forest strips along the rivers and streams as possible will greatly benefit the Shanghai metropolitan region.

Rivers and streams in Shanghai can be classified according to a management hierarchy as city, county, and village-and-town water bodies. Since the upper reaches of the Huangpu River are the drinking water sources for Shanghai City, afforestation there is a very important planning goal. The buffer widths for riparian forests should be wide enough to protect water resources from sediment and nonpoint pollution inputs from adjacent farmland. Along each side of the Huangpu River's upper reaches, the riparian forest corridor was designed to be 500 m wide, while along the upper reaches of the main tributaries of the Huangpu River, the riparian forest was designed to be 50 to 100 m wide (Johnson and Ryba, 1992; Hubbard and Lowrance, 1994; Pan and Deng, 2003). For the remaining urban and suburban rivers in Shanghai, the riparian forest was planned and designed to be 50 m wide. For county and town rivers and streams, the width of the riparian forest was designed to be 20 to 30 m and 10 to 20 m, respectively. Dianshanhu Lake is the largest freshwater lake in the Shanghai region and is another important drinking water source for the city. A 500- to 1000-m-wide lake-shore forest has been designed as a buffer for water quality protection (Fig. 14.4).

Planning for a Farmland Windbreak Network

Windbreaks or shelterbelts are very important buffer zones in agro-ecosystems. Buffer zones in agro-ecosystems are maintained to prevent soil erosion, nutrient leaching, and pesticide drift from arable fields, particularly to aquatic habitats nearby (Jenssen et al., 1994; Daniels and Gilliam, 1996). These buffer zones are also seminatural habitats supporting wildlife in farmland (Sotherton, 1984; Dennis et al., 1994). At the landscape level, buffer zones may also function as corridors in agricultural mosaics (Ma et al., 2002). Windbreaks and shelter belts are agroforests that are accepted by the farmers, particular in the delta plains area. In Shanghai, there are 2908 km^2 of farmland suitable for agroforestry. In the delta, the optimal grid size for an agroforestry network is 100 m × 100 m, in order to regulate farmland microclimate and preserve soil moisture (Li and Wan, 2002). Two or three rows of trees 6 to 10 m wide were planted along field edges, so little farmland need be devoted to the forest stands themselves.

For the design of the entire network of agroforests in Shanghai, we adopted the 100 m × 100 m grid plan. To estimate the number of tree seedlings needed to

Fig. 14.4 Riparian forest corridors network planned in the Shanghai Metropolitan region (only the corridors along main rivers and streams are illustrated)

construct this network, we calculated the amount of area that an agroforest network might occupy. Assuming that the 2908 km^2 of farmland in Shanghai can be regarded as a big square, then the area can be divided into 539 sections. Theoretically, if a big square is divided into smaller squares of $N \times N$ dimension, with W being the length of each smaller square's side, and L being the total length of the smaller square's sides, then $L = W \times N(3N - 2)$. That is to say, for the 2908 km^2 of farmland in Shanghai, the maximum length that the forest network can be is 87,048,500 m in 100 m × 100 m grids. Since the width of the windbreak greenbelt is limited to 6 m for farmland, the maximal area that would be covered by the agroforest network is approximately 522 km^2.

Pollution Isolation Forest Greenbelt Planning

There are many practices for building forest greenbelts around factories or industrial gardens to prevent pollutant diffusion. Many plants can absorb gaseous exhausts like SO_2, NO_x, and filter out particulates and reduce noise. There are many industrial parks in Shanghai, such as SINOPEC (Shanghai Petrochemical Co. Ltd.) located in Jinshan District, Baoshan Iron and Steel Co., Ltd. (Baosteel) located in the Baoshan District, and a chemical industrial park located in the Caojing area, all of which contribute a diverse array of pollutants to adjacent areas. There are increasing demands to build peri-factory or peri-industrial park forests, and not simply green-space lawn habitats, to prevent pollution diffusion from these point sources to residential areas and to protect the environment overall.

As part of Shanghai's urban forest planning program, we plan to construct forest strips around industrial parks and factories. For factories located in urban areas, the forest corridor was designed to be 600 m wide, due to limitations imposed by densely occupied land. However, for factories in suburban or ex-urban areas, the forest corridor was designed to be 800 meters wide or more. According to our calculations using Geographic Information Systems (GIS), these industrial forest corridors will probably total 143 km (Fig. 14.2).

Forest Corridor Planning for Biodiversity Conservation

Many studies provide strong evidence that habitat loss has large, consistently negative effects on biodiversity (Fahrig, 2003). Habitat loss and fragmentation have been cited widely as the major contemporary threats to biological diversity as people transform the landscape to accommodate their needs. Corridors have become popular tools for mitigating fragmentation and conserving biodiversity (Hess and Fischer, 2001). Experiments have shown that corridors reduce rates of species loss by enhancing recolonization, but only in medium-sized fragments (Collinge, 1998). Such studies will help improve our understanding of the roles of corridors when planning and designing for biological diversity conservation.

Rare animals were once found in Shanghai. These included the large Indian civet (*Viverra zibetha*), the small Indian civet (*Viverricula indica*), and the Chinese pangolin (*Manis pentadactyla*) (Huang et al., 1991). However, surveys of wildlife resources in Shanghai in 2000 have reported that these animals are no longer found (Shanghai Agriculture and Forestry Bureau, 2000). Only six wildlife species are still found in the Shanghai region: *Felis bengalensis, Arctonyx collaris, Erinaceus europaeus, Nyctereutes procyonoides, Mustela sibirica*, and *Lepus sinensis*.

To protect native wildlife still living in Shanghai, the Shanghai Agriculture and Forestry Bureau initiated a program, the "1 + 14" Wildlife Homestead Action in 2001, which included plans for protecting one large and 14 small biological conservation areas. Conservation corridors are designed to accommodate the animals' large home ranges and potential activity radii. The width of the forest corridors that transect the wildlife reserves was designed to be 1000 to 1700 m for *Felis bengalensis*, 600 to 1000 m for *Arctonyx collaris*, 200 to 300 m for *Erinaceus europaeus*, 450 to 700 m for *Nyctereutes procyonoides*, 200 to 450 m for *Mustela sibirica*, and 150 to 250 meters for *Lepus sinensis* (Da et al., 2004).

Discussion and Conclusion

Urban forest planning in Shanghai is a comprehensive and systematic program, one that has from the outset involved a multidisciplinary and interdisciplinary approach. This requires the cooperation of experts in urban ecology, urban planning, urban forestry, landscape ecology, conservation biology, landscape architecture and planning, as well as the social sciences. In particular, landscape ecology, conservation biology, and landscape architecture play very important roles in developing the urban forest plan. Constructing functional forested corridors and greenways throughout Shanghai undoubtedly will be a most important aspect of planning and design. Since the spatial pattern, width, and vegetation structure of these corridors are linked to their intended functions, their installation must be conducted with care. Funds for their long-term ecological study must also be budgeted to measure the effectiveness of these corridors for their intended purposes in case design modifications are needed in the future.

Spatial pattern in the landscape is the common denominator for the disciplines of landscape ecology and landscape architecture. Ecologists are beginning to understand the relative importance of landscape spatial structure for a variety of ecological processes. However, the application of this knowledge to landscape architecture and planning projects remains novel. The integration of landscape ecological research with landscape architectural practice provides a rich opportunity for understanding the implications of, and directing, future landscape change (Collinge, 1998). The spatial pattern and composition of corridors relies, to a certain degree, on the context of the entire landscape in a region. Details of planning and designing a corridor will be decided by factors such as topography, soil type, vegetation, and roadside or stream riparian conditions. But at the landscape level, connectivity between the corridors and habitats, as well as the corridors themselves, should be the focus, since it strengthens corridor function.

Width is the characteristic of forested corridors that is most essential in determining their functional success (Merriam and Lanoue, 1990; Bennett, 1991; Harrison, 1992; Andreassen et al., 1996; Metzger et al., 1997; Van Dorp et al.,

1997; Haddad, 1999). The widths we have established for forest corridors in Shanghai are admittedly somewhat arbitrary and may be controversial. However, since land is a very limited resource in Shanghai and fundamental research is lacking to support some recommended buffer dimensions, we still have recommended that corridors be as wide as competing land use will permit. For roadside corridors, for example, forest width design should consider the extent of the road-effect zone, which can be less than 50 m to more than 1000 m, depending on topography (upslope or downslope), wind direction (upwind or downwind), and surrounding habitat suitability (Forman et al., 1997). The recommended width should not be less than 50 m as long as no land-use conflicts or limits exist.

Vegetation structure is another important factor affecting corridor function. Here the structure refers to the vertical structure (e.g., the trees, shrubs, grass, as well as the species composition or configuration in the horizontal plane). Both structure and width allow a corridor to provide diverse habitats and more ecological niches for animals. We also recommend that native species and potential vegetation theory (Song, 2001; see also Chapter 12) should be strongly considered in urban forest construction plans.

We have presented here an initial framework for the urban forest corridor plans for Shanghai's metropolitan region. Much more fundamental research, especially in the biodiversity of Shanghai, should be carried out to support the urban planning and landscape planning process. This will allow Shanghai to be at the forefront of ecologically sound development, and to move toward becoming an eco-city in the near future. We hope past ecological research in Shanghai, as well as the pilot urban forest planning program, can act as "stepping stones" to help Shanghai become an eco-city, and thereby promote the physical and mental health of its residents by implementing more ways to live harmoniously with nature.

References

Ahern, J. (1991a) Planning and design for an extensive open space system: linking landscape structure and to function. Landscape and Urban Planning 21:131–145.

Ahern, J. (1991b) Greenways and ecology. Proceedings of Landscape/Landuse Planning Committee of the American Society of Landscape Architecture's 1991 Annual Meeting, Washington, DC. American Society of Landscape Architecture, Washington, DC, pp. 75–87.

Andreassen, H.P., Halle, S., and Anker, I.R. (1996) Optimal width of movement corridors for root voles: not too narrow and not too wide. Journal of applied Ecology, 33:63–70.

Andrews, J. (1993) The reality and management of wildlife corridors. British Wildlife 5:1–7.

Atlas of Shanghai (1997). Shanghai Scientific and Technical Publishers. Shanghai (in Chinese).

Barrett, M., Lantin, A., and Austrheim-Smith, S. (2005) Stormwater pollutant removal in roadside vegetated buffer strips. http://www.rbfconsulting.com/papers/radF5646.pdf

Beier, P., and Noss, R.F. (1998) Do habitat corridors provide connectivity? Conservation Biology 12(6):1241–1252.

Bennett, A.F. (1990) Habitat Corridors: Their Role in Wildlife Management and Conservation. Department of Conservation and Environment, Melbourne, Australia. pp. 37.

Bennett, A.F. (1991) Roads, roadsides and wildlife conservation: a review. In: Saunders, D.A., and Hobbs, R.J. (eds.) Nature Conservation 2: The Role of Corridors. Surrey Beatty, Chipping Norton, Australia, pp. 99–117.

Bennett, A.F., Henein, K., and Merriam, G. (1994) Corridor use and the elements of corridor quality: Chipmunks and fencerows in a farmland mosaic. Biological Conservation 68(2):155–165.

Binford, M.W., and Buchenau, M.J. (1993) Riparian greenways and water resources. In: Smith, D.E., and Hellmund, P.C. (eds.) Ecology of Greenways: Design and Function of Linear Conservation Areas. University of Minnesota Press, Minneapolis, MN, pp. 69–104.

Bradley, G. (1995) Integrating multidisciplinary perspectives. In: Bradley, G. (ed.) Urban Forest Landscapes: Integrating Multidisciplinary Perspectives. University of Washington Press, Seattle, pp. 3–11.

Carlson, C., Canty, D., Steiner, F., and Mack, N. (1989) A path for the Palouse: an example of conservation and recreation planning. Landscape and Urban Planning 17:1–19.

Collinge, S.K. (1996) Ecological consequences of habitat fragmentation: implications for landscape architecture and Planning. Landscape and Urban Planning 36:59–77.

Collinge, S.K. (1998) Spatial arrangement of habitat patches and corridors: clues from ecological field experiments. Landscape and Urban Planning 42:157–168.

Cook, D.I., and Van Haverbeke, D.F. (1977) Suburban Noise Control with Plant Materials and Solid Barriers, Research Bulletin EM 100, U.S. Department of Agriculture, Forest Service, Rocky Mountain Forest and Range Experiment Station, Fort Collins, CO.

Da, L.-J., Chen, K.-X. and Xin, Y.-F. (2004) The scale of ecological corridor in the urban forest of Shanghai. Journal of Northeast Forestry University, 32(4):16–18 (in Chinese with English abstract).

Daniels, R.B., and Gillia, J.W. (1996) Sediment and chemical load reduction by grass and riparian filters. Soil Science Society of America Journal 60:246–251.

Danielsen, F., Sørensen, M.K., Olwig, M.F., et al. (2005) The Asian tsunami: a protective role for coastal vegetation. Science 310:643.

DeGraaf, R.M., and Wentworth, J.M. (1986) Avian guild structure and habitat associations in suburban bird communities. Urban Ecology 9:399–412.

Dennis, P., Thomas, M.B., and Sotherton, N.W. (1994) Structural features of field boundaries which influence the overwintering densities of beneficial arthropod predators. Journal of Applied Ecology 31:361–370.

Fábos, J.G. (2004) Greenway planning in the United States: its origins and recent case studies. Landscape and Urban Planning 68:321–342.

Fábos, J.G., and Ahern, J. (1995) Greenway, Landscape and Urban Planning, Special Issue, 33(1–3).

Fahrig, L. (2003) Effects of habitat fragmentation on biodiversity. Annual Review of Ecology, Evolution and Systematics 34:487–515.

Flores, A., Pickett, S.T.A., Zipperer, W.C., Pouyat, R.V., and Pirani, R. (1998) Adopting a modern ecological view of the metropolitan landscape: the case of a greenspace system for New York City region. Landscape and Urban Planning 39:295–308.

Forman, R.T.T. (1995) Land Mosaics: The Ecology of Landscapes and Regions. Cambridge University Press, Cambridge, pp. 54–58, 146–157, 208–252.

Forman, R.T.T., and Alexander, L.E. (1998) Roads and their major ecological effects. Annual Review of Ecology and Systematics 29:207–231.

Forman, R.T.T., and Collinge, S.K. (1996) The spatial solution to conserving biodiversity in landscapes and regions. In: DeGraaf, R.M., and Miller, R.I. (eds.), Conservation of Faunal Diversity in Forested Landscapes. Chapman and Hall, New York, pp. 537–568.

Forman, R.T.T., Friedman, D.S., Fitzhenry, D., Martin, J.D., Chen, A.S., and Alexander, L.E. (1997) Ecological effects of roads: toward three summary indices and an overview for North America. In Canters, K. (ed.) Habitat Fragmentation and Infrastructure. Ministry of Transportation and Public Works and Water Management, Delft, Netherlands, pp. 40–54.

Forman, R.T.T., and Godron, M. (1986) Landscape Ecology. Wiley, New York.

Haddad, N.M. (1999) Corridor use predicted from behaviors at habitat boundaries. American Naturalist 153:215–227.

Han, J., Wu, J.S., and Allan, C. (2005) Suspended sediment removal by vegetative filter strip treating highway runoff. Journal of Environmental Science and Health, Part A: Toxic/Hazardous Substances and Environmental Engineering 40(8):1637–1649.

Hansson, L., Fahrig, L., and Merriam, G. (eds.) (1995) Mosaic Landscapes and Ecological Processes. Chapman and Hall, London.

Harrison, R.L. (1992) Toward a theory of inter-refuge corridor design. Conservation Biology 6:293–295.

Hess, G.R., and Fischer, R.A. (2001) Communicating clearly about conservation corridors. Landscape and Urban Planning 55:195–208.

Huang, J., Akbari, H., Taha, H., and Rosenfeld, A. (1987) The potential of vegetation in reducing summer cooling loads in residential buildings. Journal of Climate and Applied Meteorology 26:1103–1116.

Huang, Z.-Y., Zhou, M.-Z., Zong, Y., and Tan, H.-Z. (eds.) (1991) The list of national protected animals in Shanghai Region. Fudan University Publishers, Shanghai (in Chinese).

Hubbard, R.K., and Lowrance, R.R. (1994) Riparian forest buffer system research at the coastal plain experiment station, Tifton, GA. Water, Air, and Soil Pollution 77(3–4):409–432.

Jenssen, P.D., Maehlum, T., Roseth, R., et al. (1994) The potential of natural ecosystem self-purifying measures for controlling nutrient input. Marine Pollution Bulletin 29:6–12.

Jim, C.Y. (1999) A planning strategy to augment the diversity and biomass of roadside trees in urban Hong Kong. Landscape and Urban Planning 44:13–32.

Johnson, A.W., and Ryba, D.M. (1992) A literature review of recommended buffer widths to maintain various functions of stream riparian areas. King County Surface Water Management Division.

Li, J.-X., and Wan, Z.-H. (2002) Microclimatic effect and soil moisture change of poplar-wheat intercropping systems in Huaibei Plain. Chinese Journal of Applied Ecology 13(4):390–394 (in Chinese with English abstract).

Li, J.-X., Wang, Y.-J., Song, Y.-C., and Wang, H. (2004a) GIS-based analysis of spatial characteristics of watercourse and its water pollution in Shanghai city. China Environmental Science 24(5):632–635 (in Chinese with English abstract).

Li, J.-X., Wang, Y.-J., Shen, X.-H., and Song, Y.-C. (2004b). Landscape pattern analysis along an urban-rural gradient in the Shanghai metropolitan region. Acta Ecologica Sinica, 24(9): 1973–1980 (in Chinese with English abstract).

Linehan, J., Gross, M., and Finn, J. (1995) Greenway planning: developing a landscape ecological network approach. Landscape and Urban Planning 33:179–193.

Little, C. (1990) Greenways for America. John Hopkins University Press, Baltimore, MD.

Ma, M., Tarmi, S., and Helenius, J. (2002) Revisiting the species–area relationship in a semi-natural habitat: floral richness in agricultural buffer zones in Finland. Agriculture, Ecosystems and Environment. 89:137–148.

McGarigal, K., Cushman, S.A., Neel, M.C., and Ene, E. (2002) FRAGSTATS: spatial pattern analysis program for categorical maps. Computer software program produced by the authors at the University of Massachusetts, Amherst. www.umass.edu/landeco/research/fragstats/fragstats.html

McPherson, E.G. (1993) Evaluating the cost effectiveness of shade trees for demand-side management. Electricity Journal 6:57–65.

McPherson, E.G. (1994) Cooling urban heat islands with sustainable landscapes. In: Platt, R.H., Rowntree, R.A., and Muick, P.C. (eds.) The Ecological City, Preserving and Restoring Urban Biodiversity. University of Massachusetts Press, Boston, pp. 151–171.

McPherson, E.G., Nowak, D., Heisler, G., et al. (1997) Quantifying urban forest structure, function, and value: the Chicago Urban Forest Climate Project. Urban Ecosystems 1:49–61.

Merriam, G., and Lanoue, A. (1990) Corridor use by small mammals: field measurement for three experimental types of Peromyscus leucopus. Landscape Ecology 4:123–131.

Metzger, J.P., Bernacci, L.C., and Goldenberg, R. (1997) Pattern of tree species diversity in riparian forest fragments of different widths (SE Brazil). Plant Ecology 133:135–152.

Miller, R.W. (1997) Urban Forestry: Planning and Managing Urban Greenspaces. Prentice-Hall, Englewood Cliffs, NJ.

Naiman, R.J., Décamps, H., and Pollock, M. (1993) The role of riparian corridors in maintaining regional biodiversity. Ecological Application 3:209–212.

Oke, T.R. (1989) The micrometeorology of the urban forest. Philosophical Transactions of Royal Society [B] 324:335–349.

Pan, X.-L., and Deng, W. (2003) Advances in riparian buffers in agricultural catchments. Journal of Agro-Environment Science 22(2):244–247 (in Chinese with English abstract).

Paul, M.J., and Meyer, J.L. (2001) Streams in the urban landscape. Annual Review of Ecology and Systematics 32:333–365.

Rowntree, R., and Nowak, D.J. (1991) Quantifying the role of urban forests in removing atmospheric carbon dioxide. Journal of Arboriculture 17:269–275.

Schroeder, H.W. (1986) Estimating park tree density to maximize landscape aesthetics. Journal of Environment Management 23:325–333.

Schultz, R.C., Collettil, J.P., Isenhart, T.M., Simpkins, W.W., Mize, C.W., and Thompson, M.L. (1995) Design and placement of a multi-species riparian buffer strip system. Agroforestry Systems 29(3):201–226.

Shanghai Agriculture and Forestry Bureau. (2000) Terrestrial Wildlife Resources Investigation in Shanghai.

Shanghai Environment Protection Bureau. (2000) Annual Bulletin of Environment Quality (1996–2000), Shanghai.

Shanghai Municipal Statistic Bureau. (2002) Statistical Yearbook of Shanghai. China Statistic Press. Beijing.

Smith, D.S., and Hellmund, P.C. (eds.) (1993) Ecology of Greenways. University of Minnesota Press, Minneapolis, MN.

Song, Y.-C. (2001) Vegetation ecology. East China Normal University Press, Shanghai (in Chinese).

Sotherton, N.W. (1984) The distribution and abundance of predatory arthropods overwintering on farmland. Annals of Applied Biology 105:423–429.

Turner, M.G. (1989) Landscape ecology: the effect of pattern on process. Annual Review of Ecology and Systematics 20:171–197.

United Nations. (1999) World Urbanization Prospects: the 1999 revision. United Nations, New York.

Van Dorp, D., Schippers, P., and Van Groenendael, J.M. (1997) Migration rates of grassland plants along corridors in fragmented landscapes assessed with a cellular automation model. Landscape Ecology 12:39–50.

Vought, L.B.M., Pinay, G., Fuglsang, A., and Ruffinoni, C. (1995) Structure and function of buffer strips from a water quality perspective in agriculture landscapes. Landscape and Urban Planning 31:323–331.

Wang, S.-N. (ed.) (2001) Report of water resources survey in Shanghai. Shanghai Scientific and Technical Publishers, Shanghai (in Chinese).

Wang Z.-F., Zhu T.-Y., and Zhu, J.-W. (1985) Forest Meteorology. China Forestry Publishing House, Beijing.

Wu, J. (2001) Landscape Ecology: Pattern, Process, Scale and Hierarchy. Higher Education Press, Beijing (in Chinese).

Zube, E.H. (1995) Greenways and the US National Park System. Landscape and Urban Planning 33:17–25.

15 Management of Urban Forests in the United States

J. James Kielbaso

Definition of Urban Forestry

Many definitions of urban forestry have been proposed and written, but the one I like best was developed by the Society of American Foresters' Urban Forestry Working Group: "Urban forestry is a specialized branch of forestry that has as its objective the cultivation and management of trees for their present and potential contribution to the physiological, sociological, and economic well-being of urban society." Education of the public about tree values is implied. It broadly includes other aspects such as municipal watersheds, wildlife habitats, recreation, landscape design, recycling, and even wood fiber. Quality of life is the sum of all things that make life enjoyable, comfortable, and meaningful, including physical, mental, economic, psychological, aesthetic, and recreational benefits. It is much easier just to assume quality than to enumerate the many aspects of quality and how urban forestry contributes to them.

There have been several surveys of city tree managers in the United States over the past 20 years. A 1986 survey of 2787 cities had a 38% return. We estimated that there were 61,654,000 street trees in the 7043 U.S. cities represented. The 1986 survey of urban forestry programs in the U.S. (Kielbaso et al., 1988), therefore, allows us to draw some conclusions about common factors in those cities with urban forestry programs that come close to the ideal model. Adjustments are obviously necessary for applying these findings to China. For example, is a city's commitment to planting and maintaining trees a high priority in view of its total environmental, economic, and aesthetic assets? Also, is this commitment legal and codified? In other words, does the city have an ordinance, or other legal policy, that assigns responsibility for the planting and care of all the city's trees?

In almost all cities with good, effective tree management, the 1986 survey found a tree manager with the title of *forester* or *arborist*, who served in a subunit of government with the words *forestry* or *tree* as part of the department title. In short, a city's trees must have an advocate. When a city's tree policies are sufficiently clear, a good manager can ensure that tree spaces are vigorously defended, that tree removals are not permitted without good reason, and that new tree spaces are provided when any new projects are contemplated. So much responsibility rests on

M.M. Carreiro et al. (eds.), *Ecology, Planning, and Management of Urban Forests: International Perspectives.*

the tree manager that the city needs to be sure to hire the strongest candidate possible, define the position clearly, and then support the person. Every city with an outstanding tree program has a good spokesperson who is capable of communicating with politicians and the public about the values and the needs of the urban forest. Another component essential to getting a sound program in place is an inventory of the city's trees. This must be organized to allow the manager to locate specific trees, and it must be regularly updated. Without that information, it is unlikely that any argument in support of the program or its budget can be successful.

What values should be central to an urban forestry program? Aesthetic, economic, and environmental values begin the list. Trees are one of the important elements that tie a neighborhood together and form some sense of unity and character. They can convey a sense of well-being, and of stateliness and charm. Trees also add to the economic value of cities. Estimates have been made that landscape trees add several thousand dollars to the value of each home along a tree-lined street. Trees usually help homes sell more quickly and at higher prices than homes without trees. Multiple trees may increase the value of a home by as much as 20%, with an average value of increase between 5% and 10% (Payne and Strom, 1975). The per tree value of street trees has been variously estimated between \$544 and \$1714, and some individual trees can be valued even more highly (Kielbaso et al., 1988).

The environmental contributions of trees add other values that should be clearly emphasized for politicians and residents. With pollution in our cities and the likely prospect of intensifying global warming and urban heat islands in the future, it is important to realize the contributions that trees make to a healthy environment. As one example, the volume of carbon dioxide removed from the air by an 80-foot-tall beech tree each day is equivalent to that produced daily by two single-family dwellings (Robinette, 1972). Tree canopies have been shown to substantially reduce the cost of air-conditioning homes in warmer climates, \$147 per year for a single mobile home according to one study (Laechelt and Williams, 1974). Windbreak protection through the use of trees has been found to reduce winter heating costs as much as 20% to 30% (DeWalle, 1978). Akbari and his colleagues (1988, 1992) have estimated that, because of these factors, a well-placed tree in an urban setting may be worth as much as 14 times the value of a similar tree in a forest location in terms of its total environmental effect (McPherson, 2003). For more recent studies on energy saving, see McPherson and Simpson (2003), Simpson (2002), and Konopacki and Akbari (2002).

Trends in Tree Care

Despite these many values, the 1986 survey showed that tree care programs in the U.S. have not yet reached a desirable stage of maturity or quality. One finding was that in 1986 only 39% of responding cities had a "systematic" urban forestry program, compared with 50% in 1980. Table 15.1 describes the cities that responded to the survey and those that had systematic programs (Kielbaso, 1990). From each

Table 15.1 Survey response rates and cities with systematic tree care by U.S. population categories, 1986

Classification	No. of	Cities	Cities with systematic programs % of (responding cities)	
Population group (thousands)	cities surveyed (A)	responding (% of A)	1980	1986
Over 1000	6	33	60	
500–1000	17	53	62	44
250–499	34	53	67	39
100–249	113	55	65	32
50–99	280	49	65	42
25–49	616	44	55	44
10–24	1545	30	45	38
5–9	104	43	28	31
2.5–4.9	72	74	21	21
Total, all cities	2787	38	50	39

city we requested budget information to be able to calculate such data indices as the amount expended per capita and per tree, and the percentage of the total city budget devoted to tree care. Only 38% of the respondents were able to report how many street trees existed with any degree of assurance. That percentage suggests that the status of city tree care is rather low across the U.S. The close agreement between the percentage of cities in Table 15.1 that have systematic tree care programs (39%) and those that know the number of trees in their city is also notable.

Overall budget information on tree care programs, which allows for estimating the amount expended per capita, was provided by 71% of the responding cities (Kielbaso, 1989). So, the information in Table 15.2, which summarizes this budget analysis, bases the per capita expenditures on considerably more respondents than does the per tree estimates. Both mean and median data are presented. Generally, the mean can be seen as a goal for a relatively good program, while the median represents a fallback position, a point at which the program is at least better than half the cities that responded to the survey. The budget data provided in the survey were analyzed according to two other categories: location on public property and various types of work activities (Table 15.3). The data show that over 70% of all city tree care budgets are devoted to the three categories of planting, trimming, and removal, including stump removal. Larger cities spend less for planting and removal and more for nursery care. Cities in the Northeast and North Central regions of the U.S. spend less for trimming and more for removal, a fact greatly influenced by Dutch elm disease. Cities in the West spend considerably more for trimming and watering, and much less for removal and planting (Kielbaso, 1989).

Overall, if a city is to approach an ideal program, a more desirable balance would be to allocate about 35% to 40% of funds for pruning, 14% for removal, and 10% for planting. That would allow each of the remaining categories to increase by

Table 15.2 Mean and median annual expenditures for tree care in cities, by U.S. population, 1986

Classification population group	Per person Mean	Per person Median	$ Per tree Mean	$ Per tree Median
Over 1,000[a]	2.14	2.14	13.24	13.24
500–999[a]	1.31	1.38	9.11	7.14
250–499[a]	2.41	1.73	12.24	12.60
100–249	2.88	2.37	11.95	11.00
50–99	2.96	2.41	11.83	10.37
25–49	3.14	2.01	10.61	9.56
10–24	2.17	1.06	9.86	7.69
5–9[a]	3.29	1.10	11.98	6.00
2.5–4.9[a]	1.36	1.13	3.89	3.33
Total, all cities	2.60	1.73	10.62	8.04

[a]Responses for fewer than 12 cities in this population group. Small numbers of responses for category suggest caution when drawing conclusions from data.

Table 15.3 Budget allocated by place and work activities (percentages) in U.S. cities, 1986

Location		Work activities			
Streets	62	Trimming	30	Watering	4
Parks	26	Removal, includes stumps	28	Office	2
Public Grounds	7	Planting	14	Nursery	2
Nurseries	2	Supervision	7	Fertilization	1
Cemeteries	1	Storm work	5	Repair	2
Other	2	Spraying	4	Misc	2

an average of 1%. This would reflect the importance of regular maintenance rather than the type of crisis response in tree removals and replacement that has been typical since the Dutch elm disease epidemic in the U.S. Generally, more funds are budgeted for removal because the trees are large; therefore, several small trees can be planted for the price of one removal. When a city can move to a strong maintenance program rather than a disease- and crisis-oriented one, it is likely that many fewer removals will be required. Pruning cycles will also become shorter and less drastic, and other maintenance practices can be increased. The proper pruning of trees along streets can do much to extend their useful life and eliminate most potential problems long before they materialize.

Along with the overview of current budget priorities, the 1986 survey revealed how these American cities stand on such matters as tree ordinances, long-range plans, and record keeping. While 61% of responding cities had a tree ordinance that defined responsibility for tree care, only 13% had one that placed restrictions on the cutting of trees on private property, and only 17% had an urban forestry management plan. While 47% of the cities kept some tree records, only 11% had them computerized for easy access. It has been reported that municipal tree management budgets decreased 40% on a per person basis, and by 37% as a percentage of city budgets (items 23 and 24 in checklist listed below). Several other indicators suggest at best a status quo from our 1986 data (Tschantz and Sacamano, 1994). This and other surveys of urban forest management have not followed the same format, and might therefore be difficult to use for comparison. Most differences between surveys are not great, however.

Tree Valuation

A good place to begin an evaluation of the benefits of urban trees may be with their economic values and how these values may be determined. There are several ways that are worth considering. Some governmental units place a set, or fixed, value on trees, and then carry this value on the city records as a city-owned asset. Another governmental way to consider value is to actually account for the maintenance costs on a yearly basis and apply annuity analysis, commonly done in forest management, over the life of the tree to determine its value (Kielbaso, 1971). This, in fact, is a very accurate method, although it disregards some economic as well as intangible considerations such as aesthetics, property values, psychological benefits, and storm water control. Moderately well managed cities spent an average of $10.62 per tree per year (Table 15.2). This yields a value of $499 per tree at 23 years, about the same amount as paying $150 to plant the tree, and carrying it at 6% compound interest for 21 years (Kielbaso et al., 1988).

Several studies in the U.S. have focused on determining a property value contribution for trees. Some of these have shown that trees contribute from 5% to 15% to the value of a residential property (Payne and Strom, 1975; Morales et al., 1983; International Society of Arboriculture/Council of Tree and Landscape Appraisers, 2000). Many realtors have said that lots with trees usually sell first for home sites. From a government perspective, street trees add to the property tax base by increasing property values. There are few who consider trees a liability, except in cases of storm failures such as broken branches or actual blow-downs.

Placing values on urban trees is much more difficult than estimating wood or timber values, as is done in non-urban forestry. In cities, this type of value for trees is much lower than other kinds of values. In most instances timber value in cities should be ignored, except when salvage is necessary, as in the case of disease, insect, or blow-downs. The International Society of Arboriculture (ISA) and its cooperator, the Council of Tree and Landscape Appraisers (CTLA), have over the years developed and improved the Guide for Plant Appraisal (Kielbaso, 2001). Space does not permit a thorough explanation, but the trunk formula method is as follows:

$$\text{Appraised Value} = \text{Basic Tree Cost} \times \text{Species} \times \text{Condition} \times \text{Location}$$

where Basic Tree Cost = Installed Tree Cost + (TA_{INCR} × Unit Tree Cost per in^2 or cm^2). $TA_{INCR} = TA_A - TA_R$, where TA_{INCR} represents appraised tree increase, TA_A represents trunk area of the appraised tree, TA_R is the trunk area of a replacement size tree, and the unit tree cost is the cost per unit of the replacement size tree. The installed tree cost includes the costs of purchasing a replacement-size tree, usually 7.5 to 15 cm, and planting it. In its first seven editions, this method was rather specific to the U.S., but now in its ninth edition it is has been adapted to be more useful in other countries, with appropriate modifications (International Society of Arboriculture, 2000).

Another economic consideration is the possibility of employment in the care and management of city trees. People may be employed in the public sector to administer trees for a governmental unit, or in the private sector to care for trees on public, private, or utility properties, or to consult regarding the care of trees. There are many thriving companies in North America and Europe with which I have personal acquaintance. In the U.S., the Tree Care Industry Association (TCIA) represents major commercial arborist firms. About half of the 14,000 members of the ISA are involved in commercial work.

Environmental and Economic Benefits

There is an increasing convergence of economic and environmental values for trees. The environmental benefits are numerous. An interesting case was made by Das (1980) speaking to the Indian Agricultural Society. He suggested that a medium-sized tree of 50 metric tons, during a life span of 50 years, produced benefits of about $224,000. He included production of oxygen, animal protein, control of soil erosion and soil fertility, recycling water, sheltering birds and animals, and controlling air pollution, and placed values on each representing its respective production cost. At about the same time, a student of the noted ecologist Howard Odum stated at a meeting that they had calculated the energy that trees captured from the sun in terms of barrels of oil at $120/tree/year. Akbari et al. (1988) has modeled energy and CO_2 savings through the use of trees and albedo modifications, especially in relation to the phenomenon of the urban heat island.

Model results show that one 25-foot tall shade tree may reduce annual heating and cooling costs of typical residences by 8% to 12%, or $10 to $25; even using the low $10 value, planting could save about $1 billion per year in the U.S. (McPherson and Rowntree, 1993). In a related study, McPherson (2003) showed a 24:1 benefit-cost ratio for one species, the London plane tree, *Platanus acerifolia*, although the average of 10 species was 6.6:1, with the benefits ranked from highest to least as property value, energy savings, avoided stormwater runoff, and CO_2 removal.

Urban forestry contributes in other ways to quality of life, and Robinette (1972) was among the first to discuss these in a comprehensive way. His taxonomy of tree uses in cities included architectural, engineering, climatological, and aesthetic. The architectural uses included space articulation, screening, privacy control, and progressive realization or unfolding. The engineering uses were erosion control, noise control, atmospheric purification, traffic control, and glare and reflection control. The climatological uses included were solar radiation, wind, precipitation, and temperature controls. In addition, his aesthetic uses involved numerous artistic concepts.

Urban Heat Island Amelioration and Energy Implications

That trees can ameliorate the urban heat-island effect is perhaps the most exciting applied benefit of the urban forest (Akbari et al., 1988). Lowering air temperature in cities makes the urban environment more comfortable. In addition, lower temperature caused by more trees and green space indirectly decreases the rate at which photochemical smog pollutants, like ozone, are produced. Studies suggest that the incidence of smoggy days increases 10% with each 2.8°C increase in temperature (Akbari et al., 1992). The health of urban residents is inversely related to higher smog levels. So, anything that can be done to ameliorate the urban heat island will tend to reduce smog and improve human health as well (see Chapters 5 and 6).

Trees also help decrease energy use and CO_2 emissions, especially in regions that rely greatly on fossil fuels for cooling and heating. Anything that reduces these energy demands will reduce the amount of fossil fuel consumption. For instance, trees located strategically around buildings can reduce the amount of energy consumed for cooling by up to 50% (Parker, 1981). Studies have shown that the energy saving for cooling or heating produces a 1- to 3-year payback period for investments made in planting (Akbari et al., 1988). This is one of the most rapid payback rates for energy conservation measures. At the same time that trees directly reduce expenses, they help reduce the generation of power at the source, thus leading to a reduced output of CO_2, an important greenhouse gas. The potential to use trees to conserve energy and reduce CO_2 emissions is great in cities and has not been maximized. Akbari and colleagues (1988, 1992) estimated that 15 trees would have to be planted in a rural forest to provide the same reduction in atmospheric CO_2 accumulation as one tree in a city. Even if the ratio were somewhat lower, this fact alone should make planting urban trees a higher priority for both municipal governments and residents.

Akbari et al. (1988) conducted a study of seven cities throughout the U.S., and showed that houses built after 1980, which were assumed to have better insulation and tighter construction than houses built earlier, would save nearly 20% in annual energy costs, and about 28% at peak demand periods for cooling, by planting trees and by increasing albedo by whitewashing buildings (Table 15.4). The data in Table 15.5 enable us to more fully appreciate the savings of energy from trees and increased albedo. About 50% of the savings are attributable to trees. The location of trees near buildings to help reduce the urban heat island will have significant financial benefits to all who must pay for energy. Akbari et al. (1988) have demonstrated that planting 100 million trees near homes in the U.S. could save

Table 15.4 Average energy savings for seven U.S. cities from trees and albedo modification of buildings

	Trees change in %	Albedo change in %
Peak kW	21.7	6.4
Annual kWh	8.7	10.6

Source: Adapted from Akbari et al. (1988).

Table 15.5 Yearly savings (contributed by 100 million trees) of energy used for air conditioning in the U.S. and consequent reductions in released carbon from tree planting and albedo modification

	Residential			Small commercial			Large commercial			Total	
	Energy (%)	(10^{15} BTU)	Carbon (M Tons)	Energy (%)	(10^{15} BTU)	Carbon (M Tons)	Energy (%)	(10^{15} BTU)	Carbon (M Tons)	Energy (M Tons)	Carbon (M Tons)
Direct savings[1]	10	0.12	4	4	0.03	1	0	0.0	0	0.15	5
Indirect savings[2]	20	0.23	8	12	0.09	3	5	0.04	1	0.36	12
Total	30	0.35	12	16	0.12	4	5	0.03	1	0.51	18

[1]Direct savings are those savings resulting from modification, such as shading, of a particular building.

[2]Indirect savings are those that when modifications are made throughout an entire city, the resulting energy balance of the whole city is modified so as to further affect the individual buildings.

M tons, million tons; BTU, British thermal unit.

Source: Modified from Akbari et al. (1988).

approximately 0.25 quads (22 billion kWh) of electricity. This is worth about $2.3 billion dollars annually in 1988 dollars, and can reduce emissions of CO_2 by 9 million tons.

Air Quality and the Urban Forest

Air quality can be improved by trees through the collection of particulates and the absorption of gaseous pollutants such as ozone, carbon monoxide, sulfur dioxide, and nitrogen oxides, among others. The problem is that in large quantities, these gases also adversely affect plants. Unfortunately, a tree or other plant is not an unlimited sink for pollution gases. The concentration of ozone in urban areas increases as temperatures increase.

It has been shown that trees can reduce airborne particulates from 18 to 181 kg per tree per year. Calculations can be made to determine their environmental value. Economic values for this function alone have been estimated at over $4 per tree per year for particulate removal (McPherson, 1991). Thus, environmental and economic values overlap. Reduced pollutant gases and particulates will have significant potential health savings, as well.

McPherson (1991) showed, through computer simulation, cost–benefit results of the Trees for Tucson/Global ReLeaf project that planted 500,000 trees. Net benefits were projected to be $236 million over a 40-year period, with a benefit cost ratio and internal rate of return for all trees being 2.6 and 7.1, respectively. Per tree cooling benefits of $20.73 were projected—227 kWh ($16.34) for evapotranspirational cooling and 61 kWh ($4.39) for direct shade. In a related study in Chicago, McPherson et al. (1993) estimated air pollution mitigation in the Lincoln Park section, with 23.2% crown cover. During an average day, the benefits from these 212 hectares (525 acres) were estimated at $136, or nearly $25,000 per year for the 180 days with leaves. Nearly 75% of the benefit was due to particulate mitigation, and 20% to nitrogen dioxide, 3% to sulfur dioxide, and 1% to carbon monoxide uptake.

Because of studies like those of McPherson above, Hudson (2002), proposed that the urban forest be considered a biomass pollution shed, and more recently, a biogenic public utility. He suggested that when costs and benefits of trees are both considered, urban trees offer solid benefits beyond aesthetics. The following study for the city of Santa Maria, California, included cost-benefit estimates for the urban trees and support Hudson's evaluation (Tables 15.6 and 15.7).

Other Environmental Benefit Considerations

Urban forests may also mitigate the negative impacts of cities on soils and local hydrology in at least four ways. Trees reduce soil erosion from wind and rain.

Table 15.6 Biomass pollution shed for city of Santa Maria, California: biogenic services and benefits compared to maintenance costs of 38.9 hectares containing 24,911 trees

Biogenic services benefits—24,911 trees			
Pollutant	Filtration rate	California Control Costs	
Particulates	61 kg /day	@ $ 11.07/kg/day	$ 675.00
CO	1.8 kg/day	@ $ 2.22/ kg/day	$ 4.00
NO_2	9 kg/day	@ $ 26.67/ kg/day	$ 240.00
SO_2	45.4 kg/day	@ $ 4.41/ kg/day	$ 200.00
Total Benefits Value/Day			$1,119.00

Maintenance and production costs—24,911 trees		
Operational budget	$ 0.0391/day × 24,911 trees	$ 974.02
Liability	$ 0.0038/day × 24,911 trees	$ 94.66
Sidewalk and curb repair	$ 0.0044/day × 24,911 trees	$ 109.60
Water	$ 0.0003/day × 24,911 trees	$ 7.47
Gutter sweeping	$ 0.0001/day × 24,911 trees	$ 2.49
Total maintenance/production costs/day		$ 1,188.25

Total yearly cost/benefit analysis:		
Production costs:	365 days × 24,911 trees @ $ 1,188.25/Day	$ 433,707.60
Filtration benefits:	225 In-leaf days × 24,911 trees @ $ 1,119/day	$251,775.00
Deficit		($ 181,932.60)

Cost/benefit ratio = 1.72

58% self-sufficient

Source: Modified from Hudson (2002).

Table 15.7 Biomass-avoided energy consumption for city of Santa Maria, California: maintenance/production costs compared to energy saved benefits for 6288 trees[1]

Maintenance and production costs 6228 trees		
Operational budget	$0.0391/day × 6228 trees	$ 243.51
Liability	$0.0038/day × 6228 trees	$ 23.67
Sidewalk and curb repairs	$0.0044/day × 6228 trees	$ 27.40
Water	$0.0003/day × 6228 trees	$ 1.87
Gutter sweeping	$0.0001/day × 6228 trees	$ 0.62
Total maintenance and production costs/day		$ 297.07

Total yearly cost/benefit analysis		
Production costs:	365 days × 6228 trees @ $297.07/Day	$ 108,430.55
Energy Benefits:	225 in-leaf days × 6228 trees @ $10.00/year	$ 62,280.00
Deficit		($ 46,150.55)

Cost/benefit ratio = 1.74

57% self-supporting

[1]Assumption: 25% of the total public tree population of 24,911, or 6288, is directly reducing energy consumption costs between $10.00 and $20.00 per tree per year.

Source: Modified from Hudson (2002).

Increasing the proportion of vegetation and unpaved surfaces in urban watersheds can help avoid extremes of flooding by providing surfaces to accept and store water. Tree canopies also reduce flooding by intercepting 7% to 10% of potential runoff. Both of these functions tend to reduce the need for large sewer pipes from storm runoff (McPherson and Rowntree, 1991) and reduce the erosion rates of local streams, although most engineers would not likely reduce sewer pipe size. Urban forests may also provide tertiary treatment opportunities for wastewater, as well as places to hold storm or wastewater for aquifer recharge. The contributions to soil erosion control, flood control, sewer size control, and wastewater treatment make the urban forest a potentially great contributor to improving urban quality of life.

American Forests (2000) has a GIS-based program, "City Green," that attempts to quantify the environmental benefits of urban forests by estimating the financial implications of adding or removing trees from a landscape. Several cities, including Atlanta, Detroit, Washington, D.C., Houston, and Roanoke, Virginia, have benefited from this Geographic Information System (GIS) program. They offer examples of how cities have strengthened their local ordinances, secured additional funding, enhanced programs and staffing, and improved communication between diverse stakeholders. In addition, there are some new ways to integrate green infrastructure into communities to help meet federal clean air and water requirements. Our current research is dealing with dendroremediation, or using trees to help clean toxic soils, or brown fields.

The broader amenity values of trees and the urban forest contribute recreation opportunities, wildlife observation opportunities, and visual enhancement. Some large cities have large expanses of forested land that serve as recreational sites. Chicago has the impressive Cook County forest preserve around it. Shanghai is embarking on an unprecedented afforestation of its surrounding areas and has the opportunity to create an excellent series of wooded parks. These parks will offer opportunities to escape the artificiality and often hectic pace of the cities. The opportunity to walk, play, and observe some natural scenery among trees provides for physical and mental health benefits.

Significance of Inventories for Urban Forest Management

In most places, the urban forester does not have control of private trees, as compared to public street or park trees. We have studied street tree management in the U.S. and have found that for every tree currently growing (about 60 million) there is space for at least one more tree. These trees, which could shade hot streets and cool the microclimates of urban areas, have the potential to greatly enhance the quality of urban life through financial energy savings, CO_2 mitigation, and aesthetics. Unfortunately, at the rate that U.S. cities are planting street trees, it will require 267 years to have our streets fully planted (Kielbaso, 1990). We have concluded that there are about 10 more private trees for each street tree, or about 660 million urban trees (Kielbaso et al., 1993). In another estimate, Samson and colleagues (1992) suggested that there may be as many as 1 billion urban trees, when private yards, parks, and greenways are included, and the potential to plant 150 million to 290 million more in our cities exists. Dwyer et al. (2000), using the latest remote sensing and other advanced technologies, have concluded that there may be as many as 3.8 billion trees in urban areas.

Knowing the inventory of the urban forest is important for management, in much the same way that inventories and inventory control are important for most retail and manufacturing firms. Inventories can lead to management opportunities or challenges. With knowledge of the total tree population and its growth and mortality rates, the manager can take actions to maximize benefits or minimize liabilities, as well as evaluate outcomes of earlier decisions.

A reinventory in two U.S. cities in 1992 provides information for understanding the dynamics of the urban forest. Bowling Green, Ohio, and Lincoln, Nebraska, were remeasured after 13 and 12 years, respectively. Bowling Green increased its tree density by 41% to attain a density of 47 trees per acre (116/ha), whereas Lincoln increased its tree density by 22% (29 trees/acre; 72/ha) over this interval. The average tree growth rate for both cities at both time intervals was about 0.4 inches (1.0 cm) per year. Other studies have shown urban trees to grow less and be more stressed. Highly significant differences in growth parameters have been shown between trees growing along reasonably good street locations and in a nearby woodlot (Close et al., 1996).

Mortality in these two urban forests surprised us (36% in Bowling Green and 45% in Lincoln). Most of the mortality occurred in the small- to medium-sized trees on private property. It is commonly assumed that when trees are planted in cities, they will grow and remain there. Of the trees that died between the years of this study, 77% in Bowling Green, and 65% in Lincoln had been rated in excellent condition at the start of the study in 1980. Of the trees measured in 1992 in each city, 55% were new trees planted in the previous 12 to 13 years. Most rural forests do not have so great a turnover, although trees in the seedling and sapling sizes are abundant in rural forests. In cities, one does not find many seedlings in residential situations, nor is there the overhead competition that would be found in a forest, causing the death of understory trees. Most people cannot easily see change in the urban forest where they live. While species diversity had not changed greatly in either city between intervals, this study shows that the urban forest is much more dynamic than most would suspect (Kielbaso et al., 1993).

A Checklist for City Programs

The following list presents the data from a 1986 survey of 1062 U.S. cities in a form designed for additional practical use (Kielbaso et al., 1988). This can be a useful way to evaluate a particular city program while comparing it with others in the U.S. survey, keeping in mind that international comparisons require judgments as to applicability since government structures may vary, especially as related to ordinances, laws, and other legal situations. Tree City and Arbor Day may or may not apply, and budgets may be kept differently.

Yes/No Questions, with Yes Percentages Noted

1. Systematic management: Is the tree program managed systematically rather than on a crisis basis? (39%)
2. Is the city responsible for trees in the public right of way? (72%)
3. Is the respondent aware of any state or federal urban forestry assistance programs? (27%)
4. Are tree records kept? (47%)
5. Are tree records computerized? (11%)
6. Are trees designated as "official" trees for a block or other section of a street? (23%) [Some question this as leading to monocultures.]
7. Is there a trend toward planting small trees along city streets? (43%) [*Note*: With energy conservation concerns, the question of whether this is a positive trend should be reevaluated and larger trees used whenever possible.]
8. Does the city maintain a nursery to provide some or all planting stock? (22%)
9. Does the city have a tree ordinance that defines responsibility for tree care in the city? (61%)
10. Is there a tree preservation ordinance restricting tree cutting on private property?
11. Does city monitor tree pest problems, important in integrated pest management? (36%)
12. Are any biological controls employed against tree pest problems? (18%)
13. Are systemic treatments used, which apply pesticides directly into a tree rather than broadcast the chemical into the environment? (17%)
14. Does the city conduct educational programs regarding trees for city residents? (30%)
15. Does the tree management unit conduct educational programs for employees? (59%)
16. Does the city conduct an Arbor Day program to bring recognition to trees? (49%)
17. Does the city participate in the Tree City program? (26%)
18. Does the city participate in research relating to urban trees? (13%)
19. Is there a written or formal plan for dealing with trees in an emergency situation such as a tornado, windstorm, earthquake, typhoon, etc.? (27%)
20. Does the city have a formal written urban forestry management plan? (17%)

Numerical- or Percentage-Based Questions: Averages for Responding Cities

21. The number of full-time employees devoted to performing the work of maintaining urban trees. (7.0 employees, full-time equivalent)
22. The number of trees planted in the city in 1986, as a measure of activity of the program. (338 trees)

23. The tree care budget as a percentage of the total city budget. (0.49% mean, 0.40% median)
24. The tree care budget expressed as dollars per person. ($2.60 mean, $1.73 median)
25. The percentage of cities knowing both the number of trees and the budget allocation for tree care. (35%)
26. The tree care budget described as dollars per tree. ($10.62 mean, $8.04 median)
27. How many years of tree maintenance does the urban tree manager possess? (14.6 years)
28. The number of trees maintained per full-time employee. (3,798 trees)
29. The number of public trees in the city, contributing to the environment. (29,677 trees)
30. Is the number of trees in the city known, as basic management information? (38% know)
31. What is the number of trees per person? (0.50 trees)

These factors were chosen because urban forestry professionals generally believe they are indicators of a positive tree care program in a city; the larger the number, the more positive the relationship. Only one of the indicators—the number of trees maintained per employee—should be lower to be considered more positive. Disagreement exists on matters such as whether official trees, designated as official for a street or area, and small trees are valuable. Some professionals see such policies as having a negative environmental value. But because they do at least imply a strong, conscientious program, they are included as positive indicators in the above checklist.

These results are from the 1986 survey reported in 1988 (Kielbaso et al., 1988), based on a stratified random selection of cities. Managers were asked to respond to a lengthy questionnaire that was mailed to them three times over the course of 5 months. This also happened to be the third in a series of follow-up studies over a 12-year period. Bias in respondents from more active programs is not believed to have influenced the results, since a significant number of cities, about 60% of 1062 responding cities, having minimal programs responded. In fact, the estimates of the total number of street trees reported for this survey by managers or their subordinates were very similar to a separate survey that was conducted by qualified volunteers in randomly selected cities. In the second survey, the volunteers quantified trees along streets irrespective of any ongoing program or specific input from the city. The total numbers of street trees estimated by both studies were strikingly similar. With this concurrence, it is not believed that significant bias from active programs influenced either effort.

These surveys helped to produce the above checklist so that cities would have some sort of benchmark against which they could assess their own programs. Since so many cities were unable to provide even some of the basic information for comparisons, there is some bias in the recommendations toward those that could. The recommendations are goals to strive for, and since they represent the "better" programs, they are thought to be goals worth making the attempt to reach.

Budget, Planning, and Other Commitments to Urban Forests

The four characteristics of a city committed to urban forestry are having (1) a sound ordinance covering multiple aspects of the public/private forest, (2) a thorough inventory of city trees, (3) a professional advocate with the title of *forester* or *arborist*, and (4) being located in a forestry- or tree-oriented division of the city government. Once a city has taken those steps, the next most important need is continued funding for tree management. Depending on the situation, a desirable goal is a budget approaching an average of $2.60 per person or $10.62 per tree. However, since these averages represent comparatively sophisticated programs, attaining the medians of $1.73 per person, or $8.04 per tree, may be more realistic initial goals. Smaller communities with small budgets may find that contracting work to private companies, renting equipment, or sharing equipment with other cities may be a more sensible strategy. However, contracts must be carefully constructed so that both sides have reasonable expectations.

Only 17 % of U.S. cities have an urban forestry management plan, but such a plan is essential for management of a resource that is by its very nature long-lived and long-term. Goals must be made and prioritized so that realistic objectives may be set and measured. A long-range plan should address objectives for planting, removing, and pruning of trees, along with the training of professionals, public relations, and funding. For example, without the support of quality nurseries, a program will suffer from lack of quality tree-planting stock. Everything possible should be done to encourage and reward good nurseries and tree care companies in the community. As they succeed, general tree consciousness increases, and the city tree program will benefit.

However, many cities have not developed long-term goals or plans for managing their forest. Instead, crisis management, or management by demand is the rule. B. Hudson (personal communication) demonstrated this fallback to be inefficient. Hudson has shown that pruning on a schedule rather than on demand results in 1.35 fewer hours being spent on a per tree basis, when considered over an 11-year period—1.03 versus 2.38 hours, respectively. Specifically, it requires about twice as much time to prune each tree when done on a demand basis, rather than as a planned and organized effort.

Biodiversity Failures, Lessons, and Goals

In setting planting goals, species composition must be considered. Not enough diversity is represented by the species currently existing or being planted on city streets (Kielbaso and Kennedy, 1983; Kielbaso et al., 1988). Lessons learned from the devastations of Dutch elm disease, and more recently emerald ash borer, point to the potential dangers of low diversity tree populations, and their negative impacts on urban forest management. A conservative goal for planting diversity should be that no species make up more than 15% of the total tree population of a city. An even better goal is that no genus should account for more than 10% and no species for more than 5% of the total (Richards, 1993). Whether or not the same species should be planted on an entire segment of a street is also subject to debate. Some experts fear the suggestion of "monoculture," whereas others place more emphasis on the aesthetics and maintenance efficiency achieved when trees on the same street are similar. Good arguments exist on both sides, but the 10% and 5% standard is a good one for long-range planning, except where nonavailability of adapted species precludes such a mix. Community indices used in ecological studies can be used to describe tree diversity in a city as well. For example, population diversity may be described with the Shannon-Wiener Diversity Index (Krebs, 1984), which is as follows:

$$H^1 = p_i \times \text{Log}\ (p_i),$$

where p_i is the proportional abundance of a species, and Log (p_i) is the decimal logarithm of p_i. This has been applied to cities in a 22-city survey in the U.S. to describe and compare urban tree diversity across cities (McPherson and Rowntree, 1986). Since species composition and age structure are two readily recognized characteristics, future impacts on maintenance costs are suggested. Monotypic urban forests, by species or size, should warn of possible future insect/disease problems, or age-dependent tree failures.

Conclusion

As of 1986, municipal street tree management cost the American public approximately $425 million annually, a yearly cost of $2.60 for each city resident. This has not changed greatly during recent times. Little else is being done in our cities that has the potential to add so much to the quality of urban environments for such little cost, and with returns of as much as 19 to 1. Our urban forests have long been known by those of us in the fields of arboriculture and urban forestry to be of great value. Their values have been extolled widely as they significantly enhance urban aesthetics, economics, and ecology. Health, public safety, air quality, storm water infrastructure, historic memory, and personal satisfaction should all be added into any equation of urban quality of life, and all are affected by the condition of the urban forest. Using tree planting initiatives (like Global ReLeaf) to get people involved in improving the urban environment and its quality provides significant social as well as environmental benefits. The total physical condition of a community, its buildings, vacant spaces, and streets makes a significant difference in how members of that community feel about themselves. What we see often tells us what we are. As noted by Primak (1987), "Walking by and through trash-filled empty lots and vandalized school yards, walking along littered streets with no trees, seeing only pavement and brick when looking out the window makes people feel bad about where they live and about themselves. They know they do not matter, that their pleasure or comfort is unimportant."

Charles Lewis, Research Fellow of the Morton Arboretum, has for years been reporting on changes in people and their neighborhoods as a result of gardening (Lewis, 1990). He has reported on projects in Chicago, Philadelphia, Boston, and New York that transformed neighborhoods. His were mostly gardening examples, but larger plants such as trees should provide similar social benefits. As he said, working with plants "enhances self-image, and helps to create self-esteem.... Feeling better about himself, he feels better about where he lives." The improved attitude about self and setting is evident in his comments.

It has been estimated that 25% to 30% of all land in most cities is devoted to some form of transportation, ranging from roadways to airports and rail yards.

> If anything is to be done to improve the aesthetics of the city, the task begins with transport.... Broad tree-lined sidewalks, adjacent lawns and flower gardens, sidewalk kiosks and eating places, all combine to make the right of way serve not only the movement of vehicles but also the social and recreational needs of people and the aesthetic enhancement of the downtown area.... The redevelopment of the blighted closed-in areas of American cities could readily incorporate such islands of green along streets that have become drab and colorless. [Owen, 1969]

It is important to acknowledge that all the data presented here are from the U.S. experience. They are mostly concerned with public street trees, which are relatively easy to access, inventory, and maintain, as compared to private and woodland trees. This information cannot be directly applied in other countries without allowances for local customs and culture. It is intended here to stimulate inquiry, particularly in the case of China, with managers adopting and modifying what may be useful, and ignoring what is not relevant.

Reaching the ideal city of trees may always lie a little beyond our grasp, but we do clearly know some of the basic outlines of such a program. We can begin new programs or take one more step in the already-existing program in each of our cities. Every decision should be based not simply on the current cost of city programs, but on analyses that incorporate a view of investment in a strong future that make cities worth living in now and in the future.

References

Akbari, H., Huang, J., Martien, P., Rainer, L., Rosenfeld, A., and Taha, H. (1988) The impact of summer heat islands on cooling energy consumption and global CO_2 concentration. In: Proceedings of ACEEE 1988 Summer Study on Energy Efficiency in Buildings, Vol. 5, Asilomar, CA. American Council for an Energy Efficient Economy, Washington, DC, pp. 11–23.

Akbari, H., Davis, S., Dorsano, S., Huang, J., and Winnett, S. (1992) Cooling Our Communities—A Guidebook on Tree Planting and Light Colored Surfaces. USEPA, Policy, Planning and Evaluation (PM-221). U.S. Supt. of Documents. Washington, DC.

American Forests. (2000) City Green—Calculating the Value of Nature. Version 4.0 for ArcView. American Forests, Washington, DC.

Close, R., Nguyen, P.V., and Kielbaso, J. (1996) Urban vs. natural sugar maple growth: I. stress symptoms and phenology in relation to site characteristics. Journal of Arboriculture 22(3):144–150.

Das, T.M. (1980) The value of a tree. The Agricultural Society of India 23rd Annual General Meeting, Calcutta.

DeWalle, D.R. (1978) Manipulating urban vegetation for residential energy conservation. Proceedings of the First National Urban Forest Conference. USDA Forest Service, Washington, DC, pp. 267–283.

Dwyer, J., Novak, D., Noble, M., and Sisinni, S. (2000) Connecting people with ecosystems in the 21st century. Technical report PNW-GTR-490. USDA Forest Service, Washington, DC.

International Society of Arboriculture/Council of Tree and Landscape Appraisers. (2000) Guide for Plant Appraisal, 9th ed. International Society of Arboriculture, Champaign, IL.

Hudson, B. (2002) Is the Urban Forest a natural resource public utility? Final Report, USDA Forest Service Grant G-5–98–20–076, NUCFAC, National Urban and Community Advisory Council, Vallejo, CA.

Kielbaso, J. (1971) Economic values of trees in the urban locale. In: Proceedings of Symposium on the Role of Trees in the South's Urban Environment. University of Georgia, Athens, GA.

Kielbaso, J. (1989) City tree care programs: a status report. In: Moll G., and Ebenrech, S. (eds.) Shading Our Cities. American Forestry Association. Island Press, Washington, DC, pp. 35–46.

Kielbaso, J. (1990) Trends and issues in city forests. Journal of Arboriculture 16(3):69–76.

Kielbaso, J. (1993) Urban forestry and quality of life in cities. First Panamerican Forestry Conference, Curitiba, Parana, Brazil.

Kielbaso, J. (2001) For what it's worth. American Nurseryman. February:48–53.

Kielbaso, J., and Kennedy, M.K. (1983) Urban forestry and entomology: a current appraisal. In: Frankie, G.W., and Koehler, C. (eds.) Urban Entomology: Interdisciplinary Perspectives. Praeger, New York, pp. 423–440.

Kielbaso, J., Beauchamp, B., Larison, K., and Randall, C. (1988) Trends in urban forestry management. Baseline Data Report. International City Management Association, Washington, DC, 20:1.

Kielbaso, J.J., de Araujo, M.N., de Araujo, A.J., and Cannon, W.N. Jr. (1993) Monitoring the growth and development of urban forests in Bowling Green, Ohio and Lincoln, Nebraska. American Forests National Urban Forest Inventory, Washington, DC.

Konopacki, S., and Akbari, H. (2002) Energy Savings for Heat Island Reduction Strategies in Chicago and Houston (Including Updates for Baton Rouge, Sacramento, and Salt Lake City). Draft Final Report LBNL-49638. Heat Island Group, Lawrence Berkeley National Laboratory, Berkeley.

Krebs, C.S. (1984) Ecology: The Experimental Analysis of Distribution and Abundance, 3rd ed. Harper & Row, New York.

Laechelt, R., and Williams, B. (1974) Value of tree shade to homeowners. AFC Bulletin No.2450. Alabama Forestry Commission, Montgomery, AL.

Lewis, C.A. (1990) Plants and the community: effects of plants and gardening in creating interpersonal and community well-being. In: Proceedings: The Role of Horticulture in Human Well-Being and Social Development. Timber Press, Portland, OR.

McPherson, E.G. (1991) Economic modeling for large-scale tree plantings. In: Vine, E., Crawley, D., and Centolella, P. (eds.) Energy Efficiency and the Environment: Forging the Link. American Council for an Energy Efficient Economy, Washington, DC.

McPherson, E.G. (2003) A benefit-cost analysis of ten street tree species in Modesto, CA, USA. Journal of Arboriculture 29(1):1–8.

McPherson, E.G., Nowak, D., Sacamano, P., Prichard, S., and Makra, E. (1993) Chicago's evolving urban forest: initial report of the Chicago Urban Forest Climate Project. General Technical Report NE-169. USDA Forest Service, Radnor, PA.

McPherson, E.G., and Rowntree, R. (1986) Ecological measures and change for street tree populations. Proceedings of Third National Urban Forestry Conference, Orlando, FL. American Forestry Association, Washington, DC, pp. 65–77.

McPherson, E.G., and Rowntree, R. (1991) The environmental benefits of urban forests. In: A National Research Agenda for Urban Forestry in the 1990's. International Society of Arboriculture, pp. 45–49.

McPherson, E.G., and Rowntree, R. (1993) Energy conservation potential of urban tree planting. Journal of Arboriculture 19:321–331.

McPherson, E.G., and Simpson, J.R. (2003) Potential energy savings in buildings by an urban tree planting programme in California. Urban & Fischer Verlag.

Morales, D., Micha, F.R., and Weber, R.L. (1983) Two methods of valuing trees on residential sites. Journal of Arboriculture 9(1):21–24.

Owen, W. (1969) Transport: key to future of cities. In: Perloff, H.S. (ed.) The Quality of The Urban Environment. Resources for the Future, Washington, DC, pp. 205–230.

Parker, J. (1981) Uses of Landscaping for Energy Conservation. Department of Physical Sciences, Florida International University, Miami.

Payne, B., and Strom, E. (1975) The contribution of trees to the appraised value of unimproved residential land. Valuation 22(2):36–45.

Primak, M. (ed.) (1987) The greening of Boston, an action agenda. Report from the Boston Foundation Carol R. Goldberg Seminar. The Boston Foundation, Boston.

Robinette, G.O. (1972) Plants/People/and Environmental Quality. U.S. Dept. of Interior, National Park Service. U.S. Supt. of Documents, Washington, DC.

Richards, N. (1993) Reasonable guidelines for street tree diversity. Journal of Arboriculture 19(6):344–349.

Sampson, R.N., Moll, G., and Kielbaso, J. (1992) Opportunities to increase urban forests and the potential impact on carbon storage and conservation. In: Sampson, R.N., Hair, D. (eds.) Forests and Global Change. American Forests, Washington, DC, pp. 51–72.

Simpson, J.R. (2002) Improved estimates of tree shade effects on residential energy use. Energy and Buildings 34(10):1067–1076.

Tschantz, B., and Sacamano, P. (1994) Municipal tree management in the United States. Davey Resource Group. Davey Tree Expert Company, Kent, Ohio

16 The Urban Forest of Nanjing City: Key Characteristics and Management Assessment

Sophia Shuang Chen and C.Y. Jim

An urban forest refers to "the sum of all woody and associated vegetation in and around dense human settlements, ranging from small communities in rural settings to metropolitan regions" (Miller, 1996). This natural-*cum*-cultural vegetation entity is usually strongly influenced by human behaviors and molded by the history and fabric of urban habitats. The outstanding environmental benefits and ecosystem services of urban forests are well recognized, including contributions to aesthetics (Schroeder and Cannon, 1983; Buhyoff et al., 1984), energy savings (Dewalle et al., 1983; Rudie and Dewers, 1984), microclimatic amelioration (Simpson and McPherson, 1996), and pollution abatement (Dochinger, 1980; Beckett et al., 1998). Urban forests serve as a surrogate for and symbol of nature in human settlements, and understanding and enhancing them are high on the agenda of urban researchers and practitioners. Cities that aspire to be green, ecologically minded, and sustainable would invariably plant and maintain a exemplary urban forest.

Early urban forestry studies can be grouped into four categories: (1) species and tree size and age composition in the city (McBride and Jacobs, 1976; Profous et al., 1988; Chacalo et al., 1994); (2) tree canopy cover configuration in the city and its variation by land-use type (Rowntree, 1983; Jim, 1989; Talarchek, 1990); (3) distribution of forest patch size and number along urban-to-rural gradients (Medley et al., 1995); and (4) individual tree growth in relation to site condition based on ground surveys (Freedman et al., 1996; Jim, 1997a,b).

Urban forestry studies in mainland China have until now been mainly concerned with canopy cover configuration (Che and Song, 2001; Gao and Song, 2001; Zhou et al., 2002) and vegetation structure (Gao et al., 2002). Few studies have been based on individual trees and their interaction with urban stresses generated by dense packing of buildings, roads, pedestrians, vehicular traffic, air pollution, indifferent or improper care, and vandalism (Gilbertson and Bradshaw, 1985; Jim, 1992, 2005a,b). There is, therefore, ample scope to expand and deepen this research frontier so as to build a stronger conceptual and empirical knowledge base for improving urban forestry as a science in China. Cities in China commonly exhibit a compact mode of development with continued infilling and intensification of land uses that will impose more pressure on existing greenery stock. Since land is becoming a limiting resource, conflicts between developing land versus green space proponents are becoming increasingly acute and demand solutions. From a

M.M. Carreiro et al. (eds.), *Ecology, Planning, and Management of Urban Forests: International Perspectives.*

broader perspective, a balance needs to be found between the need for development and the need for conservation of green space and vegetation.

Many Chinese cities have adopted the modern concepts of urban forestry to guide and inform their greening strategies. However, enlightened greening policies and plans have not been matched by follow-through in effective management. Tree protection measures, for instance, are often ineffective in preserving important trees or tree clusters, including outstanding or heritage trees (Jim, 2005a,b). Many big trees had been removed to widen roads, or had been heavily pruned to resolve conflicts with juxtaposed buildings or utilities. Tree maintenance standards remain relatively low compared with the best international arboricultural practices, reflecting a chronic negligence of effective management organization and operations.

To improve and inform adaptive management efforts for China's urban forests, this chapter focuses on (1) an analysis of the key attributes of the city's tree population, and relates them to selected urban factors that are known determinants of urban tree growth; and (2) an evaluation of the pertinent aspects of urban forest management, including organization, operation, and relevant regulations. Nanjing, a principal city in east China, with a human population of 2.2 million, has been chosen for this study because it has high tree cover (about 30% of total urban area), a long urban development history, and a rapid rate of expansion and redevelopment, as have many other cities in China. Therefore, recommendations derived from this study in Nanjing, which has relatively good tree cover and tree condition, could serve as a model for a large number of other Chinese cities.

Study Area

Nanjing is located in the lower reaches of China's Yangtze River (known as the Chang Jiang River) upstream and to the west of Shanghai at latitude 32°03'N and 118°47'E, with a north-subtropical climate strongly influenced by the region's dominant monsoon regime. The mean annual temperature is 15°C, and rainfall averages 1033 mm/year. On average, a typhoon hits Nanjing one or two times per year, mainly in August. As it is situated away from the coastline, wind damage to trees is less serious compared with that in coastal cities (Jim and Liu, 1997). The study area is confined to the built-up segment of the main city area, covering nearly 130 km^2 and accommodating a population of 2.2 million in 2001. Urban growth began from a well-defined ancient core of some 2000 years of age, and has since spread to cover a much larger area. The old urban enclave, the inner city, is the most densely populated area with a high building density. The area stretching between the old urban center and the Ming Dynasty (14th century A.D.) city wall includes the majority of the city's political, economic, cultural, and social functions. Many government agencies, colleges, hospitals, and other educational or cultural institutions are located in this area. Industrial land use is mainly distributed in the area beyond the city wall to the south and the north, respectively dominated by mechanical and petrochemical industries, which have been largely developed since the 1950s. A series of large-scale residential districts have been developed around the city wall since the 1980s, and together with those in the inner city constitute the major residential areas in the city. A network composed of 117 main roads links different parts of the city, and connects the inner city with its periphery. Trees planted along these roads constitute a major portion of the urban forest in the study area.

Methods

Tree Sampling and Field Assessment

In the study area, 6527 trees, representing approximately 0.5% of the city's tree population, were sampled based on the partial inventory technique of Jaenson et al. (1992). To obtain a reliable estimate of the urban tree composition and characteristics, a two-level stratified sampling strategy was adopted. One level is the primary habitat type: residential neighborhood (RN), roadside (RS), factory (FT), institution (IT), and garden-park (GP). The other level is the trees within each subcategory within a primary habitat type (Jim and Chen, 2003). For instance, the institution habitat type was first divided into subcategories such as government agency, college, and hospital, and then the constituent trees within each subcategory were sampled at the prescribed level of intensity. The number of trees sampled in each of the primary habitat types is summarized in Table 16.1. The habitat patches were selected such that they were distributed as evenly as possible in different districts of the study area. Trees were individually measured in the field to record species, trunk diameter at breast height (DBH), crown diameter, and tree height. Tree performance assessments were based on the aggregate condition of trunk, branches, foliage, and roots, and on symptoms of disease or pest infestation. For each tree attribute, numerical scores of excellent, good, fair, poor, and dead/dying were given and ranged, respectively, from 5 to 1 points. The average performance score (APS), a summation of individual tree scores/tree count, was calculated to provide a synoptic summary of the growth condition of a entire tree population. The relative performance index (RPI), which is the APS of a tree population in a habitat type divided by the APS of the tree population of the entire city, was adopted to indicate whether a tree population for a particular habitat type had a better or worse performance compared with that of the whole city.

Table 16.1 Distribution of sampled trees in five main habitat types in Nanjing

	Roadside	Neighborhood	Factory	Institution	Garden park	Total
Tree frequency	2068	1395	1323	1105	636	6527
Percent (%)	31.7	21.4	20.3	16.9	9.7	100

Questionnaire Survey

To assess the status of the municipal tree management organization and programs, a questionnaire survey was designed that mainly used open-ended questions to facilitate a frank dialogue between the interviewer and the interviewee. Some closed-ended questions helped us to acquire more structured responses. The principal unit shouldering urban greening duties in the city is the Administrative Bureau of Landscape and Gardening of Nanjing (ABLG). For other relevant agencies, some of the questions were modified to match the specific nature of these agencies. The questionnaire survey aims to provide information on the following tree management issues in Nanjing:

- Responsibilities of major and other relevant tree management agencies
- Relationships and division of duties among tree management agencies
- Tree management approaches
- Source and expenditure of tree care funds
- Planning of tree programs
- Standards and practices of tree planting, maintenance, and protection operations
- Enforcement and implementation of tree ordinances

The survey was conducted in face-to-face interviews with a total of 10 officials in the ABLG and other relevant government departments. Their affiliated units are responsible individually or jointly for planning green space, supervising tree programs, or direct tree maintenance. Their positions ranged from department head to professional forester. For each respondent, the interview and associated discussions lasted on average of 1 to 3 hours.

Results and Discussion

Urban Forest Attributes and Associated Factors

Species Composition

A total of 108 species in 78 genera and 42 families were identified. Deciduous and evergreen species each contributed to about half of the total tree population. About 20% of the species exhibited notable ornamental attributes, such as flowers, fruits, seasonal foliage-color changes, or interesting leaf shape. The most prominent species in this group are the natives *Prunus persica, Magnolia denudata, Osmanthus fragrans*, and the exotics *Prunus yedoensis, Punica granatum*, and *Nerium indicum*. Such decorative traits have been selected at the species and cultivar levels for millennia to meet changing landscape preferences and needs. Whereas the traditional ornamentals have been cultivated for centuries, some exotics were recent introductions and hence have had a relatively brief tenure in the city. Their recent rise in popularity

Table 16.2 The top 10 species in Nanjing based on tree frequency

Species	Family	Individual frequency	Cumulative frequency percent
Platanus x acerifolia[a]	Platanaceae	967	14.8
Sabina chinensis[b]	Cupressaceae	817	27.3
Ligustrum lucidum[b]	Oleaceae	743	38.7
Magnolia grandiflora[a,b]	Magnoliaceae	455	45.7
Metasequoia glyptostroboides	Taxodiaceae	448	52.5
Sophora japonica[b]	Leguminosae	339	57.7
Cedrus deodara[a,b]	Pinaceae	335	62.9
Cinnamomum camphora[b]	Lauraceae	297	67.4
Ginkgo biloba	Ginkgoaceae	211	70.6
Trachycarpus fortunei[b]	Arecaceae	126	72.6
			72.6
Sum		4738	
Total of 108 sampled species		6527	100.0

[a]Exotic species.
[b]Evergreen species.

reflects the phenomenon of shifting landscaping fashions that are often tied to socioeconomic factors.

There were 21 species of exotic trees, mostly concentrated in the conifer families of Pinaceae, Taxodiaceae, and Cupressaceae. The notable species that have gained top-10 status include *Metasequoia glyptostroboides* and *Cedrus deodara* (Table 16.2). They were chosen for their relatively neat, compact, and easy-to-maintain crown form. Their "exotic" look in terms of symmetrical conical tree form and largely needle foliage, in a biogeographical region dominated by broad-leaved trees, also accounts for their increasing use.

The leading species, the exotic *Platanus x acerifolia* (the London plane tree), commands a frequency of 14.8% of the entire tree population. Since its introduction in the early 1900s from Europe, where it was widely planted in cities (Wasson, 2003), this deciduous broad-leaved tree has been actively planted in Nanjing. It has been adopted extensively for urban greening as avenue and park trees mainly due to its large final dimensions, fast growth rate, upright graceful form, excellent shading effects, beautiful fall-color foliage, and tolerance of air pollution, heat, drought, and poor soil condition in cities (Fig. 16.1). The species in effect has been pretested to be suitable for built-up and rather stressful urban habitats in Europe, and its introduction into Chinese cities signified the transfer of an established arboricultural practice and experience. Similarly, various Chinese species had been intentionally introduced into the Old and New Worlds since antiquity. *Platanus x acerifolia* has gained dominance in Nanjing and come to define the landscape character of some main thoroughfares and green spaces.

Species composition in many cities around the world exhibits a tendency for a small number of species to dominate the urban tree population. This trend has been reported in Athens (Profous et al., 1988), Prague (Profous and Rowntree, 1993),

Fig. 16.1 The most abundant species, *Platanus x acerifolia*, planted along the West Beijing Road in Nanjing, have matured into a pleasant green tunnel

Mexico City (Chacalo et al., 1994), Guangzhou (Jim and Liu, 2001; Jim, 2002), and Hong Kong (Jim, 1996, 2004). Nanjing was no different with the top-10 species occupying over 70% of the surveyed population (Table 16.2). The species diversity of trees in the entire city is 4.58 using the Shannon-Wiener's H index, which has been reported in urban forestry literature to range from 1.0 to 7.0 (Sanders, 1980/1981; Welch, 1994; Jim, 1996, 2001). The 30-20-10 principle, suggested by the Maryland Department of Natural Resources (Galvin, 1999), provides a measure of the magnitude of dominance attained by certain species or species groups. To avoid the situation of excessive dominance, it sets a rule of thumb that the number of trees in one family, one genus, and one species should not exceed 30%, 20%, and 10%, respectively. In Nanjing, the top-three species surpass this 10% threshold. Therefore, it is recommended that the future planting palette be modified to introduce a more diversified species composition and a less imbalanced stock of trees.

However, species diversity and population sizes do not fully characterize a city's urban forest. Patterns of planting and species distributions in different sectors of a city are also important to consider from ecological and management perspectives. Since species pattern is strongly affected by habitat conditions, it is likely that species will vary by habitat type. We found that different habitats in Nanjing did contain different species assemblages or groupings. The tree-count rankings of the top 29 most frequently found species in the five main habitat types are presented in

Table 16.3 The distribution of 29 common species[c] in five main habitat types[d] indicated by tree-count ranks[e] in Nanjing

	RS	RN	FT	IT	GP
Carya illinoensis[a]	6				
Cedrus deodara[a,b]			3	6	4
Cinnamomum camphora[b]	4	7	6		
Diospyros kaki			7		
Eriobotrya japonica[b]		8			
Firmiana simplex	9				
Ginkgo biloba				3	7
Hibiscus syriaceus		10			
Ligustrum lucidum[b]	5	1	2		5
Magnolia denudata				7	
Magnolia grandiflora[a,b]	7	2	4	5	8
Metasequoia glyptostroboides	3	3	9	4	3
Paulownia fortunei					
Platanus acerifolia[a]	1	4	5	2	
Platycladus orientalis[b]					10
Populus tomentose					2
Prunus persica					
Prunus yedoensis[a]			10	10	
Pterocarya stenoptera	10			9	
Punica granatum[a]					
Sabina chinensis[b]			1	1	1
Salix babylonica					9
Sophora japonica	2	5			6
Taxodium distichum[a]					
Toona sinensis		9			
Trachycarpus fortunei[b]		6	8	8	
Zelkova schneideriana	8				

[a]Exotic species.
[b]Evergreen species.
[c]Two of the 29 common species, *Acer buergerianum* and *Ailanthus altissima*, were not encountered in the tree survey in the five main habitat types.
[d]RS, RN, FT, IT, GP respectively refer to roadside, residential neighborhood, factory, institution, and garden-park habitats.
[e] The numbers in the table refer to the rank of species based on its frequency of occurrence within the corresponding habitat type.

Table 16.3. We found that each habitat type had its own dominant species, that some species rarely occurred in certain habitats, and others were unique to a single habitat type. For example, *Platanus x acerifolia*, an outstanding landscape species either as solitary landmark specimens or in rows, ranks first along roadsides. The interlocking branches of these mature trees arching over roads and streets have established the signature sylvan image of Nanjing city. Since it is usually planted around residential yards to serve as green screens, *Ligustrum lucidum*, an evergreen species with limited crown spread, ranked first in residential neighborhoods. *Sabina chinensis*, another popular species, occupied the top rank in the remaining three habitat types, namely factories, institutions, and garden parks. Although slow growing, this indigenous coniferous species is favored by arborists because its pagoda form and tidy crown seldom need pruning. Some species are unique or overwhelmingly concentrated in certain habitat types. For example, the famous landscape trees *Salix babylonica* and *Populus tomentosa* occurred almost exclusively in garden parks; the popular yard trees *Eriobotrya japonica* and *Toona sinensis* are frequently found in residential neighborhoods.

Tree Size and Age

The size structure of the urban forest in Nanjing tends to cluster within small size classes. Tree distributions in DBH, crown diameter, and tree height are jointly depicted in Figure 16.2, which shows that four classes have more individuals than the others. The most abundant group is composed of individuals having crown diameters of <5 m and heights ranging from 5 to 10 m. Forty-three percent of trees have a DBH of 7 to 16 cm. Fifty-four percent of trees have crown diameters of 3 to 6 m, and 37% of the trees fall in the 5- to 8-m height size class. These distributions indicate that the urban forest is young due to a high rate of new tree plantings in recent years. We predict, therefore, that tree biomass volume and complexity will

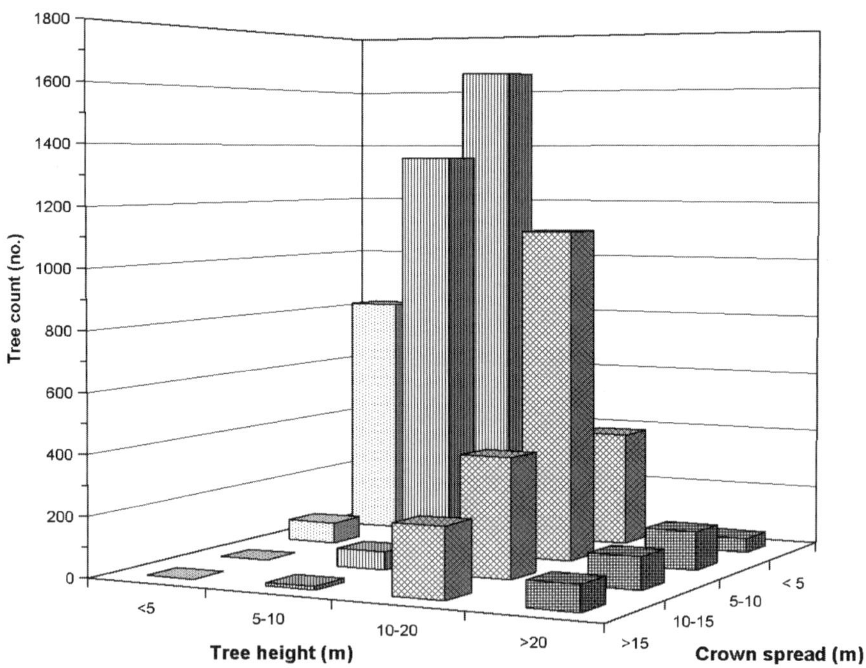

Fig. 16.2 Size structure of trees recorded in the survey

increase substantially in the future and, because of expected high growth rate of young trees, contribute greatly to environmental and ecosystem benefits. However, if these trees are not properly managed, future tree growth may also incur heavy tree care costs.

The generally young age of the Nanjing urban forest is related to rapid urban expansion and redevelopment in recent years. Ancient trees were mostly destroyed due to wars and natural disasters over the long history of the city, and only about 1000 champion trees aged over 100 years have been conserved and legally protected. Many large street trees planted in the 1920s were removed because of large-scale downtown redevelopment since the 1990s, further reducing large tree representation in the total tree population. In the last 10 years, the built-up area has expanded by about 50%, amounting to nearly 40 km^2. Many new saplings and young trees with a DBH of about 7 cm and a 1- to 2-m crown spread have been planted in these new developments. Similarly, urban renewal of some old sites has permitted the release of some land to accommodate new trees. These young trees, associated with recent development, account for the nearly 25% of small-sized trees with a DBH <10 cm, and 30% of trees with crown spread <3 m. The limited dimensions of planting sites in cramped urban spaces have resulted in widespread adoption of species with narrow crowns or small adult size. Some existing spreading specimens are often heavily pruned to avoid conflicts with adjacent buildings. In future planting programs, the dimensions and conditions of plantable sites should be assessed to optimize species selection for each site. More plantable sites of larger dimensions should be provided in new and redeveloped areas to ensure that large trees could continue to adorn the streets and green spaces of Nanjing. Otherwise, the fine landscape character afforded by large trees and their associated environmental benefits could gradually erode over the long term.

Tree Performance

The 6527 surveyed trees performed moderately well. At the two ends of the scale, both the dead/dying and excellent condition categories comprised a small proportion of the total population, 1% and 15%, respectively. The remaining three intermediate classes constituted 84% of the total (Fig. 16.3). We found 63 dead or dying trees

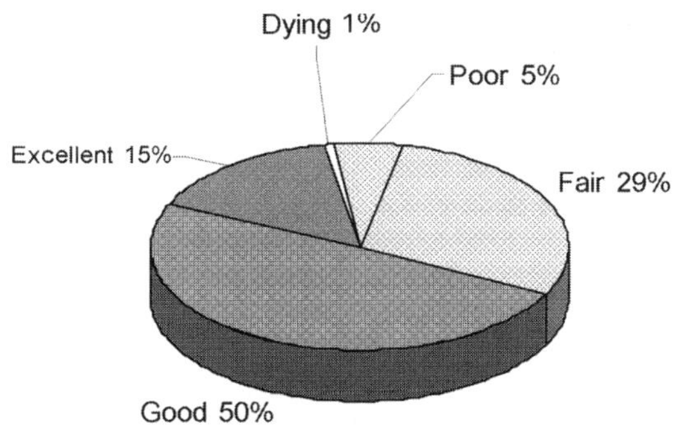

Fig. 16.3 Percentage of trees in five performance classes in Nanjing, China

that had lost most leaves or only had a few feeble and discolored leaves left. Most of them were young trees planted along roads or in residential neighborhoods, and the cause of their demise was mainly lack of care or improper planting. Some relatively large trees had died, likely due to severe pollution generated by nearby factories. In contrast, trees in excellent condition, exhibiting vigorous performance with dense foliage, balanced crown, undamaged bark, and upright trunk, either had encountered fewer disturbances from urban pressure or had stronger tolerance to stress.

Tree performance was significantly associated with tree habitat, as verified by a chi-square test ($p < .001$) performed on the 6527 trees sampled. (The chi-square test examines the statistical association between the two categorical variables of tree performance and habitat type. The variable of tree performance includes five categories of excellent, good, fair, poor, and dead/dying, and the variable of habitat type includes five categories of residential neighborhood, roadside, factory, institution, and garden park. The null hypothesis is that all categories of habitat types contain the same proportion of tree frequencies by tree performance categories.) Limited growth space, sealed ground cover, heavy vehicular traffic, air pollution, and vandalism are the common causes of tree damage and decline. The most common stress symptoms based on field observation included leaf discoloration, abnormal reduction in leaf size, twig and branch dieback, leaning trunk, and bark wounds. As illustrated in Figure 16.4, tree performances in the factory (FT), institution (IT), and garden-park (GP) habitats were better than those in roadsides (RS) and residential neighborhoods (RN). The IT lands have mostly low building density due to the inheritance of old structures and ample non–built-up land in the interstices. Also, the sizable open spaces on IT property can be guarded against development because of the nonprofit status of the land users. These open spaces are usually unpaved and suitable for tree growing. Trees are also well cared for, since a quiet green ambience is strongly desired, if not required, on IT lands. Trees in GP rank second in performance after IT, even though GP should have the best site conditions to foster tree growth. However, some GP lands are too widely paved by concrete to serve pedestrian and active recreational needs, thus disturbing tree growth to a certain degree. The FT

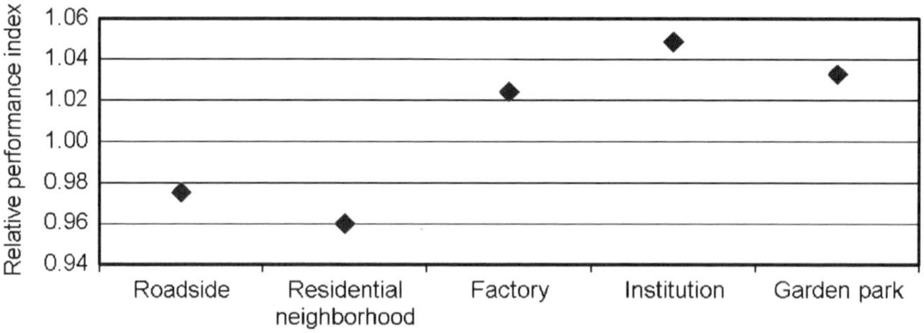

Fig. 16.4 Variations in tree performance by habitat denoted by relative performance index which ranges from 0.96 to 1.05; a value over 1.00 indicates good tree performance

trees are mainly planted along lanes and roads connecting shops, small gardens in the office area, and incidental niches suitable for trees. These trees fare relatively well because large factories usually have professional teams to care for trees, and small factories usually assign several employees to look after tree maintenance. Due to the general availability of professional tree care in Nanjing, it is perhaps somewhat surprising to find trees in industrial areas performing better than those in RN and RS. In contrast, trees associated with industrial lands in other cities usually perform poorly (Poracsky and Scott, 1999).

The shortage of effective tree management contributes to the poor tree performance in RN. Dwelling in multistory flats, the residents share the ground-level common open spaces, which are often poorly managed. After planting, the trees hardly receive any proper care by the home-owners or their management staff who is responsible for grounds maintenance. As opposed to private lands, trees in commonly held lands of the residential neighborhoods seldom receive attention from residents, and few residents would hire helpers to look after them. The sense of common ownership and hence obligation toward trees within residential lots is still rather weak. These neglected trees have to survive without professional care such as pruning, watering, and other maintenance operations. They are also widely damaged by vandalism, often aggravated by the confined open spaces that have to serve both tree growth needs and heavy recreational needs of inhabitants. The excessively high development density and poor site design of many residential areas also contribute to poor tree performance and growth.

Trees along roadsides are also beset by cramped growth conditions both below and above the ground. For RS, the long and narrow site geometry beside roads is often not a desirable habitat for growth of some plants (Forman and Godron, 1986). Trees in such narrow linear habitats are easily damaged by vehicles and pedestrians, intentionally or unintentionally. Furthermore, some urban utilities (electricity and communication lines, water and gas pipes) are commonly sited along roads over or under the ground and compete with trees for root and crown growing space (Talarchek, 1987; Jim, 1992, 1997b).

Urban Forest Management Assessment

Planting and Maintenance Responsibilities

According to the questionnaire survey and interviews of relevant officials and foresters, urban trees in Nanjing can be divided into public and private trees. Public trees mainly refer to those growing along roadsides and in public green spaces, and since they are considered part of the urban infrastructure are planted and maintained by the Greening Administrative Institute (GAI) of the districts where they are situated. The ABLG takes charge of high-level comprehensive planning of all urban trees, a responsibility that is analogous to the Department of Forestry in United States cities

(Grey, 1996). The forestry department in Nanjing, however, confines its responsibility to rural forests within municipal jurisdiction.

The separation of management responsibilities between ABLG and GAI has relegated maintenance to being a weak link in promoting the health of Nanjing's urban forest. The GAI as a public institution is funded by the municipal government, but it can obtain incomes with government permission by accepting pay for felling, removing, or transplanting nonpublic trees and undertaking greening construction projects. As a consequence, they are usually much more enthusiastic about the income-generating contract work, and spend more labor and materials on these than on public-tree duties. Therefore, public trees receive only basic maintenance and hence commonly show unsatisfactory performance. Although the ABLG is authorized to supervise GAI's service quality, it can barely implement this duty because of a lack of financial resources and administrative empowerment. This is due to the fact that the municipal subsidy is channeled directly to the GAI, which is directly administered by district governments with limited expertise to supervise GAI's tree work.

To overcome this ineffective connection in the management chain, the existing governance model needs to be overhauled. Municipal funding to GAI should be conveyed instead through a management authority with the requisite knowledge and experience, and this authority has to be the ABLG. We recommend that the duties and responsibilities of the GAI should be governed by a contract with ABLG. The contract should aim at ascertaining the scope, quality, and timeliness of all tree work in the public domain. The contract should specify a definition of work, inspection process, billing, and payment policy and procedure, etc., using the practical experience of some U.S. cities as a model (Grey and Deneke, 1986).

Since, strictly speaking, there is no private land in socialist China, private trees growing in private lots are rare. Due to the limited supply of houses with yards in Chinese cities, few yard trees are found. Compared with the public trees maintained by ABLG or GAI, the trees that belong to and are managed by enterprises, institutions, and residential neighborhood management agencies are nonpublic. Such trees are generally maintained by contractors to save expenditure in employee supervision and training. However, there are few landscape contractors and the contractor system for tree maintenance is far from mature in Nanjing. Therefore, the GAI can almost monopolize the market and set relatively high prices for maintenance operations because of the lack of competitors. Some property owners were literally forced by circumstances to resort to the illegal hiring of peasants without proper arboricultural training to prune and treat trees, frequently causing extensive damage and inducing long-term tree decline.

Within this context, new policies should be formulated to encourage the establishment of more tree maintenance companies so as to break up the virtual monopoly of tree maintenance by GAI. The in-house arboriculture staff of parks and some large enterprises can be potential competitors in the future tree service market. Also, a professional arborist association should be organized with the help of ABLG to regulate and manage this service market. Most importantly, the association should set high standards that dovetail with the best international practices, organize training of arborists, establish a continuing professional development program, and introduce a mandatory certification system for tree workers. Appropriate amendment of the tree ordinance and government policies could permit the industry to regulate and improve itself under official guidance.

Planting and Maintenance Operations

The planting plan and maintenance quality of urban forest practice in Nanjing could only be rated as adhering to a basic standard. There are no comprehensive species selection criteria and strategies, a critical bottleneck in the urban forest management regime. The ABLG provides no guidelines on specific species choices for different kinds of sites, and no overall recommended species list for the city. Although some research institutes have conducted studies on species suitability in relation to site conditions, and recommended some preferred species, these findings and recommendations have not been broadly recognized, and have seldom been incorporated into greening practices because of the gap between science and policy. The lack of a species selection strategy has already caused the extreme domination of Nanjing's urban forest by a few popular species, such as *Platanus x acerfolia, Sabina chinensis, Ligustrum lucidum*, and *Magnolia grandiflora*, that tend to perform well and are widely favored as ornamental or shade trees.

Due to institutional obstacles and inadequacies, it is premature to require a comprehensive planting plan for individual sites. However, at this stage a recommended species list for urban sites could be established based on empirical assessment of the condition of present tree stock. The list could provide guidance and hints for nonprofessional property owners who could then influence a contractor's tree planting decisions. Based on the performance assessment of the surveyed 6527 trees, a total of 64 species with APS scores ≥3.8 (the average APS of all trees is 3.73), indicating better performance in the urban environment, have been identified (Table 16.4). This list could form the basis for developing a more refined species selection matrix that includes specific recommendations for different habitats and site conditions. Meanwhile, the interim list could inform professionals and laypersons to optimize the match between species and a particular landscape situation to enhance tree performance and reduce management liability.

The quality of staff and the tree work they deliver are unsatisfactory under the current management system. The GAI is chronically short of urban forest professionals at different levels, and the same problem afflicts ABLG. The root cause of this critical shortcoming is the long-term negligence of tree maintenance in the past. The knowledge and skills of arboriculture workers are lagging behind the times and have become obsolete. They are in urgent need of updating and alignment with modern standards. For instance, topping is widely regarded by tree science as a bad and antiquated practice that should be avoided whenever possible (Shigo,

Table 16.4 Amenity tree species that perform well based on empirical assessment of tree condition in Nanjing

Indigenous species	Ornamental species: Flowering	Ornamental species: Leaf color-change	Shading species
Acer buergerianum	*Albizia julibrissin*	*Acer buergerianum*	*Aesculus chinensis*
Celtis sinensis	*Cercis gigantea*	*Acer palmatum*	*Ailanthus altissima*
Diospyros kaki	*Chimonanthus praecox*	*Bischofia polycarpa*	*Albizia julibrissin*
Fortunearia sinensis	*Chionanthus retusus*[b]	*Ginkgo biloba*	*Camptotheca acuminata*
Ginkgo biloba	*Diospyros kaki*	*Koelreuteria paniculata*	*Catalpa ovata*
Lagerstroemia indica	*Eriobotrya japonica*[b]	*Liquidambar formosana*	*Celtis sinensis*
Liquidambar formosana	*Euonymus bungeanus*	*Liriodendron tulipifera*[a]	*Euonymus bungeanus*
Photinia serrulata[b]	*Koelreuteria integrifolia*	*Metasequoia glyptostroboides*	*Ginkgo biloba*
Pistacia chinensis	*Lagerstroemia indica*	*Prunus cerasifera*	*Koelreuteria integrifolia*
Quercus chenii	*Magnolia grandiflora*[a,b]	*Pseudolarix amabilis*	*Liriodendron tulipifera*[a]
Sabina chinensis[b]	*Malus halliana*	*Taxodium distichum*[a]	*Magnolia grandiflora*[a,b]
Ulmus parvifolia	*Michelia figo*		*Magnolia officinalis*
Viburnum dilatatum	*Nerium indicum*[a,b]		*Pistacia chinensis*
	Osmanthus fragrans[b]		*Populus simonii*
	Photinia serrulata[b]		*Populus tomentosa*
	Prunus triloba		
	Prunus yedoensis[a,b]		
	Pyrus pyrifolia		
	Viburnum dilatatum		
	Ziziphus jujuba		

[a]Exotic species.

[b]Evergreen species.

1991; Harris et al., 1999), but is still widely practiced in Nanjing (Fig. 16.5). For example, topping is often performed on large *Platanus x acerifolia* along roads. Mature or semi-mature specimens have been routinely pruned by drastic head cutting of main stems, purportedly to encourage lateral development and maximize shading effects. Sometimes, the heavy pruning was conducted only to remove visual or physical obstruction of shop signs. The local arborists apparently do not realize the potential damage done to trees by topping, such as wound decay and reduced photosynthetic capacity.

Foliage removal practice provides another example. Both the International Society of Arboriculture (ISA) guidelines and the American National Standards

Fig. 16.5 Inappropriate topping of roadside trees, as shown by this *Platanus x acerifolia*, situated at West Beijing Road in Nanjing. This exemplifies a common adherence to outdated arboricultural practices in Nanjing

Institute (ANSI) A300 standard suggest that foliage removed from a mature tree within a growing season should not exceed 25%. Arborists in Nanjing are seldom concerned or are not aware of these guidelines and relevant research findings. Because urban forest operations have long been erroneously regarded as simple, and is hard and grim work with humble monetary remuneration, the green trade does not attract people of good education and quality. Only a small portion of the staff has received tertiary education in arboriculture or related disciplines. Most workers only acquired a general secondary education with little to no specialist training. The ABLG should be responsible for training and upgrading these landscape employees with the help and knowledge base of scientists and professionals in universities and research institutes. The government can work with tree specialists to formulate or upgrade a series of tree-work standards and guidelines suitable for Nanjing. They should be benchmarked against international standards, which would help to bring about improvement in tree management quality.

In addition, the common overemphasis on the acreage of green space rather than individual trees has resulted in a somewhat unexpected yet common by-product, a decrease in tree number and negligence of tree maintenance in Nanjing and other Chinese cities. Indices describing green space, such as green space ratio (GSR) and public green space area per capita, are frequently adopted as the official yardstick for evaluating the status and achievement of urban greening objectives. These macroscale numerical requirements can no doubt facilitate the growth of green areas, but they do not demand higher standards for tree density, species assemblage, final tree dimensions, planting site quality, and tree performance. A specific urban forest plan rather than a general green space plan could more holistically upgrade the entire spectrum of tree provision and management types. It should be presented as a long-term and citywide development strategy for the urban forest, accompanied by a depiction of spatial configuration of canopy coverage in all sectors of a city. The plan should be designed with a view to maximizing ecological and landscape benefits, providing an optimum species combination for the city, and contributing to the city's long-term sustainability. Based on the plan, the ABLG can furnish informed guidance for tree planting and maintenance in both public and nonpublic green spaces.

Management Regulations

In socialist China, tree ordinances cover all the land in general. Horizontally, they are categorized by the governmental departments that enforce the relevant ordinances, such as agriculture, forestry, and construction departments. Vertically, there are ordinances at national, provincial and municipal levels, as for the corresponding greening administrative ordinances (Jim and Liu, 2000). In Nanjing, over 30 regulations are implemented at the municipal level, involving the entire spectrum of tree work from tree planning to tree establishment and protection. They can be grouped into the following classes by content: (1) comprehensive law, (2) greening administrative ordinance, (3) greening protection regulation, (4) greening project management regulation, (5) greening standard and planning document, (6) obligatory planting regulation, and (7) garden city standard and management regulation.

Usually, the contents of municipal regulations are largely derived from corresponding national or provincial laws, with the addition of some particular items to cater to local conditions. However, a given item might be referred to in several regulations, and enforced through different government departments at the national, provincial, or city levels. Their respective interpretations of these regulations on the same item may differ, such as the many classification schemes for green space and many statements about the GSR for different land uses. The involvement of more than one government unit for a given duty or task does engender much bewilderment and uncertainty, and much confusion in both understanding and practice. Overall, the unnecessary duplication in different documents should be removed, the different requirements on a given item should be unified and dedicated to one regulation, and obsolete regulations and outdated clauses should be explicitly expunged. In the interest of efficient and unambiguous management of essential municipal green infrastructure, an exhaustive review and overhaul of the complex statutory and quasi-statutory documents would help to lay a firm foundation for improvements in other aspects of urban forestry work.

Conclusion

This tree survey demonstrated that the urban forest in Nanjing shares some common traits with those in Western cities. These traits include a small proportion of species occupying a large proportion of the urban tree population, and trees of small final size being commonly adopted in a high-density urban environment. However, some aspects of Nanjing's urban forest differ. Specifically, as opposed to trees in Western cities located near factories, trees in factory-industrial lands perform relatively well in Nanjing, due to a greater availability of tree care. In contrast, the lack of care significantly suppressed tree performance in some residential neighborhoods, where Western cities usually have healthy trees. This difference in tree condition may be related to corresponding differences in tree ownership patterns. An understanding of the key attributes of the urban forest in Nanjing can provide essential information for designing a comprehensive greening plan that benefits from proper species selection and improved preparation and design of planting sites. These are pertinent issues that have been neglected in Nanjing's current urban forestry program. The way that a city treats its urban forest, and nature in general, reflects to a certain degree the fundamental fabric and values of a society. Countries and cities with different social, economic, political, and environmental backgrounds tend to nurture different values and perceptions of urban vegetation.

We also suggest that the emphasis on green space area, as benchmarks for gauging the success of official greening policies, does not help the welfare of individual trees. Such a deeply ingrained bias has led to the long-term negligence of tree planting and maintenance in built-up areas. Hitherto, the tree staff and maintenance quality have been relatively poor, lacking in proper supervision, and not supported by a strong culture of excellence and achievement. The lack of incentives, career prospects, and attractive remuneration does not attract or retain good-quality workers. The horticultural industry, still in its infancy in terms of quantity and quality, could benefit from some official encouragement and support. These vexing and long-standing management problems are not specific to Nanjing. Due to similarities in administrative organization and legislative makeup, they are common issues that beset urban forestry work in most Chinese cities. Research on the key attributes of urban forests and related management of individual trees should aim at sustaining a healthy and diversified green stock. Different facets of the integrated urban forestry package (Herz et al., 2003) could benefit from a systematic overhaul to bring China's tree work on a par with international best standards.

Acknowledgments This research grant support, provided by the Committee on Research and Conference Grants of the University of Hong Kong, the NSFC and the Director's Special Grants of Nanjing Institute of Geography and Limnology of the Chinese Academy of Sciences, is gratefully acknowledged. Thanks are also due to the Nanjing municipal government for providing assistance and information on landscape planting and management, and to Prof. H.Y. Zhou of the Nanjing Forestry University for kindly providing assistance in the field survey.

References

Beckett, K.P., Freer-Smith, P.H., and Taylor, G. (1998) Urban woodlands: their role in reducing the effects of particulate pollution. Environmental Pollution 99:347–360.

Buhyoff, G.J., Gauthier, L.J., and Wellman, J.D. (1984) Predicting scenic quality for urban forests using vegetation measurements. Forest Science 30(1):71–82.

Chacalo, A., Aldama, A., and Grabinsky, J. (1994) Street tree inventory in Mexico City. Journal of Arboriculture 20:222–226.

Che, S.Q., and Song, Y.C. (2001) Extract of the remote sensing message of urban green space landscape: Shanghai city as the case study. Urban Environment and Urban Ecology 14(2):10–12 (in Chinese).

DeWalle D.R., Heisler, G.M., and Jacobs, R.E. (1983) Forest home sites influence heating and cooling energy. Journal of Forestry 81:84–88.

Dochinger, L.S. (1980) Interception of airborne particles by tree plantings. Journal of Environmental Quality 9(2):265–268.

Forman, R.T.T., and Godron, M. (1986) Landscape Ecology. John Wiley and Sons, New York.

Freedman, B., Love, S., and O'Neil, B. (1996) Tree species composition, structure, and carbon storage in stands of urban forest of varying character in Halifax, Nova Scotia. Canadian Field-Naturalist 110:675–682.

Galvin, M.F. (1999) A methodology for assessing and managing biodiversity in street tree populations: a case study. Journal of Arboriculture 5(3):124–128.

Gao, J., and Song, Y.C. (2001) Dynamic of the urban greenspace system and its relationship with landuse: case study of the southwest Shanghai city. Urban Environment and Urban Ecology 14(3):18–20 (in Chinese).

Gao, J., Song, Y.C., and Zhang, Q.F. (2002) Analysis on the characteristics of urban vegetation and its mapping based on remote sensing and geographic information system. Acta Phytoecologica Sinica 26(1):1–9 (in Chinese).

Gilbertson, P., and Bradshaw, A.D. (1985) Tree survival in cities: the extent and nature of the problem. Arboricultural Journal 9:131–142.

Grey, G.W. (1996) The Urban Forest: Comprehensive Management. John Wiley and Sons, New York.

Grey, G.W., and Deneke, F.J. (1986) Urban Forestry, 2nd ed. John Wiley and Sons, New York.

Harris, R.W., Clark, J.R., and Matheny, N.P. (1999) Arboriculture: Integrated Maangement of Landscape Trees, Shrubs, and Vines, 3rd ed. Prentice Hall, Upper Saddle River, NJ.

Herz, R., Schroeder, H., Stahre, P., Stal, Ö., and Woodward, S. (2003) Towards the integrated management of urban vegetation and infrastructure. Final Report of COST C3 Working Group Vegetation and Urban Civil Engineering, European Union.

Jaenson, R., Bassuk, N., Schwager, S., and Headley, D. (1992) A statistical method for the accurate and rapid sampling of urban street tree populations. Journal of Arboriculture 18(4):171–183.

Jim, C.Y. (1989) Tree canopy cover, land use and planning implications in urban Hong Kong. Geoforum 20(1):57–68.
Jim, C.Y. (1992) Tree-habitat relationships in urban Hong Kong. Environmental Conservation 19(3):209–218.
Jim, C.Y. (1996) Roadside trees in urban Hong Kong: part 2 species composition. Arboricultural Journal 20:279–298.
Jim, C.Y. (1997a) Roadside trees in urban Hong Kong: part 3 tree size and growth space. Arboricultural Journal 21:73–88.
Jim, C.Y. (1997b) Roadside trees in urban Hong Kong: part 4 tree growth and environmental condition. Arboricultural Journal 21:89–99.
Jim, C.Y. (2001) Species diversity of three major urban forest types in Guangzhou City, China. Forest Ecology and Management 146:99–114.
Jim, C.Y. (2002) Heterogeneity and differentiation of the tree flora in three major land uses in Guangzhou City, China. Annals of Forest Science 59:107–118.
Jim, C.Y. (2004) Characteristics of urban park trees in Hong Kong in relation to greenspace planning and development. Acta Horticulturae 643:123–128.
Jim, C.Y. (2005a) Monitoring the performance and decline of heritage trees in urban Hong Kong. Journal of Environmental Management 74(2):161–172.
Jim, C.Y. (2005b) Floristics, performance and prognosis of historical trees in the urban forest of Guangzhou city (China). Environmental Monitoring and Assessment 102:285–308.
Jim, C.Y., and Chen, S. (2003) Variations of the treescape in relation to urban development in a Chinese city. Professional Geographer 55(1):70–82.
Jim, C.Y., and Liu, H.T. (1997) Storm damage on urban trees in Guangzhou, China. Landscape and Urban Planning 38:45–59.
Jim, C.Y., and Liu, H.T. (2000) Statutory measures for the protection and enhancement of the urban forest in Guangzhou City, China. Forestry 73:311–329.
Jim, C.Y., and Liu, H.T. (2001) Species diversity of three major urban forest types in Guangzhou City, China. Forest Ecology and Management 146(1–3):99–114.
McBride, J., and Jacobs, D. (1976) Urban forest development: a case study, Menlo Park, California. Urban Ecology 2(1):14–22.
Medley, K.E., McDonnell, M.J., and Pickett, S. T. A. (1995) Forest-landscape structure along an urban-to-rural gradient. Professional Geographer 47(2):159–168.
Miller, R.W. (1996) Urban Forestry: Planning and Managing Urban Greenspaces, 2nd ed. Prentice Hall, Englewood Cliffs, NJ.
Poracsky, J., and Scott, M. (1999) Industrial-area street trees in Portland, Oregon. Journal of Arboriculture 25(1):9–17.
Profous, G.V., Rowntree, R.A., and Loeb, R.E. (1988) The urban forest landscape of Athens, Greece: aspects of structure, planning and management. Arboricultural Journal 12:83–107.
Profous, G.V., and Rowntree, R.A. (1993) The structure and management of the urban forest in Prague, Czechoslovakia. I. Growing space in metropolitan Prague. Arboricultural Journal 17:1–31.
Rowntree, R.A. (1983) Configuration of tree canopy cover in urban land uses. Geographical Perspective 51:49–53.
Rudie, R.J., and Dewers, R.S. (1984) Effects of tree shade on home cooling requirements. Journal of Arboriculture 10(12):320–322.
Sanders, R.A. (1980/1981) Diversity in the street trees of Syracuse, New York. Urban Ecology 5:33–43.
Schroeder, H.W., and Cannon, W.N. (1983) The aesthetic contribution of trees to residential streets in Ohio towns. Journal of Arboriculture 9(9):237–243.
Shigo, A.L. (1991) Modern Arboriculture: A Systems Approach to the Care of Trees and their Associates. Shigo and Trees Associates, Durham, NH.
Simpson, J.R., and McPherson, E.G. (1996) Potential of tree shade for reducing residential energy use in California. Journal of Arboriculture 22:10–18.

Talarchek, G.M. (1987) Indicators of urban forest condition in New Orleans. Journal of Arboriculture 13(9):217–224.

Talarchek, G.M. (1990) The urban forest of New Orleans: an explanatory analysis of relationships. Urban Geography 11:65–86.

Wasson, E. (chief consultant). (2003) The Complete Encyclopedia of Trees and Shrubs. Thunder Bay Press, San Diego, CA.

Welch, J.M. (1994) Street, and park trees of Boston: a comparison of urban forest structure. Landscape and Urban Planning 29:131–143.

Zhou, W.Z., Pan, J.J., Fang, S.B., and Jiang, X.S. (2002) Analysis of ecological vegetation features by thematic data in Nanjing City. Urban Environment and Urban Ecology 15(1):4–9 (in Chinese).

17
Urban Forest Structure in Hefei, China

Zemin Wu, Chenglin Huang, Wenyou Wu, and Shaoujie Zhang

Over the past 10 to 15 years, urban expansion has been occurring more rapidly than at any time in China. Today, about 41% of China's people are living in cities, compared with 26% at the end of the 1980s, and the country is still rapidly urbanizing. Thus, city greening and city beauty have become very important issues, since many Chinese cities have air pollution problems (Shao et al., 2006). Vegetation management in cities has significant potential for improving air quality and human health and well-being in China. The growth of urban areas in China provides planning opportunities for creating new urban forests to combat this air pollution problem. For these reasons, it is important to obtain a more detailed understanding of the status of urban forests in the country.

To increase our knowledge of how large a benefit could be provided by urban forests, of how we can improve and enhance both the quality and extent of urban forests, and of what management approaches best maintain urban forests, we need to understand the abundance, distribution, and structure of urban forests in cities (Kielbaso, 1988). Some research projects on urban forest structure conducted in American cities, such as Chicago and Sacramento (McPherson, 1998; Nowak, 1994a,b), have provided good models for other cities to follow. These American studies were based on obtaining structural data on their urban forests so that their benefits could be determined.

But in China, because the concept of urban forests is still new, only a few reports on this topic have been published. So to guide the future development of urban forests, we need to learn more about their current status through case studies in different cities using new methods and theory. We selected the city of Hefei for this study because it is representative of mid-sized cities in China with a noticeable urban forest component, but is projected to grow rapidly soon. Hefei has a population of one million and is projected to increase by 50% to 70% in the next 15 years. The purpose of this study was to analyze the current situation of Hefei's urban forest structure to provide basic information on its characteristics, problems, and management, and to provide a case study model for comparative studies with other cities as well.

M.M. Carreiro et al. (eds.), *Ecology, Planning, and Management of Urban Forests: International Perspectives.*

Study Area Description

Hefei is situated in southeast China 400 km east of Shanghai at 117°11' to 117° 22' east longitude and 31°48' to 31°58' north latitude. Hefei, a mid-size city and the capital of Anhui province, was built in 200 A.D. initially for defense purposes only. The city became the capital of Anhui province in 1953. Today, the city covers about 100 km^2 and has a population of one million. Hefei lies within the north subtropical climate zone. The climate typically has a moist and warm monsoon climate. Annual average temperature is 15.7°C, with an average of 2.1°C in January and 28.3°C in July; the lowest temperature recorded is –20.6°C. Mean annual precipitation is 1000 mm. The natural zonal vegetation is mixed evergreen and deciduous broadleaf forest. The major native species are *Quercus acutissima, Quercus variabilis, Platycarya strobilacea, Pterocarya stenoptera, Kerleuteria paniculata, Cyclopalanopsis glauca*, and *Liquidambar formosana*. The city is included on a list of the best green cities of China and was named as a national landscape city by the Construction Ministry in 1992. In 1994, we found that Hefei had a total of 456 species of woody plants belonging to 73 families and 170 genera. Twenty-two are coniferous species and 148 are flowering species (Wu, 1994).

The city has a grid road system with six major avenues connecting to the national highway system. There are four concentric zones around the city. The first is Ring Park, which used to be a city wall and was afforested in 1950s; the second is bounded by the first ring road, and the city area within the road, which includes the Ring Park areas, about 20 km^2 in extent; the third zone is bounded by the second ring road, and the city area within the road is approximately 100 km^2; and the fourth zone is a highway with an inner area of about 200 km^2. This last zone is planned for settlement by 2 million inhabitants in the future to meet Hefei's development goal in the new century. The study area of about 23 km^2 is centered in downtown Hefei, approximately within the area of the first ring road, and includes some university campuses located adjacent to the first ring road.

Methods of Data Collection

Within the central core area in the city, we classified the city's urban forest into five groups based on their location: street trees, parks, Ring Park, institutional grounds, and open space areas in apartment blocks (residential areas). Institutional grounds included school yards, university campuses, and government office yards. The Ring Park is the open green space surrounding the old downtown that was originally afforested at the base of city wall in the 1950s after the wall was removed. In the 1980s the green strip was rebuilt to be an open public park and became a distinct landscape type in the city. Open space areas in apartment blocks included small public green belts in neighborhoods and residential areas.

Sampling

The sample percentage of the 23 km^2 study area for each land-use category varied from 10% to 18%. We established rectangular sampling plots that varied in absolute dimensions depending on the land use group. Street tree plots were 100 m long, including both sides of the investigated road, and were spaced 900 m apart. Plots in parks were 4 m wide. Their length varied depending on the individual park, but each plot crossed from one edge of the park to the opposite side, and were located at 30-m intervals along one transect running completely across each park. Thus for each park the area sampled occupied 13% of total park area. For Ring Parks, each 300 m^2 plot (10 m long × 30 m wide) was established every 300 m internally along the long axis. There were a total of 27 samples comprising about 18% of the Ring Park area. Open space areas around apartment buildings were sampled using 2-m-wide plots, running completely across the grounds of each apartment building complex, with each plot spaced 20 m apart. For institutional grounds, 10% of the population of every tree species was randomly sampled. This tree population data set was based on records provided by the local landscape administration office. In total, 30 students participated in the project, and all the field surveys were done in both summer and fall seasons of 1998 and 1999. Sorenson's Index of Similarity ($C = 2j/a + b$, where a is the number of species in community a, and b is the number of species in community b, and j is the number of species that both communities share in common) was employed to analyze the similarity of species composition in different land-use categories.

Field Survey for Each Plot

On each plot, every tree was identified to species, and the following measurements were made: diameter at breast height (DBH) at 1.37 m above the ground, total height, height to base of live crown, crown radius, crown shape, crown condition, and health status. We established six classes of tree health based on foliage characteristics. A tree was rated as excellent (class I) if less than 5% of the crown showed dieback or leaf discoloration and the crown was symmetrical and full. Class II indicated a health rating of good, with trees exhibiting 5% to 25% dieback or discoloration; a moderate health rating (class III) indicated 26% to 50% dieback; poor health (class IV) were trees with 51% to 75% dieback; class V, dying (79% to 99%); and class VI, dead (no leaves) (Kielbaso, 1988).

Leaf area, leaf biomass, and biomass of individual trees were calculated using regression equations for deciduous urban trees (Nowak, 1994a). If no allometric equation could be found for an individual species, either the genus or family average was substituted, or biomass was computed using the separately averaged result for our hardwood and conifer species. The distribution pattern of urban forest coverage and urban forest patches was studied using aerial photography techniques (Miller, 1988).

Results

Total Tree Numbers in the Study Area

According to field survey data, we estimated that there were approximately 351,500 trees in the study area; 218,500 were broad-leaf trees, and 133,000 were conifers. We found that 61.6% of trees were in residential areas, 16% in institute grounds (university campuses and government building yards), 7.9% in the Ring Park, 8.8% in other parks, and only 5.8% along streets. Based on land-use categories, the parks (including the Ring Park) occupied only 4.7% of total land area but had 16.7% of the tree population (Table 17.1).

Tree Species Composition

We once recorded 460 woody plant species in the city, but only 85 were major tree species, including 24 conifer species and 61 broad-leaved species. The top-15 conifer species and top-20 broad-leaf species (both ranked by number of individuals) are listed in Tables 17.2 and 17.3. Among them, the top five conifer species comprised 77.5% of the conifer population, while the top 10 broad-leaved species comprised 81.8% of the broadleaf population. Nearly 72% percent of the entire tree population in the study area consisted of just 10 species: *Ligustrum lucidum, Metasequoia glyptostroboides, Cedrus deodara, Cinnamomum camphora, Sabina chinensis var. Kaizuka, Prunus cerasifera f. atropurpurea, Magnolia grandiflora, Robinia pseudoacacia, Paulownia tomentosa*, and *Juniperus formosana*.

However, using tree numbers for ranking the importance of tree species provided information on only one aspect of tree dominance. Ranking by biomass would reveal different species dominance patterns. For example, when biomass is used as an importance species criterion, then *Robinia pseudoacacia, Platanus acerifolia, Pterocarya stenoptera*, and *Bischofia polycarpa* are ranked in the top five for broad-leaved species, although they did not rank as highly when number

Table 17.1 Tree numbers in different land-use categories

	Total of land-use categories		Tree numbers	
Land-use categories	Area (ha)	% of total land use categories	Tree numbers	% of total tree population
Ring Park	42.5	1.8%	27,662	7.9
Parks	67.0	2.9%	30,743	8.8
Campus and Gov.	424.5	19.6%	56,366	16.0
Road	121.1	5.3%	20,492	5.8
Resident	1415.0	61.6%	216,277	61.5
Total*	2070.1	91.2%	351,540	100

*Not including water and farmland in the study area.

Table 17.2 Major conifer species ranked by population size

Species	Number	Percentage	Rank
Metasequonia glyptostroboides	43,229	32.5	1
Cedrus deodara	27,459	20.6	2
Sabina chinensis var. Kaizuka	17,231	13.0	3
Juniperus formosana	8110	6.1	4
Sabina chinensis cv. 'Pyramidalis'	7104	5.3	5
Pinus massoniana	7009	5.3	6
Sabina chinensis	4902	3.7	7
Platycladus orientalis	3553	2.7	8
Sabina virginiana	2300	1.7	9
Cupressus funeibris	2300	1.7	10
Taxodium ascendens	2239	1.7	11
Ginkgo biloba	1812	1.4	12
Pinus thunbergii	1780	1.3	13
Pinus taeda	1150	0.9	14
Sabina virginiana cv.'Pyramidalis'	1070	0.8	15
Other species	1752	1.7	
Total	133,000		

Table 17.3 Major broad-leaved species ranked by population size

Species	Number	Percentage	Rank
Ligustrum lucidum	92,652	42.4	1
Cinnamomum camphora	17,685	8.1	2
Prunus cerasifera f. Atropurpurea	13,386	6.1	3
Magnolia grandiflora	12,453	5.7	4
Robinia pseudoacacia	8735	4.0	5
Paulownia tomentosa	8569	3.9	6
Platanus acerifolia	7027	3.2	7
Sapium sebiferum	6634	3.0	8
Sophora japonica	6236	2.9	9
Toona sinensis	5545	2.5	10
Pterocarya stenoptera	4540	2.1	11
Brousonetia papyrifera	4193	1.9	12
Acer burgerianum	3788	1.7	13
Osmanthus fagrans	3056	1.4	14
Punica granatum	2690	1.2	15
Bischofia polycarpa	2613	1.2	16
Magnolia denudata	2331	1.1	17
Prunus persica	2305	1.1	18
Eriobotria japonica	1341	0.6	19
Sophora japonica var. Pendula	1233	0.6	20
Other species	11,528	5.3	
Total	218,540	100	

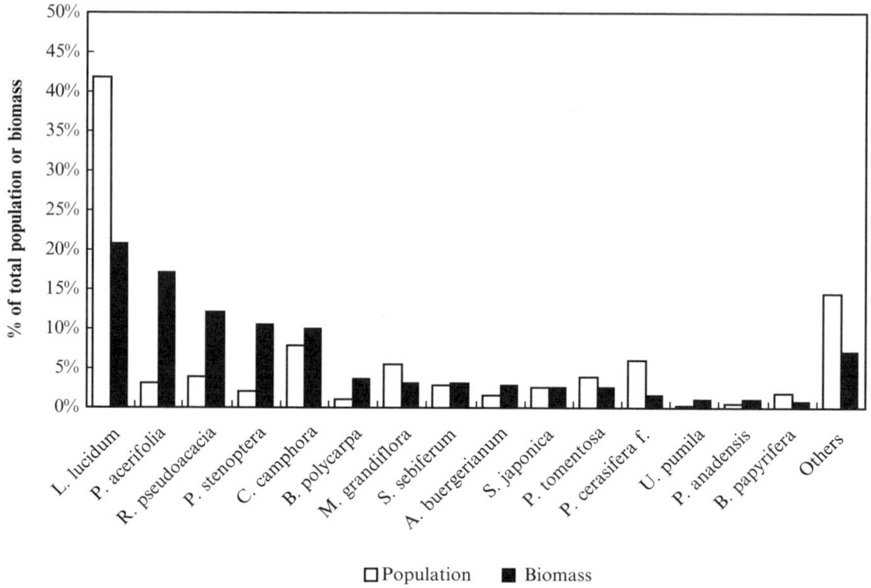

Fig. 17.1 Comparison of species dominance by number of individuals and by biomass for broad-leaved tree species in the inner 23-km^2 area of Hefei, China

of individuals was used as a criterion (Fig. 17.1). For example *Robinia* ranks third in biomass order, but ranks eighth in total number of individuals; *Pterocarya* and *Bischofia* are not included in the top-10 species ranked by number of individuals (13th and 16th, respectively), but are third and fifth in importance, respectively, when biomass is used. In any case, species in eight genera, *Metasequoia, Ligustrum, Platanus, Robinia, Cedrus, Pterocarya, Cinnamomum*, and *Bischofia*, which had the highest biomass dominance, comprised about 57.3% of tree population numbers and 73.1% of all tree biomass. Coniferous and evergreen broad-leaf trees constituted the greatest proportion of the urban tree population by individuals (37% and 37%, respectively).

Tree Species Composition by Land-Use Type

Tree species composition varied with land-use categories. Institutional yards possessed the highest species richness, with 64 species comprising 75.9% of all species found in the city. Parks had the second highest species richness, and the Ring Park and residential area followed in order. Street trees had the lowest species richness, with only 16 species recorded, and only 19.3% of all major species observed. On the other hand, the degree of species overlap between institutional grounds and parks was the highest at 32.2% similarity (Table 17.4). Similarity between species in the university campus and those in the Ring Park was 30.9%. Similarity between species in these land-use categories (institutional grounds and park) and the other three land-use types (Ring Park, resident, and street trees) ranged from 18% to 25%.

Forest Stand Characteristics by Land-Use Category

For the whole study area, the average density was estimated at 170 individuals per hectare (ha) and the average basal area at 2.86 m^2/ha. Tree density is not very low in Hefei when compared to other Chinese cities, but is much lower on a basal area per hectare basis (Wu et al., 2003). However, there are strong differences in stand characteristics of tree density and basal area for different land use types (Table 17.5). The Ring Park had the highest value for both tree density and basal area per unit land, followed by other parks. Institutional grounds ranked third for basal area per unit land but lowest for tree density. Residential areas had the lowest basal area, but ranked fourth in density.

Based on Rowntree's (1984a,b) definition of an urban forest, a site with 5 to 25 m^2/ha of tree basal area could be classified as a forest, because this basal area

Table 17.4 Species number and percent similarity between different land use categories

			% Species similarity between two different land-use types							
			Park		Ring Park		Resident		Street trees	
Item	Species number	Percent	Number	%	Number	%	Number	%	Numbers	%
Institute ground	63	74.1	37	32.2	30	30.9	23	25.3	14	17.7
Park	52	61.2			22	25.6	19	23.8	15	22.1
Ring Park	34	40.0					14	22.6	9	18.0
Resident	28	32.9							8	18.2
Street trees	16	18.8								
Whole city	85									

Note: Percent similarity determined using Sorenson's Similarity Index as described in text. Number column refers to number of species in common between the two communities being compared.

Table 17.5 Tree density (number of trees/ha) and basal area by land-use type in the study area

	Stand characteristics		
Land-use types	Average DBH	Tree density (per ha)	Basal area (m^2/ha)
Ring Park	19.2	651	18.84
Parks	18	459	11.68
Institute ground	18.8	134	3.72
Road	16.3	169	3.53
Resident	11.7	153	1.64

density would have enough biomass to influence microclimate. According to his definition, only the park and Ring Park trees in the study areas in Hefei could meet the standard of being classified as an urban forest. However, for the tree population in the study area as a whole, the existing tree population could not provide an urban forest environment at the site level, because of low stocking density.

Tree Size Structure

Forty-eight percent of the tree population in central Hefei were smaller than 10cm DBH, and occurred mostly in residential complexes. Thirty-seven percent of the tree population had a DBH of 10 to 20cm, 12% had a DBH of >20 to 30cm, 2% had a DBH of >30 to 40cm, and only 1% of population had a DBH larger than 40cm. In general, small trees accounted for a majority of the tree population in the study area. On the other hand, the distribution pattern of tree sizes in different land-use categories showed that campus, street trees, and park trees had a higher proportion of big trees (Fig. 17.2). This occurred because most of the trees in campus and institutional yards were afforested in 1950s when these areas were built. However, most trees in residential areas were newly planted in the past 10 to

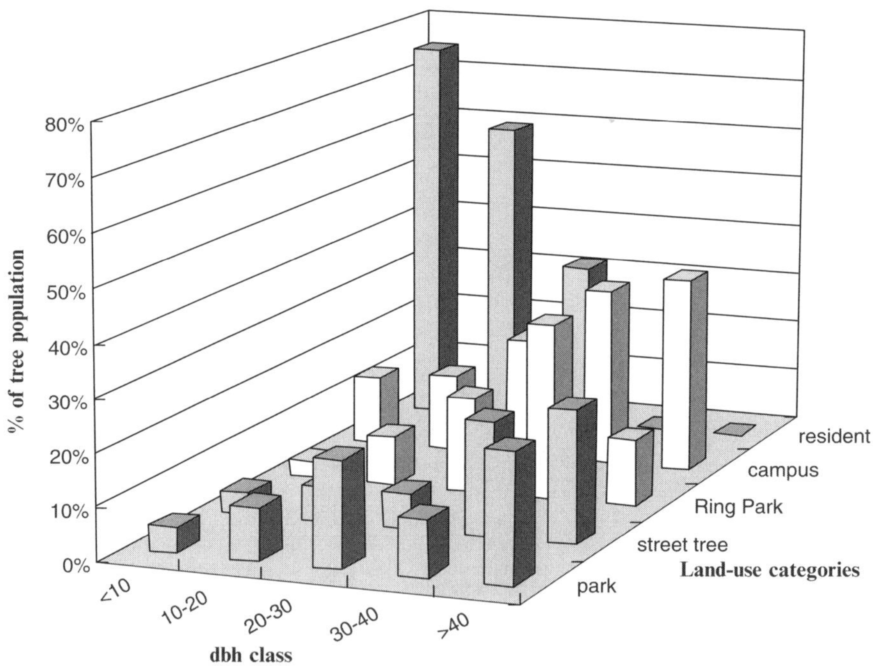

Fig. 17.2 Tree size distribution pattern in different land-use categories in the inner 23-km^2 area of Hefei, China

20 years, since the city accelerated development in that period and built more new residential areas.

Tree size distribution also varied by species. Among the 25 most important species ranked by numbers of individuals, populations of *Pterocarya stenoptera, Platanus acerifolia, Bischofia policarpa*, and *Robinia pseudoacacia* had a higher proportion of large trees (63.8%, 40.6%, 22.1%, 17.1%, respectively). These are the species that were commonly planted in 1950s to 1960s, since they were the most common species available for city greening at that time.

Tree Health Status

Tree health status is one of the most important indicators of urban forest quality. Tree health is the integrated expression of the quality of tree management, and the degree of attention paid to matching tree species characteristics and physiology with city and site condition. In Hefei, about 53.4% of the tree population in the study area was rated in good condition (40% of this number were conifers and 62% were broad-leaved), while 3.5% were dead or dying. Another 31.1% and 12% were rated in moderate and poor condition, respectively.

The land-use type with the highest proportion of dead or dying trees was the park (6.9% of its tree population), followed by the Ring Park (4.4%), and residential and street trees (both at 2.9%); the lowest dead and dying tree percentages occurred in the institutional land-use category (2.4%). We believe these patterns occur because most of the trees in parks and Ring Park were planted in the 1950s and 1960s and were such trees as *Platanus, Populus, Kerleuteria*, and *Ligustum*. These trees are all fast-growing species and they are getting old now, so their crowns are beginning to die back. Another reason may be that in parks with a more natural setting, thinning due to competition for light and growing space is occurring. However, the fact that a large part of the central city is residential, with many newly built neighborhoods and small trees in a good environment, may have skewed the data for the health condition of the total tree community.

Biomass and Leaf Area

Biomass

Using allometric equations, we estimated that there were approximately 31,741 metric tons of tree biomass in the study area, 9751 metric tons for conifers and 21,996 metric tons for broad-leaved trees. The average biomass per hectare in the study area was estimated at 15.33 metric tons. This means the broad-leaved trees (62.2% of the tree population) contributed 69% of the biomass. Tree biomass also varied by land use (Table 17.6). The Ring Park contained only 7.9% of the tree

Table 17.6 Tree biomass by land-use category

Land uses	Tree number	Biomass* and percentage for different land-use types. Whole Biomass	%	Broad leaves Biomass	%	Conifer Biomass	%	Average Metric tons/ha	kg/tree
Resident	216,277	10,392	32.7	5802	26.4	4590	47.1	7.56	48
Institute ground	56,366	7393	23.3	4884	22.2	2509	25.7	17.42	131
Park	30,743	4885	15.4	3431	15.6	1454	14.9	72.90	159
Ring Park	27,662	5395	16.9	4434	20.2	961	9.9	126.94	195
Road side	20,492	3676	11.6	3439	15.6	237	2.4	30.36	179
Total	351,540	31,741		21,990	100	9751	100	15.33	90

*Biomass in metric tons.

population and 1.8% of land area, but contributed 17% of the entire biomass, and was listed at the top position for biomass storage per hectare, with 126.94 metric tons per hectare. Parks with only 3.2% of land area ranked second, contributing about 15.4% of the total tree biomass in the study area. In contrast, residential complexes with 68.4% of the land and 61.6% of the tree population, ranked last for biomass storage per hectare, contributing about 32.7% of the whole biomass, but having a biomass density of only 7.56 metric tons per hectare. University campuses ranked fourth in biomass storage per hectare (17.4 metric tons per hectare) and ranked second for total biomass storage. Roads had a higher biomass accumulated per unit land (30.36 metric tons/ha) and contributed about 11.6% of the total tree biomass in the study area.

Biomass storage by species also varied a good deal. The average biomass per tree in the study area was 90.13 kg. Twenty species (representing 78% of the tree population) dominated the tree biomass of central Hefei by contributing about 91.2% of the entire tree biomass. About 80% of tree biomass was stored in 10 species comprising about 63.3% of tree population. We also calculated and compared the average biomass per tree for different species and found that only four species attained an average of >300 kg per tree. These were *Zelkova schneideriana* (653 kg), *Platanus acerifolia* (532 kg), *Paulownia tomentosa* (527 kg), and *Pterocarya stenoptera* (511 kg). The following species ranged between 300 and 500 kg per individual: *Ulmus pumila, Ulmus parvifolia, Bischofia polycarpa, Catalpa ovata, Robinia pseudoacacia*, and *Celtis tetrandra var. sinensis*.

By comparing biomass storage for the tree population in different DBH classes, we found that 63.7% of biomass was stored in the trees having a DBH of 10 to 30 cm. But trees with a DBH larger than 40 cm, being only 1% of the tree population, disproportionately contributed about 15.5% of the biomass storage for the entire tree population. Based on these biomass calculations, we estimated that total carbon storage by trees in the study area was approximately 21,996 metric tons, averaging 10.62 metric tons per hectare, with 63.7% of the total biomass stored in the 10- to 30-cm DBH size class.

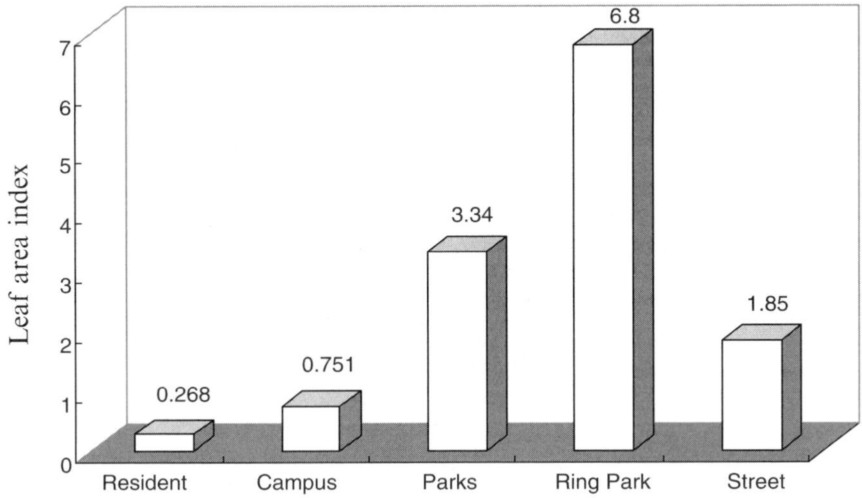

Fig. 17.3 Leaf area index for different land-use categories in the inner 23 km^2 area of Hefei, China

Leaf Area and Leaf Area Index (LAI)

There was a total leaf area of 1.345 million m^2 for broad-leaved trees in the study area, with an average of 61.5 m^2 per individual tree. The LAI achieved was 0.65 m^2 leaf area/m^2 ground area. Not unexpectedly, the average leaf area per tree differed by species. *Paulownia tomentosa* had the highest leaf area on a per tree basis, attaining values of up to 390 m^2. This was followed in succession by *Bischofia polycarpa, Liquidamber formosana, Firmiana simplex*, and *Quercus acutissima*, all trees having more than 200 m^2 of leaf area per tree. The 15 major broad-leaved species that dominate urban forest biomass contributed about 90% of total leaf area in Hefei's central districts. *Ligustrum lucidum* was listed in first place, followed by *Magnolia grandiflora, Paulownia tomentosa*, and *Platanus acerifolia*. The LAI of the urban forest by land–use category also varied, with the Ring Park having the highest value (6.38 m^2/m^2), and residential neighborhoods having the lowest value (0.268) (Fig. 17.3).

Urban Forest Coverage Analysis

We analyzed aerial photographs to calculate the amount of total forest coverage by land-use category. The total coverage by trees was 26.2% in the study area. We also mapped the distribution pattern of urban forest coverage for this central district of Hefei (Fig. 17.4). We found that land having less than 30% tree canopy coverage occupied 30% of the study area, and that 33% of the land had tree coverage of less than 5%. This low coverage area was located in the commercial zone in the city center. The Ring Park, covering only 20% of the land area, had more than 50% of the central city's tree canopy coverage (Fig. 17.4).

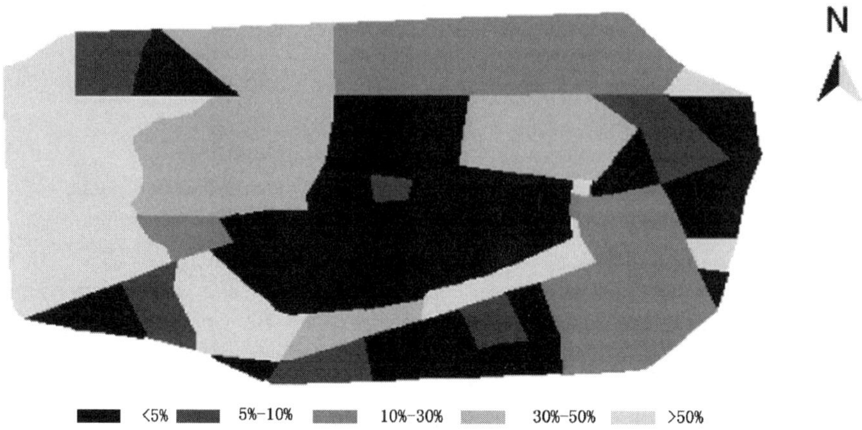

Fig. 17.4 The pattern of urban forest coverage in the study areas of Heifei, China

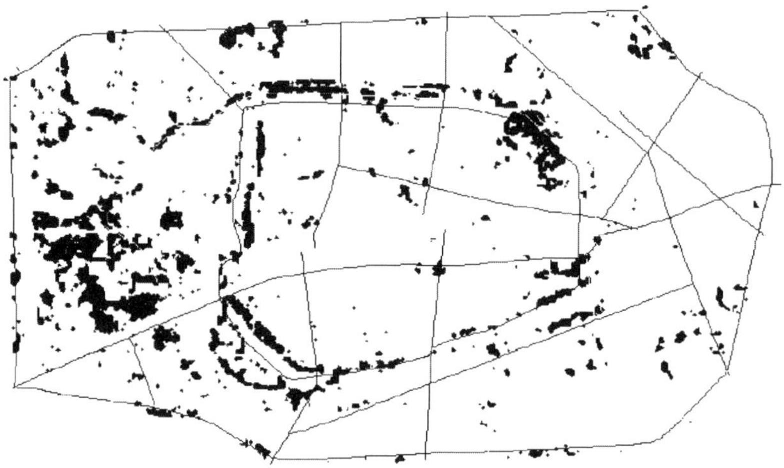

Fig. 17.5 The distribution pattern of urban forest patches greater than 0.011 hectare in size in Hefei, China

Table 17.7 Frequency distribution pattern of forest patch size classes inside the first ring road area

Urban forest Patch size class	Urban forest patch numbers	% of total patches
Small (<500m^2)	198	48.5
Medium (501 m^2–2000 m^2)	99	24.3
Large (2001 m^2–10,000 m^2)	76	18.6
Extra large (10,001 m^2–50000 m^2)	29	7.1
Huge (>50,001 m^2)	6	1.5

Urban forest patches, which we defined as land units with over 30% tree canopy coverage, consisted of 408 individual patches with a total area of 161.2 ha. The largest patch was 12 ha and the smallest was 0.011 ha (which was the smallest area that could be identified from aerial photographs). More patches were distributed in the western part of this city study area than in the eastern section, and more were located in the Ring Park due to intersection by roads through the park (Fig. 17.5).

Urban forest patches were divided into five size classes: small, medium, large, extra large, and huge (Table 17.7). In the study area 48.5% of the patches were small (<500 m^2), and patches larger than 1 ha were only 8.6% of the total. Only six patches had areas over 5 ha each. The patch size categories were based on area, but in nature heterogeneous habitats exist where boundaries between two different ecosystems are detectable. This is called the ecological edge. In general, the width or extent of an ecological edge for a forest site is considered to be about five times the height of its constituent trees. In Hefei, since the Ring Park had a more typical urban forest environment, the average height of its component trees (11 m) was used as the basis for an exercise in calculating ecological edge length and density. We assumed an average width of the ecological edge to be about 55 m wide. On this basis then, the minimum patch area needed to support an inner core environment with habitat conditions that differ from the edge should be at a minimum 9800 m^2, or approximately 1 ha. Based on this calculation, there were only 38 urban forest patches (comprising 9.3% of the total) in Hefei that were large enough to support inner habitat conditions distinct from edge conditions. Although these 38 patches were 9.3% of the total number, their inner habitat area comprised 19.7% of the total urban forest area of the city.

Appraisal of Hefei's Urban Forest with Recommendations for Improvement

For the entire central study area, the existing trees cannot produce a distinct forest environment as an urban forest should do because of the low basal area per hectare. But some land-use categories, such as parks and the Ring Park with higher basal areas per hectare, have forest stand characteristics and can be considered urban forest communities.

In the study area, the average DBH of trees is low, since there were fewer big trees, and since almost 48% of the trees were smaller than 10 cm DBH. This can be explained by the fact that a large proportion of the city's trees were planted in recent decades.

Several tree species dominated the tree population, and the city has not paid a great deal of attention in diversifying its species stock. This has led to a rather monotonous landscape environment and is also a source of future instability for the urban forest community. The city needs to pay more attention to stock diversification for future plantings.

Urban forest patches are distributed unevenly in the study area, with the western sections having more trees than those in east. The eastern section of Hefei contained industrial fields and has been rebuilt to be commercial and residential squares in the recent years. More trees should be planted in this eastern section in future.

Most of the forest patches in Hefei were classified as small. Given our tree species, urban forest patches of roughly square or circular shapes that attain a size of 1 ha may be able to support inner habitat conditions, which is important for supporting and improving the ecological services functions of the forest for the city's people.

Establishment of city plazas and landscaped avenues has become more popular in recent years. Therefore, species with small crowns and flowers were selected as street trees, resulting in decreased environmental benefits that can be derived from the city's street tree population. So city plazas and streets should improve their planting design and diversify their planting palettes with different tree species that vary in size. Priority should be given to planting deciduous tree species that grow to larger sizes.

Trees in residential areas contribute less to the improvement of the city's environment than might be expected. Since residential areas constitute such a large proportion of the city's total land use, it provides an important opportunity for improving the urban forest where it can make a large difference to people's health and comfort. If 15 new trees for per hectare were added in residential areas, the city's total tree population would increase by 6%.

The most important objective for Hefei city now is to build an urban forestry program to meet increasing demands for tree maintenance.

References

Kielbaso, J.J. (1988) Trends in urban forestry management. Baseline Data Report, Vol. 20, No. 1. International City Management Association, Washington, DC, p. 30.

McPherson, E.G. (1998) Structure and sustainability of Sacramento's Urban Forestry. Journal of Arboriculture 24:174–189.

Miller, R.W. (1988) The Urban Forest. Prentice-Hall, Englewood Cliffs, NJ.

Nowak, D.J. (1994a) Urban forest structure: the state of Chicago's Urban forest. Northeastern Forest Experiment Station, General Technical Report NE-18. U.S. Department of Agriculture, Washington, DC, pp. 3–18.

Nowak, D.J. (1994b) Atmospheric Carbon Dioxide Reduction by Chicago's Urban forest. Northeastern Forest Experiment Station, General Technical Report NE-18. U.S. Department of Agriculture, Washington, DC, pp. 95–114.

Rowntree, R.A. (1984a) Ecology of the urban forest—introduction to part I. Urban Ecology 2:1–11.

Rowntree, R.A. (1984b) Ecology of the urban forest—introduction to part II. Urban Ecology 2:229–243.

Shao, M., Tang, X.Y., Zhang, Y.H., and Li, W.J. (2006) City clusters in China: air and surface water pollution. Frontiers in Ecology and the Environment 4:353–361.

Wu, Z. (1994) The flora of Hefei city. Research report of Forestry Institute, Anhui Agricultural University (in Chinese).

Wu, Z., and Wu, W. (2003) The analysis of urban forest landscape pattern in Hefei city. Journal of Applied Ecology 14:2117–2122 (in Chinese).

18
Forests and Forestry in Hesse, Germany: Meeting the Challenge of Multipurpose Forestry

Rolf Schulzke and Sebastian Stoll

The state of Hesse is located in the center of the Federal Republic of Germany. Forty-two percent or 9000 km^2 of Hesse is covered with forests. Such high forest coverage was not always the situation. At the end of the 18th century forests were devastated due particularly to overexploitation for fuel wood and forest grazing. Forest cover in Hesse had been reduced to 25%. The existing forests were described as stands with many gaps, the trees only being usable as fuel wood, and unsuitable for use as lumber. The soil was exhausted. At that time foresters recognized the situation to be critical. They demanded that forests should be managed and used according to explicit plans and in a sustainable manner. Consequently, a principle of sustainability was developed and became the credo of sound forest management (Hessisches Ministerium für Umwelt, Landwirtschaft und Forsten, 2000). Over the course of time, the demand for supplying timber and other demands were made on these forests. As urbanization increased, recreational and watershed protection functions became increasingly more important, especially in the last few decades. These functions are not limited to urban forests. Forests surrounding towns and cities, although not classified as urban forests in the narrow sense of the definition, are nevertheless supposed to fulfill productive recreational and protective functions—functions that can often conflict. This chapter describes a system that demonstrates how multipurpose forestry can work to create an environment that also helps connect urban and rural areas into a more integrated system.

Current Forest Conditions in Hesse

Forests cover 42% of the state of Hesse and the Rhineland Palatinate. Hence these two regions exhibit the highest proportion of forest cover of the 16 German states. The spatial distribution of these forests, however, is rather uneven, with forest cover by districts ranging from less than 20% to more than 60%. For several years now, forest area has slightly increased due to the fact that the number of farms is decreasing and farmers show growing interest in afforestation programs. Forty percent of these forests are owned by the state, 35% by communities (among them the municipalities of Frankfurt, Kassel, and Wiesbaden), and 25% by private forest

M.M. Carreiro et al. (eds.), *Ecology, Planning, and Management of Urban Forests: International Perspectives.*

owners. In Hesse, 418 of 426 municipalities own their own forests (Hessisches Ministerium für Umwelt, Landwirtschaft und Forsten, 2000). The forests in Hesse are characterized by a high proportion (54%) of indigenous broad-leaved tree species. Hessian forests show a standing average volume of 319 m^3/hectare (ha), with an annual increment of approximately 11 m^3/ha. Therefore, silvicultural management is a necessity to tend the forests. A very important part of the annual felling is derived from beech (*Fagus sylvatica*). Hessian beech is still an important export, even as far as China where the round timber is converted to sliced and peeled veneer.

However, today Hessian forestry is also facing serious problems. We do not have a situation as bad as in that at the end of the 18th century, when natural disasters, like flooding and erosion, leading to famine and unemployment, occurred due in large part to forest destruction. But for now, the forests adjacent to the conurbations of Kassel and the Rhein-Main Region surrounding Frankfurt are in a very deteriorated condition. About 25% of the Hessian population lives in the Rhein-Main region, covering only 7% of the state's area, but being one of the most industrialized parts of Germany and perhaps of Europe. Consequently, land development pressure is high. Since 1900 about 25% of the original forests have been converted to other land uses (Hessische Landesforstverwaltung, 1999). On the other hand, there is an increasing need to maintain the protective and recreational functions of the forest, as demanded by the public. Finding solutions for these land-use conflicts is a difficult challenge for multipurpose forestry management.

Forest Act

The Forest Act of Hesse, issued in 2002 and amended in 2002, states the basic responsibilities of forest owners and establishes the criteria for sound and sustainable forestry (Hessisches Ministerium für Umwelt, Ländlichen Raum und Verbraucherschutz, 2002). Forest owners have to manage their forest not just for their welfare, but also for everybody's welfare in a sustainable, informed, and methodical manner according to the basic principles of forestry and landscape conservation. By doing so, productive, protective, and recreational functions can be maintained. Characteristics of sound and sustainable forest management are the following:

- Long-term existence and sustainability of forest production
- Conservation of forest ecosystems as habitat for maintaining a diversity of plant and animal species
- Avoidance of large area clear-cuts
- Selection of site-adapted tree species
- Use of forest reproductive material that maintains and enhances genetic diversity
- Avoidance of pesticide use

- Care in tending and regenerating forests, as well as in harvesting and removing forest products, so as not to disturb the capacity of the forest soil to sustain forest production (e.g., by avoiding soil compaction)
- Application of forest stand and forest soil protection techniques

The forest act has established on a regional level the forestry framework plan. The whole system includes the National Forest Program on the state level and the Forest Management Plan on the local level. Adherence to these plans is obligatory in order to guarantee the multipurpose functions of these forests (Bundesministerium für Verbraucherschutz, Ernährung und Landwirtschaft, 1975). A forest program, which must be put into concrete terms for the region, has to be elaborated for the country. The aim is to contribute to the improvement of living conditions and economic development. However, the different interests of forest owners and the public often need to be reconciled. Goals of forestry development and measures need to be clearly articulated in order to deal with conflicting interests. The goals and measures derived from the forestry planning framework may have impact on other economic and sociopolitical sectors that will need to be considered and integrated into the regional development plan (e.g., highway construction). Every federal, state, and municipal authority, as well as other public planning commissions, are obliged to consider forestry goals when considering land use plans that may affect the forests.

For every state- and community-owned forest a detailed, mid-term (10-year) management plan has to be submitted to safeguard the sustainability and multipurpose function of each forest. These plans have to be approved by the forest divisions of the regional councils. For private forests larger than 100 ha, the same regulation is compulsory. The Forest Act has strict regulations as to the procedure and obligations following legal or illegal clear-cutting of forests. The reforestation of clear-felled areas is a legal obligation for all forest owners. Permission is necessary for the conversion of forest land to other land uses. Such conversions have to be denied in cases where there the greater public interest is to conserve the stands. Permission to convert forest land shall be governed by the proof of compensatory afforestation in the neighborhood. This regulation was established to provide disincentives to clear-cut and remove forests from the landscape (Bundesministerium für Verbraucherschutz, Ernährung und Landwirtschaft, 1975; Hessisches Ministerium für Umwelt, Ländlichen Raum und Verbraucherschutz, 2002; Regierungspräsidium Kassel, 1997).

There are legal requirements concerning the structure and tasks of the Hessian Forest Administration. Being owned by the Hessian Ministry of the Environment, Rural Development, and Consumer Protection, the Hessian State Forest Service is subdivided into two units: Hessen-Forst and the Forest Authority (Hessisches Ministerium für Umwelt, Ländlichen Raum und Verbraucherschutz, 2002). The operational unit called Hessen-Forst is entrusted with the management of the state-owned forests. This state-owned company has its own budget that allows it to be more flexible and to work according to economic principles. Hessen-Forst not only manages the 340,000 ha of state forests, but also offers services to communal and private forest owners who manage an additional 430,000 ha of forest land. Both communal and private owners in particular like to rely on the existing organizational structure and ask for managerial support for which they pay. However, there is no obligation for Hessen-Forst to do so. The authoritative unit includes the Hessian ministry and its forestry department. They supervise the operational unit (i.e., the State Forest Enterprise) and, together with its subordinate bodies at the regional and district levels, steer forest politics and enforce legal regulations.

Multipurpose Forestry

Socioeconomic development has led to a change in the goals of forest management in many parts of the world. For a long time, the production of wood was the main concern. Today, there is general agreement that our forests serve productive, protective, and recreational functions, often simultaneously, although, depending on local conditions, one or more of these functions is more dominant. Social and protective functions have gained increasing importance especially in conurbations. Therefore, forest management has to be conducted in such a way that the goals for multiple purposes can be achieved. The choice of tree species, for example, should be matched to the existing site-specific objectives. To be in a position to characterize protective and recreational functions in a transparent and comprehensible manner, the instrument of forest function surveying and mapping was introduced as an essential part of the forest inventory process. For each forest area the protective and recreational functions have been regularly assessed and mapped for more than 20 years. This has been done according to criteria that are the same for the entire Federal Republic of Germany. The introduction of the surveying and mapping of forest functions includes a definition of those functions, a description of the effects striven for, and recommendations for continued management (Arbeitsgemeinschaft Forsteinrichtung, 1974). A two-level classification shows whether and to what extent a function influences forest management. The results are both written down in the forest management plan and fixed on a special map describing the various forest functions and their subcategories.

These functions, which will dictate the types of silvicultural measures applied to the system, are identified, assessed, and mapped in the following categories:

- Water conservation forest: maintains clean ground and surface water, increases ground-water recharge, and controls surface runoff, thus reducing flood risk.
- Soil protection forest: protects neighboring areas from the negative effects of wind, water, and snow erosion; rock falls; and landslides.
- Protective forest against climatic impacts: protects settlements, agricultural areas, and recreational facilities against the negative impacts of various climatic phenomena.
- Antipollution forest: reduces the harmful or irritating impacts of noise, dust, and air pollutants.
- Forests under nature protection: offer habitats to a great variety of species or represent a rare forest association.

- Recreational forest: primarily functions to improve human health by providing opportunities for relaxation and low-impact leisure activities.

In case the implementation of plans have negative impacts on forests, the relevant authorities have to consider the findings of these scientific assessments. They are obliged to demonstrate whether the destruction of forest could be avoided and, if damage is inevitable, how damage to the forest could be minimized.

The sustainable fulfillment and improvement of the various functions can only be guaranteed, if the forests are stable, vigorous, and healthy. Site-adaptation of the stands and sound management techniques are prerequisites. Therefore, silvicultural methods have changed from a plantation orientation to methods that mimic a more natural forest structure (see Chapter 19). The consequences are that forest management now aims to do the following:

- Establish forests that contain a species composition, vertical structure, and processes that are close to nature.
- Promote mixed stands.
- Prefer natural regeneration under the canopy of the mature stand (avoidance of clear-felling).
- Produce high-quality timber.

Experience gained by forest owners and managers has proven that these methods have ecological and economic advantages. Economic benefits are derived from the production of timber and nonforest products using silvicultural methods that take advantage of natural processes of tree regeneration instead of establishing row plantations. In addition, allowing natural processes to dominate leads to other ecological benefits, since these forest show improved resistance against biotic and abiotic threats and offer better opportunities for protective and recreational functions (Hessische Landesforstverwaltung, 1999; Hessisches Ministerium für Umwelt, Landwirtschaft und Forsten, 2000). However, these favorable conditions cannot always be realized in a short time. Sometimes planting a nurse crop is necessary, and admixed species need silvicultural treatment. Existing forests, although not representing the favorable setup in large areas, offer good opportunities for silvicultural improvements, such as enrichment planting or advanced planting of admixed species. Sometimes planting could be avoided since successional processes, for example, could be used as a nucleus for the development of stable forests (see Chapter 23).

Case Study

Seventy-six percent of Hessian forests must fulfill at least one protective function. Ninety-nine percent of the forests in the vicinity of Kassel are characterized on average by more than three levels of protective functions. More than six functions are observed in 10% of the municipality's forest area. These figures reflect the importance of forests in serving the well-being of our citizens and in protecting agricultural land and urban infrastructure. To minimize the negative impacts, activities resulting in the conversion of forest land should be located outside very important forests. In cleared areas, afforestation programs have had to be planned. Consequently, the municipality of Kassel has submitted an application to declare its neighboring forests as protected forests, as allowed by the Hessian Forest Act. This declaration has to be made by the regional forest authority. If accepted, then all activities that do not comply with sustainable forest management prerogatives and the maintenance of forest cover are forbidden. Conversion of forests to other kinds of land use will not be allowed (for example, settlements or quarries). But unlike conservation areas, where any human activity is forbidden, forest management is an essential part of the protection program. In the meantime, 77% of these forests are still declared recreational forests, where recreational objectives have to be taken into account while preparing forest management plans.

Although forests cover large areas of Hesse, there are parts of the country where the proportion of forests is quite low, and their size does not ensure that protective functions will be provided. To overcome these shortfalls, it is understood that forest area should be increased wherever suitable. Increasing forest area can be promoted by afforestation, which means establishment of new forests on land that was formerly agricultural land. Because afforestation creates new additional forests, financial subsidies will be granted to the forest owner. Nevertheless, permission must be granted to authorize afforestation activities, and it can only be refused for special reasons that will be discussed later.

Before carrying out afforestation activities, several administrative and silvicultural issues must be addressed. The forest authorities should direct afforestation activities to areas and sites with the following characteristics:

- The proportion of forest area is below average.
- Maximum benefit for the public and nature, as well as private benefit, could be achieved.
- Conditions allow high forest productivity.
- Landscape scenery and recreational functions will be improved.
- The conservation of species and biotopes could be improved (e.g., by planting riverine forests or where biotopes can be linked by forest networks).
- Protection against pollution and damage caused by climatic phenomena is necessary (e.g., by the establishment of shelter belts or plantations alongside highways and industrial plants).
- Soil protection, especially erosion control, could be improved (e.g., on slopes above settlements and roads, in areas surrounding lakes and rivers, or in large arable areas exposed to wind erosion).

Afforestation should not occur in areas and sites where

- species and biotope conservation could be placed at a disadvantage;
- landscape scenery and recreational functions would be spoiled, e.g., at view points or in areas of historic importance; or
- the area is essential for agricultural production and structure.

In 1994, guiding principles for identifying areas for future afforestation activities were established by the regional councils after intensive discussion with the relevant stakeholders (farmers, nature conservation authorities, planning departments). Before these meetings, the district forest authorities of the region of Kassel had reported in 1993 that up to 50,000 ha of nonforest land would be suitable for additional afforestation. The 1994 task was to take into account the interests of other stakeholders (Regierungspräsidium Kassel, 1997). A crucial point during the discussion was the size versus number of the forest areas to be created. It became obvious that a number of smaller patches, as long as they did not fall below a minimum size, would be of greater value than a big block of forest. As a rule of thumb, the diameter of individual forest areas should not be smaller than one or two heights of a mature tree that would grow under the conditions at that site. This dimension is the minimum size for silvicultural activities and allows formation of a specific forest climate. The positive effects of wind-breaks reach up to 20 to 25 times the height of the trees. Therefore, with regard to protective functions, it is advisable to establish several smaller belts of tree plantations than one big block. Due to the interests of other stakeholders, it proved to be much more difficult to find areas for large-scale afforestation than to gain acceptance for several smaller plots.

As a result of this procedure, 16,000 ha were identified where afforestation should have priority over other categories of land use (Regierungspräsidium Kassel, 1997). However, this proposal is binding on the authorities. In addition, there is always the opportunity for a landowner to apply for an afforestation permit even if the area was not identified as being a priority for forested land use. Up until now realization of afforestation potential has not been very extensive. Farmers are reluctant to afforest their land. Beside requiring areas for food production, farmers need areas on their land to serve other nature conservation functions, or for taking advantage of the government's set-aside program for arable land that takes land out of production for economic reasons.

Conclusion

It should be emphasized that every tree planted and every new forest established will improve the living conditions of people. Over time, the primary goal of forestry, to produce timber, has been enlarged to include protective and recreational functions that are gaining in importance. Forest management, particularly in and near urban areas, has to consider silvicultural techniques that serve the goals of multipurpose forestry. Eventually, the ecological and economic benefits of such forests will strengthen the acceptance of forest management for multiple functions. For example, revenues from timber production will allow financing investments for recreational functions. Forests in and around urban areas are important for creating a natural environment that connects urban and rural areas, offering both recreational benefits and employment. They fulfill social, ecological, and productive functions. Interdisciplinary planning and legal instruments are necessary to guarantee the existence of forests of particular sizes and spatial distribution. Forest function surveys and mapping are suitable tools for supporting forestry interests in the planning process. Although forest management has become a multidimensional field and challenge, the applied approaches and methods being developed to meet multipurpose forestry goals should provide us with some optimism that we will be able to meet this challenge of guaranteeing the existence of forests to serve the manifold expectations people associate with them.

References

Arbeitsgemeinschaft Forsteinrichtung. (1974) Leitfaden zur Kartierung der Schutz- und Erholungsfunktionen des Waldes. [Working Group on Forest Planning. Introduction to the mapping of protective and recreational forest functions.]

Bundesministerium für Verbraucherschutz, Ernährung und Landwirtschaft. (1975) Bundeswaldgesetz vom 2.Mai 1975 in der Fassung vom 27.7.1984. [Federal Ministry of Consumer Protection, Food and Agriculture. Forest Act of the Federal Republic of Germany issued May 2, 1975, amendment of July 27, 1984.]

Hessische Landesforstverwaltung. (1999) Gefährdung der Wälder im Rhein-Main Gebiet, Mitteilungen der Hess. Landesforstverwaltung, Band 35. [Hessian Forest Service: Situation of the forests in the Rhine-Main Area. Communication of the Hessian Forest Service.]

Hessisches Ministerium für Umwelt, Landwirtschaft und Forsten. (2000) Werden und Wandel, Hessens Forstwirtschaft auf dem Weg in das 3.Jahrtausend, Mitteilungen der Hess. Landesforstverwaltung, Band 37. [Hessian Ministry of Environment, Agriculture and Forestry: Development and Change, Hessian Forestry on the way into the 3rd Millennium, Communication of the Hessian Forest Service.]

Hessisches Ministerium für Umwelt, Ländlichen Raum und Verbraucherschutz. (2002) Hessisches Forstgesetz vom 10 November 1954 in der Fassung vom 18 Juni 2002. [Hessian Ministry of Environment, Rural Area and Consumer Protection: Hessian Forest Act issued 10th November 1954, amendment of 18th June 2002.]

Regierungspräsidium Kassel. (1997) Forstlicher Rahmenplan Nordhessen. [Regional Council of Northern Hesse: Forest Framework Planning for the Region of Northern Hesse.]

19 Experiences in the Management of Urban Recreational Forests in Germany

Michael Jestaedt

Managing forests often entails sustaining an ecosystem to meet different cultural expectations. These expectations or objectives can be determined by the site, by the owner, or by society. The production of timber for income generation is usually an important objective, one that can readily be quantified. At the same time, a forest provides other benefits that cannot be quantified as easily. These include such forest functions as microclimate amelioration, protection of water and soil resources, air pollutant filtration, and noise reduction. In a densely populated country forests provide a refuge for many endangered species that cannot survive in an urban or agriculturally dominated landscapes. Forests in rural areas are primarily managed for timber and ecological protection functions. However, because of their aesthetic appeal and contrast with city hardscapes, forests have become the most important areas for public recreation in the crowded areas of Central Europe. Consequently, management goals for forests depend on their geographic location.

While objective and scientific criteria can be used to manage a forest for timber production and for ecological protection, the recreation value of a forest is determined by subjective criteria like beauty, diversity, and emotional demand. Criteria must also be found for distinguishing forest management goals from management of public gardens or wooded parks. Parks differ from forests in their natural stocking and management, whereas management for timber will always be included as goals for urban forests. A park, however, is a landscaper's creation, and serves to satisfy aesthetic and emotional goals. Therefore, natural successional forces are not determinants of species composition in a wooded park, as they would be in a forest, due to continual human intervention in keeping plant communities in the same condition.

Characteristics of Urban Forests

Historic Development of Forests in Central Europe

After the glacial period, Central Europe was almost exclusively covered with forests. With colonization, deforestation increased to ensure agricultural production. As a consequence, forests were mostly reduced to sites that were not usable for

M.M. Carreiro et al. (eds.), *Ecology, Planning, and Management of Urban Forests: International Perspectives.*

agriculture because of unsuitable soils or orography. Urban areas thus developed in regions with productive soils or in places that offered especially favorable transportation possibilities. Forests were only useful for the supply of timber or for feeding cattle (acorn, grass, leaves). Consequently, forests were relegated to marginal border areas (Küster, 1999). When forests were to be saved in urban areas, a specific authority was necessary to stop settlement. For example, kings did not allow settlement in certain forests that were called king's forests or *banforests* and were used by the king for national interests and for hunting. Consequently, the forests that remain today in many urban areas were *banforests* in former times (for example, the Dreieichenhain near Frankfurt, the Reichswald near Nürnberg, and the Kingsforest near Cologne).

At the end of the 19th century, due to increasing industrialization and a growing population density, the negative impact of a lack of urban forests became obvious. Some far-sighted local authorities began afforestation programs, like that in Munich for protecting ground water. Authorities in Cologne decided to plant a green belt around the city using an area formerly devoted to fortifications that had become useless after the First World War. In the Ruhr region, many landfills were recultivated and afforested (see Chapter 23).

Ecological Impact

Because of numerous human influences, urban forests differ from natural or unmanaged forests. One of the most important influences is the large number of people that visit urban forests in Germany (Table 19.1). The forest of the city of Frankfurt in the state of Hessia (6000 hectare [ha]) has 6 million visitors every year (Ebert, 1993). Such a high number of visitors obviously has a high impact on this forest ecosystem. Permanent disturbances, noise, deposits of rubbish, and other

Table 19.1 Visitors to forests in different federal states of Germany

Federal States, 1982	Inhabitants (in millions)	Stock area (ha)	Visits per ha per year	Visits per inhabitant
Schleswig-Holstein	2.01	129,000	373	24
Hamburg	1.43	3,700	4251	11
Niedersachsen	5.50	944,100	101	17
Bremen	0.58	100	134,600	23
Nordrhein-Westfalen	13.50	752,000	467	26
Hessen	4.42	806,000	126	23
Rheinland-Pfalz/Saarland	3.74	780,800	125	26
Baden-Württemberg	7.11	1,231,900	185	32
Bayern	8.50	2,283,400	93	25
West-Berlin	1.67	7,000	5,333	23
Federal Republic	48.56	6,936,000	168	24

Source: Lösch, 1980.

forms of pollution are the consequence. Soil compression and erosion in steep areas result. The risk of forest fires increases, as many smokers are not cautious. However, large fires do not generally occur, because fires are soon discovered by visitors and successfully extinguished.

Urban forest patches are also highly fragmented by roads, pipelines, and electrical wires that run both above and below ground. Consequently, the typical forest microclimate that can form beneath a closed canopy cannot develop. Average temperatures and the risks of storm damage will therefore rise. The influence of fragmentation, due to roads and infrastructure on the integrity of forest habitat, is generally underestimated by the public, because the resultant negative effects take several years to appear.

Urban forests are also exposed to increasing pollution. The discharge of dust, sulfur dioxide (SO_2), and nitrogen in various chemical forms (up to 40 kg nitrogen per hectare per year in this region) changes soils greatly. Pollution-sensitive and endangered species disappear, community composition of biotopes change, and the gene pool of the remaining populations can be influenced (Gregorius et al., 1985; Scholz, 1986). When trees are stressed by air pollution, soil loss, and soil compaction, their ability to resist attacks by biotic parasites decreases, and further susceptibility to abiotic damage increases (Rehfuess, 2000). Stabilizing feedback within the forest ecosystem can become compromised. Forest dieback can be easily recognized close to many cities in Germany.

Protection of Urban Forests

The existence of forests close to cities must be guaranteed over the long term because of their importance in preserving the quality of life of residents in the area. Forests are too often destroyed for the construction of supply lines, roads, and land for building or for agricultural use, even if other areas could be used. Ecological arguments are rarely taken into consideration when economic profits are to be made quickly. Land ownership patterns also contribute to forests being preferentially developed over other land-use or land-cover types. Large development projects can be more easily realized when many owners do not have to be consulted or appeased, as is the case for many German forests. Long-term protection of forests and their functions, therefore, must be ensured by diverse political authorities that make planning decisions (see Chapter 18).

A legal framework for forest protection exists in almost every country. Forest law in Germany includes regulations that prohibit deforestation and rules of declaration of *banforests* or recreation forests. Deforestation in urban areas is always linked with obligations of compensatory afforestation, where developers are obliged to afforest another adequate area in the same landscape to compensate for forested area lost. If these regulations are followed, developers will refrain from carrying out most projects, because adequate sites are hard to find in conurbations and because land prices are very high. Principles of preservation, however, also have to be introduced into the governmental planning system. Conservation laws and laws for the protection of water and soil may prohibit illegal operations. Yet the most effective kind of protection over the long term is the citizens' knowledge about the importance of forests for their personal welfare and quality of life. If this environmental awareness is promoted, economic and political decision makers will plan projects in and near forests more carefully.

Developing Urban Forests for Recreation

Results of Opinion Polls

With growing industrialization, proportionately more people live in urban areas and become exposed to various forms of stress from jobs, increased exposure to noise and pollution, and lack of regular exercise. As shorter workweeks evolved during the social evolution of the last century, people have more leisure time. For those who wish to spend more time outdoors, forests can be very suitable locations for various forms of recreation and relaxation (Ruppert, 1971). Indeed, 70% of the German population prefers to relax in a landscape that is characterized by forests (Bichelmaier, 1969). Opinion polls and surveys are needed to help urban forest managers maintain forests in good condition, while responding to this high intensity public demand.

Zundel and Völksen (2002) have conducted and interpreted a number of surveys over the last 30 years. Often interpretation is difficult because survey results can be very contradictory. Results may depend on the following factors:

- Whether people were questioned in the forest or in the city
- Whether they were given exact and clear definitions of the terms used
- Whether the surveys were carried out in the same season

The following principles about frequency of visits and about the prefered types of recreation forests can be distinguished:

- Old forests are preferred.
- Mixed forests are considered more attractive; mixtures of hard- and softwood species and trees of different ages are considered more beautiful.
- Women prefer open stands without a dense understory.
- In summer, hardwood species are preferred, but in winter softwood species are preferred.
- Older people visit forests more often than younger people do.
- A forest should vary in its structure and offer a sense of adventure.
- Recreation equipment and clear marking of footpaths are desired.
- Small-scale harvesting of timber does not disturb most visitors.

Development of a Concept for Guiding Visitors

In every forest there are sensitive areas that should be visited as rarely as possible, if at all. To determine the location and the size of rare or sensitive biotope habitats, an inventory of the forest's biotopes should be conducted so that they are not destroyed by visitors (Arbeitskreis Forstliche Landespflege, 1984). In many cases it is necessary to regulate the population size of game stock (roe deer, wild boar). For this purpose, undisturbed zones should be established where visitors should be kept out.

If managers are to guide visitors to certain locations of the forest and steer them away from others, then visitor estimates first need to be obtained. Additionally, their means of transport to the forest should also be ascertained, whether they take public or private transportation and also where they can find parking lots. For example, if people have no information about parking, they are less likely to visit a forest. Furthermore, the type of visitors that use the forests should be determined. Are they walkers, bikers, horseback riders, or, if there is enough snow in wintertime, cross-country skiers? Conflicts may emerge between the different groups of visitors, because they make different demands on the quality and shape of the trails. For example, bikers are not easily convinced to use the same trail as walkers. On the one hand, intensive horseback riding makes the trails impassable for walkers, but on the other hand, walkers and horseback riders destroy the cross-country ski runs. Consequently, path layouts must direct the different groups so that conflicts can be avoided. This can be achieved by designating specific trails for single groups and sign-posting them accordingly. However, this is usually not enough for separating the different user types. Not only do trails have to be made desirable for one group, they also have to be made unattractive for other groups. We have found that rough gravel is normally avoided by horses, while loose sand deters walkers and bikers. Undisturbed forest areas can be designated and protected by diverting visitors on suitable trails away from those areas, by intentionally putting certain sections in bad condition for travel, or by blocking and afforesting open views through the woods that invite rogue trail formation.

Roads and Paths in Recreation Forests

When large crowds use a nearby urban forest, a higher road density is needed to accommodate them, compared to a forest managed for purposes other than recreation. In Central Europe, depending on site conditions, a road density of 20 to 30 m per ha is strived for, but in densely populated areas 50 to 70 m per ha is often the case. Such high road densities result in high maintenance costs. Although not every path has to be passable for vehicles, a high proportion of roads that are passable for automotive traffic must still be built. When a forest is visited by large numbers of people, ambulances and police cars often have to be called. Also, transport of timber or construction material into or out of the forest must be facilitated, and staff responsible for supervision and repair work have to access all areas quickly.

To reduce the negative effects of high road density on these forests, roads should be paved with permeable tarmac. For ecological reasons, asphalt use should be the exception. Also road cover material that does not produce too much dust during dry periods should be used. To make orientation in forests for visitors easier, hiking trails have to be marked well, and should always lead back to the starting point. Connections to other trail systems should be attempted. Paths for horseback riding should be of adequate length; stabilization with gravel is not necessary, but the branches of the trees must be removed up to a height of at least 3 m.

Silvicultural Approaches for Creating and Maintaining Healthy Recreation Forests

Through various silvicultural means, a forest area can be developed so that it suits the rational as well as the emotional expectations that forest visitors associate with a walk in the forest. The forest offers the city residents a sense of freedom as they encounter a natural environment that contrasts with the mechanized urban environment. As a consequence, it is important to influence the aesthetic attractiveness of a forest (Salisch, 1911; Ruppert, 1971).

For both ecological and aesthetic reasons, the stands of a recreation forest should always have a well-developed structure. In existing old stands that do not meet these expectations, the canopy can be opened a little so that natural subcanopy tree regeneration can be enhanced. This occurs since light-demanding and shade-tolerant trees exist together in the understory. In the first stage, the femel-cut regeneration approach (small patch cuts of 30 to 40 m) is used to create stands that at first favor growth of shade-tolerant species. A few years later, the patches are enlarged to favor light-requiring species. This ultimately results in multispecies stands of different ages, with greater vertical structure within the canopy and therefore a plenter, rather than even-aged, forest structure. If it is necessary to transform large monotypic stands to a mixed species and age structure, then clear cutting must be avoided. Cultivating an understory of shade-tolerant tree species or planting below the canopy with species desired for future forest composition may help in such cases. These stands can then gradually be transformed into multilayered, mixed species stands that can then regenerate naturally. Old stands that are not able to regenerate any longer can be dealt with similarly.

When afforesting an area close to a city, it is advantageous to cultivate first a nurse crop of fast-growing species, like poplar or alder. The desired future dominant species can be planted in the understory later and will grow faster once the nurse-crop species are removed. In this way a visitor can quickly get the impression of experiencing a forest, although the aesthetic optimum has not yet been reached. In addition, stand structural diversity is established quickly (Arbeitskreis Forstliche Landespflege, 1994).

To enhance visitor enjoyment, forests should have viewing areas overlooking the landscape. Older landmark trees should be carefully tended in these sites. Views of particular special natural formations, like boulders, should not be blocked. Meadows and clearings should be kept open. Forest edge habitats should be formed in a way that trees and shrubs will grow in a terraced structure.

Facilities for Recreation

From the public's perspective, the attractiveness of every forest area can be increased with simple recreational facilities. At the same time, such facilities can be used to steer and concentrate visitors to certain forest locations and not others. Many years ago, Ruppert (1960) developed numerous suggestions for so doing that are still valid today. First, facilities and structures that orient and welcome visitors must be set up. Orientation boards showing the trail system must indicate distances, the average walking time, and distinctly label different kinds of trails. For example, trail types include those that focus on natural history education or others more suited for keeping fit and where simple sports equipment can be used. This eliminates uncertainties in a visitor's mind that are caused by a lack of local knowledge.

On forest fringes, parking lots (stabilized areas in the shade) have to be constructed. Sports and game equipment can be built on playgrounds. In doing so, it should be considered that not only children but all age groups have to be attracted. The purpose here is to concentrate many visitors at forest fringes. Picnic and barbecue areas placed at forest fringes can serve this same purpose. Most of the litter generated would also be concentrated here, making trash pickup easier. Moreover, such siting lowers the risk of forest fires, which increasingly start as uncontrolled fires that are lit for food preparation. Meadows, which can be used for ballgames, are also extremely attractive to the public. However, they have to be mowed regularly.

Typical equipment that is needed along trails includes benches that are placed at regular intervals and at scenic locations. Simple huts can serve as shelters against sudden rain. A natural lake or an artificially constructed one increases the attractiveness of all sites. Extremely popular are wildlife game parks using local species (red deer, fallow deer, roe deer). Here the visitor can watch the animals in their natural environment. Outside these types of parks, it is quite difficult to watch these animals because they often change behavior to become night active to avoid humans.

Building materials for recreational facilities should be natural looking and match the landscape. Wood and natural stones that are taken from this landscape are preferred to artificial materials like concrete, steel, or plastic. Once these structures are built, they must be maintained, and the litter near them must be cleared away immediately. Damage to equipment and trails has to be repaired as soon as possible. Otherwise littered or run-down conditions provoke vandalism and prevent visitors from using such establishments, and so cannot fulfill their intended purpose of concentrating and directing visitors.

Supporting Ecological Functions in Recreation Forests

The ecological importance of urban forests cannot be estimated highly enough. In comparison to its surroundings, a forest is a natural ecosystem with a high diversity of valued species. Especially at the forest fringe zones you can find refuges for many species that cannot survive in open landscapes. Every step that serves to protect and develop forest fringes also affects the open landscape. A lot of experiments have been conducted to determine how best to support species diversity of our urban and other forests (Arbeitskreis Forstliche Landschaftspflege, 1984). This starts with preserving old stands, leaving dead timber to decompose in the stand, preparing nesting places for birds, and creating places where bats can spend the winter. The maintenance and tending of natural stream biotopes and the artificial construction of special biotopes that endangered species depend on are among the many objectives for urban forest management. In the course of managing forests, there are many possibilities for positively influencing the ecological quality of our landscapes.

Because of forests, erosion is minimized or avoided in the greater landscape. The best protection for areas that are endangered by erosion is extensive afforestation. It is therefore of special importance to avoid management plans and follow-through activities that increase the likelihood of soil erosion. Clear-cutting on steep slopes must be avoided. When timber is removed, suitable skidders with broad tires should be used, so soil compression is reduced.

Silvicultural approaches can also influence the collection and protection of ground water. Because of high usage rates for ground water in Central Europe, there is only a little accrual of ground water during the plant-growing season. However, it is possible to increase the level of ground water in winter by cultivating hardwood rather than softwood tree species, because softwood trees intercept more rainfall. Studies have been conducted to demonstrate that ground water levels are influenced by tree species. Because forest soils exhibit intensive and deep rooting of trees, and a large amount of soil humus created over time, rain is able to percolate to lower soil horizons more easily and surface water flow is thereby reduced. To protect the soil, herbicides and insecticides have to be minimally used. Also any operations that compress or permanently disturb the upper soil horizons must not be allowed so that rainwater penetration can occur. Machines that might endanger the quality of subsoil water (through leaking motor oil and hydraulic oil) must not be used.

Compared with an open landscape and built-up areas, a special climate develops in a forest because of tree transpiration, causing the air above forests to be cooler and more humid (Röhrig, 1980; Mayer, 1992). Since wind exchange occurs between the forest and neighboring residential areas, living conditions in these settled areas have more comfortable microclimates. For this reason, it is often useful to promote a high transpiration rate from a forest and increase advection rates via turbulence to neighboring communities. This ecological service function (microclimate moderation) can be enhanced by constructing a forest with many different canopy types and heights that increase air turbulence. Mixtures of light-demanding and shade-tolerant trees also enhance forest functions under changing environmental conditions.

Because of its large canopy and leaf surface area, forests filter large amounts of pollutants from the air. Evergreen softwood trees accumulate relatively more particulate dust and poisonous materials, like heavy metals from the air. While this improves air quality, it must be remembered that these poisonous materials may accumulate in the plants and soil and may become a serious problem to the ecosystem. Forests also are able to reduce noise pollution and may be established around noisy industrial plants for this purpose. Forests are also able to cover or hide unsightly buildings or industrial plants. If forests are planted for these reasons, we recommend that fast-growing trees and a high proportion of the evergreen trees be planted.

Management Operations Issues

Safety Precautions

Because of the many roads cutting through these recreation forests, the built-up areas along the forest fringe, and the many facilities for visitors, all of which maximize visitor use, it is extremely important to focus on public safety management. Trees must not endanger people, buildings, and traffic. Jurisdictional rules demand high staff expenditures to maintain public safety. All trees adjacent to built-up areas, roads, parking lots, and frequently used hiking trails have to be examined for their stability twice a year. Trees that might be a danger have to be removed immediately by well-trained employees using special equipment including climbing gear.

Timber Harvesting in Recreational Forests

Although timber harvest can generate important income for recreational forests, it should be done as inconspicuously as possible. Clear-cutting and concentrated patch cuts must not occur. Harvesting of some trees of exploitable size does not attract attention, as long as it is spread over a large area. After thinning, timber must be removed carefully, and damaged forest roads need to be restored as soon as possible. In case bark beetles are a problem, which is normally the case in highly stressed urban forests, material that facilitates the further development and spread of such insects should be removed (i.e., using a shredder). In the past, visitors did not feel disturbed by the timber harvest in recreational forests (Hanstein, 1967; Zundel and Roether, 1978). Nowadays, urban populations in Germany understand little about the ecosystem services provided by forests due to their growing alienation from natural processes and ignorance of a forest's economic importance. Only long-term public education by forest administrations can help reverse these growing, counterproductive public trends (Krafft, 2002).

Economic Issues

Of course, forest management success depends not only on ecological conditions and the desires of the population, but also on political decision makers. Budget quantity and timing cycles must make adequate funds available to managers when problems arise, as well as for general maintenance. However, since the benefits of forests close to cities are not easily quantified in monetary terms, the political leadership in Germany up to now has not been supportive of preventing development in and near urban forests (Ruppert, 1971).

Economic yields from the sale of timber, income from utility permissions (e.g., electrical supply lines), and from rents and leasing do not cover all the costs of maintaining these forests. In 2001, the city of Frankfurt spent 1100 Euros per hectare of urban forest, but the direct economic yields came to only 300 Euros per ha. A deficit of 800 Euro had to be covered. But what has to be mentioned is that these expenses included the maintenance of a large information center and the costs of six large forest playgrounds with supervisors. Forty-eight percent of the costs incurred for forest management were due to building and maintaining recreational facilities. The amount of 150,000 Euros is spent on garbage disposal for the urban forest system every year (Stadt Frankfurt, 1993). To place these costs in perspective, Ebert (1993) determined the distribution of costs for different public areas on a per city inhabitant basis each year and found the following:

Tending of public gardens and parks: 42 Euros
Zoological garden: 9 Euros
Urban forests: 7 Euros

So, compared with other similar public institutions involving green space, the per capita annual expense of managing Frankfurt's urban forest was actually quite modest. Despite these figures it has still not been easy to convince political decision makers to purchase more areas for planting forests close to the city. Therefore, economically accounting for quality of life and environmental benefits of urban recreation forests and communicating this to city council members and the city's residents is an educational strategy that must also be adopted by forest managers, if these forests are to be maintained and increased over the long term.

References

Arbeitskreis Forstliche Landespflege. (1984) Biotoppflege im Wald –Ein Leitfaden für die forstliche Praxis. Kilda-Verlag, Greven.

Arbeitskreis Forstliche Landespflege. (1994) Waldlandschaftspflege. Hinweise und Empfehlungen für Gestaltung und Pflege des Waldes in der Landschaft. Ecomed–Verlag, 2. Auflage, Landsberg.

Bichelmaier, F. (1969) Die Erholungsfunktion in der Raumordnung. Verlag Paul Parey, Hamburg and Berlin.

Ebert, W. (1993) Multifunktionale Forstwirtschaft in Frankfurt am Main. Unveröffentlicher Vortrag.

Gregorius, H.R., Hattemer, H.H., Bergmann, F., and Müller-Stark, G. (1985) Umweltbelastung und Anpassungsfähigkeit von Baumpopulationen. Silvae Genetica 34:230–241.

Hanstein, U. (1967) Über die Gewohnheiten, Ansichten und Wünsche der Waldbesucher. Allg Forstzeitschrift 22:H.27.

Krafft, U. (2002) Wald, Forstwirtschaft und Holz –Wahrnehmung in der Gesellschaft. Forst and Holz 57:614-617.

Küster, H. (1999) Geschichte der Landschaft in Mitteleuropa. C.H.Beck-Verlag, München.

Lösch, G. (1980) Typologie der Waldbesucher. Betrachtung eines Bevölkerungsquerschnitts nach dem Besucherverhalten, der Besuchermotivation und der Einstellung gegenüber Wald. Dissertation, University of Göttingen.

Mayer, H. (1992) Waldbau auf Soziologisch-Ökologischer Grundlage. Gustav-Fischer-Verlag, Stuttgart.

Rehfuess, K.E. (2000) Anthropogene Veränderungen von Waldböden – Folgerungen für die Bewirtschaftung. Forst and Holz 55:3–8.

Röhrig, E. (1980) Waldbau auf ökologischer Grundlage, Band 1. Verlag Paul Parey, Hamburg and Berlin.

Ruppert, K. (1960) Der Stadtwald als Wirtschafts- und Erholungswald. BLV Verlagsgesellschaft, München.

Ruppert, K. (1971) Zur Beurteilung der Erholungsfunktion stadtnaher Wälder. Dissertation, University of Göttingen.

Salisch, H.V. (1911). Forstästhetik. 3. Auflage, Berlin.

Scholz, F. (1986) Luftverunreinigungen- ein Umweltfaktor, der Forstgenetik und Forstpflanzenzüchtung vor neue Aufgaben stellt. Allgemeine Forstzeitschrift 41:1288–1294.

Stadt Frankfurt. (1993) Stadtwald in Zahlen. Forstamt der Stadt Frankfurt.

Zundel, R., and Roether, V. (1978) Verhalten und Wünsche von Erholungssuchenden in waldreichen Ferien- und Wochenenderholungsgebieten. Unveröffentlicht.

Zundel, R., and Völksen, G. (2002) Ergebnisse der Walderholungsforschung. Verlag Dr. Kessel, Oberwinter.

20
Modeling the Social Benefits of Urban Parks for Users

Giacomo Secco and Grazia Zulian

Monitoring quality of life has become an extremely crucial subject for discussion in both developed and developing countries. There are two main topics that stem from these debates: defining concepts such as quality of life, environmental quality, and sustainability; and developing methods and approaches to assess the quality of a living environment. These issues have been synthesized well by Irene Van Kamp (2003, 2004), who remarked on the need for a conceptual multidisciplinary and shared framework for approaching environmental quality and quality of life issues. She also pointed out the strong link between conceptual frameworks and assessment tools.

Here we focus on assessing the quality of public urban greening as a service for the population, since it is considered an important resource for maintaining the quality of the urban environment. We approach this issue in the same way as Pacione (2001, 2003), who stated that quality of life is the result of complex interactions between spatial, structural, and social factors. On the whole, it is recognized that the achievement of a high quality of life is correlated with the satisfaction of human needs, these being primary (food, housing, health, education) and secondary (recreation, higher education). Meeting these needs is strongly linked to social, economic, and cultural contexts, but, at the same time, it is recognized that the availability of public services is a key aspect of living quality, and that it is important to establish methods for evaluating the extent to which services can supply people's needs (Pacione, 2001; van Kamp, 2003).

Pacione (2001) proposed the concept of "collective consumption" (coined by Castells in 1979) for indicating the provision of an adequate supply of services to a person. Collective consumption refers to all of the goods and services that are free or not part of the market economy and that are provided for the community. Public services are crucial elements of a complex society. They provide strategic means of achieving social solidarity and fundamental human rights, because they are concerned with the social and geographic redistribution of provisions, without focusing on the monetary market. They are, in our opinion, essential for exercising citizenship and democratic rights. Dwelling places, education, health care, public transportation, culture, and recreational facilities are all common benefits and, when provided as community services, they represent a guarantee of shared projects and experiences.

M.M. Carreiro et al. (eds.), *Ecology, Planning, and Management of Urban Forests: International Perspectives.*

Our specific research interest is the development of a system that evaluates services provided to a city dweller by public urban green spaces, paying particular attention to the issue of green-space accessibility. Specifically, in this chapter we propose a model for analyzing the distribution of social benefits derived from one specific type of urban forest patch—public green spaces equipped for recreational activities. Urban and neighborhood parks, public gardens, and playgrounds play a crucial role in enhancing the quality of the urban environment by providing opportunities for people to meet, play, and otherwise benefit from publicly shared open space.

The multiple roles of urban forests and green spaces demonstrate the need for interdisciplinary approaches for evaluating their quality and capacity to provide benefits. Generally speaking, when analyzing urban greening contributions to people's quality of life, the relative surface area of green spaces in the city and public access to green space are often considered (assessed as total number of square meters (m^2) of green space that the public can access, for example). These coarse indicators of public benefit may be appropriate at the national or regional levels, but are not sufficient for estimating their value and ability to benefit the public at the local level. Our method for assessing the benefits of urban greening for the community is based on a "needs assessment" line of research, which emphasizes that public services ought to be located *where they are most needed* and not only where they are lucrative (Pacione, 2001). The method can also be used to analyze other community services, such as health services, and cultural or educational facilities. Our long-term goal is to provide a model that can be used to inform and improve the process of urban planning for the allocation of basic public services, particularly those involving health care, educational and cultural opportunities, and recreational facilities. In this chapter, we describe our method and evolving model for evaluating park availability to the residents of a city.

Our Conceptual and Methodological Approach

The Social and Spatial Structure of Urban Green Spaces and Parks

To optimally benefit a city's residents, public urban green space should be considered and assessed as a system with an interconnected network of different categories of green space. From a social point of view, we can identify two main groups of green areas (Gobster, 2001; Kit Campbell Associates, 2001):

- Spaces with clearly defined and diversified functions and facilities, for example:
  - Urban and neighborhood parks and gardens
  - Playgrounds
  - Green corridors

- Spaces characterized by zones where the focus is on the equal distribution of natural (or semi-natural) resources; these green spaces lack a designated purpose and are often located in areas that cannot be reached by many users

Here we focus on public urban parks and gardens that are essential for the well-being of people as environmental, landscape, and social resources. These are locations where people can meet, benefit from being in the open air, and play. Strategies for the management of these kinds of public green spaces (Tjallingii, 1996; Schrijnen, 2000) should take into account the following factors and their interrelationships:

- Greening
  - Planning, management, and maintenance of green spaces
  - Needs of the communities (children, elderly people, less favored groups)
- Mobility of users and available transportation system
  - Residential area planning

Therefore, a balanced and efficient management of the system involves broader urban policies and a need for methods of analyzing all the aspects involved in the provision of green space. In fact, the *quality* of the green-space system should be measured not only using coarse quantitative indices (for example, m^2/inhabitant), but also by assessing the distribution of sites, the quality of supply capacity, and their degree of integration with the social context of their neighborhoods.

Constructing a Model for Describing a Park's Social Profile

The main goal of our approach is to develop a methodology for analyzing the benefits of urban green areas based on their social functions. The model is constructed in two stages:

1. Analysis of site capacity to provide social services
2. Spatial analysis of site accessibility

In this section we illustrate these stages of analysis and describe how this model was applied in the towns of Padova (Padua) and Piove di Sacco (Italy).

Generally speaking, we begin our analysis by studying the relationship between people and parks in the town. Therefore, when assessing green-space provision within an urban context, it is important is to assess people's needs, perceptions, and desires. For instance, using questionnaires, interviews, and focus groups, Coles and Bussey (2000) assessed the value of urban woodlands based on the social meanings they had for their users. Burgess et al. (1988a–c) studied attitudes of park users by organizing focus groups and providing structured questionnaires and open interviews to set up a framework for green-space planning. Oguz (2000) conducted a survey on three parks in Ankara (Turkey) to explore the characteristics of park users and user satisfaction, and to report problems and requirements. Paul Gobster (1991, 1998, 2001) examined user perception and activities, using both qualitative and quantitative techniques. Finally, De Vries (2002) and De Vries et al. (2001) integrated spatial analyses and social surveys to analyze the relationships between urban greening and quality of life measures.

To examine the urban green space needs of a local community, we conducted two social surveys in Padova. In 2001, we interviewed 300 parks users about their activities in the parks, as well as their habits, mobility, and socioeconomic status. During the summer of 2002, we carried out a second survey in Padova in which 550 questionnaires were used to evaluate the distances traveled to and from the parks by different groups of people (Zulian, 2003a,b). The data gathered were then used to set the quantitative parameters of our model with respect to urban context and people's needs.

Site-specific conditions, traditions, and activities vary greatly depending on individual perception and a park's sociocultural context. Based on a review of the literature (Gobster, 1991, 1998, 2001; Di Fidio, 1993; Goossen and Langers, 2000; Kit Campbell Associates, 2001; Gambino, 2002; Park and Open Spaces Association of Japan, 2002), as well as the results of our social surveys, we formulated the following criteria for determining a site's potential for providing quality services:

- All sites provide at least a minimum degree of service to the entire population.
- The ability of an area to attract its users is tightly linked to
  - the type and the extent of services it can provide, and
  - the distance from home.
- Specific factors, external to the park, are also important for understanding the relationship between people and parks; these include
  - organization of the transportation infrastructure and public transportation system,
  - urban zoning, and
  - presence of impassable barriers (e.g., rivers, train tracks).

User age is also a significant variable in determining a park's quality for users, since it affects user mobility, autonomy and the attitudes concerning the level and type of services in the parks, and specific needs concerning activities and facilities. That's why we classified the users into eight age groups that varied with respect to user autonomy in reaching the park, their degree of physical mobility, and the types of recreational activities and facilities desired. All of these factors play a role in determining how users perceive park quality. Table 20.1 describes the main characteristics of the age groups used in our study.

Our model has two main components. The first rates site capacity to provide services and social opportunities, and the second analyzes spatial accessibility to parks and gardens. In the following subsections we describe these two model components in more detail. Then we present two case studies in which the method has been applied.

Table 20.1 Characteristics of the age groups used in our study

Age	Description
<5	Young children: completely dependent on parents; visit the park accompanied by adult who continually watches and assists child in play
5–11	Children in their first years of school: limited autonomy and mobility
12–14	Young teenagers: visit the parks in groups; need equipment for sports and group activities; mobility still limited, but biking allows greater autonomy
15–18	Older teenagers: less limited autonomy; frequently participate in sports and recreational activities
19–40	Young adults: often with young children; some visit the parks alone or in groups for recreational and sports activities
41–60	Middle-aged adults: not limited in mobility or autonomy; mostly need to relax and benefit from open air; use sports trails, especially during weekends and in summer
61–75	Retired adults: greater free time to visit parks, often for less active recreational activities (socializing, reading the paper, visiting with friends); mobility problems can occur
>76	The elderly: often have major mobility problems, need green space near home

Rating Site Capacity

We assume that each park has a specific social value, one that varies according to the user groups considered. To evaluate the site's provision of services the following concepts were developed:

- Nominal value
  - Assessed using the elements that identify the type of service and that are essential for providing the service (for parks and gardens the nominal value is based on usable green area)
- Added value
  - Changes in the nominal value of a site due to the presence of specific facilities and contextual factors
  - Equipment and facilities increase the value of the site according to their quantity and type
  - Contextual factors inside or surrounding the site change the value of a site; if conditions are favorable, value is added to the nominal value (>1); if unfavorable, value is subtracted (<1); three main factors are considered in this category: accessibility, recreational opportunities, and safety
- Social value: the value assigned to a site that considers the amount of green area, the facilities, the equipment, and contextual factors; social value is expressed in service units; Figures 20.1 and 20.2 show the factors that are used to rate the social value of parks.

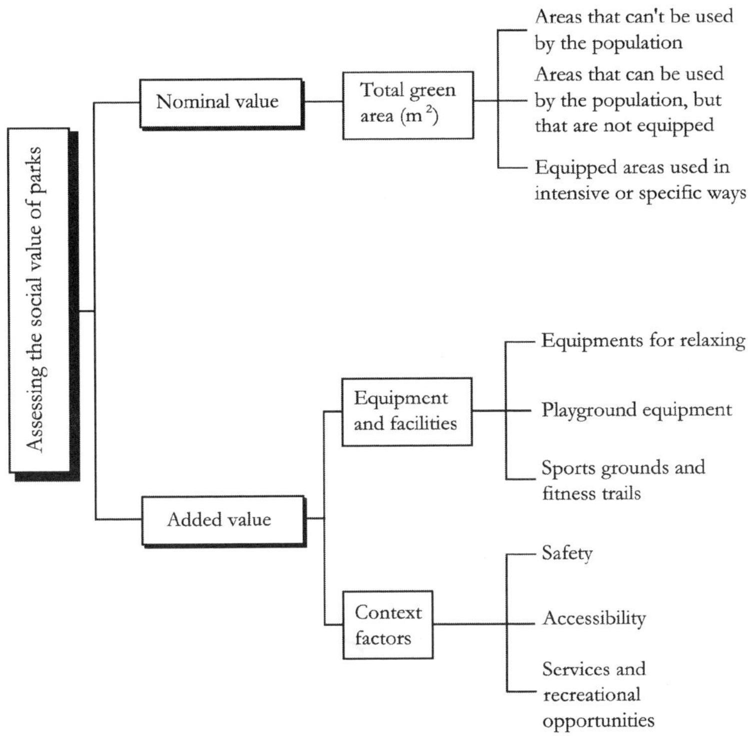

Fig. 20.1 Factors used to assess the social value of parks

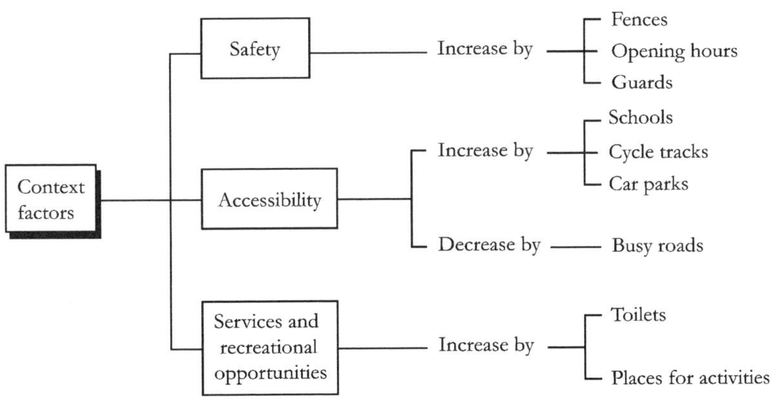

Fig. 20.2 Contextual factors that increase or decrease the social value of parks

- Service unit (S.U.): measure of the social value, corresponding to the value of 1 m^2 of green space area without facilities; the S.U. varies in relation to the presence of facilities and equipment (counted and then assessed using their weighting coefficients), as well as to contextual factors (estimated as percentage values and assessed using their weighting coefficients)

Parks and public gardens play an important role in the urban environment as a place for social, recreational, and cultural exchange. All the activities, their organization, and their effectiveness in enhancing the quality of the surroundings and of the broader urban environment depend on an interrelated ensemble of circumstances. We assume that the park's social value depends on its facilities and on particular contextual factors that have different meanings depending on users needs, desires, and concerns. Therefore, we established specific weighting coefficients for each factor by age group in our quantitative model of calculating a site's capacity to provide services (Table 20.2). These coefficients were determined by analyzing the social surveys proposed in Padova and St. Etienne (Zulian, 2003a,b). Before this model can be transferred to other cities, these coefficients may have to be recalculated to reflect each community's feelings, needs, desires, and concerns about urban greening and neighborhood quality of life. We welcome and encourage applied studies using our model in other cities so as to improve our understanding of how parks are used and understood by people in different cultures and environmental contexts.

As shown in Tables 20.2 and 20.3, we classified three main types of green areas, different types of equipment for recreational activities, and three contextual factors.

- Green areas:
- Areas that cannot be used by the population due to their morphological characteristics or vegetation type
  - Areas mainly serving environmental and landscape functions

Table 20.2 Weighting coefficients for assessing green areas and equipment

	Coefficients by age group (years)							
Park elements	<5	5–11	12–14	15–18	19–40	41–60	61–75	>76
Green area (m^2)	0.6	0.8	1.0	1.0	1.0	1.0	0.8	0.6
Play grounds (m^2)	1.3	1.3	0.0	0.0	0.0	0.0	0.0	0.0
Unusable areas	0.1	0.1	0.1	0.1	0.1	0.1	0.1	0.1
Benches	10.0	20.0	30.0	60.0	90.0	90.0	180.0	200.0
Tables	25.0	50.0	100.0	250.0	225.0	225.0	450.0	500.0
Rubbish bin	17.5	12.5	12.5	20.0	20.0	20.0	22.5	25.0
Complex game	1500.0	1500.0	0.0	0.0	0.0	0.0	0.0	0.0
Single game	600.0	420.0	0.0	0.0	0.0	0.0	0.0	0.0
Exercise trail	0.0	60.0	180.0	600.0	600.0	240.0	30.0	0.0
Playgrounds	0.0	600.0	4000.0	4000.0	2600.0	600.0	0.0	0.0

Table 20.3 Percentage increase to nominal scores for different contextual factors

Factors	Characteristics			Increase
Accessibility				
Cycle tracks	Proximity (within			4%
Schools	a defined distance)			4%
Car parks	Proximity (within			4%
Bike parks	a defined distance)			4%
Busy roads	Absent within a defined distance			4%
Safety				
	Absent			0%
			Not effective	3%
Fences		Open	Effective	18%
	Present		Without guards	42%
		With opening hours	Guards	60%
Services				
Restrooms				12%
Places for activities				3%

- Areas that can be used by the population, but that are not equipped
  - Areas having primary functions that are general
  - Areas important for adults, teenagers, and those who do not need particular equipment
- Equipped areas used in intensive or specific ways
  - Areas having specific primary functions
  - Areas important for specific user groups
  - Areas having facilities and activities that should be determined by the target user group's characteristics

To estimate a green area's social value for adults and teenagers (the user groups that can mainly use it), we assigned to it a value equivalent to its area in square meters. The weight coefficients chosen for the other user groups vary according to their propensity to profit from them. Elderly people, for example, do not fully benefit from green areas lacking benches or other such basic equipment. To assess the social value of recreational and sport facilities, we counted and weighted them according to the importance estimates for the different user classes. We considered the following facilities:

- Recreational facilities
  - Benches, tables, and rubbish bins
  - Those items mainly important for people who are just looking for a place to sit or walk (elderly people, senior)

- Games
  - Simple games, important for young children (ages 0 to 6 years), such as sandboxes, swings, teeter-totters, slides, merry-go-rounds
  - Complex games, providing play activities for older children (ages 7 to 12); for example, play structures (play forts, play houses, complex sliders), gyms
- Sports equipment
  - Exercise trails, where exercises are combined with walking or jogging (we count the number of exercise stations); playgrounds for noncompetitive, non-agonistic activities (we count the number of play grounds)

We assessed the following contextual factors:

- Accessibility
  - All the elements of the park and it surroundings that allow user access
  - Biking/pedestrian trails, bike racks, streets with little traffic, schools within a determined distance; these factors are extremely important for the user classes that are less mobile
- Safety
  - All the elements of the park and it surroundings that are involved with a serene (easy-going) use of the park
  - Controlled and patrolled areas; those with restricted hours of operation, fences
- Recreational possibilities and general services
  - All the facilities of the park that accommodate the practice of cultural and recreational activities (meeting rooms, for example)

The social value of a park also varies according to the percentage of scores set for the contextual factors. Table 20.3 synthesizes the percentages for these scores by category, and Table 20.4 shows the differences in their weighting coefficients in the model by age group.

Finally, to evaluate the highest potential social value for a particular urban park, we need to identify the functional characteristics of that park that can ideally supply all the user groups (Table 20.5), calculate its current and highest possible social value (with respect to its circumstances), and then compare

Table 20.4 Weighting coefficients for assessing value of contextual factors for a park by age group

	The greatest percentage variation by age group (years)							
Factor	<5	5–11	12–14	15–18	19–40	41–60	61–75	>76
Safety	60.0	60.0	30.0	24.0	24.0	24.0	42.0	60.0
Accessibility	14.0	20.0	20.0	20.0	20.0	20.0	16.0	12.0
Services	15.0	12.0	10.5	7.5	7.5	7.5	12.0	15.0

Table 20.5 Functional characteristics of a park that can supply all the user groups and representation of their social value

Categories	Aspects	Amount	Notes
	Total area	80,000	Total dimensions
Lawn areas (m^2)	Zone with lawn areas	75,000	Zones equipped for sports and leisure activities
	Playgrounds	5000	Zone reserved for children only
Facilities (units)	Benches	50	
	Tables	12	Simple equipment
	Trash Cans	20	
	Complex games	5	3 play structures, 2 gyms
	Simple games	20	1 sandbox, 5 swings, 8 teeter-totters, 3 slides, 3 merry-go-rounds
	Exercise trail	24	Number of exercises
	Sports grounds	2	Basketball, volleyball
	Safety	100	Controlled areas, hours of operation, enclosures
Contextual factors (score in percentage)	Accessibility	100	Biking/pedestrian trails, bike racks, streets with low traffic, school within 150 m
	Services	100	Restrooms, meeting rooms

each real park with the ideal one. The functional and structural characteristics of the "ideal park" must be determined at the start of the study when choosing which elements will be used to assess the social value of parks in a particular urban context.

The park's social value for each user group can be represented graphically (Fig. 20.3). The social value is expressed as ratio with the total green area of the park (numerator). We can then compare the following:

- Social value assessed for the eight age groups (each sector)
- Social value of each group with the social value of a park that is able to supply all the users' needs (the gray circle in Fig. 20.3)

In this example (Fig. 20.3), the social value of the park is highest for users in the 5- to 11-year age group and is 2.33 times greater than the value of the park based on green area alone. We can then compare the level of service among age classes or among different parks using the scale proposed in Table 20.6.

Spatial Analysis of Parks

Ideally, the availability of parks and their recreational equipment should provide equal opportunities for a city's residents. However the distribution of public services, including parks and public gardens, depends on urban morphology, pre-existing urban structures, and political decisions both past and present. Simple

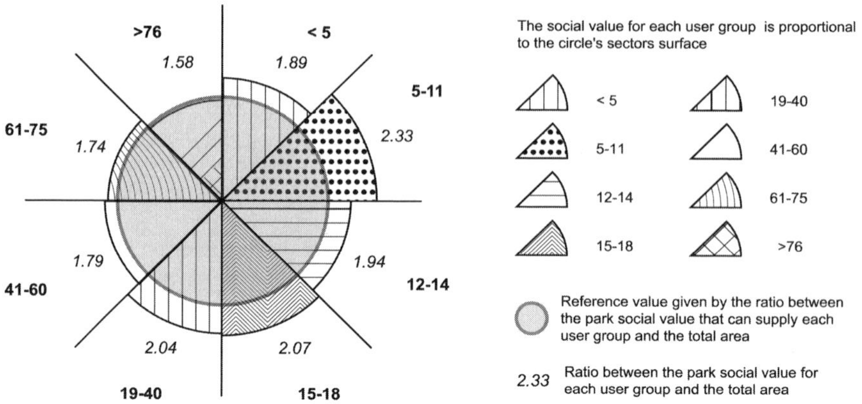

Fig. 20.3 A graphic representation of the social value of a park assessed for each user group by age. In this example, the social value of the park for children between ages 5 and 11 is the highest (2.33 times greater than the park's value based on area alone)

Table 20.6 The ratio of a park's social value to its the green area without equipment or contextual factors

The ratio of a park's social value to its green area	Score
>1.9	Very good
1.7–1.9	Good
1.2–1.6	Scarce
<1.2	Insufficient

quantitative indices (e.g., m^2 of public space per person) for evaluating the availability of urban public services do not sufficiently capture the following factors:

- The pattern of parks across a town
- The type of services offered in each park and their spatial distribution across town
- The suitability of a park's opportunities according to the needs of potential users
- The urban neighborhoods that need or are deficient in public parks

To identify the zones that need public parks, the second component of our model involves capturing the existing spatial interaction between public parks and their users and expressing them as indices that are more context specific. Evaluating a site's available services, its location in the local landscape, and the location of potential users facilitates identification of the zones or neighborhoods that are most deficient in services for each user group. Proposals can then be developed to bring park services into better alignment with the needs and demographics of the residents in that section of the city.

Spatial interaction represents the *estimated flow between locations*, a movement of people, freight, or information (Rodrigue et al., 2006). Models of spatial interaction measure flows and predict the consequence of changes in the conditions generating flows. Each spatial interaction is composed of an origin-and-destination pair, and flows between them are a function of the attributes of the location of origin, the attributes of the destinations, and the level of friction on flow rates imposed by the distance between the origin and the destination. So, in order to develop a spatial interaction model to assess the flow between people and parks, we need to consider the characteristics of the people (origins), the attributes of parks (destination), and the cost of the friction.

A further consideration in constructing spatial models is the degree of spatial desegregation in data, and the spatial units used to map results. These must be defined for each study. Handy and Niemeier (1997) define spatial desegregation as the spatial level (the scale) in which flows are measured. Therefore, accessibility can, in fact, be assessed by working at the individual level (Hanson and Schwab, 1987), at the household level, at the census block level, and so on. The smaller the zone or scale, the greater the desegregation and the precision in the measurements. This scaling issue in the field of zone analysis is better known as the modifiable area units problem (MAUP), and arises "from the imposition of artificial units of spatial reporting on continuous geographical phenomena resulting in the generation of artificial spatial patterns" (Heywood, 1998). The MAUP occurs when data cannot be measured as a single point, but need to be "contained" in spatial units in order to be measured or represented (as socioeconomic or epidemiological data, for example). So the main problems are known as the *scale effect*, that is, the variation in numerical results due to the number of zones used in the analysis. Moreover, the *zonation effect* is the variation in numerical results arising from the grouping of small areas into larger units (Armhein, 1995). For more information about approaches and methods for analyzing accessibility in the social sciences, see Bailey and Gatrell (1995), Kwan (1998, 1999), and Miller (2006).

To analyze the interaction between people and parks, the following aspects were taken into account in our study:

- Residents
  - Localized at the household level (the residence is assumed as the point of departure/arrival of trips)
  - Divided into age groups and according to their behaviors and needs (see Table 20.2)
- Parks
  - Characterized using the social value calculated for each age group to assess the capacity of parks to attract the users
- The flow estimated between the two locations as the distance between origin (households) and destination (park entrances)

  - Flow assessed using Euclidean distance has been assumed to be an appropriate trade-off for estimating the flow cost at the urban level of analysis (Apparicio et al., 2003)
- Due to MAUP problems, results were represented at the census block level

To model user mobility behaviors, we employed an inverse logistic function (Lando and Zanetto, 1991), modified to attribute the value of 1 to the origin (Secco, 1994) and to increase correlation with the experimental data. The equation is as follows:

$$f(d)=\frac{1+K}{K+e^{\alpha d^{0.5}}} \tag{1}$$

where:

- d represents the distance between the park and the resident
- α and K are the parameters of the function, specific for each user group, varying with the social value of the parks as follows:
  - α:
    - The size parameter of the function; it determines the inflection point's location
    - As α decreases, the inflection point moves to higher values of d
  - K: the shape parameter of the function
- e: the base of the natural logarithm

To obtain results that relate to the real behavior of the population, we need to adjust those parameters using experimental data. After analyzing the data we gathered using the social surveys, we found the following:

- The shape of the function is almost constant as the social value of the service and population age is varied (e.g., the ratio of the distance d_{10} where the function is reduced to 10% and the distance d_{70} where it is reduced to 70% is always 3.5).
- A relative user index can be calculated. The y-axis unit represents $f(d)$ or the relative fruition of the park (fruition = use index, the ratio of real users to potential users of the park determined by calculating the ratio between the fruition at the distance d and the fruition at distance 0). Figure 20.4 represents the function obtained by plotting data for age group 0 to 11 years for a middle-sized park in Padova.
- The amplitude of the function varies positively and strictly as the logarithm of park social value increases. The higher the social value of the park, the greater its capacity to attract the users.

We also found that the function of distance traveled with age is a bell-shaped curve, with the less mobile user groups (children and the elderly) having lower values for distance traveled, as expected. By interpolating the experimental data

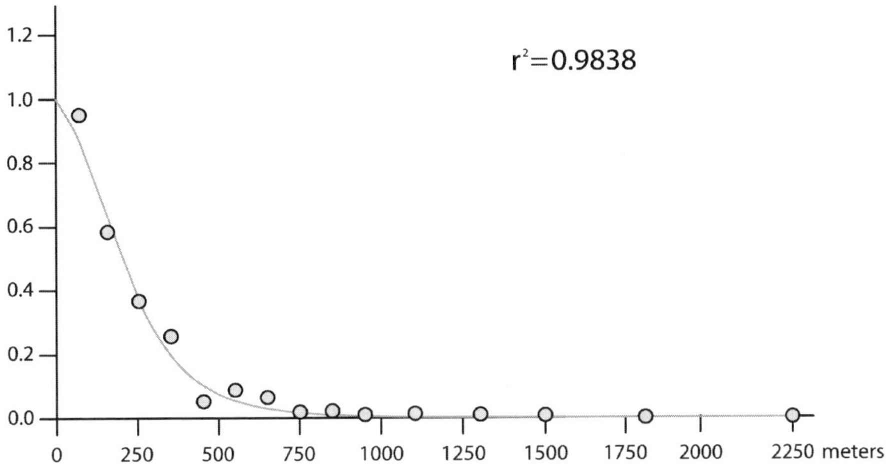

Fig. 20.4 An example of a distance function based on data from the 0- to 11-year age group

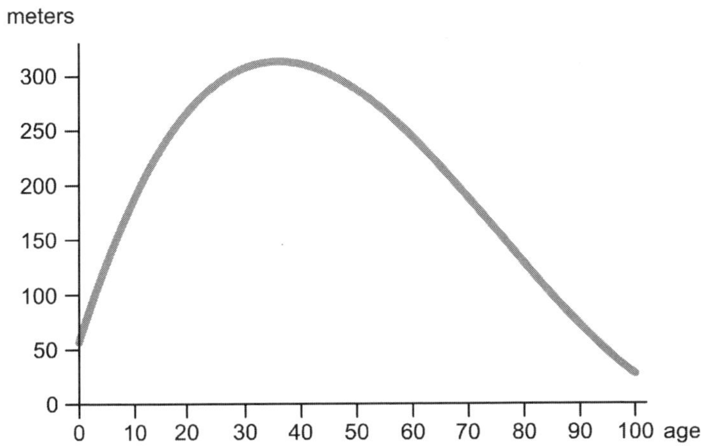

Fig. 20.5 An example of how the distance function varies with age of the park user

gathered from the surveys, we obtained a model of the variation according to the age of users (Fig. 20.5). To calculate the services offered to the population, the social value of each park was allocated among all citizens, giving each one an amount proportional to a distance function. The function parameters are adjusted in relation to the user age to account for their different mobility.

Considering the allocation of each park (park r) among all residents, the next steps are followed:

- Calculate $w_{r,s}$, a weighting factor that varies depending on the distance between a person's home (or point of origin) and the nearest entrance of park r:

$$w_{r.s} = f_c(d_{r.s}) \quad (2)$$

where:
$d_{r.s}$ is the distance between park r and the home
$f_c(d)$ is the distance function, calibrated for the user group c.

- Assuming that $w_{r.tot}$ is the aggregated weighting factor of all citizens, the quota $q_{r.s}$ of park r allocated to its citizen*s* is as follows:

$$q_{r.s} = Q_{r.c} \frac{w_{r.s}}{w_{r.tot}} \quad (3)$$

where $Q_{r.c}$ is the social value of the park for user class c.

- The contextual value of park r is given by the sum of the values allocated to each individual:

$$Q_{r.cont.} = \sum_{i=1}^{n} q_{r.i} \quad (4)$$

where n is the number of residents.

- The aggregated quota of service for a person s is given by the sum of quotas derived from all parks:

$$q_s = \sum_{j=1}^{np} q_{j.r} \quad (5)$$

- where np is the number of parks.

Results of Applying the Model in Two Case Studies

Applications of the method are presented in two case studies. The goal for the first study was to analyze the social impact of the renovation of a riverbank tract in the town of Padova (Italy). The second study assessed the services offered by parks and gardens in Piove di Sacco (Italy) (Fig. 20.6).

In 2003 to 2004, a study was conducted in Padova to assess the social impact of the possible renovation of the riparian areas of a portion of the urban hydrographic system. The main goals of our project were to optimize the following factors:

- Providing public access to the riparian zones, the riverbanks, and nearby green areas
- Designing a system of pedestrian/biking trails that connects to the surrounding biking network
- Providing management criteria for the riparian areas (which in Italy are controlled by a state body)

Fig. 20.6 A map of Italy showing the locations of the city of Padova and the town of Piove di Sacco

- Providing for a wide range of activities:
  - Recreational activities of mild impact that do not require equipment or disturb the landscape
  - Recreational activities such as sports activities to be concentrated along already equipped axes

The particular morphology of the hydrographic system renders the plan important at the district level, the urban level, and the suburban level. Figure 20.7 shows the town of Padova and the structure of the river network system, and Figure 20.8 shows the location of the parks and public gardens in the planning zone.

For general improvement of the site's condition, particular attention was paid to the following issues:

- Management of the areas reserved for bike and pedestrian trails
- Provision of adequate park equipment
- Restoration of the autochthonous (native) vegetation
- Management of riverbank vegetation to mitigate bank erosion

Herein, we describe the analysis methods selected for the social components of the model. In addition to a study of the street network, urban structure, and the

Fig. 20.7 The riverbank planning zone area within Padova, Italy

typology and structure of the vegetation, the social value was assessed for each area of public green space for all of the user groups using the methods discussed previously. The social value was assessed for the already existing and equipped areas, and also for the newly created sites. Our analysis also went as far as to simulate the social value of each site for each of the different management strategies proposed. The graphic summary was included in the cartography at the end for reference to the operations discussed during phases of general design and precise planning of the individual interventions. Figure 20.9 is an example of a map created for the discussion of the proposed plan. In addition to the traditional thematic cartography, we presented the graphics showing the real social values of parks and the social value estimated after the interventions (Fig. 20.10).

In Padova the entire method could not be applied because some data were lacking. To present an example of the spatial analysis model, we include our study of the urban greening of the municipality of Piove di Sacco, a small urban area located

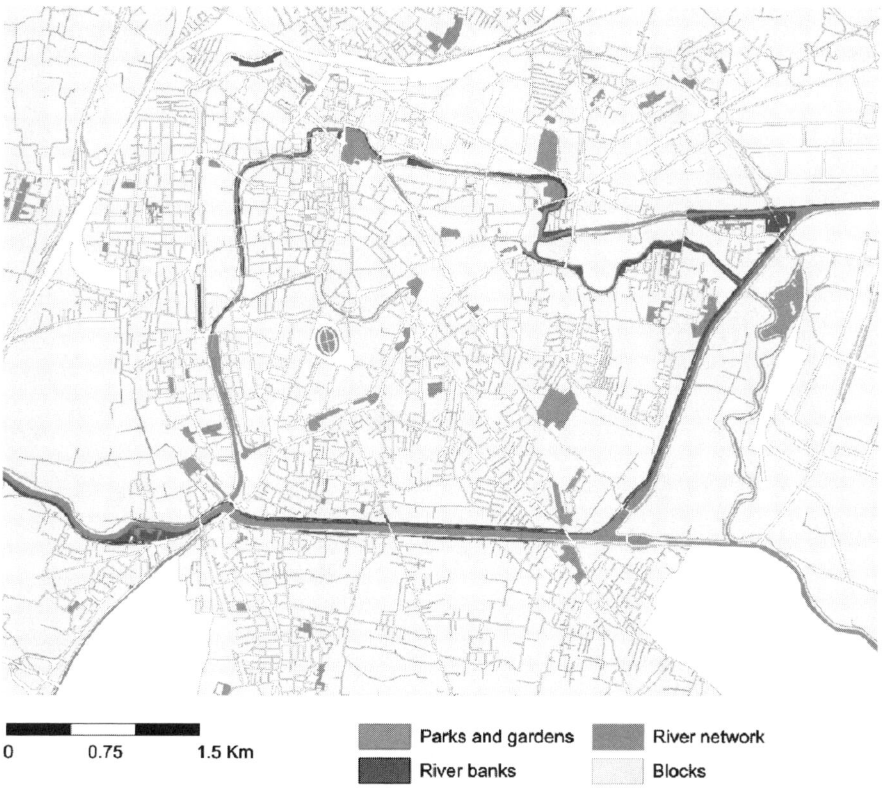

Fig. 20.8 A close-up of the planning zone showing the distribution of parks and public gardens within the city of Padova, Italy

near the city of Padova. The municipality manages 515,384 m^2 of public greens, 138,125 m^2 of which we found to be equipped parks. According to Italian norms, Piove di Sacco should have 420,000 m^2 of equipped greens, that is to say, 200% more than it currently offers.

To point out the main areas deficient in services and to plan a strategy of intervention, the two phases of our methodology were applied (evaluation of sites' ability to provide services and a spatial analysis of the accessibility of the parks). Among all 40 of the existing parks, only one site was able to provide quality services for all of the user groups: the Bosco di Pianura (about 50,000 m^2). Moreover, the attractiveness of this park was further decreased since it was located near an industrial zone and was far from the city center. The other areas were not able to satisfy the demands of the existing residential population, specifically those of the user groups classified as weaker in terms of mobility (0 to 5 years and >70 years). This is shown in Figure 20.11, where the same zones service fewer children ages 0 to 5 years than the entire population.

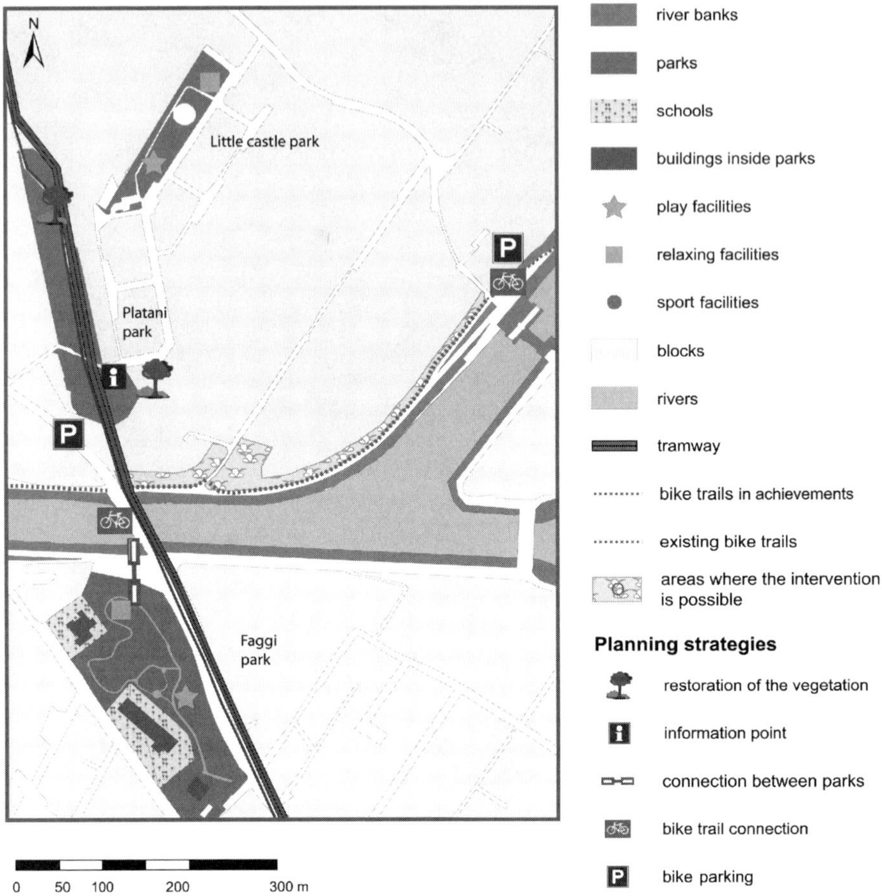

Fig. 20.9 Specific management strategies for Padova, Italy

Our analysis indicates that Piove di Sacco needs to take action to avoid patterns of inequality in the distribution of urban green spaces. One strategy that could be adopted is taking into account public open spaces that are available and their local context (land use, population density, main building typologies, lines of action stated in the general town planning scheme). Our modeled output has been suggested as another strategy for avoiding this undesirable situation. First, we recommended improvement of services in the existing sites by increasing facilities and equipment. We also proposed that 20 existing areas be equipped with facilities for the weaker user groups. Moreover, we identified 10 free open spaces (adding 98,700 m^2 to the town's green space) that could be converted into new public parks and equipped to satisfy specific user groups. Lastly, we suggested improving the social value of two existing parks by adding new facilities and extending them to a total area of 42,000 m^2.

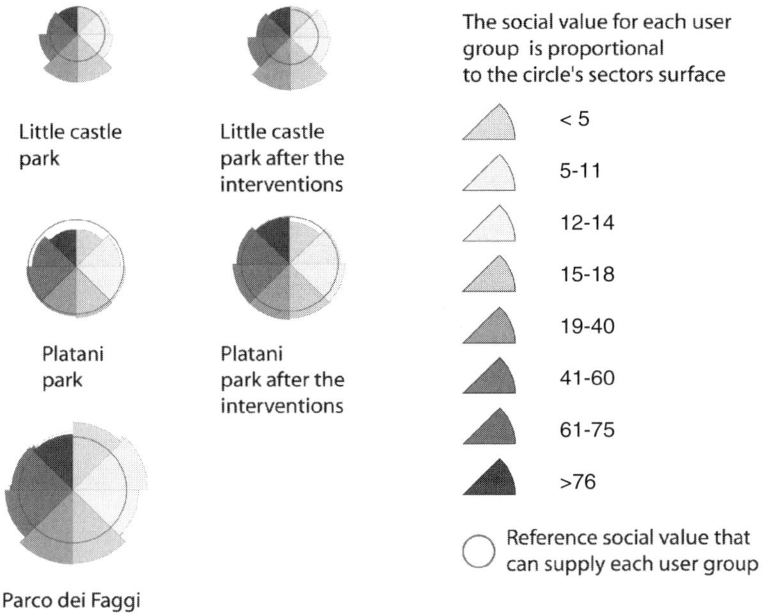

Fig. 20.10 Graphics representing the real social values of parks and the social value estimated after the interventions for the parks represented in Figure 20.9

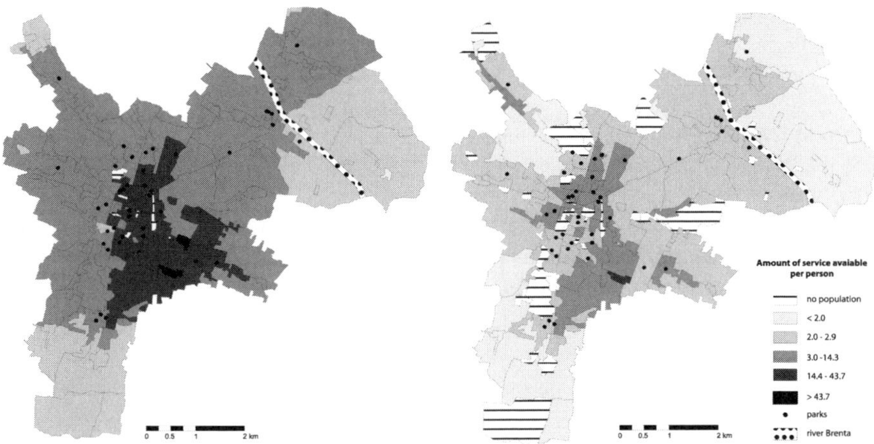

Fig. 20.11 Comparisons of services offered to the total population (left) and to children (right) in the town of Piove di Sacco, Italy

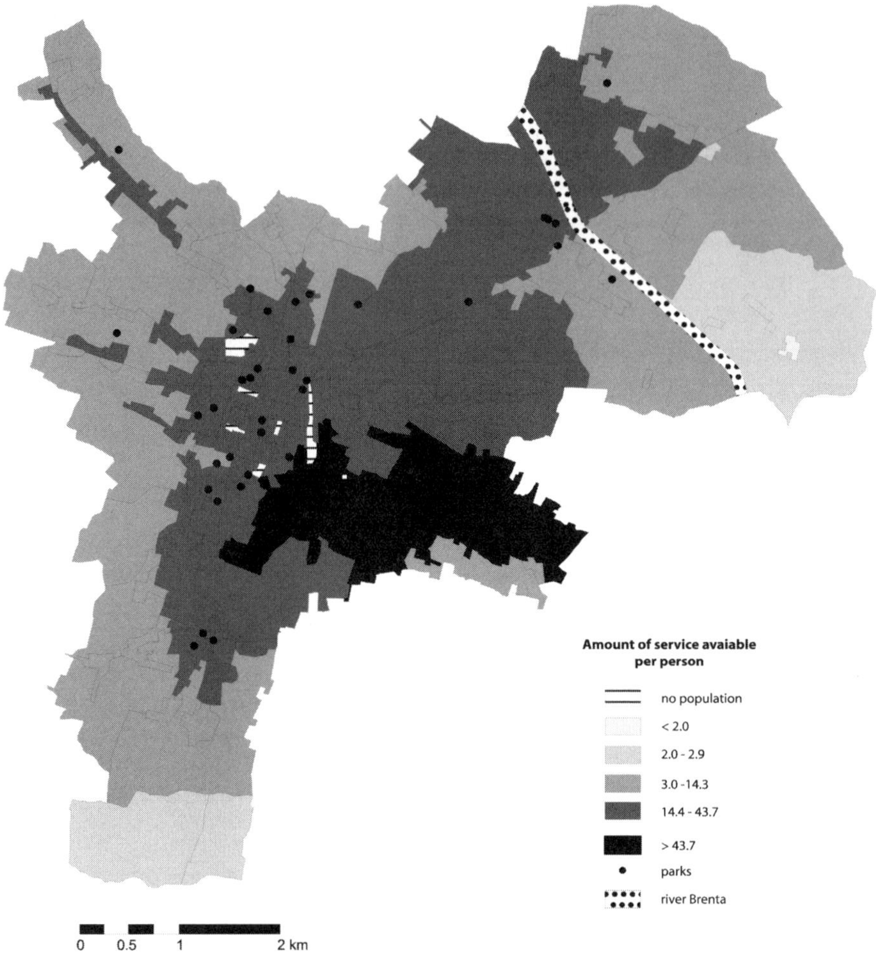

Fig. 20.12 Modeled simulation of the improved services offered to the total population of Piove di Sacco, Italy by increasing equipment in the parks. Compare with Figure 20.11, left panel

This new system could satisfy all users groups, changing the services available and its spatial dynamics. The areas in the northeast and in the center of town would provide more services for children and teenagers. The eastern area would also improve, but still not sufficiently to satisfy three residential settlements planned in the near future (Fig. 20.12).

Conclusion

The urban public greening system is an ecological, economic, and social resource for society. However, the presentation of multifunctional and multiscalar dynamics to describe such a system's value has proven to be very difficult, since the planning and the management of a site is very complex and in a constant state of flux. Therefore, to develop a sustainable planning policy for green-space resources, it is very important to use a holistic, integrated, and multidisciplinary approach. The proposed method described herein predominantly focuses on the social aspects of green-space value that deal with the dynamics between parks and public gardens and users. It is clear that a close examination of the demands, behaviors, and attitudes of a population is imperative for the balanced organization of a service that, in densely urbanized areas, can acquire a wide range of social functions and benefits. These include outdoor recreational activities, participation in cultural encounters and initiatives, and the regaining of a relationship with the natural world (despite great human impact), not only through visitation but also through environmental education and horticulture.

Our method confronts these planning and management issues by considering and attempting to quantify two important aspects of a green-space network's social value: the ability of these sites to provide services and their spatial dynamics. We continue to refine and improve our method and model. For example, the criteria for differentiating user groups need to be revised. Although age does prove to be a good descriptive variable and a datum that is also easy to collect and to work with in disaggregated terms, it is not always sufficient for taking into account the behaviors of each individual. Talen (2002), Talen and Anselin (1998), and Pacione (1986) based their analyses on the socioeconomic characteristics of a population, while others have focused their attention on differences due to ethnicity. With respect to the ability to provide services, currently this is measured almost exclusively using functional characteristics, without evaluating the importance of vegetation or fauna. We also recommend that the spatial analysis method for determining accessibility should be revised to examine the importance of different transportation networks and infrastructure as well.

References

Apparicio, P., Shearmur, R., Brochu, M., and Dussault, G. (2003) The measure of distance in a social science policy context: advantages and costs of using network distances in eight Canadian metropolitan areas. Journal of Geographic Information and Decision Analysis 7(2):105–131.

Armhein, C. (1995) Searching for the elusive aggregation effect: Evidence from statistical simulations. Environment and Planning [A] 27(1):105.

Bailey, T.C., and Gatrell, A.C. (1995) Interactive Spatial Data Analyses. Longman, Harlow, Essex, England.

Burgess, J., Harrison, C.M., and Limb, M. (1988a) Exploring environmental values through the medium of small groups: 1 theory and practice. Environment and Planning [A] 20:309–326.

Burgess, J., Harrison, C.M., and Limb, M. (1988b) Exploring environmental values through the medium of small groups: 2 illustrations of a group at work. Environment and Planning [A] 20:457–476.

Burgess, J., Harrison, C.M., and Limb, M. (1988c) People, parks and urban green: a study of popular meaning and values for open spaces in the city. Urban Studies 25:455–473.

Coles, R.W., and Bussey, S.C. (2000) Urban forest landscapes in the UK, progressing the social agenda. Landscape and Urban Planning 52:181–188.

De Vries, S. (2002) Heath benefits of more natural living environment. Paper presented at the IUFRO European Regional Conference: Forestry Served Urbanised Society. Agosto, Denmark, Copenhagen, pp. 27–30.
De Vries, S., Verheij, R., and Groenewegen, P. (2001) Nature and health, the relationship between health and green space in people's living environment. Proceedings of the 12th ELRA Congress, Cultural Events and Leisure Systems: Sense of Identity and Sense of Space, April 18–22, pp. 72–86.
Di Fidio, M. (1993) Architettura del Paesaggio, Criteri di Pianificazione e Costruzione. Pirola Editore, Milan.
Gambino, R. (2002) Le acque come struttura portante del verde. In: De Biasio, C.L. (ed.) Il Sistema del Verde Urbano, Elementi di Riconversione Ecologica della Città. Il Poligrafo, Padua, pp. 63–87.
Gobster, P.H. (1991) Forest vegetation in urban parks: perceptions of inner city children. In: Vander Stoep, G.A. (ed.) Proceedings of the 1991 Northeastern Recreation Research Symposium, April 7–9, New York, State Parks Manegement and Research Institute, University of Massachussetts, Amherst, MA.
Gobster, P.H. (1998) Urban parks as green walls or green magnet? Interracial relations in neighbourhood boundary parks. Landscape and Urban Planning 41(1):43–55.
Gobster, P.H. (2001) Neighbourhoods—open space relationships in metropolitan planning: a look across four scales of concern. Local Environment 6(2):199–212.
Goossen, M., and Langers, F. (2000) Assessing quality of rural areas in the Netherlands: finding the most important indicators for recreation. Landscape and Urban Planning 46:241–251.
Handy, E., and Niemeier, L. (1997) Measuring accessibility: an exploration of issues and alternatives. Environment and Planning [A] 29:595–613.
Hanson, S., and Schwab, M. (1987) Accessibility and urban travel. Environment and Planning [A] 19:735–748.
Heywood, I. (1998) Introduction to Geographical Information Systems. Addison Wesley Longman, New York.
Kit Campbell Associates. (2001) Rethinking Open Space. Open Space, Provision and Management: A Way Forward. Scottish Executive Central Research Unit, Edinburgh. http://www.scotland.gov.uk/cru/open_space.pdf.
Kwan, M.-P. (1998) Space-time and integral measures of individual accessibility: a comparative analysis using a Point-based framework. Geographical Analysis 30(3): 191–216.
Kwan, M.-P. (1999) Gender and individual access to urban opportunities: a study using space-time measures. Professional Geographer 51(2):210–227.
Lando, F., and Zanetto, G. (1991) La dinamica territoriale dell'industria veneta. In: Bernardi, R., Zanetto, G., and Zunica, M. (ed.) Il Veneto. Diversità e Omogeneità di una Regione—Emergenze Territoriali e Socioeconomiche. Pàtron Editore, Bologna, pp. 11–58.
Miller, H.J. (2006) Place-based versus people-based accessibility. In: Levinson, D., and Krizek, K.J. (eds.) Access to Destinations: Rethinking the Transportation Future. Elsevier, New York. http://www.geog.utah.edu/~hmiller/research.html.
Oguz, D. (2000) User's surveys of Ankara's urban parks. Landscape and Urban Planning 52:165–171.
Pacione, M. (1986) Quality of life in Glasgow. Environment and Planning [A], 18, 1499–1520.
Pacione, M. (2001) Urban Geography: A Global Perspective. Routledge, London.
Pacione, M. (2003) Urban environmental quality and human wellbeing—a social geographical perspective. Landscape and Urban Planning 65:19–30.
Park and Open Spaces Association of Japan (POSAJ). (2002) Park and Recreation Administration in Japan. POSAJ, Tokyo.
Rodrigue, J.-P., Comtois, C., and Slack, B. (2006) The Geography of Transport Systems. Routledge, New York. http://people.hofstra.edu/geotrans/.
Schrijnen, G. (2000) Infrastructure networks and red-green patterns in city regions. Landscape and Urban Planning 48:191–204.

Secco, G. (1994) "*Per una valutazione del rischio ambientale da fitofarmaci nel Veneto*". A. Ge.I., Gruppo di Lavoro Per una mappa del rischio ambientate in Italia Materiali quattro, 19–46, Department of Geography, University of Padova (Italy).

Talen, E. (2002) Pedestrian access as a measurement of urban quality. Planning Practice and Research 17(3):257–278.

Talen, E., and Anselin, L. (1998) Assessing spatial equity: an evaluation of measures of accessibility to playgrounds. Environment and Planning [A] 30:595–613.

Tjallingii, S.P. (1996) Ecological conditions. IBN-DLO, The Institute for Forestry and Nature Research, Wageningen, Netherland.

Van Kamp, I., Leidelmeijer, K., Marsman, G., and de Hollander A. (2003) Urban environmental quality and human well-being—towards a conceptual framework and demarcation of concepts; a literature study. Landscape and Urban Planning, 65:5–18.

Van Kamp, I., van Loon, J., Droomers, M., and de Hollander, A. (2004) Residential environment and health: a review of methodological and conceptual issues. Reviews on Environmental Health 19(3–4): 381–401.

Zulian, G. (2003a) Le attitudini dell'utenza di parchi e giardini pubblici: il caso di Padova. In: Leone, U. (ed.) Aree Dismesse e Verde Urbano. Pàtron Editore, Bologna, pp. 105–118.

Zulian, G. (2003b) Un'analisi del servizio offerto agli utenti dal verde urbano: metodi e applicazioni. Ph.D. dissertation, Università di Padova, Dipartimento di Geografia, G. Morandini.

21
Potential Leaf Area Index Analyses for the City of Toronto's Urban Forest

W.A. Kenney

One commonly used measure of urban forest structure is canopy cover, or simply the proportion of the city that is covered by tree canopies as visualized from above. This measure is intuitive and relatively easy and inexpensive to measure. However, canopy cover only represents the urban forest in two dimensions and fails to recognize differences in species and tree condition. Many of the benefits that cities derive from their urban forest can vary directly with the total leaf area of the forest, for example, reduction in air temperature, sequestering gaseous pollutants, and carbon sequestration (Nowak, 1994a). In addition, urban forests play a role in moderating urban forest climate through shading buildings, people, and hard surfaces, and through evapotranspirational cooling and the windbreak effect (McPherson et al., 1988; Akbari and Taha, 1992; McPherson and Rowntree, 1993; Brown and Gillespie, 1995). Increasing leaf area will increase shading and evapotranspiration, and could also have a direct impact on the windbreak effect. Similarly, storm water attenuation (Xiao et al., 1998) will be affected by leaf area, as well as other factors.

Consequently, an estimate of the total leaf area of the urban forest would be more informative than simply an estimate of its canopy cover. The most commonly used measure of leaf area is the leaf area index (LAI), defined as the total leaf surface area per unit land area. While LAI offers some indication of the urban forest's ability to provide services to the community as outlined above, urban forest management and planning requires some indication of the potential that exists to expand leaf area. A low canopy cover, in itself, says little about the potential for that area to support a tree canopy. For example, a portion of a city may be found to have a canopy cover of 20%; urban forest managers may consider expanding this to say 30%. Without some indication of the carrying capacity or potential to support additional trees within the city as a whole or in specific areas, managers are unaware of the practical possibilities of reaching their goals. The potential leaf area index (PLAI) is a measure of the carrying capacity of an urban area and is a function of the amount of available growing space and its configuration (Kenney, 2000). Armed with an estimate of the LAI and PLAI, urban forest managers and planners can more effectively address the protection and enhancement of this important resource. This chapter reports on a study whose purpose was to estimate the LAI and PLAI for the City of Toronto, Ontario.

M.M. Carreiro et al. (eds.), *Ecology, Planning, and Management of Urban Forests: International Perspectives.*

Methods

This project was part of a larger research initiative to estimate the role of Toronto's urban forest in mitigating greenhouse gases using the United States Department of Agriculture's (USDA) Forest Service, Urban Forest Effects (UFORE) model. Toronto has an area of approximately 63,000 hectares (ha) and a population of approximately 2.5 million. For the purposed of the UFORE study and to estimate the PLAI and LAI for the city's urban forest, the city was stratified into broad land-use classes as outlined in Table 21.1. A total of 211 circular plots, each with an area of 400 m^2, were randomly distributed among these eight land-use classes proportionally to the amount of tree cover (Nowak et al., 2003).

Field work was conducted during the months of June, July, and August, 2000. The coordinates of each feature within the plot were surveyed using a transit. Features such as trees, individual shrubs, utility poles, light standards, traffic signs, sewer grates, and fire hydrants were recorded as points. The vertices of polygon features such as shrub beds, turf, wild grass, soil, mulch, water, buildings, asphalt, cement, rock, wood, and other impervious surfaces were surveyed to locate precisely the position and shape of each feature. One hundred percent of each plot was allocated to polygon themes. Additionally, any infrastructural element that was outside of the plot but whose zone of influence (through buffer zones, discussed below) fell within the perimeter of the plot was also recorded.

A Geographical Information System (GIS) theme was created for each feature category present (trees, fire hydrants, asphalt, buildings, turf, etc.). Each theme was then categorized as either soft surface suitable for tree planting (grass, soil, herbaceous, wild grass, duff, and mulch) or hard surface unsuitable for planting (asphalt, cement, wood, rock). Seldom is 100% of a soft surface, as defined above, available for planting. For example, personal preferences with respect to urban trees are extremely variable. While social factors can be limiting with respect to tree establishment, they are ephemeral with respect to the time frames discussed for strategic management and may periodically shift with changing ownership and attitudes

Table 21.1 The distribution of plots by land-use class and the number of survey plots in each

Land-use class	Area of land-use class (ha)	Number of plots
Commercial	1,427.8	9
Industrial	11,727.6	12
Institutional	2,627.0	11
Low-density residential	13,870.8	42
Medium-density residential	15,618.1	49
High-density residential	4,261.8	14
Open area	12,698.8	63
Special purpose	971.3	11
City total	63,203.2	211

Table 21.2 Average buffer widths derived from a survey of six cities

Feature	Buffer width (m)
Sidewalk, driveway, patio, etc.	2.0
Building	2.5
Fence	1.5
Fire hydrant	3.1
Power transmission tower	19.0
Other utilities	1.5
Railway bed	4.5
Stop sign	11.7
Storm sewer	1.5
Street light	4.1
Street sign	2.0
Traffic light	22.5
Water	2.0

Source: Adapted from Duffy (1999).

within any planning horizon. For this reason, they were not taken into consideration when determining the actual plantable space present in a given plot.

The actual proportion of the soft surface that is available for planting is also limited by the adjacent infrastructure. Growing space available for tree planting will be reduced due to buffers placed on certain surfaces and infrastructure to allow for proper tree growth and to avoid interference with the feature in question. Such interference can include reduced visibility of street signs, impaired access to emergency utilities such as fire hydrants, or physical interference through crown growth into buildings or root growth damaging sidewalks. Duffy (1999) conducted a survey of six Ontario cities to investigate the typical buffer widths applied to various features with respect to tree planting. The average buffer widths for each feature shown in Table 21.2 were applied to their respective themes in the GIS. These buffered themes were then cut from the soft surface theme to produce a new one indicating the available plantable space. Below-ground utilities (e.g., natural gas mains, hydroelectric, cable television, telephone, etc.) were not considered in the calculation of plantable space. While these do pose very real constraints on tree establishment and growth, the data necessary to locate these features were not available.

Calculation of Potential Leaf Area Index

To estimate the amount of leaf area that could potentially be supported in a given plot, a procedure adapted from Duffy (1999) was utilized. This procedure involves "manually" positioning trees of known size and leaf area within the GIS in such a

Table 21.3 Prototype tree parameters used in this study: these sizes are considered representative of urban trees in general

Prototype tree size	Total height (m)	Crown length (m)	Crown width (m)	Shading coefficient	Leaf area (m^2)
Small	5	4	3	0.80	26
Medium	10	8	6	0.80	172
Large	20	15	12	0.80	410

way that leaf area would be maximized within the area available for planting in each plot. Duffy identified four classes of prototype trees based on cluster analysis of 57 species commonly planted in Ontario's urban forests. Tree data accumulated in the plots for the current study indicated that the prototype tree sizes used by Duffy (1999) were somewhat larger than those present in Toronto. For this reason, the prototype trees used in this study were somewhat smaller, and compressed into three size classes of known height, canopy width, and leaf area (Table 21.3). The available planting space identified in the GIS was filled with trees from these three size classes in such a way that all stems were in available plantable space and their canopies did not overlap, but could overhang unplantable space.

Buildings represented a special case with respect to placing trees. The "no-planting" buffer width applied to all buildings was 2.5 m. Given this buffer, the crowns of medium-sized prototype trees could potentially overlap structures by 0.5 m and large-sized trees could overhang by 3.5 m. A 0.5-m overhang for medium-sized trees was considered acceptable and in practical terms would result in only minor pruning to the tree. A 3.5-m overhang for large-sized trees was considered acceptable for small buildings (less than or equal to three stories or 10 m) as the bulk of the canopy would overhang the structure rather than contact it directly. However, for buildings taller than three stories, large trees were placed so that their canopies did not contact the building.

To achieve the highest possible leaf area per plot, available growing space was filled preferentially with large trees first, then medium-size trees, and finally small trees. Again, this planting strategy may not be practical or desirable in all situations, but the goal of this exercise was to establish a benchmark for the maximum leaf area that a plot could potentially support. Leaf area of the prototype trees was calculated using a relationship developed by Nowak (1994b, 1996) for estimating the leaf area of open-grown deciduous urban trees:

$$Y = (e^{-4.3309+0.2942\,H\,+0.7312\,D\,+5.7217\,Sh\,-\,0.0148\,S\,+0.1159})$$

where Y is the total leaf area (m^2), e is the natural logarithm, H is the crown height (m), D is the crown diameter (m), Sh is the shading coefficient (a measure of how much light is transmitted through a canopy), and S is the an estimate of crown surface area, $\pi D(H + D)/2$.

These prototype trees were assigned a shading factor of 0.80. This value is lower than that of the hardwood average of 0.83 (Nowak et al., 2002). This value is justified as it is a more conservative estimate of the average leaf area of the commonly used urban trees, given that many of the denser shade trees (such as *Juglans* sp., *Fagus* sp., and *Celtis occidentalis*) are not very common in the city and other more open trees (such as *Gleditsia triacanthos*) are abundant. The total leaf area (single-sided) of all prototype trees on each plot were summed to give a potential leaf area for the entire plot. These were then divided by the area of the plot (400 m^2) to give the PLAI in units of m^2 of leaf area per m^2 of ground area.

Results and Discussion

The average existing LAI and PLAI for each land use category are illustrated in Figure 21.1. As would be expected, the highest average PLAI values were found in open areas, institutional areas, and special-purpose areas (e.g., exhibition grounds). While the same is true for LAI, institutional and special-purpose land uses had extremely large standard errors, which is most likely due to the very broad definition of these land-use types and the variability in the extent and configuration of nonplantable space.

Residential areas were found to have increasing values of PLAI as the building density decreased across the three categories (residential high density to residential low density). Average LAI values for these land-use categories were relatively uniform. The difference between the LAI and PLAI in the medium-density residential and low-density residential land-use types suggests that there is potential to increase leaf area in these two land-use types. Conversely, there appears to be little opportunity to increase leaf area in high-density residential areas.

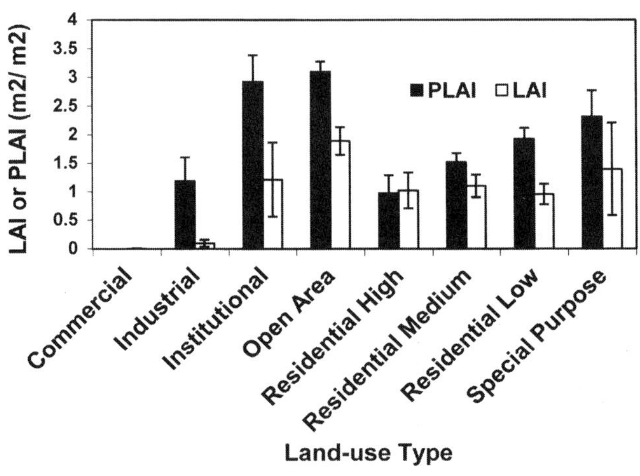

Fig. 21.1 Leaf area index (LAI, open bars) and potential leaf area index (PLAI, solid bars) by land-use category for the City of Toronto. Error bars represent standard errors. LAI and PLAI units are in m^2 of leaf surface area (one side)/m^2 of ground surface area

Industrial areas were found to have an average PLAI much higher than the corresponding LAI. Many of the industrial areas surveyed could be considered light industry and, as such, are characterized by buildings immediately surrounded by parking areas, which are in turn surrounded by substantial open space. In many cases this open space was dominated by turf with relatively few trees. This suggests that this type of industrial site could represent a significant opportunity for increasing leaf area. Perhaps corporate property owners, in collaboration with their employees, could be encouraged to become more actively involved in the stewardship of the portion of the urban forest found on their property.

Interestingly, all of the field sample plots surveyed in commercial areas had PLAI values of 0 and only a single commercial field plot was found to contain a very small amount of existing leaf area. This is primarily due to the very intensive use of most commercial areas. They either consist of street-side shops with little if any plantable space between the building and the curb, or strip malls and larger shopping centers dominated by buildings and parking areas. Therefore, expanding leaf area in commercial areas may be relatively challenging, since it seems necessary to increase PLAI by increasing planting space itself, rather than simply take advantage of existing but underutilized planting space.

The observed high PLAI levels throughout most of the city are encouraging from the standpoint of increasing Toronto's urban forest. With the exception of the old City of Toronto, virtually all areas of the amalgamated Toronto area could support more leaf area as indicated by the difference between the existing and potential leaf area density. As mentioned earlier, this is an optimistic estimate, as not all of the areas with high PLAI values can practically support the number of trees predicted. For example, we did not take such factors as personal preference into consideration. Some homeowners may prefer not to have any trees on their property. Similarly, some industrial sites, while currently exhibiting a substantial gap between the LAI and PLAI, may be reserving space for future expansion. Consequently, our estimates must be seen as a reflection of the existing situation at the time of sampling without a measure of landowner preference.

The ratio of LAI to PLAI provides an indication of the extent to which the potential carrying capacity of an area is currently met. In this study, some areas had LAI values that exceeded PLAI (i.e., the ratio of LAI/PLAI exceeds 1), indicating that the existing LAI exceeds the potential. While this seems intuitively unreasonable, it possibly stems from a combination of a number of factors. The selection of the prototype trees may be conservative because it underrepresents large trees such as silver maples *(Acer saccharinum)* that dominate some older parts of the city. Since all the prototype trees were deciduous, the conifers included in the LAI calculations were also underrepresented. Since conifers have a higher shading factor (Nowak et al., 2002), the use of 0.80 as a shading factor may also have been overly conservative and low. In older parts of the city, we suspect that our estimates of PLAI represent the current situation but the LAI actually reflect historical growing conditions. Large well-established trees may have developed in areas of higher PLAI. Subsequent road and sidewalk widening and "in-fill" housing redevelopment may have reduced the PLAI while the existing trees are still surviving in the more restricted space. The next generation of trees in these areas may have a more difficult time reaching their potential.

Our method of locating trees in the GIS was intended to mimic a reasonable scenario in the urban environment. As such, we did not allow crowns of the trees "planted" in the GIS to overlap. However, this likely underestimated conditions that could be found in natural forested areas where the level of canopy overlap is much greater. The presence of natural forests sites in our study would then be expected to have LAI/PLAI values greater than 1.

Figure 21.2 illustrates the distribution of plots with LAI/PLAI ratios greater than 1. With the exception of the plots in the northeast corner of the city, most of these are located in older residential areas with well-established trees.

Figure 21.3 shows plots with LAI/PLAI ratios of less than 0.5, indicating areas with significant potential to expand leaf area. These plots tend to be on the edge of the city where building densities are lower. The plots on the northeast corner in this instance represent active or old agricultural areas (part of the "open space" land-use type) that have not yet been developed, but do not have significant tree cover.

Just as canopy cover provides only limited information about urban forest structure, estimates about surface cover percentages are also limited in their usefulness. For example, the amount of soft surface is unlikely to consistently represent

Fig. 21.2 The distribution of plots in the City of Toronto with LAI/PLAI ratios greater than 1. Ratios greater than 1 indicate plots in which the existing LAI exceeds the potential or PLAI. See text for discussion. With the exception of the plots in the northeast corner of the city, most of these plots are located in older residential areas with well-established trees

Fig. 21.3 Distribution of plots in the City of Toronto with LAI/PLAI ratios less than 0.5 indicating plots in which the existing LAI is less than 50% of the estimated PLAI

the carrying capacity for trees, since the configuration of soft surfaces and their association with utilities and other features will have an impact on the available growing space and the size of trees that can be supported. For example, a 1-ha plot of good soil that is circular or square can support much greater leaf area than 1 ha of similar good soil that is in a tree lawn 0.5 m wide and 20 km long.

Conclusion

An understanding of the distribution of existing leaf area will provide urban forest managers with information that will aid them in planning for the long-term supply of ecological and socioeconomic benefits to their communities. While canopy cover is intuitive and easy to determine, this measure fails to provide sufficient information about the condition of the urban forest canopy. With the development of computerized tree inventory systems and GIS, it is possible to account for species and tree condition and to develop a mechanism to measure the urban forest canopy in three dimensions. For example, species variation can be incorporated through the use of shading factors as a component in the leaf area algorithm applied in the UFORE model and in this project. An estimate of the LAI based on ground samples can also incorporate an assessment of the condition of the tree crowns in the sample plots. Extrapolating these tree condition estimates to the urban forest as a whole will provide a more realistic measure of the state of the urban forest canopy than just canopy cover.

By estimating a measure of carry capacity, such as PLAI, forest planners and managers will be able to identify factors that limit not only the existing leaf area but also the potential leaf area of parts of their forest. For example, what might be the implications of a new planning policy that changes building density on the urban forest? How might we expand the urban forest in an area by improving the amount and configuration of the growing space? The concept of leaf area density encourages the protection of larger trees where possible. This does not just mean planting species that have the genetic potential to become large, but it also requires a management regime that carries them in to maturity. While many municipalities have by-laws or ordinances to protect trees, how many have regulations that protect growing space in such a way that established trees will remain healthy and reach their genetic potential to contribute to a healthy, livable community?

While many municipalities are focusing considerable attention on assessing canopy cover, the usefulness of such estimates is questionable. This example of the application of LAI and PLAI analysis to the urban forest of Toronto illustrates how a more critical look at the city's tree canopy can provide direction for urban forest management and planning. However, this approach of estimating PLAI outlined here needs further refinement before it is likely to be widely adopted. The manual location of each of the prototype trees in each of the plots in the GIS is time-consuming. The development of an efficient GIS optimization routine to accomplish this task automatically is needed. The current approach estimates LAI and PLAI for the sample plots, and the values are then extrapolated to the entire land-use type represented by those plots. A more efficient method for creating a GIS layer of PLAI and the ratio of LAI/PLAI will improve the likelihood that this approach to assessing urban forests will be applied in the field.

Acknowledgments The author would like to thank the Ontario Ministry of Environment for the support of a larger study entitled, "The Role of Urban Forests in Greenhouse Gas Reduction" (contract ENV99 4691, June 2001) that made this work possible. Much of the text for this chapter is derived from the unpublished final project report. Special thanks go to Christina Idziak and Colin Anderson for their assistance in collecting and analyzing field data and with the preparation of the manuscript. Jennifer McKay, Jennifer Raffin, and Wendy Strickland provided valuable assistance in data collection and analysis.

References

Akbari, H., and Taha, H. (1992) The impact of trees and white surfaces on residential heating and cooling energy use in four Canadian cities. Energy 17:141–149.

Brown, R.D., and Gillespie, T.J. (1995) Microclimate Landscape Design: Creating Thermal Comfort and Energy Efficiency. John Wiley & Sons, New York.

Duffy, N.M. (1999) Design limitations to potential leaf area density in urban forests. M.Sc.F. thesis, Faculty of Forestry, University of Toronto.

Kenney, W.A. (2000) Leaf area density as an urban forestry planning and management tool. Forestry Chronicle 76(2):235–239.

McPherson, E.G., Herrington, L.P., and Heisler, G. (1988) Impacts of vegetation on residential heating and cooling. Energy and Buildings 12:41–51.

McPherson, E.G., and Rowntree, R.A. (1993) Energy conservation potential of urban tree planting. Journal of Arboriculture 19(6):321–331.

Nowak, D.J. (1994a) Urban forest structure: the state of Chicago's urban forest. In: McPherson, E.G., Nowak, D.J., and Rowntree, R.A. (eds.) Chicago's Urban Forest Ecosystem: Results of the Chicago Urban Forest Climate Project. Gen. Tech. Rep. NE-186. U.S. Department of Agriculture, Forest Service, Northeastern Forest Experiment Station, Radnor, PA, pp. 3–18.

Nowak, D.J. (1994b) Air pollution removal by Chicago's urban forest. In: McPherson, E.G., Nowak, D.J., and Rowntree, R.A. (eds.) Chicago's Urban Forest Ecosystem: Results of the Chicago Urban Forest Climate Project. Gen. Tech. Rep. NE-186. U.S. Department of Agriculture, Forest Service, Northeastern Forest Experiment Station, Radnor, PA, pp. 63–81.

Nowak, D.J. (1996) Estimating leaf area and leaf biomass of open-grown deciduous urban trees. Forest Science 42(4):504–507.

Nowak, D.J., Crane, D.E., Stevens, J.C. and Hoeh, R.E. (2003) The Urban Forest Effects (UFORE) Model: Field Data Collection Manual. USDA Forest Service North Eastern Research Station.

Nowak, D.J., Crane, D.E., Stevens, J.C., and Ibarra, M. (2002) Brooklyn's Urban Forest. USDA Forest Service General Technical Report NE-290.

Xiao, Q., McPherson, E.G., Simpson, J.R., and Ustin, S.L. (1998) Rainfall interception by Sacramento's urban forest. Journal of Arboriculture 24:235–243.

22 Spatial and Temporal Change of Urban Vegetation Distribution in Beijing

Jun Yang, Peng Gong, and Jinxing Zhou

Urban vegetation is one of the important infrastructural components of any urban ecosystem. The existence of well-distributed and abundant vegetation cover in cities can provide many benefits for city dwellers. The most obvious benefit to the public is aesthetic. Trees, shrubs, and lawns add natural color, shape, and texture to the rectilinear concrete and asphalt surfaces in cities and conceal unpleasant spots from view (Miller, 1997). Urban vegetation can also generate direct economic benefits by producing timber, fruits, fuel wood, cut flowers, and many other goods (Kuchelmeister and Braatz, 1993). However, urban vegetation also supplies other more indirect benefits by providing environmental services that enhance quality of life in cities (see Chapter 16). Urban vegetation can filter air pollutants, sequester CO_2, shade and shelter homes from sun and wind, intercept urban runoff, and provide habitat for wildlife (Miller, 1997; Xiao et al., 1998; McPherson and Simpson, 1999; Nowak et al., 2000).

Various studies have been conducted to quantify the benefits and costs of urban vegetation (McPherson, 1997, 2000, 2003; McPherson et al., 1999; Nowak and Dwyer, 2000). Among their conclusions was the realization that environmental services provided by urban vegetation are strongly related to the structure of that vegetation. The type of vegetation, size, and spatial distribution of urban vegetation are structural factors that mattered most to the degree of benefit obtained. However, urban vegetation management in China places major emphasis on the absolute quantity of green space, such as total area of urban vegetation, total canopy cover, and per capita vegetation cover, and overlooks other structural attributes of urban vegetation as indicators of improvement in urban greening plans. Among those neglected structural characteristics is the spatial distribution pattern of urban vegetation. One good example of why and how this neglect comes about during planning and management is the misrepresentation of statistics to artificially report improvement in vegetation cover. To boost the statistics of vegetation cover in the city, some municipal governments combine the small amount of green cover existing in the densely populated central city districts with the high vegetation cover in rural areas at the urban fringe to get a high average urban canopy cover. Accepting such simplistic indexes as a basis for planning green space does not provide incentive to create more green space in densely

M.M. Carreiro et al. (eds.), *Ecology, Planning, and Management of Urban Forests: International Perspectives.*

populated areas where the public needs it most (see Chapter 9). These simple statistics are also used because they are relatively easy to obtain as compared to data dealing with spatial distribution. Studies that shed light on the spatial distribution of vegetation in cities can help to counter the problem mentioned above and improve the effectiveness of urban greening practices for cities in China and other parts of the world.

Beijing is China's second largest metropolitan area, with a population of 14.23 million in 2002. From 1991 to 2002, the population grew by 30% (Beijing Municipal Bureau of Statistics, 2005) and the total built-up area increased by about 60% (Che et al., 2005). The negative environmental outcomes of such rapid growth have recently caused Beijing's officials to reorient and devote more resources to making Beijing a more sustainable eco-city. Interest in enhancing the green space of the city is one of these objectives. However, the statistical finding that vegetation cover in Beijing City had reportedly risen rapidly in the last decade was met with some skepticism. According to the Beijing Municipal Bureau of Parks (2004), the city's canopy cover (i.e., the percentage of ground surface covered by the vertical projections of tree canopies, shrubs, grass, and other vegetation) increased from 28% of the city's built-up area in 1991 to 32.83% in 1997 and 36.54% in 2001. The built-up area in Beijing is defined as land that has been developed, covered with buildings and other man-made surfaces, and has basic public transportation and utility services. The municipal government has plans to increase the area of green space to 40% and the canopy cover to 45% of the built-up area by 2010 (Beijing Municipal Bureau of Parks, 2004). These numbers are striking when one considers that the average tree canopy cover in urban areas in the United States is only 27.1% (Dwyer et al., 2000). In addition, Beijing has a population density of 11,164 people per km^2 in its built-up area compared to an average urban population density of 928 people per km^2 in the U.S. in 2000 (U.S. Census Bureau, 2004). So it is reasonable to ask: How real was this increase? Because of the way statistics can be used to mask other kinds of trends, knowledge of where green space is distributed throughout the city must be determined to answer this question, and not simply data on total coverage.

To address this issue, this study had two main objectives: (1) to determine the actual percentage of vegetation in urban Beijing at different time periods, and (2) to determine the spatial distribution of urban vegetation and detect the spatial and temporal changes from 1991 to 2002.

Materials and Method

Study Site

Beijing is the capital city of China. The center of Beijing lies at 39° 92' N and 116° 46' E. It has a population of 14.23 million as of 2002. The climate of Beijing is a typical continental monsoon climate with four distinctive seasons.

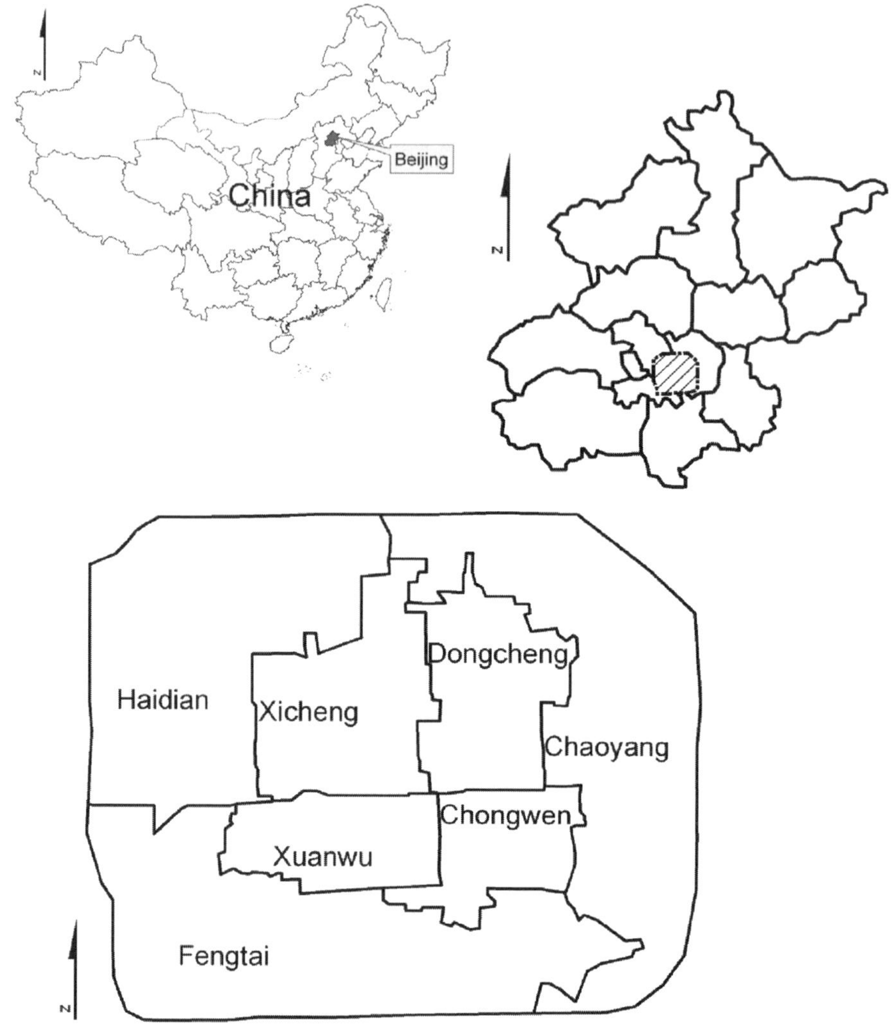

Fig. 22.1 The study area. The upper left map shows the position of Beijing in China. The middle map shows the administrative area of Beijing, and the shaded area is the study area. In bottom map, the outermost boundary is the fourth ring road for the city. The area contained inside is the central city of Beijing, which includes four central city districts and parts of three suburban districts

Most precipitation is concentrated in July and August. The frost-free period is about 180 days. The area administrated by the Beijing municipal government is 16,807 km^2. Traditionally, the four central city districts, Xuanwu, Xicheng, Chongwen, and Dongcheng, plus parts of three suburban districts, Haidian, Fengtai, and Chaoyang, are viewed as the central city (Fig. 22.1). The area of the central city is about 305 km^2 and constitutes the major portion of the city's built-up area. This study focused on the central city area where about 4.5 million people live. So the ecosystem services provided by vegetation are likely to be proportionately greater here than in many rural locations, since such a large number of people benefit.

Method

Multidate satellite images of Beijing were analyzed to extract vegetation information for the central city of Beijing. Satellite images are very useful for exploring spatial patterns in the urban environment. Their relatively large coverage enables studying the land cover/land use at an entire city scale. Furthermore, repeated coverage of the same place allows the user to catch temporal changes in these features. In this study, satellite images covering the study area were acquired. Then a spectral unmixing method was used to extract urban vegetation from the images (Small, 2001). The unmixed vegetation fraction was assumed to equal the canopy cover rate of vegetation because satellite images were taken above the canopy of vegetation during the growing season. The vegetation fraction derived for each pixel was the ratio of the area of vertical projection of the vegetation canopy on the ground to the total ground surface area inside that the pixel. A pixel in this image corresponds to a square area of 30 m by 30 m on the ground. Finally, the spatial and temporal changes of vegetation were detected using an image differencing method (Hayes and Sader, 2001).

Data Acquisition and Preprocessing

For this comparative study we acquired two Landsat 5 Thematic Mapper (TM) satellite images taken on May 16, 1991 and May 6, 1997, and one Landsat 7 Enhanced Thematic Mapper Plus (ETM+) image taken on May 22, 2002. All those images covering the scene numbered 123/32 in the Landsat WRS-2 system, which included the entire Beijing administrative area and parts of its neighboring provinces. These images were carefully chosen from hundreds of images, because they were taken in the same month, thereby reducing seasonal differences in vegetation phenology that could influence interpretation of the data. All images were taken on days that had zero cloud cover above the city, which minimizes the influence of clouds on classification results. Besides the satellite images, one 1:50,000 topographic map and one 1:25,000 street map of Beijing were also purchased as ancillary data for this analysis. The two Landsat 5 TM images were registered to the Landsat 7 ETM+ image. Geometric and radiometric corrections had been conducted on the Landsat 7 image by the U.S. Geological Survey, Earth Resources Observation and Science data center. No atmospheric correction was conducted in this study, because this information was unavailable for Beijing. Also, atmospheric correction may introduce new uncertainties to the unmixing results. After preprocessing, the studied area was clipped from the rectified satellite image for further analysis.

Spectral Linear Unmixing

The linear unmixing method (Small, 2001) was used in this study to derive the vegetation fractions from the three images. Linear unmixing is thought to produce a more accurate representation of the actual vegetation distribution in the urban environment than traditional classifying and Normalized Difference Vegetation Index (NDVI) methods (Small, 2001). The linear mixing model assumes that the spectral reflectance of a pixel can be described as a linear combination of end-member spectra, for example:

$$R(\lambda) = f_1 E_1(\lambda) + f_2 E_2(\lambda) + \cdots + f_n E_n(\lambda) \tag{1}$$

where $R(\lambda)$ is the observed reflectance at wavelength λ, the $E_i(\lambda)$'s are the end-member spectra, and f_i is the corresponding fraction of the n end members contributing to the observed reflectance. For b spectral bands that the satellite sensor has, the linear mixing model is composed of b linear equations and can be written as

$$Ef = r \tag{2}$$

where E is a b-by-n matrix in which the columns are the n end-member spectra, b's are the spectral bands, and f is the vector of end-member fractions, describing the observed reflectance vector r. To solve for f, the number of bands must be equal to or greater than the number of end members. The least-squares solution for this linear mixing model, in case of uncorrelated noise, is given by the following equation:

$$f = (E^T E)^{-1} E^T r \tag{3}$$

where E^T is the transposed matrix of E. The result is a set of end-member fractions for each pixel.

In this study, four end members were chosen based on the dimensionalities of the data. The dimensionalities of the data were observed by conducting a minimum noise fraction (MNF) transformation and displaying the transformed data in the feature space. The vertices of the image data in feature space are potential end members. The four end members chosen from feature space for this study were *vegetation, low-albedo area* (e.g., house, road, shade), *high-albedo area* (e.g., bare soil, pavement, cement), and *water bodies*. All image processing was conducted using the PCI Geomatica (V8.2) software package. Vegetation fraction maps were produced by using the Matlab (V6.1) software package. The vegetation fractions resulting from the unmixing process were compared to the statistics from the Beijing Bureau of Municipal Parks. There are no official vegetation statistics for the entire study area since the bureau tallied it by district. So the vegetation cover information of four central city districts, Xuanwu, Xicheng, Chongwen, and Dongcheng, were obtained from the vegetation fraction map and compared to the official data in 1997 and 2002. The official data by district for 1991 were not available.

Criteria for Detecting Change in Vegetation Cover

After conducting spectral unmixing, the resulted vegetation fraction maps of 1991, 1997, and 2002 were paired, and image differencing (Hayes and Sader, 2001) was conducted on each pair. An arbitrary threshold of 50% of value change was set to determine change or no change in vegetation cover, since smaller changes in vegetation cover are hard to detect accurately (Foody and Boyd, 1999). Since the goal of this study was to explore large trends in vegetation cover and distribution in the central city area over a decade, the ability to detect small and subtle changes is not critical. Therefore, adopting a high threshold value of 50% was justified, and provides conservative estimates of change. The decision was made to place a pixel in the changed land cover class category, if the vegetation fraction inside that pixel had changed by 50% between two dates. A positive change meant that the vegetation had increased during this period. If negative, then the vegetation had decreased. The changes were highlighted and exported to one image. Here the image only showed where significant vegetation cover changes had occurred between the two dates, not the exact percentages of change. The accuracy was visually inspected by comparing the change images with the raw TM images. For reasons of time and budget limitations, ground-truth data were not specifically collected for accuracy assessment purposes for this study.

Results

To construct vegetation fraction maps (Fig. 22.2), the abundance of vegetation in each pixel was represented by the gray value assigned to that pixel. The brightness of the pixel represents the percentage of vegetation cover. The brighter the pixel, the higher the percentage of vegetation cover that exists in that pixel. A dark pixel means that there is no vegetation in that pixel or the percentage of vegetation cover was too low to be detectable using the current image and methods. The results showed that the center of city in 2002 had higher vegetation cover than in 1991.

The mean vegetation fractions for the study area and the root mean square error (RMSE) value are shown in Table 22.1. Vegetation fractions give the estimated vegetation cover in the study area. The result showed an overall increase of vegetation cover from 1991 to 2002 in the study area.

The unmixing results for 2002 were compared to another ground study conducted in the same area in summer 2002 (Yang et al., 2005). The classification results of the satellite image were verified by a stratified random sampling of 204 plots throughout the central city. A total vegetation cover of 24.5% was calculated by the on-the-ground study conducted by Yang et al. (2005), while the estimate from the satellite image study was 21%. There was no significant difference between the two numbers, indicating that our satellite classification method was reasonable and sound. For 1991 and 1997, no field surveys were conducted, but the results are comparable to 2002 because all images were interpreted using same method.

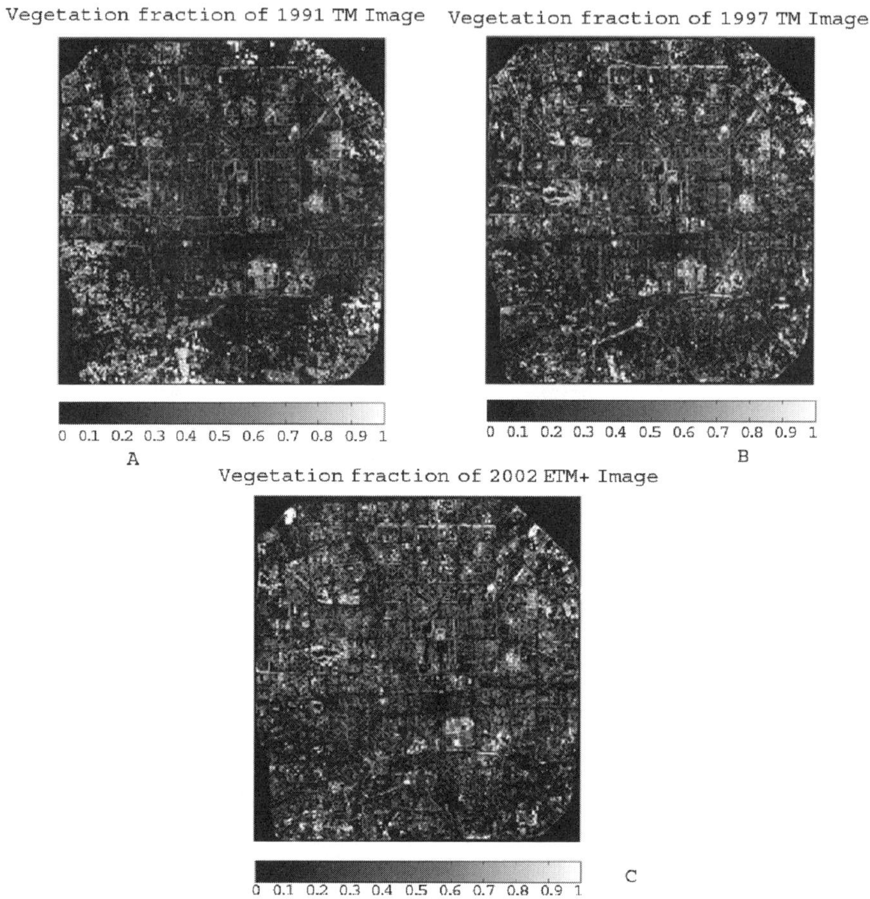

Fig. 22.2 The vegetation fraction maps for 1991 (A), 1997 (B) and 2002 (C) in the central area of Beijing within the fourth ring road boundary. The vegetation fraction is shown as different gray levels in the scale, where 1 represents pure vegetation (lighter) and zero represents no vegetation (darker) in that pixel

Table 22.1 Mean vegetation fractions and roots mean square error (RMSE) distributions in study area for 1991, 1997, 2002; standard errors denoted in parentheses

Year	Vegetation fraction (%)	RMSE (%)
1991	0.17 (0.26)	0.05 (0.03)
1997	0.20 (0.24)	0.07 (0.06)
2002	0.21 (0.22)	0.08 (0.06)

To estimate the goodness of fit in the model, we used the RMSE, which is a measure of the amount of the spectral reflectance not accounted for by the model. A low RMSE value for one pixel means the unmixing results for that pixel was less uncertain. The overall RMSE of the unmixing results was less than 10%. This low RMSE value showed that the unmixing results were within an acceptable range (Small, 2001).

To determine whether official estimates of large vegetation change in the city over a decade period were detectable from our satellite imagery analysis, we compared out vegetation fraction results with official data of canopy cover rate in the four central city districts (Table 22.2).

The results showed that the vegetation fractions estimated by unmixing are 10% to 30% lower than official data. A paired t-test showed that this difference between the official data and the estimated vegetation was significant ($p < .05$). The possible reasons for this discrepancy are discussed below.

Change detection was conducted on the vegetation fraction maps by using image differencing techniques. We found that major changes occurred from 1991 to 2002 (Fig. 22.3).

Table 22.2 Comparison of estimated vegetation fraction values with official records of the canopy cover rates in four central city districts

	1997		2002	
District	Official record	Estimated	Official record	Estimated
Xuanwu	0.25	0.17	0.22	0.16
Chongwen	0.32	0.23	0.31	0.24
Dongcheng	0.27	0.22	0.26	0.23
Xicheng	0.27	0.22	0.26	0.22

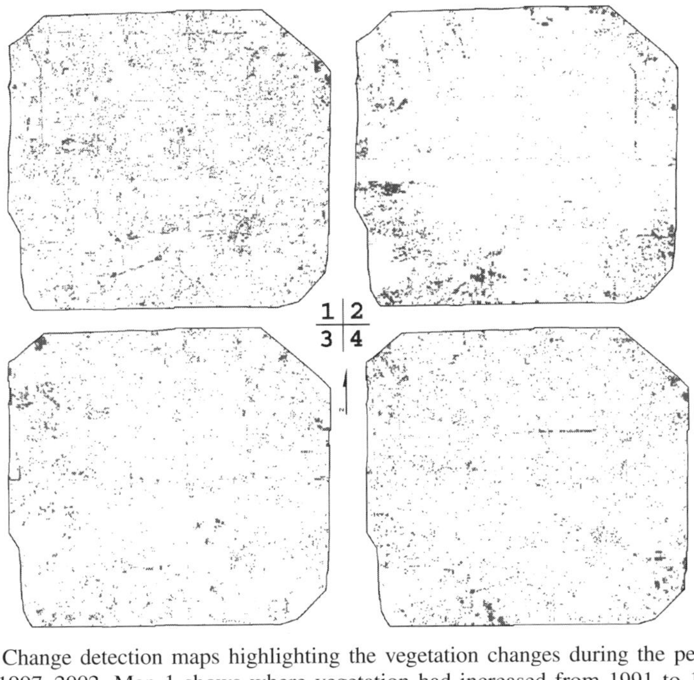

Fig. 22.3 Change detection maps highlighting the vegetation changes during the period 1991–1997 and 1997–2002. Map 1 shows where vegetation had increased from 1991 to 1997 in the central city of Beijing. Map 2 shows the places where vegetation had decreased from 1991 to 1997. Map 3 shows where new vegetation had been planted from 1997 to 2002. Map 4 shows the places where vegetation cover has been lost during 1997 to 2002

Most of the vegetation decrease happened at the urban fringes, especially at the corners of the central city. The increase of vegetation mainly happened in the central districts. Those patterns were more obvious during the period from 1991 to 1997 than from 1997 to 2002.

Discussion

Our estimated canopy cover for the central area of Beijing differed from the official records. Possible reasons for this discrepancy include the following: (1) The spectral unmixing model is not sensitive to small vegetation habitats inside a pixel. Small (2001) found that uncertainty in quantifying the vegetation fraction of less than 0.2 was high. It was possible that some trace vegetation components in pixels were not detected. (2) The method used by the Beijing Bureau of Municipal Parks tends to overestimate the canopy cover of vegetation. For example, the cover area of street trees was calculated by multiplying the total length of each street where street trees were planted by 1.5 m. In most streets in Beijing, this method overestimates the real canopy cover of street trees because the canopies of most street trees are small due to heavy pruning practices (Yang and McBride, 2003; see also Chapter 16), and there are many gaps on streets where trees are dead or not planted at all. However, the general trends of greater or lesser vegetation cover by district determined by our analysis did agree with those shown in the official records.

Two major trends were observed from the change maps (Fig. 22.3). First, from 1991 to 2002 vegetation in urban Beijing increased, with most of that increase occurring from 1991 to 1997. Second, most of vegetation cover increase occurred in the central four districts, while at the same time a significant amount of vegetation disappeared in the three suburban districts at the urban fringe. This resulted from the rapid expansion of the central city toward its outer periphery, a classic urban sprawl phenomenon. The built-up area in Beijing increased from about 390 km^2 in 1991 to 640 km^2 in 2002 (Che et al., 2005). The high density in the central four districts made it impossible to expand the built-up area through infilling, so most development happened in part of the suburb districts that is included in fourth ring road. The increase in built-up area caused the loss of farmland and orchards at the urban fringe. For example, at the southeast corner of the central city area shown in map B in Figure 22.2, 158 ha of farmland and orchards were converted into a large residential and high-tech industrial park in 1994. The newly developed areas have less vegetation cover due to high-density development and the unwillingness of the developer to allocate expensive land for green space (see Chapter 9).

From 1997 to 2002, there were no major changes in the total vegetation cover in the central city of Beijing, even though the distribution of urban vegetation had changed. The most obvious change was the construction at the northwest corner of the study areas of Wanquan Park, which covers an area of 5.5 ha. At the same time, in the four central city districts, many small vegetation patches disappeared, while few large patches appeared. This observation showed that the highly controversial practice of demolishing and rebuilding old residential areas in central city districts had a negative impact on the distribution of vegetation cover in the central city. To make room for wider roads and high-rise apartment buildings, existing large trees in these older residential areas were either removed or relocated. This practice resulted in a decrease of small patches of canopy cover all over the center area. The municipal government tried to compensate for the loss of canopy cover by constructing large green spaces where land is available. In total, 51 patches of green space with an area larger than 1 ha were constructed between 1999 and 2002 in the study area (Beijing Bureau of Municipal Parks, 2004). This is reflected in the large patches of new vegetation that can be seen in Figure 22.3, with Wanquan Park being the largest among them.

The information on the spatial and temporal changes of urban vegetation in Beijing revealed in this study is useful for urban vegetation management. Our results showed that the effort that Beijing municipal government put into urban greening was detectable. The millions of dollars invested in urban greening from 1991 to 2002 did improve the canopy cover in the central city by 3% to 4%. This was certainly a great accomplishment, considering the high population density, high impervious cover, and the rapid development in Beijing. On the other hand, the results from this study also sound an alarm for future urban vegetation planning and management. The planned 45% canopy cover may be hard to realize in Beijing, at least in the center districts of the city. As demonstrated by this study, there were large amounts of vegetation cover, mainly farmlands and orchards, disappearing at the urban fringes. The municipal government in Beijing should realize that to improve the sustainability and environmental quality of life in the area, it is necessary to protect and improve existing vegetation cover at the same time as new green space is constructed at great cost. New urban planning tools and regulations need to be instituted to deter the takeover of valuable farmland by developers, farmland that may be needed to provide greater food security for the city's residents in the future.

In this study, a spectral linear mixing model was used to analyze multitemporal satellite images of Beijing and produced meaningful quantitative area estimates of vegetation and its change inside the city. Also, the change detection by image differencing highlighted the hot spots that have undergone change over the past decade. However, this methodology also has its limitations. There is high uncertainty associated with estimation of area for vegetation patches less than 20% of pixel area. If this method is to be used in future studies, hyperspectral images or high-resolution satellite images should be used to increase the accuracy of the unmixing results.

Acknowledgments This research was supported by the fund from Forest Science Division, University of California at Berkeley. The authors want to thank Dr. Joe McBride, who provide valuable suggestions and help on this project.

References

Beijing Municipal Bureau of Parks. (2004) http://www.bjbpl.gov.cn/.

Beijing Municipal Bureau of Statistics. (2005) http://www.bjstats.gov.cn/.

Che, W., Zhang, Y., Li, J.Q., et al. (2005) Multi-functional storage of rainwater in urban area. Water and Wastewater Engineering 31:25-29 (in Chinese, abstract in English).

Dwyer, J.F., Nowak, D.J., Nobel, M.H., and Sisinni, S.H. (2000) Connecting People with Ecosystems in the 21st Century: An Assessment of Our Nation's Urban Forests. USDA Forest Service, Syracuse, NY.

Foody, G.M., and Boyd, D.S. (1999) Detection of partial land cover change associated with the migration of inter-class transitional zones. International Journal of Remote Sensing 20:2723–2740.

Hayes, D.J., and Sader, S.A. (2001) Comparison of change—detection techniques for monitoring tropical forest clearing and vegetation regrowth in a time series. Photogrammetric Engineering and Remote Sensing 67:1067–1075.

Kuchelmeister, G., and Braatz, S. (1993) Urban forestry revised. Unasylva 44:3–12.

McPherson, E.G. (1997) Quantifying urban forest structure, function, value: the Chicago Urban Forest Climate Project. Urban Ecosystems 1:49–61.

McPherson, E.G. (2000) Expenditure associated with conflicts between street tree root growth and landscape in California. Journal of Arboriculture 26:289–297.

McPherson, E.G. (2003) A benefit-cost analysis of ten street tree species in Modesto, California, U.S. Journal of Arboriculture 29:1–5.

McPherson, E.G., and Simpson, J.R. (1999) Carbon dioxide reduction through urban forestry: guideline for professional and volunteer tree planters. USDA, Forest Service, Pacific Southwest Research Station, Albany, CA.

McPherson, E.G., Simpson, J.R., Peper, P.J., and Qingfu, X. (1999) Benefit-cost analysis of Modesto's municipal urban forest. Journal of Arboriculture 25:235–248.

Miller, R.W. (1997) Urban Forestry: Planning and Managing Urban Greenspace. Prentice Hall, Upper Saddle River, NJ.

Nowak, D.J., and Dwyer, J.R. (2000) Understanding the benefits and Costs of urban forest ecosystems. In: Kuser, J.E. (ed.) Handbook of Urban and Community Forestry in the Northeast. Kluwer Academic/Plenum, New York, pp. 11–22.

Nowak, D.J., Kevin, L.C., Rao, S.T., Sistia, G., Luley, C.J., and Crane, D.E. (2000) A modeling study of the impact of urban trees on ozone. Atmospheric Environment 34:1601–1603.

Small, C. (2001) Estimation of urban vegetation abundance by spectral mixture analysis. International Journal of Remote Sensing 22:1305–1334.

U.S. Census Bureau (2004) 2000 Census of population and Housing, Population and Housing Unit Counts PHC-3–1, United States Summary, Washington, DC.

Xiao, Q., McPherson, E.G., Simpson, J.R., and Ustin S.L. (1998). Rainfall interception by Sacramento's urban forest. Journal of Arboriculture 24:235–244.

Yang, J., and McBride, J. (2003) A unique technique for street tree planting in Beijing. Arboricultural Journal 27:1–10.

Yang, J., McBride, J., Zhou, J., and Sun, Z. (2005) The urban forest in Beijing and its role in air pollution reduction. Urban Forestry and Urban Greening 3:65–78.

23
Long-Term Observations of Secondary Forests Growing on Hard-Coal Mining Spoils in the Industrial Ruhr Region of Germany

Henning Haeupler

The Ruhr region is a large industrial region in northwest Germany with very dense settlements and other man-made environments, resulting from very intense hard-coal mining of earlier times. This region was the former "black country" in Germany in the sense that Nerys Jones describes (see Chapter 8). This black material consists of hard-coal mining waste, a substrate that becomes dry and hot in the summer, up to 80°C on the surface (personal observation), but very wet in winter and spring. This waste has been piled into hills, the largest of which are 22 to 100 m high, upon a landscape that is otherwise relatively flat. Most of these waste heaps are reafforesting, with only a very few still actively being formed. Because of their close association with highly developed areas, we consider these emerging forests to be one type of urban forest that we are calling "industrial forests." We consider this forest type to be important in this region, since it covers an area of almost 10,000 hectares (ha).

Nearly all of the 22 large heaps had been planted by landscape managers (not ecologists), and therefore the plant species consisted of exotic ornamental woody plants, nearly all being species of shrubs and trees available in garden centers and tree nurseries (species lists are found in Schwiederowski, 1994). Because these waste piles exhibit extreme soil conditions of a substrate type called "Berge" (Wiggering and Kerth, 1991; Weiss et al., 2005), nearly all planted trees and shrubs, exotic as well as native ones, exhibit a high rate of die-back after transplantation. So the purpose of the project we describe here is to understand the ecology of the piles that are becoming naturally vegetated to learn how we can accelerate the process of secondary succession with species that can persist on these disturbed sites.

Site Descriptions and Methods

Forests have been developing on these waste heaps over the last 50 years. Today the older stands have a canopy almost completely dominated by birch (*Betula pendula* and *B. pendula x aurata*) with a subcanopy consisting of young oaks (*Quercus robur*) and a few beech (*Fagus sylvatica*). However, we cannot predict which type of climax forest will develop on these sites in the future. It is assumed

M.M. Carreiro et al. (eds.), *Ecology, Planning, and Management of Urban Forests: International Perspectives.*

by Hurtienne (1990; cited by Jochimsen, 1991a) that a deciduous beech-oak forest (former Fago-Quercetum, today Periclymeno-Fagetum) will develop after the stage dominated by birch-oak forest (Betulo-Quercetum roboris). The most natural stand that currently exists (on a former mine called Hugo III) is a birch forest dominated by *Betula pendula* with only a few other woody species, and with only a small amount of oaks and beeches along the periphery. For comparative purposes, finding natural reference forests that have been unaffected by humans within the last 50 years has been difficult, since many waste areas have been transformed into new industrial or residential settlements in this region.

Recently a new research project (Industrial Forests of the Ruhr Region) has been established by the Landesanstalt für Ökologie, Bodenordnung, und Forsten (LÖBF), a national institution for scientific research and administration in ecological affairs. The goal of this project is to investigate the natural succession of vegetation and plant population dynamics in these industrial forests using permanent plots. For experimental design details, see Weiss (2003) and Weiss et al. (2005). This project includes studies of the soil and forest structure, and floristic, vegetational, and faunistic observations.

We investigated plant community development on three abandoned coal mines (Zollverein, Rheinelbe, and Alma) that were at three different stages of succession:

1. Zollverein: pure waste; an early successional stage (Fig. 23.1)
2. Rheinelbe: scrub dominated by *Buddleija davidii* (Fig. 23.2) or tall perennial herb stands (Fig. 23.3)
3. Alma: forests dominated either by *Betula pendula* that arose spontaneously (Fig. 23.4), and one old but planted stand of false acacia (*Robinia pseudacacia*) (Fig. 23.5)

Fig. 23.1 Early successional stage community growing on pure coal waste at the former coal mine Zollverein. Zollverein has been a world heritage site since 2002. (Courtesy of C. Kert.)

Fig. 23.2 A slightly older sere consisting of scrub community dominated by *Buddleija davidii* and growing on former mine, Rheinelbe. (Courtesy of P. Keil.)

Fig. 23.3 Perennial herb community at former coal mine Alma. (Courtesy of C. Kert.)

Since the successional trajectory of a site is partly determined by which seeds arrive at a site and which seeds can then germinate and grow there, we collected seeds and examined the composition of the seed bank at different soil horizons over time. Therefore, to understand seed bank species composition and dynamics, we constructed special traps to collect diaspores deposited (shown in Weiss et al., 2005) at different depths in the soil. Species composition

Fig. 23.4 Nearly 50-year-old spontaneous birch (*Betula pendula*) stand at former coal mine Rheinelbe. (Courtesy of C. Kert.)

Fig. 23.5 Nearly 50-year-old planted stand of false acacia (*Robinia pseudoacacia*) at former coal mine Zollverein. (Courtesy of C. Kert.)

and abundance of diaspores in the seed bank were also determined by sieving the soil and germinating diaspores. These detailed seed bank observations were conducted in the first 3 years of the project in May (spring) and September (autumn).

Vegetation relevés (percent coverage categories by species using the Braun-Blanquet method; discussed in Knapp, 1984) and floristic observations were made every year for 6 weeks during the growing season in June and July. However, we did not use the usual six to seven classes of "degrees of coverage" described in the Braun-Blanquet method. Instead, plant coverage was estimated directly in 5% increments. We also used the frequency-method of Raunkiaer (1934) as another plant abundance measure. These methods are described in more detail in Knapp (1984). Initial results for this project have been reported by Haeupler et al. (2003) and Weiss et al. (2005). In this chapter we also present results from two unpublished theses (Kert, 2001; Schürmann, 2001).

The degree of hemeroby for the species found at these sites were also determined. Hemeroby represents the level of dependency of plants on human influence (Steinhardt et al., 1999). The classification categories from greatest to least degree of naturalness are as follows: ahemerobe (natural biocenosis); oligohemerobe (close to natural—limited wood removal, limited pasturage); mesohemerobe (semi-natural— clearing, occasional ploughing); β-euhemerobe (relatively far from natural—fertilizers, lime, pesticides applied); α-euhemerobe (far from natural—drained, fertilized, ploughed); polyhemerobe (biocenosis destroyed and covered with impervious material); and metahemerobe (artificial—biocenosis destroyed).

Results

The percentage of diaspores of different plant species in different soil horizons in spring (Fig. 23.6, top) and in autumn (Fig. 23.6, bottom) at Alma II varied greatly. In May and September quite different percentages of diaspores of the same taxa and different species combinations were detected at each depth horizon (0 to 2 cm, 2.1 to 5 cm, 5.1 to 10 cm). Total number of diaspores was low, and species richness varied at each depth at each sampling period (Fig. 23.7). However, both diaspore density and diaspore species richness were greater in September than in May, and highest in the top 0 to 2 cm of soil in September. These dynamic changes in diaspore species composition and density suggest that a permanent seed bank does not yet exist at this site.

We also observed a high degree of year-to-year variability in the vegetation at the Alma I site (Table 23.1). Only 11 taxa of the 51 plant species found in this plot were observed growing in all 3 years. Eight new taxa arrived in the second year, and 22 additional taxa arrived in the third year. However, some taxa also disappeared from the site during the course of this study—three taxa in the second year and four in the third year. Four taxa were seen only in the second year of this study. Therefore, during the first two successional stages both the plants and their diaspores in the soil were very dynamic. Therefore, it is evident that the seed bank during the pioneer stage (I) is not persistent, even though these successional stages persist for a period of more than 15 years (personal observation). This suggests that the species composition of the ecosystem was not self-sustainable and still

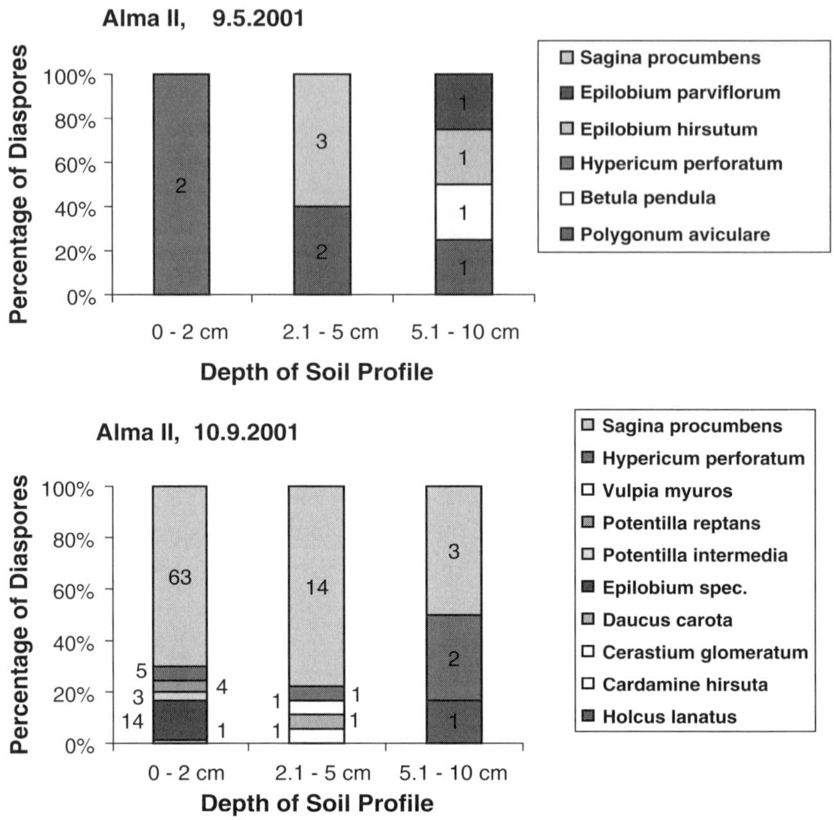

Fig. 23.6 Top: The percentage of diaspores of different species in varying depths of soil in the early successional plot at Alma in the spring. The number of diaspores is low and entered as values in the bar columns. *Bottom*: The same in autumn. Comparison of both graphs show that species composition of the soil seed bank differs and is very dynamic even within a single year, showing that this plot does not have a permanent seed bank

changing. However, in spite of this dynamic, species richness increased steadily in all plots in the 3 years of the study (Fig. 23.8).

While species richness is important, the kinds and quality of species during succession are also determinants of future species trajectories in these communities. Surprisingly, we found 19 species in these sites that are classified as rare and threatened species in North Rhine–Westphalia, four that are threatened within Germany, and five additional species that are expected to become endangered soon (Table 23.2). Therefore, these pioneer stages of succession on the waste heaps are, without doubt, of remarkably important floristic interest in the region. In addition, even though any one site may be inhabited for only a few years by these species, these waste heaps may play a role at the landscape level by serving as stepping stones for colonizing the next new waste heaps.

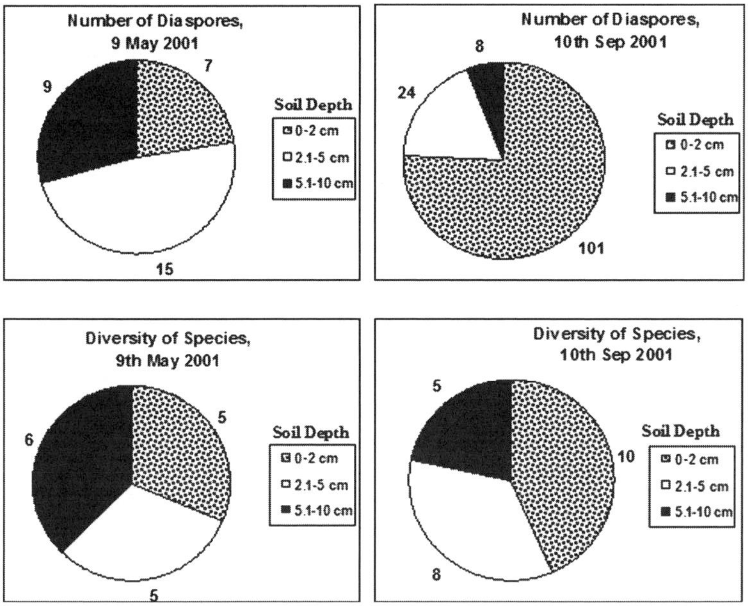

Fig. 23.7 Distribution of diaspore density (top panels) and species richness (bottom panels) at different depths in the soil. Left: May, late spring. Right: September, early fall. Both diaspore density (m^{-2} basis) and diaspore species richness were greater in September than in May and highest in the top 0 to 2 cm of soil in September

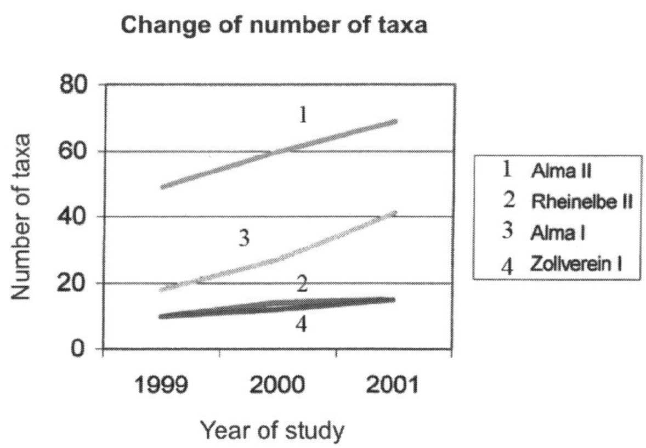

Fig. 23.8 Changes in plant species richness during the first two stages of succession (initial and shrub stage) on all three former mines. At all sites species number increased within two years but change was most evident at the Alma I and II sites

Table 23.1 The species dynamics of early stage vegetation in the Alma plots in first 3 years of study

Occurring all 3 years	Appearing	Disappearing	Occurring only in year 2
	In Year 2	In Year 2:	
Agrostis stolonifera	*Acer* sp.	*Echium vulgare*	*Cerastium semidecandrum*
Anagallis arvensis	*Cirsium vulgare*	*Equisetum arvense*	*Epilobium hirsutum*
Atriplex prostata	*Epilobium ciliatum*		*Robinia pseudoacacia*
Daucus carota	*Holcus lanatus*	*Melilotus* sp.	
Festuca rubra	*Poa annua*		*Sonchus oleraceus*
Plantago major	*Salix caprea*		
Polygonum aviculare	*Taraxacum officinale*		
Senecio inaequidens	*Tussilago farfara*		
Solidago gigantean			
Sonchus asper	In Year 3:		
Tripleurospermum perforatum	*Achillea millefolium*	In Year 3:	
	Artemisia vulgare	*Carex hirta*	
	Betula pendula	*Chenopodium polyspermum*	
	Calamagrostis epigejos	*Chenopodium rubrum*	
	Carex muricata		
	Cerastium glomeratum	*Elymus repens*	
	Conyza canadensis		
	Crepis tectorum		
	Dactylis glomerata		
	Dittrichia graveolens		
	Echinochloa crus-galli		
	Epilobium tetragonum		
	Hieracium pilosella		
	Hypericum perforatum		
	Lactuca serriola		
	Leontodon saxatilis		
	Pastinaca sativa		
	Poa compressa		
	Populus nigra var. *"italica" hybrid*		
	Potentilla reptans		
	Vulpia myuros		

Figure 23.9 shows the plant community's degree of hemeroby at each of the sites. As seen in the top three panels of Figure 23.9, the percentage of poly-, α-euhemerobic, and β-euhemerobic (synanthropic) plants decreased over the 3-year period. The percentage of native species (oligohemerobes) increased and the

Table 23.2 The list of species registered in the Red Data Books of Germany (BRG)/ North Rhine-Westphalia (NRW) and in the Ruhr-Region (WT)

	Sites			Red list status		
Taxon	Alma	Rheinelbe	Zollverein	NRW	WT	BRG
Asplenium trichomanes	-	-	x	*	3	3
Carex disticha	x	-	x	*	*	3
Carex rostrata	-	-	x	3	*	2
Carex vesicaria	-	-	x	3	3	2
Carlina vulgaris	x	-	-	*	3	2
Centaurium erythraea	x	x	-	V		
Centaurium pulchellum	x	x	x	3	3	*
Cynosurus cristatus[1]	-	x	-	V		
Cyperus fuscus	x	-	-	*	3	3
Dianthus armeria armeria	x		-	3	3	3
Dianthus deltoides	x	-	-	3	3	1
Galium saxatile	x	-	-	*	*	3
Geum rivale	x	-	-	3	3	3
Hieracium pilosella	x	x	x	V		
Hypericum humifusum	-	x	-	*	3	*
Leontodon hispidus	-	x	-	*	3	3
Myosotis ramosissima	x	-	-	*	3	2
Polystichum aculeatum	x	-	-	*	0	D
Potamogeton crispus	x	-	-	3	3	3
Potentilla argentea	x	-	-	*	3	3
Potentilla supina	x	-	-	*	2	3
Pyrola minor	x	-	-	3	2	0
Rubus nemorosoides	-	x	x	*	3	D
Rhamnus carthatica[2]	x	-	-	*	3	3
Sanguisorba officinalis	-	-	x	*	2	2
Schoenoplectus lacustris	-	-	x	*	3	3
Scrophularia auriculata	x	-	-	*	-	2
Securigera varia	x	-	-	*	3	*
Verbena officinalis	x	x	-	*	*	3
Veronica anagallis-aquatica	x	-	-	*	*	3
Σ 30	Σ 21	Σ 8	Σ 9			

[1]Probably brought in in seed mixtures.
[2]Probably planted.
x, occurrence in Alma plots, Rheinelbe and Zollverein.
The degrees of endangerment: 0 = extinct; 1 = extremely endangered, near extinction; 2 = highly endangered; 3 = vulnerable; V = decreasing; D = not known; * = no endangerment.
Source: after Gausmann et al. (2005).

percentage of polyhemerobes (mostly occurring on very unnatural sites) decreased with successional stage from Alma I to Rheinelbe III, and from Zollverein to Zollverein III in the bottom three panels (Fig. 23.9). This indicates that succession is progressing toward more natural vegetation over time. In the forests at Rheinelbe

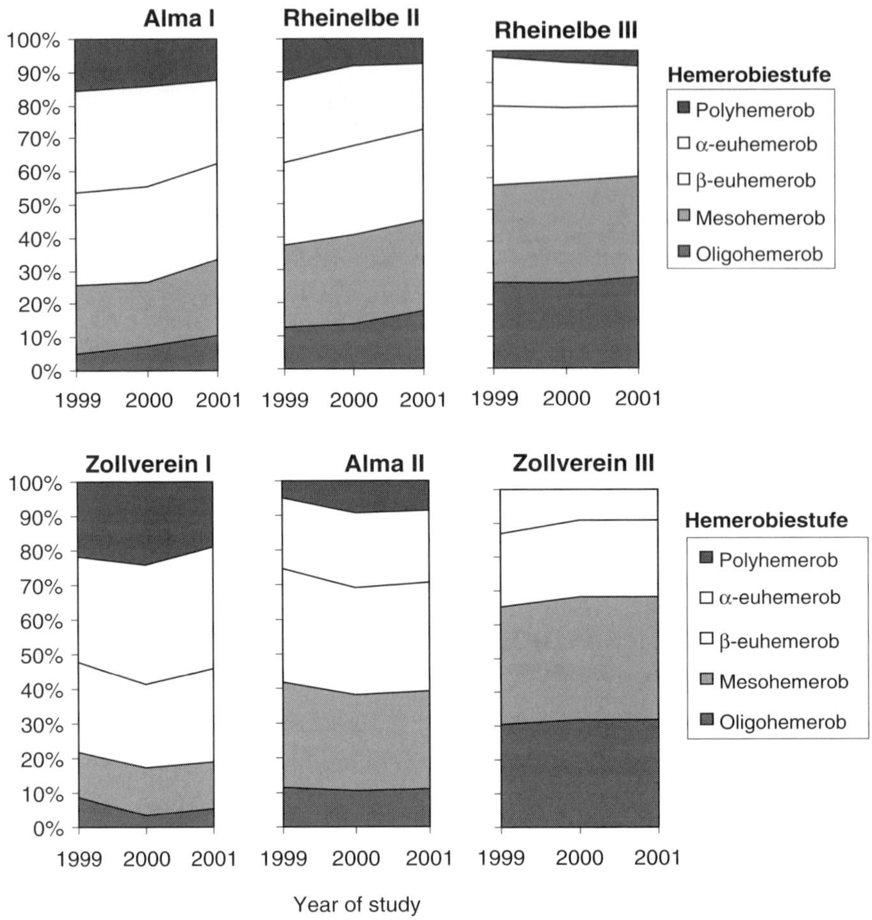

Fig. 23.9 Top: Degree of hemeroby of plants in sites at different stages of succession. The degree of human influence on the species decreases from poly- to oligohemerobic classes (top to bottom in the legend). I, initial stage; II, shrub stage; III, woodland. In the upper row the dynamic of change in hemerobic species diminishes gradually from stage I to stage III. In the lower row, the change between years is not that great at any one site, although between-site variation is great. Level of hemeroby after Frank and Klotz (1990). Bottom: The same as in the top, but in autumn. Species composition is very dynamic between years even at different soil depths. The percentage of diaspores and species composition differs between spring and fall

III and Zollverein III, we observed the percentage of meso- and oligohemerobic plants to be much higher (compare the two panels on the right in Fig. 23.9). This indicates that the forests have more native species than early successional sites with their many alien plant species.

We observed other patterns of change during succession. While the pioneer stages, including the shrubland, were species rich, the birch forests were very species poor. What happened in between these successional stages? At what stage along the successional continuum does species richness decline? These are a few of the many questions we wish to address by establishing and observing long-term permanent plots. This particular project has not been given any time limit, a very new aspect of funding scientific work in Germany. We hope that in perhaps 10 to 15 years we not only will learn what the species composition will be on these sites, but also will have results that might enable us to recommend improved methods and species for vegetating these disturbed patch types. The experiments of Jochimsen (1991b) reveal that the rich stage of tall native herbs could be established for a long time. We have observed this on other waste heaps, too, such as the one on "Hoppenbruch" (Schwiederowski, 1994). However, succession is delayed considerably in the birch forests. In the birch forests, the soil substrate has changed from pH 8 or 9 (very alkaline) to pH 4 or 3 (or lower, very acid) by weathering of pyrite (see articles in Wiggering and Kerth, 1991). The birch forests are not the historical climax vegetation in the Ruhr region, and we should expect climax forests consisting of beech or oaks. In 2004, the first signs of climax community development were detected in the herbaceous layer beech seedlings emerged. This provides hope that in time these birch stands will be succeeded by a more diverse forest.

Conclusion

Like other people in urban areas worldwide, the public in the Ruhr region needs and deserves a green environment. The question is, What is the best way to accelerate a sustainable greening of this highly altered landscape? The planting of exotic ornamental woody plants, especially in monocultures, is, indeed, not the right way and has not been successful. All these exotic stands (and those planted with indigenous trees and shrubs as well) have proven to be very problematical, because of soil erosion and the heavy die-back of many plants on this extremely difficult substrate of coal mining waste heaps (see Schwiederowski, 1994). This has made repeated supplementary planting necessary. The dry substrate, with its very low water-retaining capacity, requires irrigation. All these management efforts are expensive.

Taking advantage of the natural establishment of plants may provide a more inexpensive and sustainable means of reforesting these sites. Jochimsen (1991b) proposed sowing a mixture of native plants from ruderal sites nearby (mostly aesthetically pleasing members of the alliance *Dauco-Melilotion*). This plant community can establish itself on coal mining waste heaps without the aid of humans but needs time to develop (in Alma nearly 6 to 10 years). After the perennial herbs have established, forest succession can also naturally occur later and without much, if any, cost. This can be seen in the more than 10-year old experimental fields of Jochimsen (1991b).

This approach would not apply to waste places or other urban settings where there are not many mature native trees in the neighborhood to serve as seed sources. In such situations reafforestation can be realized only by planting native species in the way Miyawaki (see Chapter 12) has successfully established in Japan. In the case of China, since it has one of the richest floras in the world, there is no need to collect new species from around the world, as some have advised, and Chinese landscape architects and plant nurserymen should remember the rich biotic heritage of their large country for providing plant materials and seed sources for greening their human-dominated environments.

References

Frank, D., and Klotz, S. (1990) Biologisch-Ökologische Daten zur Flora der DDR. 2. Aufl., Wiss. Beitr. Univ. Halle, vol 32.

Haeupler, H., Kert, C., and Schürmann, M. (2003) Industriewal Ruhrgebiet–Floristisch-vegetationskundliche Betrachtungen. IÖR-Schriften 39:159–167.

Jochimsen, M. (1991a) Ecological aspects of plant settlement on coal mining heaps. In: Wiggering, H., and Kerth, M. (eds.) Coal Mining Waste Heaps—Impact on and Change of a Densely Populated Industrial Region. Vieweg, Braunschweig/Wiesbaden pp. 155–162.

Jochimsen, M. (1991b) Reclamation of coal mining heaps based on natural succession. In: Wiggering, H. and Kerth, M. (eds.) Coal Mining Waste Heaps—Impact on and Change of a Densely Populated Industrial Region. Vieweg, Braunschweig/Wiesbaden, pp. 189–194.

Kert, C. (2001) Untersuchung im Rahmen eines Langzeitmonitorings auf Restflächen des Bergbaus im Ruhrgebiet. 2. Vegetationsdynamik. Unpublished thesis, Ruhr-University Bochum.

Knapp, R. (ed.) (1984) Sampling methods and taxon analysis in vegetation science. In: Handbook of Vegetation Science, part 4. Junk, The Hague, Boston, Lancaster.

Raunkiaer, C. (1934) The Life Forms of Plants and Statistical Plant Geography: Being the Collected Papers of Raunkiaer. Translated from Danish by H. Gilbert-Carter. Oxford University Press, London, pp. 201–282.

Schürmann, M. (2001) Untersuchung im Rahmen eines Langzeitmonitorings auf Restflächen des Bergbaus im Ruhrgebiet. 1. Flora und Samenbank. Thesis, Ruhr-University Bochum.

Schwiederowski, C. (1994) Vergleich von Gehölzbeständen auf künstlich begrünten und sich selbst überlassenen Bergehalden. Am Beispiel des Landschaftsbauwerks Halde Hoppenbruch und der Halde Holzheide. Thesis, Ruhr Universität Bochum.

Steinhardt, U., Herzog, F., Lausch, A., Müller, E., and Lehmann, S. (1999) Hemeroby index for landscape monitoring and evaluation. In: Pykh, Y.A., Hyatt, D.E., Lenz, R.J. (eds.) Environmental Indices—System Analysis Approach. EOLSS, Oxford, pp. 237–254.

Weiss, J. (2003) Industriewald Ruhrgebiet. Freiraumentwicklung durch Brachesukzession. LÖBF-Mitteilungen 1/03, pp. 55–59.

Weiss, J., Burghardt, W., Gausmann, P., et al. (2005) Nature returns to abandoned industrial land: monitoring succession in urban-industrial woodlands in the German Ruhr. In: Kowarik, I., and Körner, S. (eds.): Wild Urban Woodlands. New Perspectives for Urban Forestry. Springer, Berlin, Heidelberg, New York, pp. 143–162.

Wiggering, H., and Kerth, M. (eds.) (1991) Coal Mining Waste Heaps—Impact on and Change of a Densely Populated Industrial Region. Vieweg, Braunschweig/Wiesbaden, (German with English abstracts).

24
Selection of Pollution-Tolerant Trees for Restoration of Degraded Forests and Evaluation of the Experimental Restoration Practices at the Ulsan Industrial Complex, Korea

Chang Seok Lee and Yong Chan Cho

In many parts of Korea, forest vegetation near industrial complexes began to show symptoms of decline in the 1970s, when active industrialization was begun (National Institute of Environmental Research, 1981; Lee, 1992). These symptoms also appeared in urban areas that have experienced chronic air pollution since the 1990s (National Institute of Environmental Research, 1990; Kim, 1991, 1994; Lee, 1992; Rhyu, 1994; Rhyu and Kim, 1994). Given that industrialized and urbanized areas are continuously expanding in Korea, forest areas showing such symptoms are likely to increase in the future. Symptoms of vegetation decline are due to the indirect impacts of soil acidification as well as to the direct effects of air pollutants on the plants themselves (Kitajima, 1988; Kim, 1991; Rhyu and Kim, 1994; Freedman, 1995; Gunn, 1995). Vegetation decline induces structural simplification and functional weakening of plant communities, and consequently leads to negative effects on other biotic communities, such as animal and microbial communities (Smith, 1990; Freedman, 1995). Restoration of degraded ecosystems, therefore, is urgently required to prevent the spread of such pollution damage (Gunn, 1995).

The Society for Ecological Restoration (2002) defines restoration as "the process of assisting the recovery of an ecosystem that has been degraded, damaged, or destroyed." Ecological restoration is the return of an ecosystem to a close approximation of its condition prior to a disturbance (National Research Council, 1991), and it can also be defined as the re-creation of naturalistic and self-maintaining ecosystems (Berger, 1993).

When we seek to restore an ecosystem damaged by environmental pollution, we can achieve this goal by preventing or reducing further pollution from being produced at the source, or adapting to the situation by planting species tolerant to pollutants (Bradshaw, 1992; Gunn, 1995, 1996; Dobson et al., 1997). Species tolerant to environmental pollution are those that can persist through growth and reproduction, or even expand their distribution range in the polluted environment, while pollution-sensitive species disappear from such environments. These processes consequently lead to changes in species composition at the community level (Barrett and Bush, 1991).

This chapter reports on a study that had four goals. The first was to select tolerant plant species by comparing their growth rates in polluted and control sites. Additional comparisons were made by growing plants in polluted soil transported from the industrial complexes, with and without treatments to reduce acidity (addition of dolomite).

M.M. Carreiro et al. (eds.), *Ecology, Planning, and Management of Urban Forests: International Perspectives.*

The second was to investigate amelioration effects of dolomite and sludge treatment on the acidified forest soil in Ulsan industrial park. The third was to evaluate the effects of dolomite and sludge application on leaf growth of *Quercus serrata, Alnus firma*, and *Ligustrum japonicum* that were planted for restoration. The fourth was to investigate the processes of plant community development after the trees were planted.

The Study Area

The Ulsan industrial complex is located on the southeastern coast of the Korean peninsula and is representative of many industrial complexes in Korea (Fig. 24.1). Construction of this industrial complex was begun in the late 1960s and is still expanding. The industrial activity in Ulsan is dominated by heavy chemical industries, including the petrochemical industry, that emit 60,000 metric tons of sulfur dioxide (SO_2) annually. These SO_2 emissions result in mean atmospheric SO_2 concentrations of 0.03 parts per million (ppm) and a maximum daily mean of 0.10 ppm. Such severe air pollution has not only caused forests to degrade to grassland and even bare ground (Fig. 24.1), but also caused the soil to acidify. Mean soil pH at the industrial site was as low as 4.1, as compared to that of the unpolluted site, 20 km south, which was 4.8 (Kim et al., 1996). The Ulsan site also had a low base cation content (Ca^{2+} and Mg^{2+}), which was from one-third to one-half that of unpolluted healthy areas nearby, and had an Al^{3+} content that was two to three times higher than that of unpolluted areas (Kim et al., 1996).

The Yeocheon industrial complex on the southern coast of the Korean peninsula (Fig. 24.1) was built in the early 1970s. Since then the air pollutants that originate from the many industrial facilities have induced visible changes in the spatial distribution and structure of nearby vegetation. Just as in Ulsan, industrial activities focus on the heavy chemical industry, including petrochemicals. The major pollutant is also SO_2 with emissions of 5000 metric tons per year, with an average atmospheric SO_2 concentration of 0.02 ppm and an annual mean maximum daily value of 0.06 ppm. Such severe air pollution not only caused vegetation to degrade from forest to grassland or bare ground, but also led to soil acidification with pH ranging from 3.5 to 6.7, with a mean of 4.4 (Kim et al., 1996). Moreover, due to such acidification, the soil contains a Ca^{2+} and Mg^{2+} content that is from one-third to one-half that of control reference areas and an Al^{3+} content two or three times higher than control reference areas (Kim et al., 1996).

Mt. Dotjil, chosen as the target restoration area for this study, is the most severely damaged area in the Ulsan industrial complex. This mountain is surrounded by industrial facilities on its eastern, western, and southern slopes, and by landfill and urban areas on its northern slope. As a result, this mountain has continuously experienced severe environmental pollution since the late 1960s when the industrial complex began operation. Vegetation patches occur interspersed with bare soil, and consists of grassland and several tree plantations (Fig. 24.2). Our restoration focused on areas of bare soil and grassland having low ecological quality in both structure and function. Land with vegetation coverage below 20% was classified as

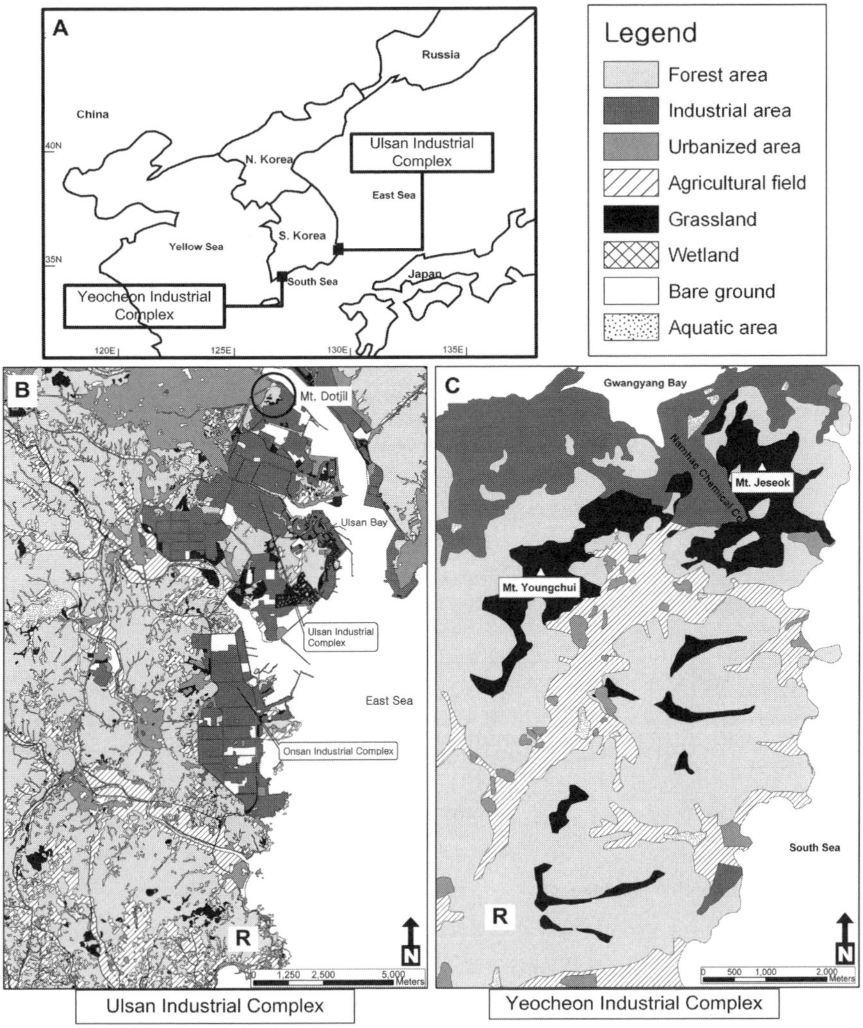

Fig. 24.1 **(A)** Regional map of the Korean peninsula showing location of the research sites at the industrial complexes of Ulsan and Yeocheon. **(B)** Land-use and land-cover pattern maps at the Ulsan site. **(C)** The Yeocheon site

bare ground (hereafter BG). The BG patch type was divided into two groups depending on the importance values of the grass, *Miscanthus sinensis*, or the vine, *Pueraria thunbergiana* (kudzu). Grassland was also divided into two types, one dominated by *M. sinensis* (hereafter GG) and the other by several forbs (grassland forb, hereafter GF). *M. sinensis* maintains a dense ground cover and thereby inhibits other plants from establishing in their matrix. Control reference sites were chosen in an unpolluted area more than 20 km south of the Ulsan industrial complex (Fig. 24.1). Ambient SO_2 of this area was less than half that of the polluted area

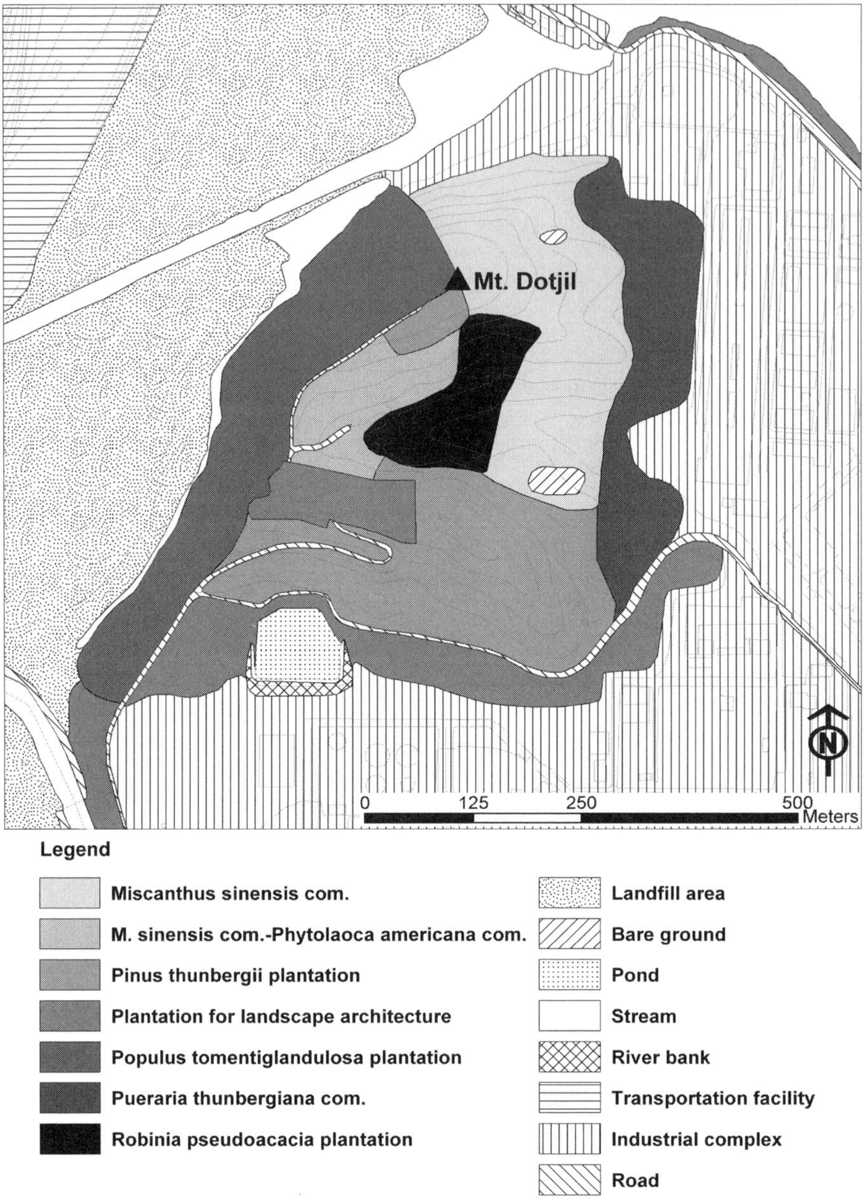

Fig. 24.2 A landscape ecological map, including land-use pattern, on Mt. Dotjil designated as the target area for restoration and its surrounding area in the Ulsan industrial complex

(Park, 1991). This control reference area also exhibited a better environmental condition than that of the polluted area in the physicochemical properties of soil, as mentioned earlier.

Methods

Selection of Tolerant Plant Species

Field Experiment

To select tolerant plants, we carried out two experiments: one involving on-site transplants and the other involving pot cultures grown in a greenhouse. In one study, sample plants obtained from seeds collected in the forest near the polluted site were transplanted into polluted and unpolluted sites. In the second study, plants were grown in pots containing polluted or amended soils. Soil characteristics of the study site and soils placed in pots are shown in Tables 24.1 and 24.2, respectively. The polluted site was located 300 m from the pollution source, and directly exposed to air pollutants. The control site was approximately 20 km from the industrial pollutant source, and protected from the direct effects of air pollutants because of topographic conditions (Fig. 24.1).

For the field study, 3-year-old healthy seedlings of 11 tree and shrub species were planted in April 1995. There were 10 replicate seedlings per species used in this on-site transplantation study. In the polluted site, sample plants were planted in bare ground formed naturally due to vegetation loss from severe pollution damage, and bare ground that resulted from the removal of the existing vegetation dominated by the grass, *M. sinensis*. Control sites were also prepared by removing existing forest vegetation. Seedlings were planted regularly in rows at an interval of 0.5 m. To reduce bias due to planting location, the order of plant placement in each row was random.

After transplantation, seedling growth response and performance was measured as leaf area (LA) and biomass (B), estimated by multiplying the square of stem diameter by plant height (D^2H) (Long and Hallgren, 1987). Measurements were

Table 24.1 Soil chemical properties in the experimental transplantation sites near the Ulsan and the Yeocheon industrial complexes; data are the mean values of three samples per site ± (1 standard deviation [SD])

Site	Organic matter (%)	pH	T-N (%)	P	K^+	Ca^{2+}	Mg^{2+}	Al^{3+}
				(ppm)				
				Ulsan				
Reference	14.0	4.76	0.75	27.5	102.3	267.1	132.9	243.4
	(3.0)	(0.84)	(0.17)	(3.5)	(25.8)	(95.9)	(38.8)	(15.1)
Polluted	10.8	4.08	0.43	14.5	104.2	151.1	45.8	448.6
	(5.2)	(0.26)	(0.12)	(2.8)	(23.7)	(33.9)	(25.9)	(25.7)
				Yeocheon				
Reference	13.9	4.95	0.97	24.3	112.3	254.3	132.9	265.7
	(3.3)	(0.75)	(0.26)	(4.7)	(25.9)	(94.9)	(55.2)	(17.5)
Polluted	10.4	4.28	0.95	23.9	95.9	145.3	31.2	475.3
	(2.0)	(0.36)	(0.24)	(5.2)	(31.5)	(78.5)	(28.0)	(23.4)

Table 24.2 Chemical properties of soil used to select tolerant plants in the pot culture; the polluted soil was collected near the Ulsan and the Yeocheon industrial complexes; amended soil was treated with dolomite; data are the mean values of three samples ± (1 SD)

Site	Organic matter (%)	pH	T-N (%)	P	K^+	Ca^{2+}	Mg^{2+}	Al^{3+}
				(ppm)				
				Ulsan				
Polluted	14.4	4.93	0.82	29.2	93.7	264.5	121.9	233.7
	(4.6)	(0.64)	(0.19)	(4.8)	(27.8)	(65.9)	(36.2)	(26.7)
Amended	15.2	4.43	0.37	16.7	94.2	108.1	51.2	425.6
	(4.3)	(0.36)	(0.14)	(3.9)	(25.3)	(38.9)	(35.7)	(36.3)
				Yeocheon				
Polluted	16.6	4.36	1.02	23.8	94.8	271.5	44.0	313.9
	(5.4)	(0.68)	(0.23)	(5.1)	(16.3)	(64.7)	(27.1)	(41.5)
Amended	15.4	5.56	1.07	22.9	94.7	1318.2	44.0	248.6
	(3.3)	(0.74)	(0.29)	(4.7)	(21.5)	(265.6)	(27.1)	(34.9)

carried out at 1-month intervals for 6 months from April to September and performed for three individuals per species. We determined the growth of leaf area for each species from measurements of these three individuals. The growth index of biomass was derived from the difference of biomass obtained during the same experimental period—6 months. Tolerance ratios (tolerance ratios of leaf area, Tr_{LA} and biomass, Tr_{B}) were calculated for each species using the ratios of growth coefficients in the polluted sites to those in the controls:

$$Tr_{LA} = \frac{\text{Growth coefficient of leaf area in the polluted site (or pot)}}{\text{Growth coefficient of leaf area in the control site (or pot)}} \quad (1)$$

$$Tr_{B} = \frac{\text{Growth coefficient of biomass in the polluted site (or pot)}}{\text{Growth coefficient of biomass in the control site (or pot)}} \quad (2)$$

We obtained a tolerance index (*Ti*) for plant growth by averaging the mean of the two tolerance ratios described above:

$$Ti = \frac{Tr_{\mathrm{LA}} + Tr_{\mathrm{B}}}{2} \quad (3)$$

We regarded a species with a *Ti* value greater than 1 to be a pollution tolerant species.

Pot Experiment

Polluted soil was transported from the Ulsan and the Yeocheon industrial complexes for the pot experiments. A portion of the soil was amended with dolomite ($Ca{\bullet}Mg\ (CO_3)_2$) to increase soil pH. Dolomite raises soil pH and increases available Ca^{2+} and Mg^{2+} due to the following chemical reactions in soil solution (Kreutzer, 1995):

Initial chemical reaction: $Ca \bullet Mg(CO_3)_2 + 2H^+ \rightarrow 2HCO_3 - + Ca^{2+} + Mg^{2+}$

Second reaction: $2HCO_3^- + 2H^+ \rightarrow 2CO_2 + 2H_2O$

Net reaction: $Ca \bullet Mg(CO_3)_2 + 4H^+ \rightarrow Ca^{2+} + Mg^{2+} + 2CO_2 + 2H_2O$

Sample plants were cultivated in plastic cylindrical pots with a diameter of 15 cm and a height of 30 cm. Each pot contained 3 kg of soil. For soil amelioration, each pot received 2970 g soil amended with 30 g dolomite. The pot experiment was performed using the same procedures over the same period as those of field experiment. For the pot experiment, 3-year-old healthy seedlings of 10 tree and shrub species were planted in April 1995. One sample plant was planted in each pot, and there were five replicate pots per species.

Soil Analysis

The following soil properties were measured: pH, organic matter, and total sulfur (S), total nitrogen (N), phosphorus (P), K^+, Ca^{2+}, Mg^{2+}, and Al^{3+} content. Soil pH was measured with a bench-top probe after mixing the soil with distilled water (1:5 ratio, weight per volume) and filtering the extract (Whatman No. 44 paper). Organic matter (OM) content was obtained by measuring the loss after ignition for 4 hours in a muffle furnace at 400°C. Total nitrogen was measured using the micro-Kjeldahl method (Jackson, 1967). Available P was extracted in 1N ammonium fluoride (pH 7.0) and exchangeable K^+, Ca^{2+}, Mg^{2+}, and Al^{3+} content were extracted with 1N ammonium acetate (pH 7.0 for K, Ca, and Mg, and pH = 4.0 for Al) using inductively coupled plasma (ICP) atomic emission spectrometry (Shimadzu ICPQ-1000) as described in Allen et al. (1986). Total S was determined by a method of Bardsley and Lancaster (1960).

Restoration Practice and Evaluation of the Effects

Design for Restorative Treatment

Mt. Dotjil was divided into three vegetation categories as follows: BG (≤20% of ground vegetation cover); GG (>20% ground vegetation that was dominated by *M. sinensis*); and GF (>20% ground cover that was dominated by the vine *P. thunbergiana* and other forb species). A total of 36 plots (2 m × 2 m) were established; there were 12 plots in each vegetation category (three plots for each of the following treatments: dolomite addition, sewage application, mixture of dolomite and sludge, and control). Chemical characteristics of the

Table 24.3 Chemical properties of sewage sludge used as a substrate ameliorator in this experiment; data were obtained from the mean values of five samples and values in parenthesis indicate standard deviation

Environmental factors	Content
Organic matter (%)	77.0 (4.7)
Total nitrogen (%)	3.2 (0.6)
Available phosphorus (%)	0.4 (0.1)
K (ppm)	46.9 (15.3)
Ca (ppm)	14,026.8 (1,840.6)
Mg (ppm)	1,567.4 (198.9)

sludge are given in Table 24.3. Dolomite, sewage sludge, or the 1:1 volume per volume (v/v) mixture of both was applied once at a rate of 1000 kg ha^{-1} following Kim et al. (1996). A total of 27 three-year-old seedlings (nine each for *Q. serrata, A. firma*, and *L. japonicum*) were planted in each plot in March 1996.

Evaluation of Soil Restoration Treatments

The effects of the soil treatments on restoring healthier soil characteristics were evaluated. Soil samples of the top 10-cm horizon were collected in September 1997 at five random points in each plot, pooled, air-dried at room temperature, and sieved through a 2-mm mesh. Methods and procedures of soil analysis are the same as those applied to diagnose the difference in properties between polluted and unpolluted sites. Soil characteristics (pH, OM, N, P, K, Ca, Mg, and Al) in the four treatment plots (control, dolomite, sludge, and mixed) were compared with one-way analysis of variance (ANOVA) and Tukey's honestly significant difference (HSD) test at $\alpha = .05$ (Statistical Analysis System, 2001).

Leaf Growth of *Q. serrata, A. firma, and L. japonicum*

In September 1997, after two growing seasons, relative growths of total leaf area in the dolomite, sludge, and mixed plots were measured for *Q. serrata, A. firma*, and *L. japonicum*. The percent relative growth of each species was determined by calculating the fraction of the total leaf area in a particular treatment plot to the total leaf area in the control plot and multiplying by 100. Total leaf area of each species in each plot was calculated by multiplying the number of leaves by the mean area of five randomly selected leaves that were measured with a leaf area meter (Ushikata X-Plan 380). The relative growth of the total leaf area in the four treatment plots was then compared with ANOVA and HSD at $\alpha = .05$ (Statistical Analysis System, 2001).

Vegetation Analysis

A total of five reference plots (20 m × 20 m) were established in a relatively undisturbed forest approximately 20 km south of Mt. Dotjil (Fig. 24.1). The purpose of designating reference plots was to compare them, as a target community, with the restored plots in Mt. Dotjil. Neither soil amelioration nor planting was conducted in the reference plots.

Vegetation data were collected twice in the BG, GG, and GF stands and the restored stands in Mt. Dotjil in 1998 and 2002 and once in the reference plots in 2002. Five plots were surveyed for each stand. Plots (2 m × 2 m) were used for BG, GG, and FG stands, and 5 m × 5 m plots were used in the restored stands. All the plant species were identified following Lee (1985) and Korea Plant Name Index (2004). Dominance of each species was estimated with the ordinal scale of Braun-Blanquet (1964), converted to cover classes (1 for <5% up to 5 for >75%), and subjected to detrended correspondence analysis (DCA) ordination (Hill, 1979). As a measure of species diversity and dominance, a rank abundance curve (Kent and Cocker, 1992; Lee et al., 2002; Magurran, 2004) was constructed and the Shannon-Wiener diversity index (H') was calculated for each stand type.

We determined and compared the heights and biomass of *M. sinensis* in 1998 as a measure of preemptive dominance of this species in the GG plots and in 2002 in the restored and reference stands. To do this, five subplots (20 cm × 20 cm) were established randomly in each of 1998 and 2002. Plant height was determined using measuring tape and by averaging values for five randomly chosen stems in each subplot. Above-ground vbiomass was harvested, dried for 48 hours at 80°C, and weighed. The heights and biomass of *M. sinensis* among plots were compared with ANOVA and HSD at $\alpha = .05$ (Statistical Analysis System, 2001).

Results

Tolerant Species Selected by Transplant Experiment

Eight of the 11 species tested in a transplant experiment in the Ulsan industrial area qualified as tolerant species with a *Ti* value greater than 1 (Table 24.4). The five oaks, which dominate nearby native forests, are endemic species in Korea. *Styrax japonica* forms a dense pure stand around the polluted industrial complex naturally, and two broadleaf evergreen shrubs, *L. japonicum* and *Poncirus trifoliata*, are included among the eight tolerant species (Table 24.4). Only three species were found to be tolerant in the conditions at the Yeocheon industrial area (Table 24.4).

Table 24.4 Tolerance index of species based on three seedlings transplanted to the Ulsan and the Yeocheon sites

	Ulsan			Yeocheon		
	Tolerance ratio		Tolerance	Tolerance ratio		Tolerance
Species	Leaf area	Biomass	index	Leaf area	Biomass	index
Quercus aliena	0.84	3.40	2.12	1.60	1.00	1.30
Q. serrata	1.32	2.43	1.88	0.80	0.94	0.87
Q. acutissima	1.18	1.50	1.34	0.92	1.19	1.06
Poncirus trifoliata	1.10	1.45	1.28	–	–	–
Q. dentata	1.20	1.33	1.27	0.57	0.71	0.64
Q. mongolica	1.23	1.27	1.25	0.98	0.72	0.85
Ligustrum japonicum	0.98	1.40	1.19	1.00	0.88	0.94
Styrax japonica	1.14	1.09	1.14	1.26	1.03	1.15
L. obtusifolium	1.49	0.29	0.89	–	–	–
Alnus firma	1.09	0.67	0.88	0.93	0.29	0.61
Celtis sinensis	0.72	0.11	0.42	1.00	0.75	0.88
Eurya japonica	–	–	–	0.59	0.09	0.34

Note: Species are ranked from the most to least tolerant. Growth coefficient of leaf area was obtained from growth equation that shows a change of leaf area measured at 1-month intervals for 6 months from April to September, 1995, in the field. Growth index for biomass was obtained from the difference of biomass obtained during the same experimental period—6 months. The tolerance ratios were calculated as the ratios of growth coefficients of leaf area and biomass in the polluted site to those in the control site. The tolerance index is the mean of the two tolerance ratios calculated for leaf area and biomass. Species with a tolerance indexes >1 are considered good candidates for future restoration experiments in industrially polluted sites.

Tolerant Species Selected from Pot Culture

We found that four of 10 species were tolerant to polluted soil collected from the Ulsan and the Yeocheon industrial complex when grown in pot cultures (Table 24.5). Two native oaks (*Q. aliena* and *Q. mongolica*) and *S. japonica*, which were selected as tolerant species in the transplantation experiment, are included among the four tolerant species (Table 24.5). *A. firma*, a nitrogen-fixer, was found to tolerate soils, but it is a nonnative introduced from Japan. However, this species has been used for large-scale afforestation programs in the southern province in Korea to prevent landslides and has now become naturalized in Korea (Cho and Lee, 1998; Lee and Cho, 1998).

Soil Characteristics and Leaf Growth

Dolomite neutralized soil, and increased Ca and Mg significantly. Particularly, reduction of mobile Al was noted in the neutralized soil. Sludge also improved soil fertility with elevated OM, N, and P. Combination effects from a mixture of both

Table 24.5 Tolerance index of species grown as seedlings in pots containing polluted soils collected at the Ulsan and the Yeocheon industrial complexes

	Ulsan			Yeocheon		
	Tolerance ratio		Tolerance	Tolerance ratio		Tolerance
Species	Leaf area	Biomass	index	Leaf area	Biomass	index
Alnus firma	8.21	0.28	4.25	2.71	0.64	1.68
Styrax japonica	1.50	1.60	1.50	0.26	0.87	0.57
Quercus aliena	1.47	1.10	1.29	2.03	1.89	1.96
Q. mongolica	1.40	0.74	1.07	1.32	0.65	0.99
Ligustrum obtusifolium	1.46	0.49	0.98	0.52	0.49	0.51
Q. serrata	0.56	1.30	0.93	2.28	13.00	7.63
Q. acutissima	0.73	0.33	0.53	1.00	0.45	0.73
L. japonicum	0.54	0.37	0.46	0.86	0.37	0.62
Q. dentata	0.39	0.36	0.38	0.63	1.40	1.02
Celtis sinensis	0.25	0.29	0.27	0.74	0.15	0.45

Note: Species are ranked from most to least tolerant. Growth coefficient of leaf area was obtained from growth equation that shows a change of leaf area during cultivation from April to September, 1995, in green house. Growth index of biomass was calculated as the difference in biomass obtained during the same experimental period—6 months. The tolerance index is the mean ratio of the growth coefficients for leaf area and biomass in polluted soil to those in soil ameliorated with dolomite.

ameliorators were seen in pH and OM (except GG plots), N (except GG plots), P, Ca, Mg, and Al contents (Fig. 24.3).

The leaf growth of *Q. serrata* increased in all treatments (dolomite, sludge, and mixed) in the BG ($p<.001$) and GF ($p<.01$), but not in GG plots ($p=.05$). *L. japonicum* expanded its leaves significantly with all treatments in BG ($p<.001$), GF ($p<.01$), and GG plots ($p=.05$). The effects of soil amelioration on *A. firma* were not significant ($p=.05$) in GG and GF plots. The dolomite and mixed treatments increased the leaf growth of *A. firma* slightly in BG plots, but much less so than *Q. serrata* and *L. japonicum*. No statistically significant variation was found among the effects of dolomite, sludge, and mixed treatments across all species (Fig. 24.4).

Vegetation Structure, Composition, and Diversity

From 1998 to 2002, the height and total biomass of *M. sinensis* decreased from 170.3 ± 5.5 cm to 124.0 ± 14.4 cm and from 812.2 ± 162.0 $g \cdot m^{-2}$ to 198.9 ± 60.7 $g \cdot m^{-2}$, respectively (Fig. 24.5). On the other hand, height and total biomass of *M. sinensis* in the reference stands were 47.0 ± 13.7 cm and 25.0 ± 19.1 $g \cdot m^{-2}$, respectively (Fig. 24.5).

Ordination of stands left in a natural condition was carried out to evaluate the likelihood of natural recovery of the degraded vegetation (Fig. 24.6). The GF stands were restricted at the left lower corner on a plane formed by axes I and II without any relation to the survey years 1998 and 2002 (Fig. 24.6). The GG stands showed a trend

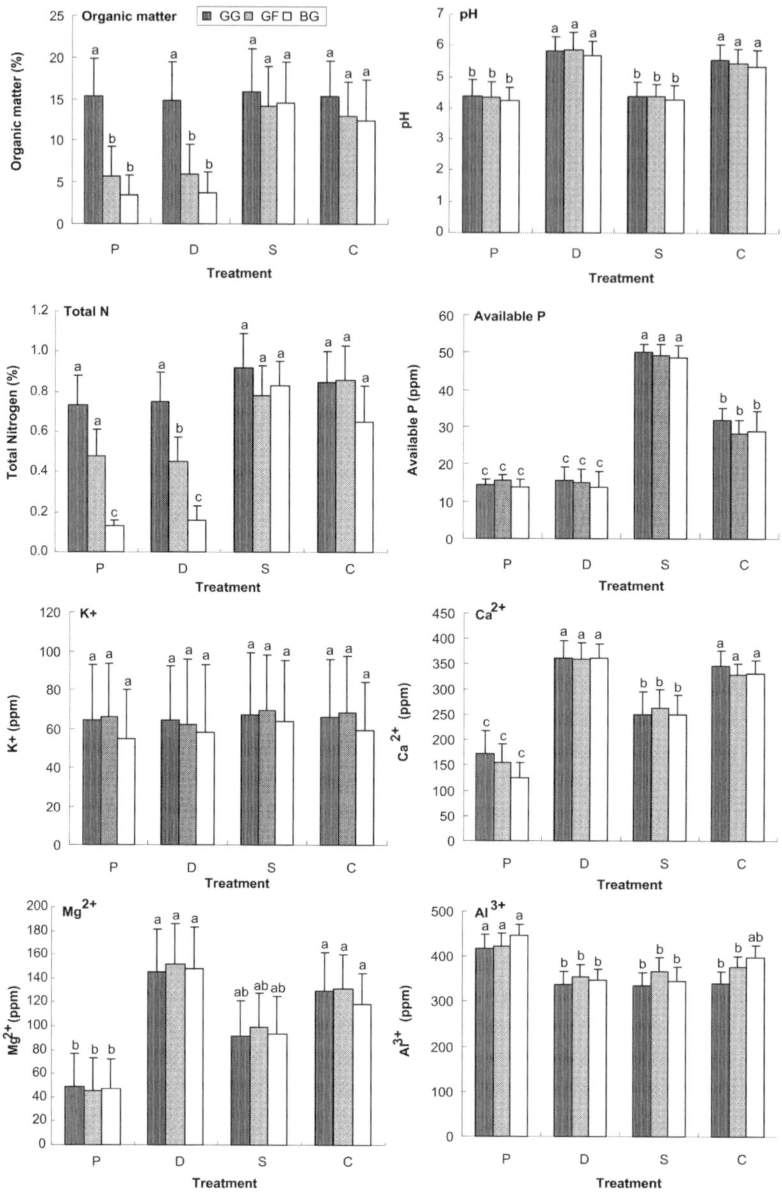

Fig. 24.3 Effects of control (C, $n = 5$), dolomite (D, $n = 5$), sludge (S, $n = 5$) and mixed (M, $n = 5$) treatments on soil characteristics in bare ground (BG), grassland (GG), and forb land (GF) plots in the Ulsan industrial park. Each bar was expressed with mean and standard error of mean. Tukey's honestly significant difference (HSD) test was conducted on each of the parameters that show a statistically significant difference among the three types of treatments at $\alpha = .05$; the means with the same alphabetical character (in superscript), for each parameter, were not different from each other

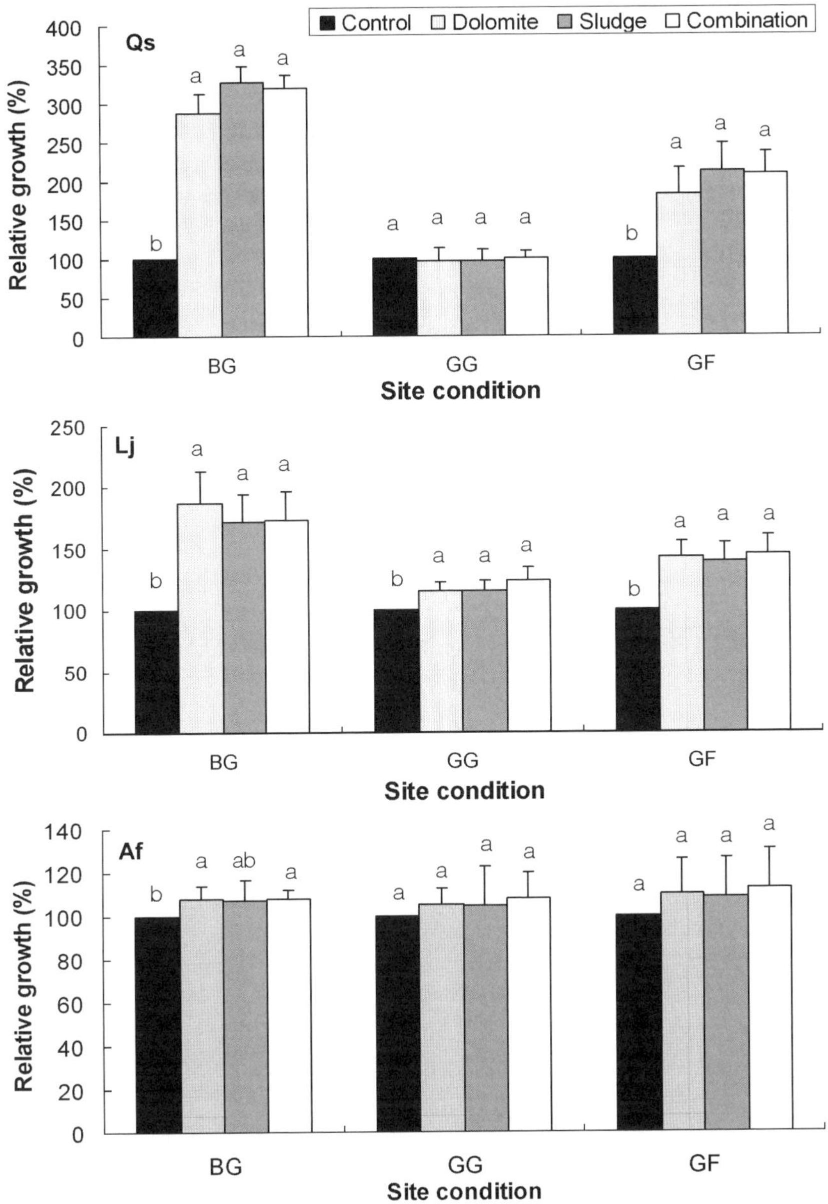

Fig. 24.4 Growth response of *Quercus serrata* (Qs), *Ligustrum japonicum* (Lj), and *Alnus firma* (Af) to control, dolomite, sewage sludge, and mixed treatments in the bare ground (BG), grassland (GG), and forb land (GF) plots

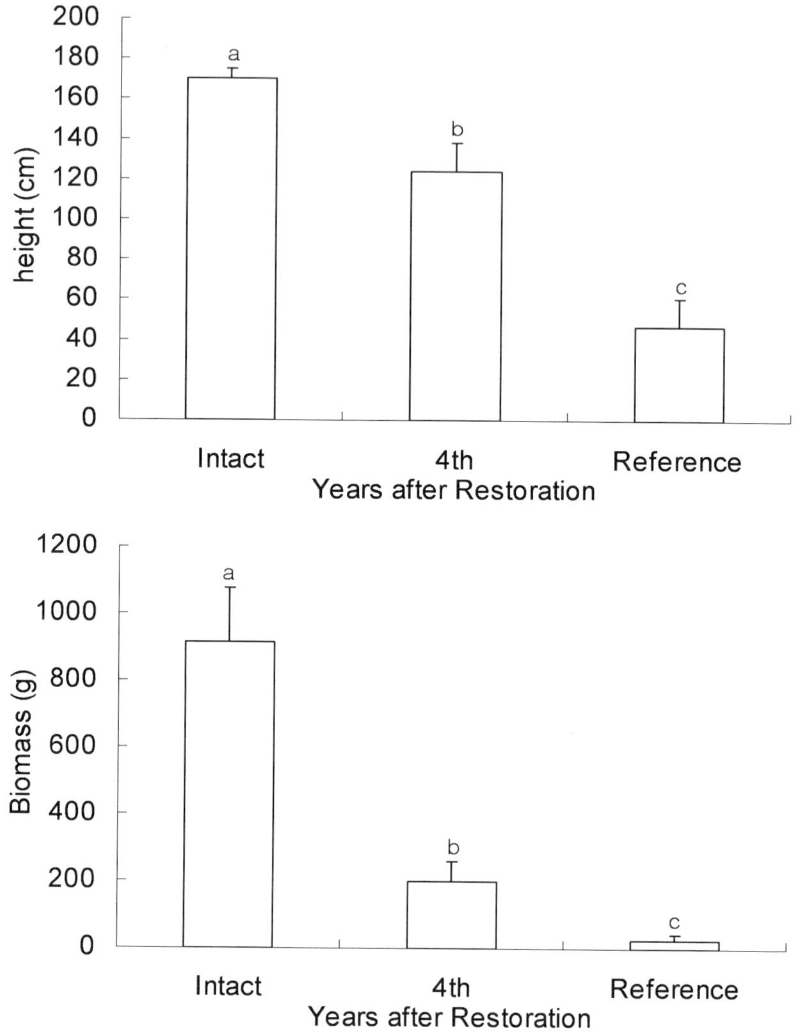

Fig. 24.5 Heights and biomass productions of *Miscanthus sinensis* in grassland (intact), the restored (4th year), and reference stands

to be arranged on the right side on the plane without any remarkable difference in species composition in the different survey years. The BG sites were arranged close to GG stands. The naturally restored stands, those expressed as the former bare ground sites (FB), were distributed between GG and GF stands. Such results imply that bare ground was replaced by GG stands dominated by *M. sinensis* and GF stands dominated by *P. thunbergiana* depending on the site. In reality, *M. sinensis* and *P. thunbergiana* are the first invaders on concave and convex micro-topography, respectively. But both GG stands and GF stands did not show any change in species

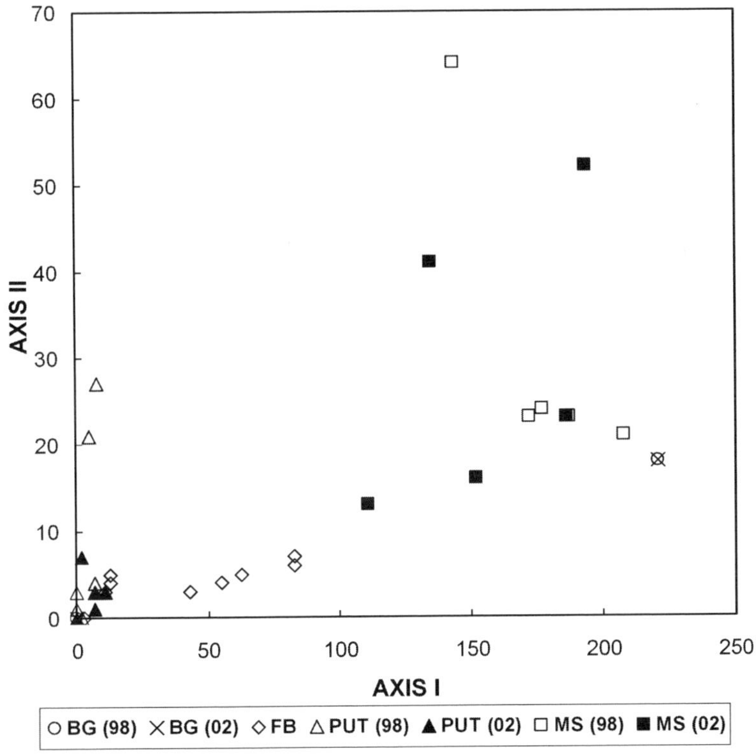

Fig. 24.6 Detrended correspondence analysis (DCA) ordination of several degraded vegetation, which were left in natural condition without any restorative treatment. Data from 1998 and 2002 indicate years that vegetation survey was carried out. BG, bare ground; GF, forb grassland dominated by *Pueraria thunbergiana* (PUT); GG, grassland dominated by *Miscanthus sinensis* (MS)

composition during the past 4 years (Fig. 24.6) and their composition was similar to that of approximately 10 years ago as well (Lee, 1992).

In ordinations carried out to assess restorative effects, the stands where tolerant species were introduced were arranged between sites left in the degraded state, such as BG, GG and GF, and the reference sites (Fig. 24.7). Moreover, the restored stands tended to move closer to the reference stands over time.

Rank-abundance relationships (Fig. 24.8) revealed two trends in species diversity. First, species richness responded positively to the restorative treatment. Second, the degree of dominance, determined by the steepness of the curves, declined in response to the restorative treatment. In response to the restorative treatment, the relative abundance of average species was higher, indicating a more even distribution of space occupancy. As a consequence, both richness and evenness increased with time since restoration. The Shannon-Wiener index was also higher in the restored stands and increased over time, reflecting the same results observed in the species rank-abundance relationships.

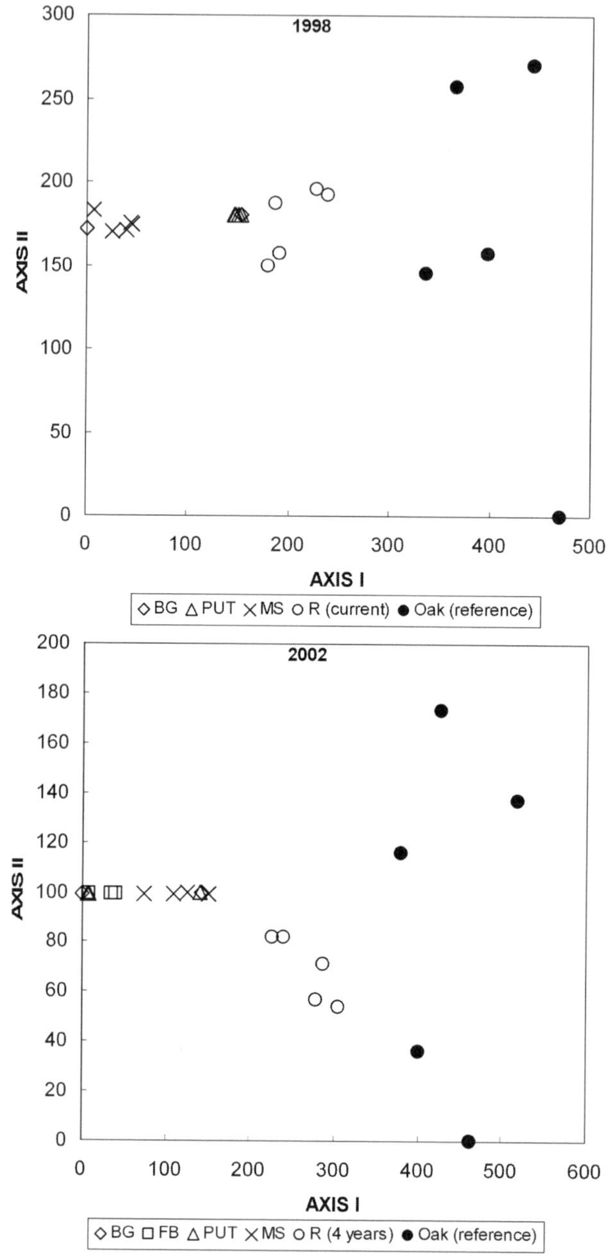

Fig. 24.7 DCA ordination of vegetation including degraded (BG, FB, GF, and GG), restored, and reference stands. BG, bare ground; GF, forb land dominated by *Pueraria thunbergiana*; GG, grassland dominated by *Miscanthus sinensis*; FB, former bare ground; R (current), the restored stands surveyed in 1998, current year of restoration; R (4 years), the restored stands surveyed in 2002, passed 4 years after restoration; Oak (reference), oak stands designated as the reference stands

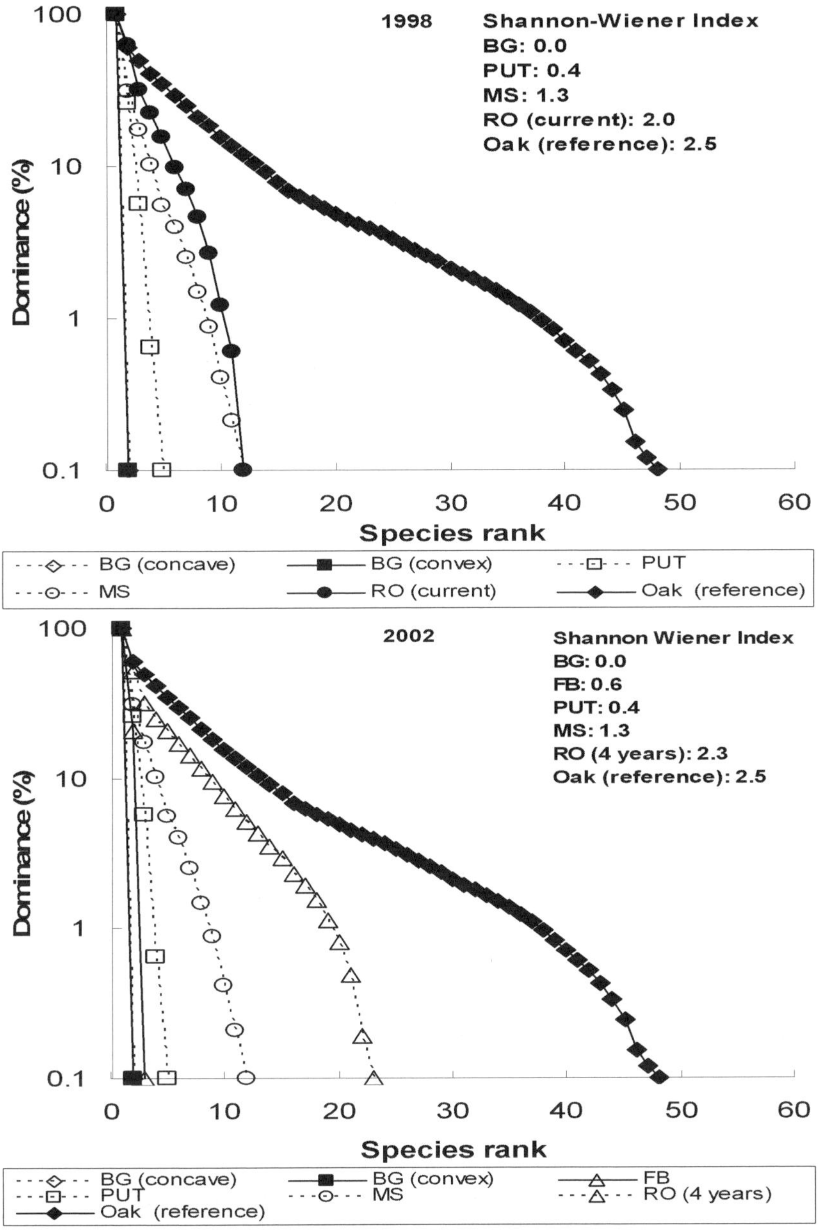

Fig. 24.8 Rank-abundance curves of vegetation including degraded (BG, FB, GF, and GG), restored, and reference stands (Abbreviations as in Fig. 24.7)

Discussion

Comparisons of Species Tolerance to the Different Polluted Sites

Some species *(Q. aliena, Q. acutissima, S. japonica)* were found to be tolerant to polluted site conditions in the field trials in both Ulsan and Yeocheon. The number of species selected for tolerance in each industrial site differed. This may be due to the different properties of the natural and polluted environments in the two regions (Table 24.1) (National Institute of Environmental Research, 1990). This difference in the number of tolerant species selected in each site suggests that restoration plans must take into consideration regional characteristics. A "one size fits all" restoration plan will likely not succeed. Similarly, the numbers of tolerant species selected in the pot culture trials using soils from each region was low (50%). *Q. aliena* and *A. firma* were found to tolerate conditions in the pot culture trials in both regions. Soil differed in soil physical and chemical properties and degree of pollution. Even the degree to which soil amendments could improve soil conditions for plant growth differed in each region (Table 24.2). Species growth responses in the transplantation field trials and the pot experiments generally agreed with one another. This result implies that both methods for screening trials for tolerant species were equally valid for this region.

Effects of Dolomite and Sewage Sludge

The reduction of Al^{3+} in the sludge plots suggests the sludge is a chelating agent for Al^{3+} (de la Fuente et al., 1997; Wong and Swift, 2003). The mixed application of dolomite and sludge resulted in both neutralization and fertility improvement of soil. Effects of sludge application on OM, N, and P were likely masked by robust production of *M. sinensis* in the GG plots; therefore, we do not recommend sludge application in the lands with dense grass cover. Meanwhile, a concurrent treatment with dolomite and sludge is recommended for soil amelioration in the BG and GF plots.

Although dolomite and sludge contributed to ameliorating the acidified soil, there are some serious concerns for land application of dolomite and sewage sludge due to the potential for contamination of ground water and eutrophication (Kaupenjohann et al., 1987; Freedman, 1995; Kreutzer, 1995). By stimulating mineralization of soil organic matter, dolomitic liming causes ground water pollution by increasing nitrate release from the soil (Kaupenjohann et al., 1987; Kreutzer, 1995). We, therefore, recommend restricting use of those soil ameliorators.

Leaf growth by *Q. serrata* and *L. japonicum* was likely enhanced by reduction of acidity (dolomite plots), addition of N and P (sludge plots), or both (mixed plots). Reduction of Al^{3+} toxicity is also a strong possibility (Ulrich,

1980). In reality, many tree species are sensitive to the concentrations of free Al^{3+} in the soil solution (Hutchinson et al., 1986). Aluminum phytotoxicity results in rapid inhibition of root growth due to impedance of both cell division and elongation (Blamey and Edwards, 1989). This results in reduced volume of soil explored by the root system and direct interference with uptake of ions such as calcium and phosphate across the cell membrane of damaged roots (Kochain, 1995). Soil nutrients such as P, Ca^{2+}, and Mg^{2+} exacerbate the problem of inefficient nutrient uptake due to restricted root growth and root damage (Sumner et al., 1991).

Additions of undecomposed plant materials such as prunings to acid soils often increase soil pH, decrease Al^{3+} saturation, and improve conditions for plant growth generally (Hoyt and Turner, 1975; Asghar and Kanehiro, 1980; Ahmad and Tan, 1986; Bessho and Bell, 1992). Similarly, addition of plant residue composts, urban waste compost, animal manure, and coal-derived organic products to acid soils increase soil pH, decrease Al^{3+} saturation, and improve conditions for plant growth (Hue and Amien, 1989; Alter and Mitchell, 1992; Hue, 1992). The recycling of these waste products for soil amelioration has a double benefit for both the environment and the economy, provided that the waste materials are not contaminated with harmful impurities. These humic substances confer metal binding and pH buffering capacities, which are important determinants of the pH of the treated soil (Wong and Swift, 2003).

No significant effects of dolomite on *A. firma* (Fig. 24.4) suggest that *A. firma* is tolerant to acidified soil (Lee et al., 2004). Also, *A. firma*, an N-fixer (Torrey, 1978; Kalaskustkii and Pariiskay, 1983), may grow well in infertile soil. This was likely a reason for no significant effect of sludge treatment on the leaf growth of this species (Fig. 24.4). N-fixing plants have been used for enhancing soil fertility in revegetation projects elsewhere (e.g., Gunn, 1995; Winterhalder, 2000), and we recommend planting *A. firma* as an initial step for revegetating in Mt. Dotjil.

Vegetation Development

Establishment of *Q. serrata, L. japonicum*, and *A. firma* with dolomite and sludge reduced the dominance of *M. sinensis* (Fig. 24.5). The reduction of *M. sinensis* may encourage recruitment and establishment of other species (Winterhalder, 2000). In the polluted area, if a forested ecosystem is being affected, then the tree stratum is generally impacted first and is "stripped" away. As trees decline, shrubs and then the ground vegetation are affected. This sequential death of different vertical strata of the vegetation, known as a "peeling" or "layered vegetation effect" (Gordon and Gorham, 1963; Woodwell, 1970), led to growth of a dense grass mat in this area. This grass mat also appears during the recovery process of areas made barren by severe pollution damage (Winterhalder, 2000). These mats are usually pollution tolerant (Cox and Hutchinson, 1980; Archambault and Winterhalder, 1995) and can become very dense (Winterhalder, 2000). *M. sinensis* was also classified as a tolerant species (Lee et al., 2004).

Plants that form mats, thickets, or simply dense stands often inhibit species change in succession by monopolizing resources such as light, nutrients, and water, and resisting replacement by other species. Thickets typically inhibit succession (Tolliver et al., 1995; Nakamura et al., 1997) or plant growth (Allen and Allen, 1988; Walker, 1994), as was demonstrated in the inhibition model of Connell and Slatyer (1977). The demise of a thicket generally means that succession can resume (del Moral and Bliss, 1993). Plants that form mats or thickets declined over time after restoration, and hence the decline of problematic species such as *M. sinensis* was evaluated as signifying a successful restoration (Buckley et al., 2002).

As was shown in Figs. 24.6 and 24.7, there was no progress, or very little, in vegetation development through natural recovery in the control plots. Meanwhile, plots with the restorative treatment revealed noticeably different trajectories, along with enhanced species diversity (Fig. 24.8), toward the target point (reference vegetation).

Evaluation of Restoration Success

One way of setting a baseline and measuring restoration success is to define the normal biological integrity of a system and then to measure deviations from it. Integrity implies an unimpaired condition, or the quality or state of being complete or undivided. Biological integrity is defined as "the ability to support and maintain a balanced, integrated, adaptive biological system having the full range of elements and processes expected in the natural habitat of a region" (Karr, 1996, 1998; Society for Ecological Restoration, 2002). To practice evaluation on a restored system, ecological attributes of the system are compared with those of the undisturbed system. In our study, we decided to compare species composition and biodiversity of the restored stands with undamaged healthy stands of reference areas. Species composition of stands dominated by restorative treatment resembled those of the reference stands (Fig. 24.7), and their diversity increased (Fig. 24.8). In contrast, stands without any restorative treatment not only showed different species composition from that of the reference stands (Fig. 24.7), but also had a lower biodiversity (Fig. 24.8). Consequently, the restorative treatment increased both biological integrity and ecological stability and thereby met the restoration goal (Aronson et al., 1993; Karr, 1996, 1998; Society for Ecological Restoration, 2002).

Recommendations for Restoration

Continual growth of the human population and human land uses leads to declines in the quality of environment. Further, the natural landscapes that provide many ecosystem services are rapidly being converted to agriculture, industrial and urban sites, and even wasteland. The biodiversity and habitability of the planet are now more threatened than ever before. Therefore, it is imperative that degraded land be rehabilitated and that adjoining natural landscapes be protected. However, it is clear that degradation thresholds have been crossed in many habitats, and succession alone cannot restore viable and desirable ecosystems without intervention (Rietkerk et al., 1997). As was shown in our results (Fig. 24.6), natural succession is protracted, and thus restoration actions must promote natural processes and direct development along desirable trajectories.

To restore degraded ecosystems, in particular those degraded by pollution, we have applied soil ameliorators including lime (Lee et al., 1998; Winterhalder, 2000; Edmeades and Ridley, 2003). Although those soil amendments contributed to improving the polluted environment and thereby achieved successful revegetation (Winterhalder, 2000), they may cause other problems, such as ground water pollution and eutrophication (Kaupenjohann et al., 1987; Freedman, 1995; Kreutzer, 1995). In this respect, we recommend planting tolerant plants or applying fertilizer to plants rather than applying soil amendments as a restorative treatment in all cases.

References

Ahmad, F., and Tan, K.H. (1986) Effect of lime and organic matter on soybean seedlings grown in aluminum-toxic soil. Soil Science Society of America Journal 50:656–661.

Allen, E.B., and Allen, M.F.(1988) Facilitation of succession by the monomycotrophic colonizer Salsora kali (Chenopodiaceae) on harsh site: effects of mycorrhizal fungi. American Journal of Botany 75:266–267.

Allen, S.E., Grimshaw, H.M., and Rowland, A.P. (1986) Chemical analysis. In: Moore, P.D., and Chapman, S.B. (eds.) Methods in Plant Ecology. Blackwell, Oxford, pp. 285–344.

Alter, D., and Mitchell, A. (1992) Use of vermicompost extract as an aluminium inhibitor in aqueous solutions. Community Soil Science Plant Analysis 23:231–240.

Archambault, D.J.P., and Winterhalder, K. (1995) Metal tolerance in Agrostis scabra from the Sudbury, Ontario area. Canadian Journal of Botany 73:766–775.

Aronson, J., Floret, C., Le floc'h, E., Ovalle, C., and Pontainer, P. (1993) Restoration and rehabilitation of degraded ecosystems in arid and semi-arid lands. A review from the South. Restoration Ecology 1:8–17.

Asghar, M., and Kanehiro, Y. (1980) Effects of sugar-cane trash and pineapple residue on soil pH, redox potential, extractable Al, Fe and Mn. Tropical Agriculture 57:245–258.

Bardsley, C.E., and Lancaster, J.D. (1960) Determination of reserve sulfur and soluble sulfates in soils. Proceedings of Soil Science Society of America 24:265–268.

Barret, S.C.H., and Bush, E.J. (1991) Population processes in plants and the evolution of resistance to gaseous air pollutants. In: Taylor, G.E., Pitelka, L.F., and Clegg, M.T. (eds.) Ecological Genetics and Air Pollution. Springer-Verlag, New York, pp. 137–165.

Berger, J.J. (1993) Ecological restoration and non-indigenous plant species: a review. Restoration Ecology 1:74–82.

Blamey, E.P.C., and Edwards, D.G. (1989) Limitations to food crop production in tropical acid soils. In: van der Heide, J. (ed.) Nutrient Management for Food Crop Production in Tropical Farming Systems. Institute for Soil Fertility, Haren, The Netherlands, pp. 73–94.

Bradshaw, A.D. (1992) The biology of land restoration. In: Jain, S.K., and Botsford, J.W. (eds.) Applied Population Biology. Kluwer Academic Publishers, Dordrecht, pp. 25–44.

Braun-Blanquet, J. (1964) Pflanzensoziologie. Grundze der Vegetaionskunde. Springer-Verlag, Wien.

Buckley, P., Ito, S., and McLachlan, S. (2002) Temperate woodlands. In: Perrow, M.R., and Davy, A.J. (eds.) Handbook of Ecological Restoration. Cambridge University Press, Cambridge, UK, pp. 503–538.
Cho, H.J., and Lee, C.S. (1998) Ecological diagnosis and development of ecological management system of urban forest on Mt. Hwangryung in Pusan, Korea. Korean Journal of Ecology 21:779–789.
Connell, J.H., and Slatyer, R.O. (1977) Mechanisms of succession in natural communities and their role in community stability and organization. American Naturalist 111:1119–1144.
Cox, R.M., and Hutchinson, T.C. (1980) Multiple metal tolerances in the grass *Deschampsia cespitosa* (L.) Beauv. From the Sudbury smelting area. New Phytologist 84:631–647.
de la Fuente, J.M., Verenice, R.R., Jose Luis, C.P., and Luis, H.E. (1997) Aluminum tolerance in transgenic plants by alteration of citrate synthesis. Science 276:1566–1568.
del Moral, R., and Bliss, L.C. (1993) Mechanisms of primary succession: Insights resulting from the eruption of Mount St. Helens. Advances in Ecological Research 24:1–66.
Dobson, A.P., Bradshaw, A.D., and Baker, A.J.M. (1997) Hopes for the future: restoration ecology and conservation biology. Science 277:515–522.
Edmeades, D.C., and Ridley, A.M. (2003) Using lime to ameliorate topsoil and subsoil acidity. In: Rengel, Z. (ed.) Handbook of Soil Acidity. Marcel Dekker, New York, pp. 297–336.
Freedman, B. (1995) Environmental Ecology: The Ecological Effects of Pollution, Disturbance, and Other Stresses, 2nd ed. Academic Press, San Diego.
Gordon, A.G., and Gorham, E. (1963) Ecological aspects of air pollution from an iron-sintering plant at Wawa, Ontario. Canadian Journal of Botany 41:1063–1078.
Gunn, J.M. (ed.) (1995) Restoration and Recovery of an Industrial Region. Springer-Verlag, New York.
Gunn, J.M. (1996) Restoring the smelter damaged landscape near Sudbury, Canada. Restoration and Management Notes 14:129–136.
Hill, M.O. (1979) DECORANA—a FORTRAN Program for Detrended Correspondence Analysis and Reciprocal Averaging. Section of Ecology and Systematics, Cornell University, Ithaca, New York.
Hoyt, P.B., and Turner, R.C. (1975) Effects of organic materials added to very acid soils on pH, aluminum, exchangeable NH4 and crop yields. Soil Science 119:227–237.
Hue, N.V. (1992) Correcting soil acidity of a highly weathered Ultisol with chicken manure and sewage sludge. Community Soil Science and Plant Analysis 23:241–264.
Hue, N.V., and Amien, I. (1989) Aluminum detoxification with green manures. Community Soil Science and Plant Analysis 20:1499–1511.
Hutchinson, T.C., Bozic, L., and Muñoz-Vega, G. (1986) Responses of five species of conifer seedlings to aluminum stress. Water, Air and Soil Pollution 32:283–294.
Jackson, M.L. (1967) Soil Chemical Analysis. Prentice-Hall, New Delhi.
Kalaskustkii, L.V., and Pariiskay, A.N. (1983) Nitrogen-fixing symbiosis of actinomycetes with plants. Biology Bulletin 9:171–185.
Karr, J.R. (1996) Ecological integrity and ecological health are not the same. In: Schulze, P. (ed.) Engineering within Ecological Constraints. National Academy, Washington, DC, pp. 100–113.
Karr, J.R. (1998) Rivers as Sentinels: using the biology of rivers to guide landscape management. In: Naiman, R.J., and Bilby, R.E. (eds.) River Ecology and Management: Lessons from the Pacific Coastal Ecosystem. Springer-Verlag, New York, pp. 502–528.
Kaupenjohann, M., Hantschel, R., Zech, W., and Horn, R. (1987) Ergebnisse von Düngungsversuchen mit Magnesium and vermutlich immissionsgeschädigten Fichten [Picea abies (L.) Karst.] im Fichtelgebirge. Forstwiss Centralbl 106:78–84.
Kent, M., and Cocker, P. (1992) Vegetation Description and Analysis: A Practical Approach. CRC Press, Boca Raton, FL.
Kim, J.H. (1991) Forest decline syndrome affected by air pollution. In: Proceedings of Symposium Held by the Korean Society of Landscape Architecture and The Forestry Research Institute of Korea, pp. 3–25 (in Korean).

Kim, J.H., Mun, H.T., Cho, D.S., and Lee, C.S. (1996) Selection and breeding of tolerant species and bio-indicator to air pollution and acid rain. A research report for development of the advanced technology. Ministry of Environment of Korea, Seoul.

Kim, J.W. (1994) Ecophysiological responses of Quercus mongolica stand to air pollutants and a model for its productivity affected by ambient ozone. Ph.D. thesis, Seoul National University (in Korean with English abstract).

Kitajima, K. (1988) The current status of the forest decline and its research in Central Europe. Japanese Journal of Ecology 38:269–277.

Kochain, L.V. (1995) Cellular mechanisms of aluminum toxicity and resistance in plants. Annual Review of Plant Physiology Plant Molecular Biology 46:237–260.

Korea Plant Name Index. (2004) http://www.koreaplants.go.kr:9090/.

Kreutzer, K. (1995) Effects of forest liming on soil process. Plant and Soil 168–169:447–470.

Lee, C.S. (1992) A study on restoration of vegetation destroyed by air pollution. A research report for Post-Doc. Fellowship Program funded by Scientific Research Foundation of Korea, Seoul (in Korean with English abstract).

Lee, C.S., and Cho, H.J. (1998) Landscape structure and ecological restoration of Mt. Hwangryung in Pusan, Korea. Korean Journal of Ecology 21:791–797.

Lee, C.S., Kim, J.Y., and You, Y.H. (1998) Amelioration of soil acidified by air pollutant around the industrial complexes. Korean Journal of Ecology 21:313–320.

Lee, C.S., Lee, K.S., Hwangbo, J.K., You, Y.H., and Kim, J.H. (2004) Selection of tolerant plants and their arrangement to restore a forest ecosystem damaged by air pollution. Water, Air, and Soil Pollution 156:251–273.

Lee, C.S., You, Y.H., and Robinson, G.R. (2002) Secondary succession and natural habitat restoration in abandoned rice fields of central Korea. Restoration Ecology 10:306–314.

Lee, T.B. (1985) Illustrated Flora of Korea. Hyangmoonsa, Seoul.

Long, S.P., and Hallgren, J.E. (1987) Measurement of CO_2 assimilation by plants in the field and the laboratory. In: Coombs, J., Hall, D.O., Long, and Oscurlock, J.M.O. (eds.) Techniques in Bioproductivity and Photosynthesis, 2nd ed. Pegamon Press, Oxford, pp. 61–94.

Magurran, A.E. (2004) Measuring Biological Diversity. Blackwell, Boston.

Nakamura, F., Yajima, T., and Kikuchi, S. (1997) Structure and composition of riparian forests with special reference to geomorphic site conditions along the Tokai River, northern Japan. Plant Ecology 133:209–219.

National Institute of Environmental Research. (1981) Studies on the ecosystem around the industrial complexes. NIER, Seoul.

National Institute of Environmental Research. (1990) A study on the evaluation and the prevention of damage to plant communities by air pollution: a comparison of the structure and productivity of plant communities around industrial complexes and an unpolluted area. NIER, Seoul.

National Research Council. (1991) The Restoration of Aquatic Ecosystems: Science, Technology, and Public Policy. National Academy Press, Washington, DC.

Park, W.C. (1991) A study on the selection of adaptable tree in air pollution area. Journal of Korea Air Pollution Research Association 7:55–65 (in Korean with English abstract).

Rhyu, T.C. (1994) Mechanism of forest decline syndrome of Pinus rigida stands by acid pollutants and its recovery in Seoul Metropolitan region. Ph.D. dissertation, Seoul National University.

Rhyu, T.C., and Kim, J.H. (1994) Growth decline of pitch pine caused by soil acidification in Seoul metropolitan area. Korean Journal of Ecology 17:287–297.

Statistical Analysis System. (2001) PROC User's Manual, 6th ed. SAS Institute, Cary, NC.

Smith, W.H. (1990) Air Pollution and Forests, 2nd ed. Springer-Verlag, New York.

Society for Ecological Restoration. (2002) The SER Primer on Ecological Restoration. Society for Ecological Restoration International, Arizona.

Sumner, M.E., Fey, M.V., and Noble, A.D. (1991) Nutrient status and toxicity problems in acid soils. In: Ulrich, B., and Sumner, M.E. (eds.) Soil Acidity, Springer-Verlag, Berlin, pp. 149–182.

Tolliver, K.S., Colley, D.M., and Young, D.R. (1995) Inhibitory effects of Myrica cerifera on Pinus taeda. American Midland Naturalist 133:256–263.

Torrey, J.G. (1978) Nitrogen fixation by actinomycete-nodulated angiosperms. BioScience 28:586–592.

Ulrich, B. (1980) Die Wälder in Mitteleeuropa: Meβergebnisse ihrer Umweltbelastung. Theorie einer Gefährdung. Prognose ihrer Entwicklung. Allgemeine Forst-Zeitschrift 35:1198–1202.

Walker, G.P.L. (1994) Geology and volcanology of the Hawaiian Islands. In: Kay, E.A. (ed.) A Natural History of the Hawaiian Islands. University of Hawaii Press, Honolulu, pp. 53–85.

Winterhalder, K. (2000) Landscape degradation by smelter emissions near Sudbury, Canada, and subsequent amelioration and restoration. In: Innes, J.L., and Oleksyn, J. (eds.) IUFRO Research Series 1. Forest Dynamics in Heavily Polluted Regions. CABI Publishing, Oxon, UK, pp. 87–119.

Wong, M.T.F., and Swift, R.S. (2003) Role of organic matter in alleviating soil acidity. In: Rengel, Z. (ed.) Handbook of soil acidity, Marcel Dekker, New York, pp. 337–358.

Woodwell, G.M. (1970) Effects of pollution on the structure and physiology of ecosystems. Science 168:429–433.

25 Restoration Planning for the Seoul Metropolitan Area, Korea

Chang Seok Lee, An Na Lee, and Yong Chan Cho

The unintended negative consequences of urbanization are large problems in both developed and developing countries. For a long time now, people have considered the natural environment as a resource to be exploited. Recently, however, our attitude toward nature, in general, and urban green space, in particular, has begun to shift profoundly as the negative environmental and social effects of our industrialized economies have become increasingly evident. In an urban environment, green space is indispensable for the well-being of residents due to its diverse ecological functions, such as pollution filtration, conservation of biodiversity, and climate control (Grey and Deneke, 1986; Gilbert, 1991; Miller, 1998). However, the portion of an urban area occupied by the natural environment decreases, and the portion occupied by an artificial environment expands as urbanization proceeds. This situation increases environmental stress due to an imbalance between pervious natural surfaces and impervious built surfaces. As a consequence, the ecological functions of nature that reduce such stresses are not taken advantage of by society (Taoda, 1979; Smith, 1990; Freedman, 1995; Miller, 1998).

Natural ecosystems are homeostatic, but severe environmental pollution caused by rapid population growth and industrialization prevents them from sustaining their structure and function within normal bounds where they can recuperate from disturbance and stress unassisted (Innes and Oleksyn, 2000). In Korea, vegetation has begun to exhibit symptoms of decline near industrial complexes and large cities (National Institute of Environmental Research, 1981; Lee, 1992; Lee et al., 2004, 2007). Acidification of soils has caused loss of basic cations, such as Ca^{2+} and Mg^{2+}, and increases in toxic Al^{3+}, further degrading the resiliency of natural ecosystems in this region (Kim, 1991, 1994; Lee, 1992; Rhyu, 1994; Rhyu and Kim, 1994). Restoration of degraded ecosystems, therefore, is urgently required to ameliorate the effects of such pollution damage (Gunn, 1995).

In urban areas, human influences are especially pervasive, and some ecosystem types have been eradicated or greatly altered. Therefore, no large restoration project can be considered adequate in a big city without an attempt to restore large areas. However, the vast majority of restoration projects are small, because most people recognize that restoration at the level of the landscape is too costly (Noss, 1991). But landscape restoration need not be prohibitively expensive. Landscape-scale restoration, with the aid of human labor, can rely on the natural

M.M. Carreiro et al. (eds.), *Ecology, Planning, and Management of Urban Forests: International Perspectives.*

recovery processes of ecosystems, and thus be more cost-effective. Restoration projects should be expanded from the local and community levels to the regional landscape level to achieve goals of creating a more pleasant environment for people, as well as creating habitats for sustaining greater biotic diversity in urban areas. Recognizing spatial context is key to carrying out landscape-scale restoration projects. Therefore, obtaining a regional land database is required (Noss, 1991).

This chapter reports on a study that had two goals. The first was to obtain information on the degree of ecological degradation of soil and vegetation structure in Seoul's metropolitan area, and to determine the city's land-use patterns. The second goal was to synthesize such information and use it to suggest a landscape level restoration plan to reduce ecological degradation in the city.

The Study Area

Seoul, the capital of South Korea, is located in the central Korean Peninsula and covers 605 km^2 of land (126°46'15" to 127°11'15" E longitude, 37°25'50" to 37°41'45" N latitude; Fig. 25.1). The elevation of the study area ranges from 20 to 800 m above sea level (Fig. 25.2). The parent rock of the mountainous areas around Seoul consists mostly of granite and gneiss, and in the flat land beside rivers and streams consists of alluvium (Fig. 25.3). Soil in these areas has been classified into the Suam, Osan, Asan, and Anryong series, which developed on gneiss and granite bedrock (Seoul City, 1997a). The climate of Seoul is continental, with warm and moist summers, and cold and dry winters. From 1971 to 2000 the mean annual temperature was 12.2°C and the mean annual precipitation was 134.4 cm (Korea Metrological Administration, 2001).

The mountainous vegetation of Seoul consists of four major plant communities distributed along an elevational gradient, with the *Pinus densiflora* community in the mountain peaks, the *Quercus mongolica* community in the upper slopes, the *Carpinus laxiflora* community in the lower slopes, and the *Zelkova serrata* community in the mountain valleys (Lee, 1997). Alder (*Alnus japonica*) stands have remained in the plains and valleys of lowlands that have escaped urbanization (Seoul City, 1997b, 1998; Lee et al., 2002b). Much of the natural landscape in the Seoul metropolitan area has disappeared due to extensive deforestation for fuel, building material, and other purposes during the 20th century (Yim, 1989). The human population of Seoul has increased from 2.4 million in 1960 to 10.3 million as of 2005 (Seoul City, 2005). During this period, the percentage of green space has decreased from 70% in 1960 to 24% in 1998, mostly to accommodate housing (Yim, 1989; Kim and Choe, 1997; Seoul City, 1998). Seoul City has designated most of the forested mountains as green-belt zones in its suburban areas in an attempt to prevent further loss of green space. Under the city's current green-belt ordinance, no commercial, industrial, or urban development is permitted in these forests (Kim and Choe, 1997).

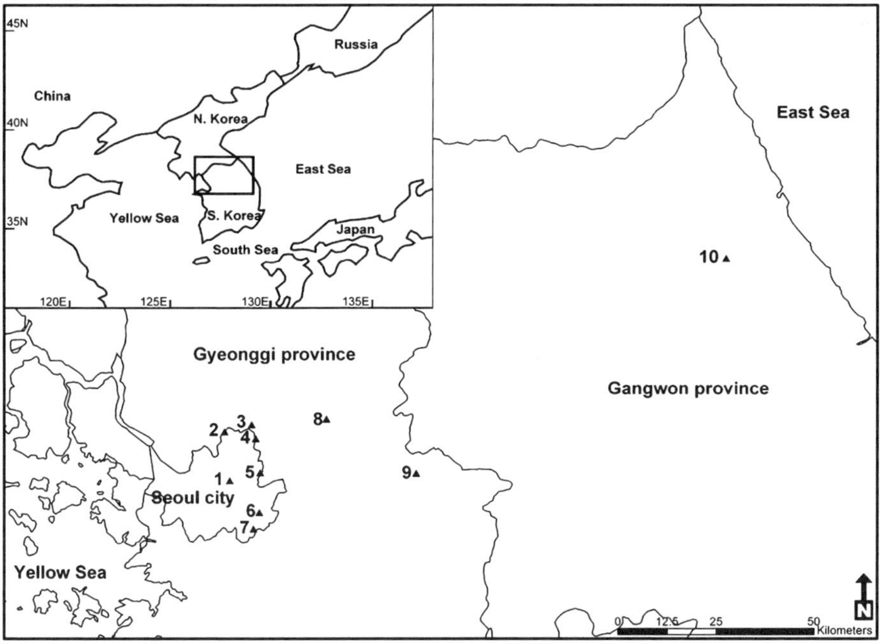

Fig. 25.1 A map shows the study areas. Area 1; areas 4, 5, and 6; areas 2, 3, and 7; area 8 and 9; and area 10 were classified as urban center, inner urban boundary, outer urban boundary, rural, and natural areas, respectively. 1: Mt. Nam, 2: Mt. Boukhan, 3: Mt. Surak, 4: Mt. Bulam, 5: Mt. Acha, 6: Mt. Daemo, 7: Mt. Cheonggye, 8: Mt. Cheonma, 9: Mt. Yongma, 10: Mt. Jumbong

Methods

Monochrome aerial photographs (1:5000 scale) taken in the winter of 1996 were used to identify vegetation types and landscape boundaries. Vegetation types and landscape elements identified on these photographs were confirmed in the field. All of the landscape elements in the urbanized areas were confirmed by visiting all the blocks divided by roads with a width of more than 8 m. The identified landscape attributes were overlain on the 1:5000 scale topographical maps. Patches smaller than 1 mm (25 m^2) on this map were excluded from this study due to the uncertainty of their size and shapes (Nakagoshi et al., 1992). Mapping was carried out using ArcView Geographic Information System (GIS). Landscape ecological analyses of the maps were determined with ArcView GIS software (Environmental System Research Institute, 1996). To evaluate the effect of fragmentation on the vegetation patches, the landscape level fractal dimension was obtained by applying Fragstats 3.3 (McCarigal and Marks, 1995). Fractal dimension values vary between 1 and 2. Values approaching 1 occur when shapes in the landscape are simple, linear, and geometric, such as squares, and values approach 2 for mosaic shapes with highly convoluted perimeters.

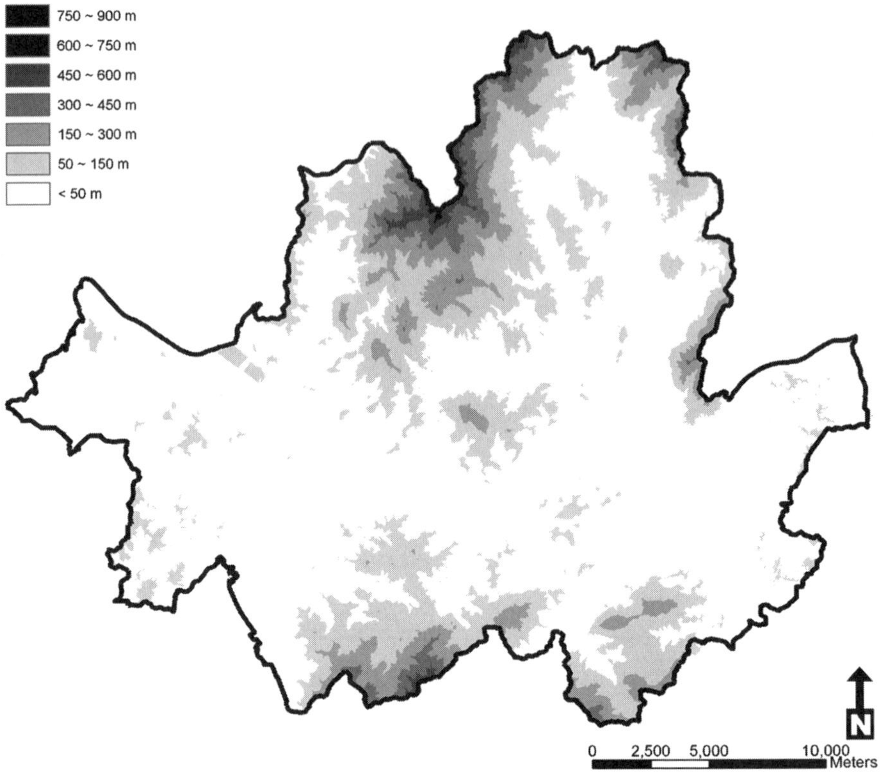

Fig. 25.2 A map showing the elevational topography of the Seoul metropolitan area

Soil samples were collected from 320 grids, measuring 2 km × 2 km, throughout the entire area of Seoul (all 605 km^2). Soil properties measured were pH, Ca^{2+}, Mg^{2+}, and Al^{3+} content. Soil pH was measured with a bench top probe after mixing the soil with distilled water (1:5 ratio, weight per volume) and filtering the extract through Whatman No. 44 paper. Exchangeable Ca^{2+}, Mg^{2+}, and Al^{3+} concentrations were measured after extraction with 1N ammonium acetate (pH = 7.0 for Ca and Mg, and pH = 4.0 for Al) and using inductively coupled plasma (ICP) atomic emission spectrometry (Shimadzu ICPQ-1000) described in Allen et al. (1986). Total sulfur (S) was determined by the method of Bardsley and Lancaster (1960).

Vegetation data were collected in the urban center (Mt. Nam), the inner urban boundary (Mts. Daemo, Bulam, and Acha), the outer urban boundary (Mts. Surak, Boukhan, and Cheonggye), a rural area (Mts. Cheonma and Youngmun), and a natural area (Mt. Jeombong) (Fig. 25.1). Vegetation survey was conducted in 55 plots, with five plots in each of the following sites: Mts. Nam (Mt. N hereafter), Daemo (Mt. D), Bulam (Mt. Bl), Acha (Mt. A), Surak (Mt. S), Boukhan (Mt. Bk), Cheonggye (Mt. Cg), Cheonma (Mt. Cm), Youngmun (Mt. Y), and 10 plots in Jumbong (Mt. J). The size of each plot was 20 m × 20 m.

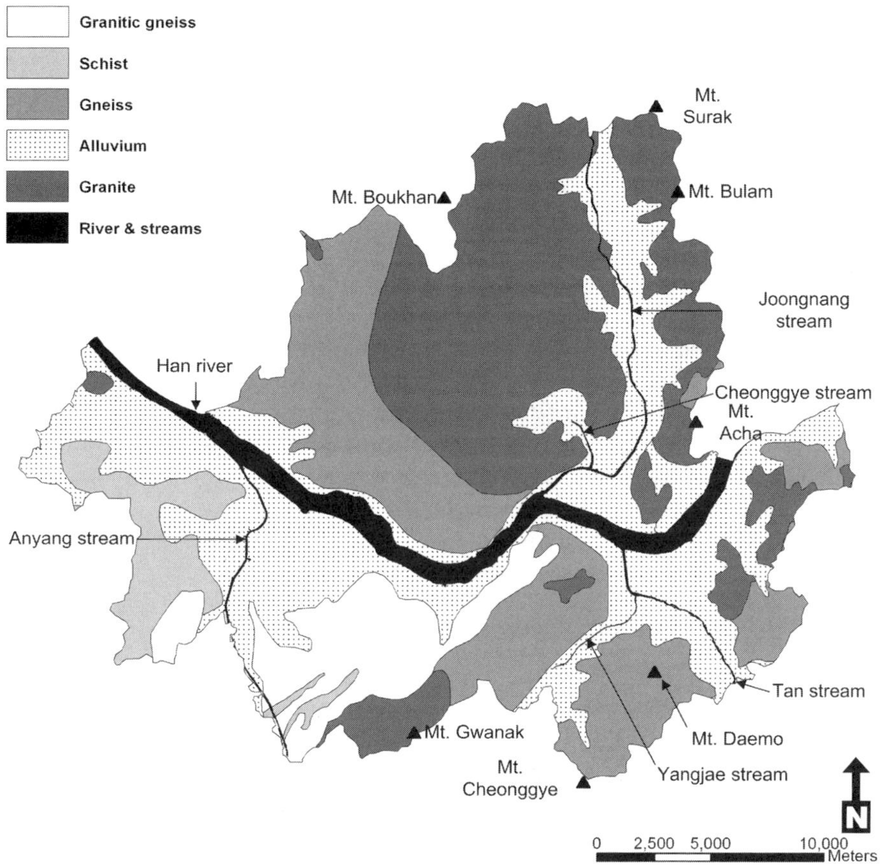

Fig. 25.3 A map showing the parent rock of the Seoul metropolitan area

All the plant species in each plot were identified using the Korean Plant Name Index (2004). For major tree species, stem diameters (at breast height for mature trees or at stem base for seedlings and saplings) were measured and sorted by diameter classes. The vegetation survey was conducted by applying the phytosociological procedure of Braun-Blanquet (1964). Dominance of each species in each plot was estimated by ordinal scale (1 for $\leq 5\%$ up to 5 for $>75\%$), and each ordinal scale was converted to the median value of percent cover range in each cover class. Relative coverage was regarded as equivalent to the importance value of each species. Relative coverage in percent was determined by dividing the cover fraction of each species by the summed cover of all species in each plot and then multiplying by 100. A matrix of importance values for all species in all plots was constructed and used as input for ordination using detrended correspondence analysis (DCA) (Hill, 1979). To describe and compare species diversity and dominance among sites, rank abundance curves (Kent and Cocker, 1992; Lee et al., 2002a; Magurran, 2004) were plotted. The Shannon-Wiener diversity index (H') (Magurran, 2004) was also calculated for each stand in each site.

Results

Landscape Structure

Once the landscape ecological map was generated for Seoul, it became apparent that secondary forests, plantations, and agricultural fields were restricted to the city's fringe, while the urban center had little vegetation (Fig. 25.4). Moreover, vegetation in the urban center was of low ecological quality, since most was fragmented into small patches and consisted of species introduced by landscape architects without ecological consideration.

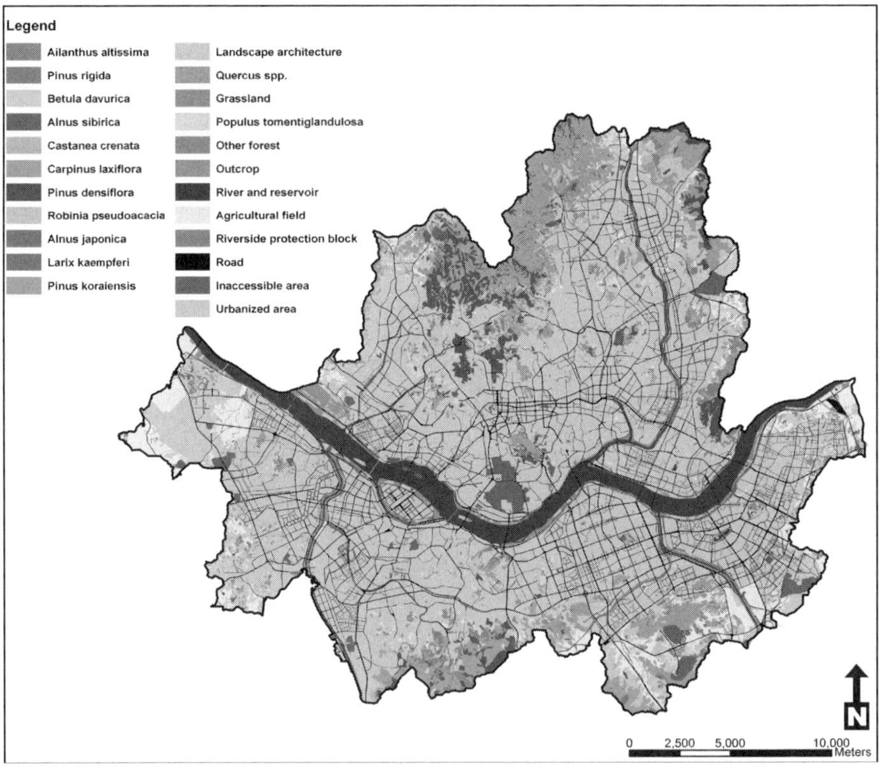

Fig. 25.4 A map showing distribution of landscape elements in the Seoul metropolitan area (redrawn from Seoul City, 2000a)

Table 25.1 Landscape elements identified by classifying the landscape ecological map of Seoul, Korea, based on aerial photographs taken in 1996

Landscape element type	Number	(%)	Area	(%)
Secondary forest				
Ailanthus altissima	10	0.04	3.66	0.01
Carpinus laxiflora	7	0.03	3.73	0.01
Betula davurica	24	0.09	28.39	0.04
Alnus japonica	26	0.09	13.60	0.02
Pinus densiflora	746	2.68	2141.02	3.27
Quercus spp.	1580	5.67	5971.07	9.12
Other forest	209	0.75	263.29	0.40
Subtotal	2602	9.34	8424.76	12.86
Plantation				
Larix leptolepis	42	0.15	54.79	0.08
Alnus hirsuta	125	0.45	59.63	0.09
Castanea crenata	202	0.73	293.20	0.45
Pinus koraiensis	338	1.21	270.37	0.41
Populus tomentiglandulosa	393	1.41	468.07	0.71
Pinus rigida	837	3.00	1326.48	2.03
Plantation for landscape architecture	2246	8.06	1808.43	2.76
Robinia pseudoacacia	2128	7.64	4467.95	6.82
Subtotal	6311	22.65	8748.92	13.36
Agricultural field	1595	5.73	3253.03	4.97
Urbanized area				
Road	3500	12.56	5439.58	8.31
Urbanized area	10,845	38.93	31957.40	48.80
Subtotal	14,345	51.49	37,396.98	57.11
Grassland	2108	7.57	1859.95	2.84
Others				
Outcrop	58	0.21	84.88	0.13
Riverside block	179	0.64	147.35	0.23
Inaccessible area	155	0.56	1207.41	1.84
Aquatic system	505	1.81	4364.75	6.66
Subtotal	897	3.22	5804.39	8.86
Total	27,858	100	65,488.03	100

Table 25.1 summarizes the landscape element types identified from the map, the number of patches for each type, and their areas. The element types in decreasing order of dominance by area were urban areas (57.1%), secondary forests (12.9%), plantations (13.4%), miscellaneous (8.9%), agricultural fields (5.0%), and grasslands (2.8%). Element types in decreasing order by the number of patches were urban areas (51.5%), plantations (22.7%), secondary forests (9.3%), grasslands (7.6%), agricultural fields (5.7%), and miscellaneous (3.2%). When the rankings by area and the number of patches were compared, plantations and grasslands were ranked higher in number than in area, indicating that these landscape elements

Table 25.2 Shape indices of each landscape element identified from the land cover map of Seoul

Landscape element	Fractal dimension
Secondary forest	1.29
Plantation	1.27
Agricultural field	1.28
Urban areas	1.51
Grassland	1.42
Others	1.32

were more fragmented than the others. Urban areas and grasslands exhibited the highest fractal dimension (FD) values, indicating that the perimeters of these patches were more complex than those of the others (Table 25.2).

Spatial Differences in Soil Properties

Soil pH tended to be lower in plots in the urban fringe than in plots within the urban center, although the difference was not statistically significant due to high degree of variation among grids (Fig. 25.5). Mean soil pH did not vary greatly by grid region (central, intermediate, and marginal grids; Table 25.3). Soil Ca^{2+} and Mg^{2+} concentrations followed the pH trends (Table 25.3), but total S and Al^{3+} concentrations were higher in the urban fringe than in the city center (Table 25.3). Most of these chemical properties of soil are strongly related to soil acidification and to each other (Table 25.4).

Vegetation Structure

Mongolian oak (*Quercus mongolica*) stands are the most widely distributed and are representative of late successional communities in Korea (Krestov et al., 2006). The DCA ordination (Fig. 25.6) showed that stands in the urban center (Mt. N) were clustered in the lower left corner of the graph, with stands in the natural area (Mt. J) on the opposite end of axis I. The stands in the outer urban boundary (Mts. Bk, C, and S) and the rural stands (Mts. Cm and Y) were intermediate in distance between the most urban stands (Mt. N) and the most natural stands (Mt. J) along axis I. Urban stands and sites within the inner urban boundary (Mts. A, B1, and D) were distributed in gradient fashion parallel with axis II. Thus, these results showed that species composition in the urban center was similar to those in inner urban boundary and differed greatly from those in the natural areas, with outer urban and rural sites being intermediate in species composition.

Species richness was lowest in the urban center (Mt. N), followed in increasing order of richness by stands in the rural, inner urban boundary, outer urban boundary,

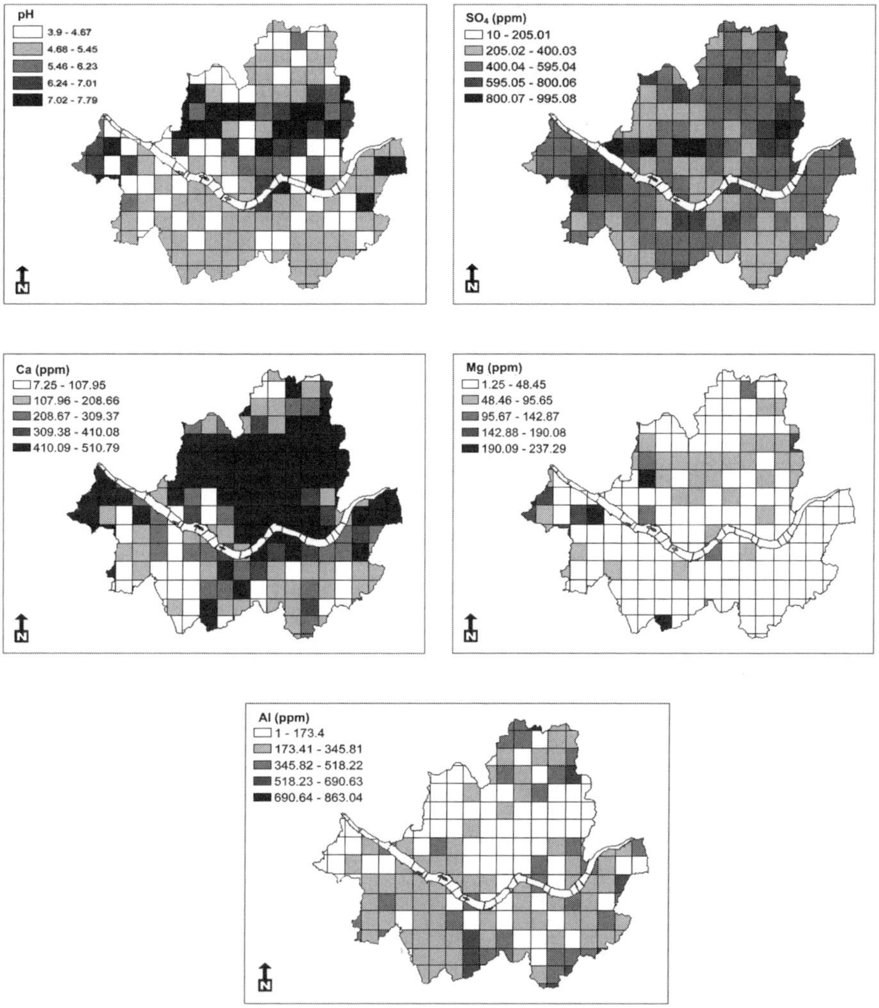

Fig. 25.5 Spatial distribution of several soil properties, such as pH, SO_4, Ca^{2+}, Mg^{2+}, and Al^{3+} in the Seoul metropolitan area

Table 25.3 A comparison of soil environmental factors among central, intermediate, and marginal zones in Seoul; values in parentheses indicate ± 1 standard deviation

	pH	SO_4	Ca^{2+}	Mg^{2+}	Al^{3+}
Central	5.73	262.3	411.5	46.9	145.8
	(1.17)	(201.6)	(127.6)	(39.1)	(139.0)
Intermediate	5.32	281.9	318.5	41.2	203.8
	(1.02)	(155.9)	(172.5)	(39.2)	(156.1)
Marginal	5.23	282.6	271.3	39.4	254.1
	(0.96)	(120.4)	(177.9)	(46.7)	(189.6)

Table 25.4 Pearson *r* correlation values between soil environmental factors

	pH	SO_4	Ca^{2+}	Mg^{2+}	Al^{3+}
pH		−0.29*	0.84**	0.49**	−0.60**
SO_4			−0.13	−0.14	0.28*
Ca				0.54**	−0.57**
Mg					−0.40**
Al					

*p<.05; **p <.01.

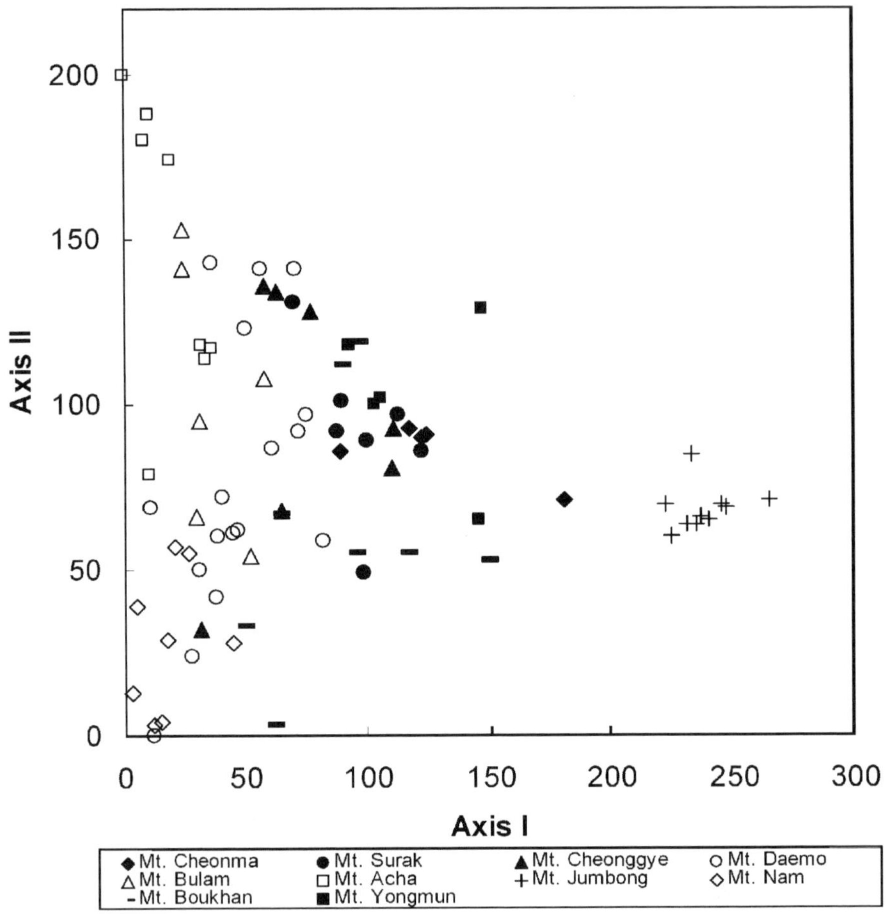

Fig. 25.6 Stand ordination of the Mongolian oak forest established in urban and suburban areas around Seoul

and natural areas (see x-axis in Fig. 25.7). Species diversity in these stands, as measured by the Shannon-Wiener index, showed somewhat different trends. The Shannon-Wiener index did not reveal significant differences in diversity among stands in the urban center, inner and outer urban boundary, and rural areas, but did

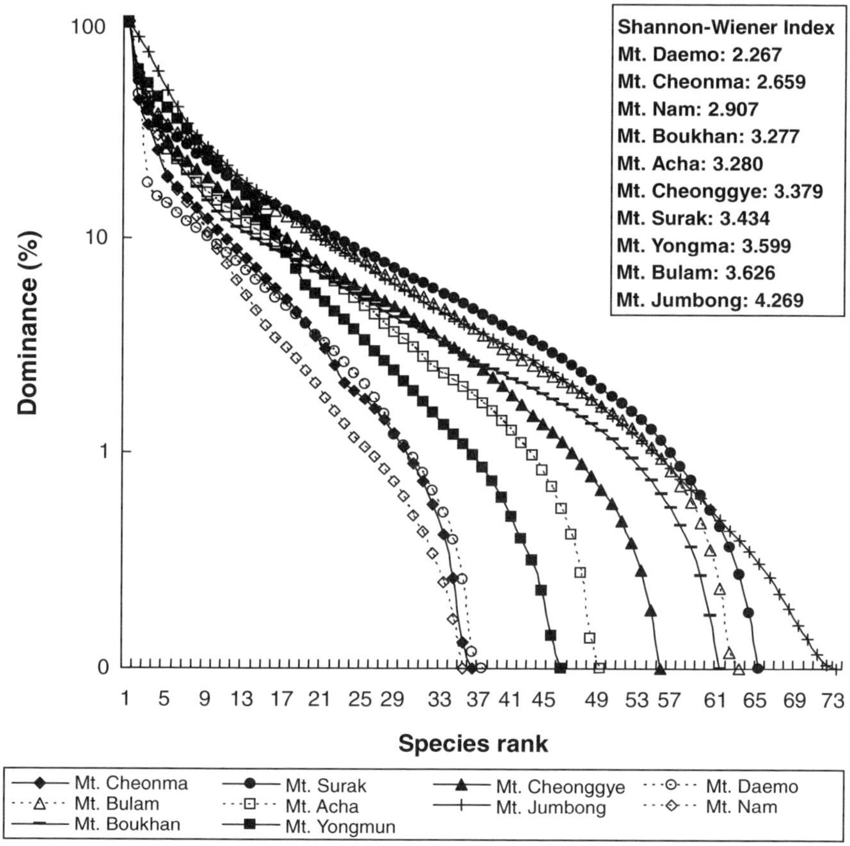

Fig. 25.7 Rank-abundance curves of the Mongolian oak forests established in 10 study areas

show that diversity in the natural areas (Mt. J) was a good deal higher than those of the other areas (Fig. 25.7).

Vegetation stratification, which was expressed as coverage of canopy trees, subcanopy trees, shrub, and herb layers, differed among study areas (Fig. 25.8). Except for the natural area, all stands had low relative coverage by the herb layer. Stands in urban center exhibited higher relative coverage by the subcanopy tree layer than in other areas.

Tree Size-Class Distribution

In urban Mt. Nam, the diameter class distribution of major trees in these Mongolian oak stands revealed that oaks dominated the larger diameter classes and even exhibited a high frequency in smaller size classes (Fig. 25.9). However, *Sorbus alnifolia*

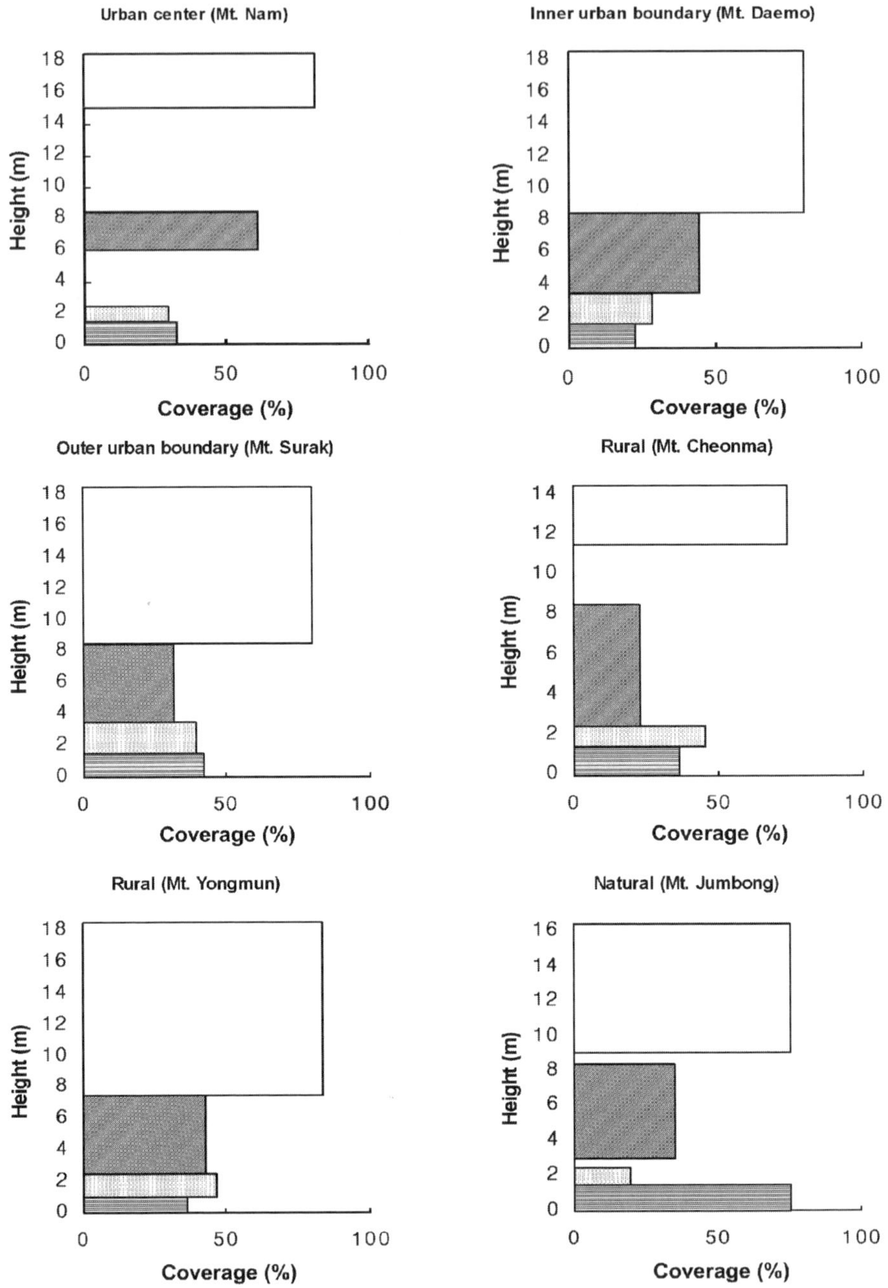

Fig. 25.8 Canopy profiles of the Mongolian oak established in each study area

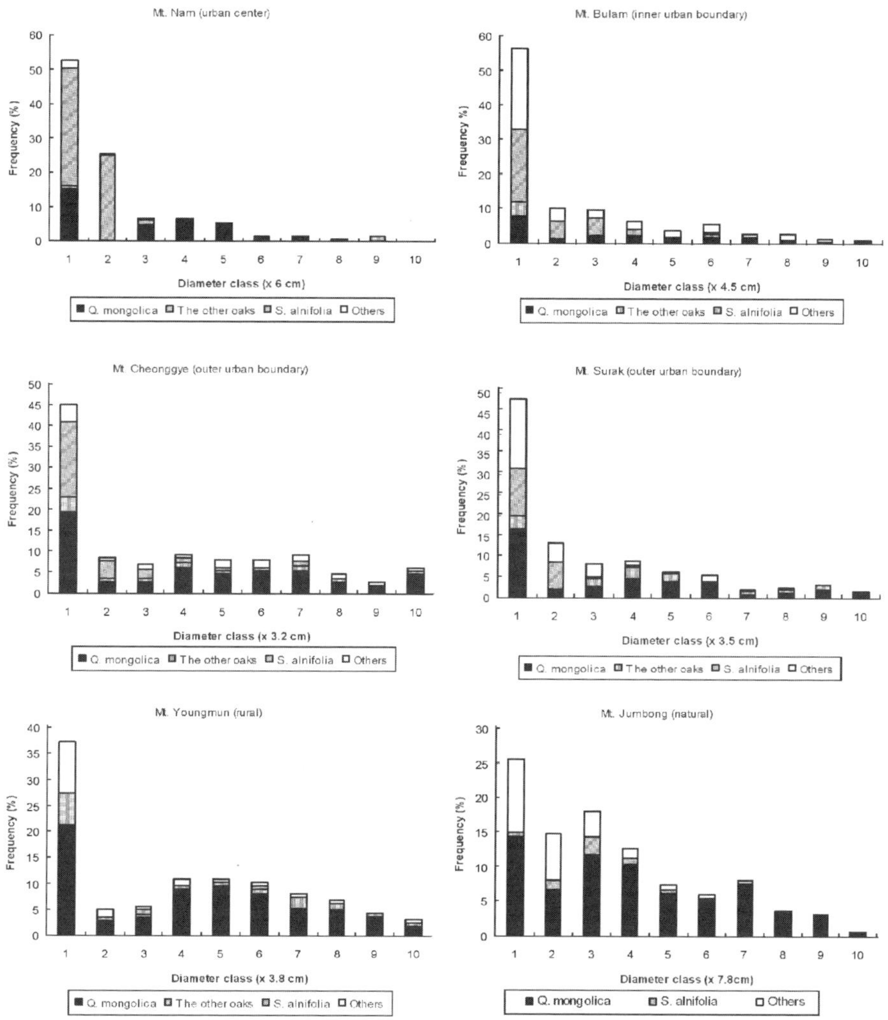

Fig. 25.9 Frequency distribution of diameter classes of major tree species composed of the Mongolian oak forests established in urban and a natural area center inner and outer urban boundaries, and rural areas around Seoul

(Korean mountain ash) dominated the smaller diameter classes in stands at Mt. Nam. Mongolian oak stands located on the inner and outer urban boundaries exhibited similar patterns to that of Mt. Nam except for a lower occupancy by *S. alnifolia* in the smaller diameter classes. In the rural and natural areas, Mongolian oak dominated all diameter classes. From these results, it is expected that Mongolian oak stands in the rural and natural areas would be maintained continuously as Mongolian oak communities, whereas urban oak stands would be replaced by *S. alnifolia*. Considering that Mongolian oak stands are the representative vegetation of the late successional stage in the Korean peninsula (Krestov, 2006), this successional trend can be interpreted as being retrogressive (Barbour et al., 1999).

Strategy for Ecological Restoration—A Geographic Information System Planning Exercise

Ecological maps of the different landscape element types in Seoul show that natural vegetation is restricted to the urban fringe, whereas artificial vegetation and impervious surfaces (e.g., buildings, roads) were relatively more evenly distributed (Fig. 25.10). Uneven distribution of vegetation becomes a starting point for creating a distinctive urban climate dominated by a heat-island effect. Further, the effect induces temperature inversions that trap pollutants over cities for extended periods and thus aggravates environmental stress to both people and other species in the urban area (Henry and Dicks, 1987; Nichol, 1996; Miller, 1998; Bonan, 2002). Soil acidification and vegetation decline syndrome are also appearing in the urban center and the urban boundary areas in Seoul. Such an uneven distribution of natural vegetation as seen in Seoul can be considered a cause of various phenomena related to ecological degradation discussed earlier.

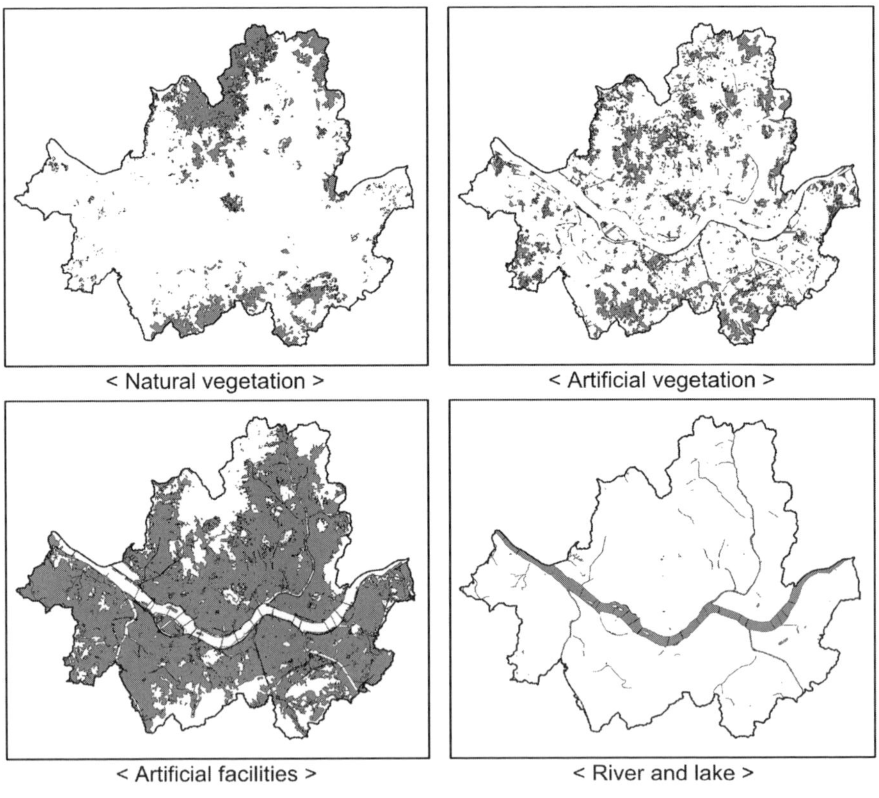

Fig. 25.10 Distribution of major landscape elements in the Seoul metropolitan area

To solve these problems, Seoul needs a vegetation restoration plan based on establishing a green network system that uses sound landscape ecological principles for its design and development. Piecemeal restorations that do not consider the entire landscape context are too fragmented in time and space to offer the most satisfactory results. As a first step, we have to decide the location of green networks and use restoration methods that are sensitive to the spatial context of each landscape element. As part of this exercise, we postulated enlarging existing green elements using 100- to 400-m buffer widths at 100-m intervals from existing forest vegetation, rivers, and streams. As baselines for the green network, we designated the 10 largest patches as core forest vegetation. The Han river, and the Jungnang, Wooi, Bulgwang, Anyang, Tan, and Yangjae streams were chosen as essential baseline river and streams for forming green networks throughout the city (Fig. 25.11). Core forest vegetation in our design includes both natural vegetation and artificial plantations, because plantations located on the urban boundary were restored to be similar to natural vegetation (Lee et al., 2004). Although areas along rivers and streams have low ecological potential, they do provide connectivity across the landscape, and so we included them as core vegetation.

In this GIS mapping exercise, we found that a 100-m buffer extension enhanced the connectivity of many forest fragments in the Seoul landscape (hypothetical green pathways 1 to 20; Fig. 25.12). A 200-m buffer enlargement

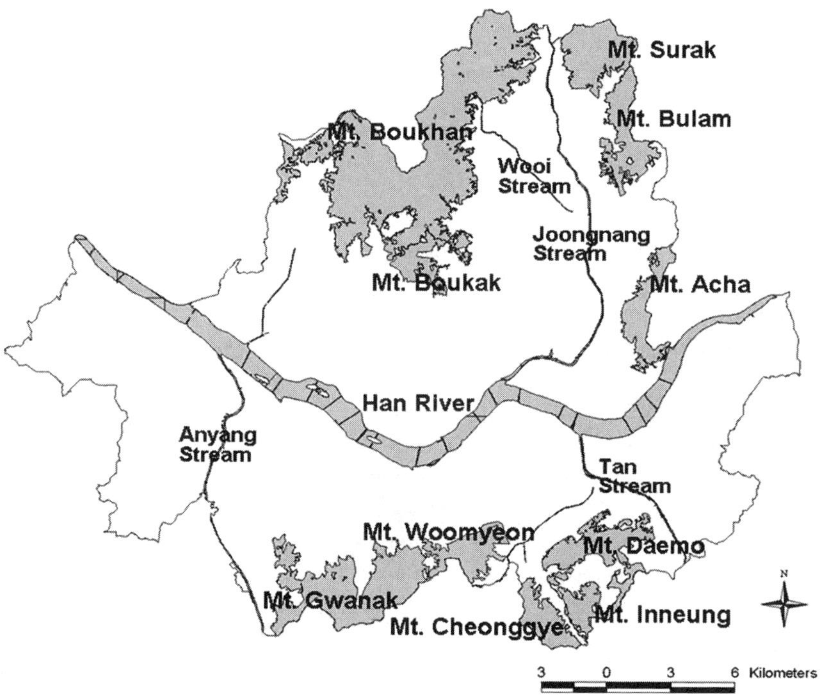

Fig. 25.11 A map showing the core green spaces

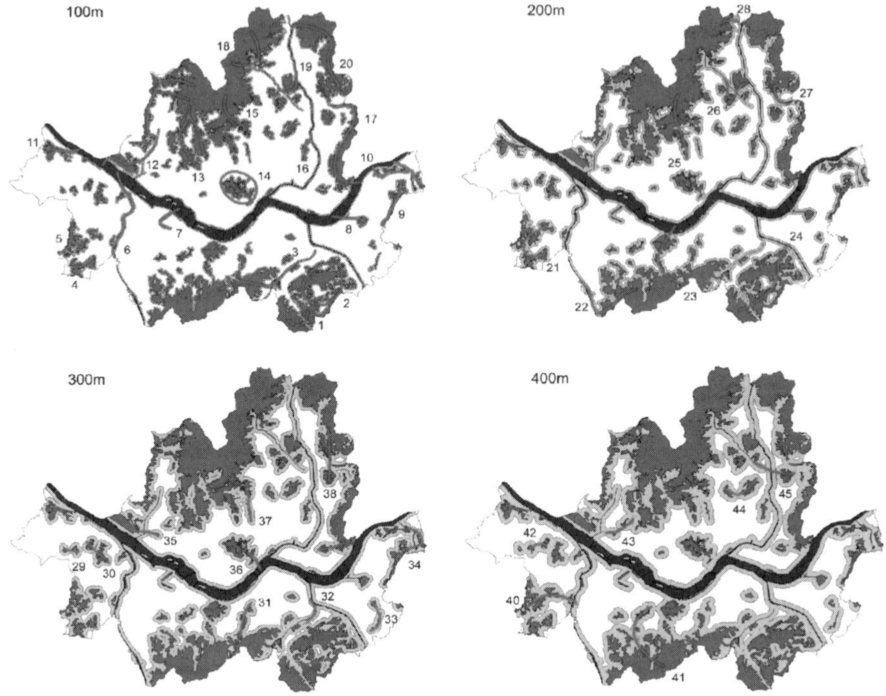

Fig. 25.12 Maps showing restoration plan hypothetically extended with 100-, 200-, 300-, and 400-m widths from the existent forest and riparian vegetation to determine direction of green network. Numbers indicate the order that new greenery space will be created

greatly improved connectivity not only between forests (hypothetical green pathways 21, 25, and 27), but also between forests and streams (hypothetical green pathway 22 to 24, 26, and 28; Fig. 25.12). Both 300- and 400-m buffer enlargements restored not only near-complete connectivity among fragmented forest patches, the Han river, and several streams, but also resulted in an even distribution of green space throughout the city (Fig. 25.12). Such connections among fragmented patches not only can enhance biodiversity conservation and ecological functioning of the entire urban ecosystem, but also can improve many environmental problems (Conine et al., 2004).

This exercise also revealed the directional aspects of the green network that could be created in specific parts of the city. Up to now, Seoul city has carried out an environmental policy to increase the dimensions of green areas and improve their quality through several campaigns, such as the Project for Creation of the Urban Environmental Forest, the 10 Million Trees Planting Campaign, and the Campaign to Create School Forests. However, most of these greening projects did not follow landscape-level, ecological restoration strategies such as ours. If ecological restoration projects are to

Table 25.5 List of recommended plant species for restoring footpaths and riparian areas in Seoul, Korea

Location	Primary trees	Secondary trees	Shrubs	Herbs
Sidewalks	*Zelkova serrata*	*Celtis sinensis*	*Callicarpa japonica*	*Disporum smilacinum*
	Carpinus laxiflora	*Styrax obassia*	*Euonymus alatus*	*Ainsliaea aceriflora*
	Quercus serrata	*Magnolia sieboldii*	*Ligustrum obtusifolium*	*Carex siderosticta*
	Alnus japonica	*Styrax japonica*	*Rosa multiflora*	*Carex okamotoi*
			Spiraea prunifolia var. *simpliciflora*	*Pyrola japonica*
Riparian areas	*Alnus japonica*	*Acer ginnala*	*Salix gracilistyla*	*Phragmites communis*
	Salix koreensis	*Salix* spp.	*Rosa multiflora*	*P. japonica*
	Ulmus parvifolia	*Populus* spp.	*Spiraea pruniflora* var. *simpliciflora*	*Carex* spp.
	Zelkova serrata		*S. integra*	*Impatiens textori*
	Fraxinus rhynchophylla		*Viburnum sargentii*	

solve our large-scale urban environmental problems and achieve desirable levels of biological conservation, such systematic, landscape-level strategies will inevitably be required.

At the scale of the patch to be restored, plans can be based on information about the potential natural vegetation (Table 25.5) (Seoul City, 1997, 1998; also see Chapter 12). To maintain good connectivity in the landscape, we opted to focus our first restoration plans on river and stream banks, including sidewalks that run alongside them. To restore sidewalks, we chose two methods depending on available width (Fig. 25.13). When space is narrow, we suggest transplanting street trees from the roadside to the center of these riparian paths. Furthermore, we suggested increasing vegetation volume by introducing subcanopy trees, shrubs, and herbs on both sides of the sidewalks. When sidewalks were wide, we applied similar methods, but increased the proportion of tall trees.

To restore our urban river, which has low ecological quality, we selected a stepwise restoration process, in which the urban river recovers natural river features gradually by imitating the form of natural rivers (Fig. 25.14). Our natural river reference models were those within the civilian control zone (CCZ), which have not been disturbed by people since the Korean War (about 50 years) (C.S. Lee, unpublished data). In the initial restoration stages, we would plan to introduce plants on the existing riparian concrete blocks to promote greater safety during flooding events. To realize such a restoration practice, we can use a soil ameliorator called Terra Cottem to improve the physicochemical properties of the planting bed (Lee et al., 2004).

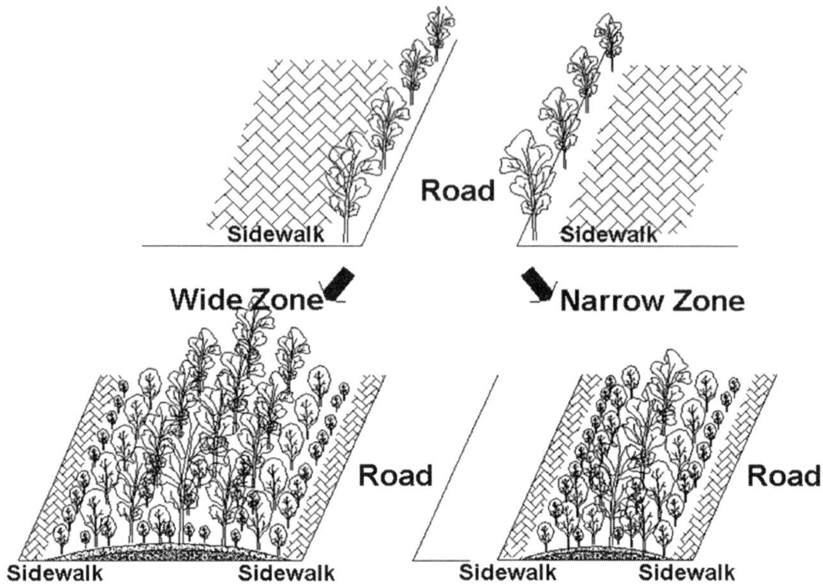

Fig. 25.13 A restoration plan to improve ecological quality around the roadside

Discussion

The Effects of Low Vegetation Cover on Urban Climate

The climate of a given location may be considered at three scales: macroclimate, mesoclimate, and microclimate (Hiesler and Herrington, 1976). Cities do not exert an influence on the macroclimate, since this occurs at a scale of hundreds of square kilometers. Mesoclimate is determined by smaller scale variation caused by topographic features, water bodies, and impervious cover. By affecting albedo and evapotranspiration in urban areas, vegetation reduces air temperature in urban areas and, therefore, influences urban mesoclimate (Parker, 1989; Souch and Souch, 1993). Microclimate variation occurs at the scale of hundreds of square meters down to a few square centimeters and can also be caused by variation in elevation in the tens of meters (Hiesler and Herrington, 1976). For example, a forest patch has a different microclimate than an adjacent clearing.

Cities are often referred to as urban heat islands, with the urban center having the highest temperatures. This is primarily due to the low amount of vegetation in urban center compared to the suburbs and beyond (Figs. 25.4 and 25.10). Cities also use large amounts of energy and emit this energy as waste heat, further exacerbating the urban heat-island effect. Buildings, asphalt, and concrete absorb solar radiation, and emit long-wave radiation that heats the atmosphere (Akbari et al., 1990). Moreover,

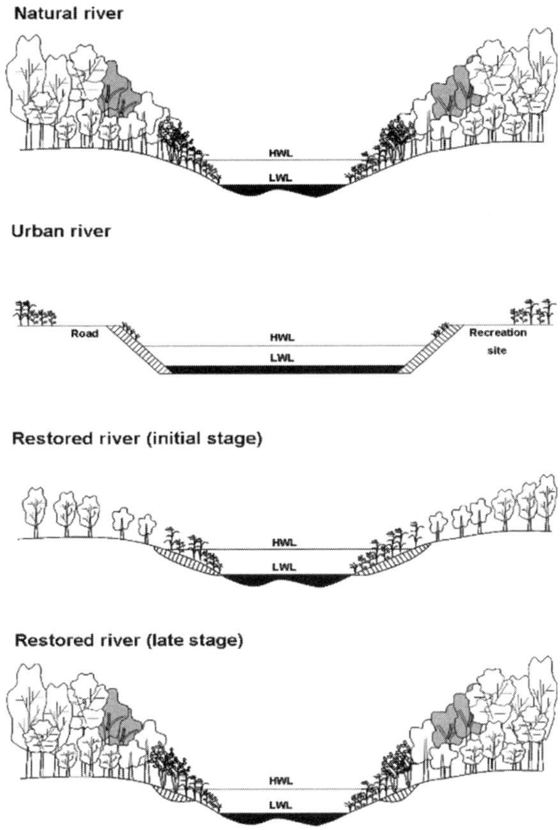

Fig. 25.14 A plan to restore the urban river, which was transformed to utilize and control the water. This method was designed by imitating natural rivers within the civilian control zone (CCZ), which have remained without any artificial disturbance for about 50 years since the Korean War

those artificial structures hold heat for extended periods. This heat moves from buildings, asphalt, and concrete to the cool air as the air temperature decreased after sunset to form atmospheric temperature inversion (warm air over cold air) (Miller, 1998; Seoul City, 2000b; Chiras, 2001). On the other hand, trees and other vegetation use large amounts of solar energy to evaporate water via transpiration to cool leaf surfaces. Evaporative water loss via transpiration also reduces air temperatures in urban areas. Trees and other vegetation can also contribute indirectly to temperature reduction by reducing urban energy consumption by intercepting and using solar energy, and by reducing building energy demand through shading and reducing wind speed. Therefore, low vegetation coverage in cities results in larger temperature gradients between urban areas and suburbs or beyond (Akbari et al., 1992). Indeed, Seoul's heat island effect is very large (Lee, 1985), as the temperature difference between the urban center and boundary was about 11°C (Fig. 25.15).

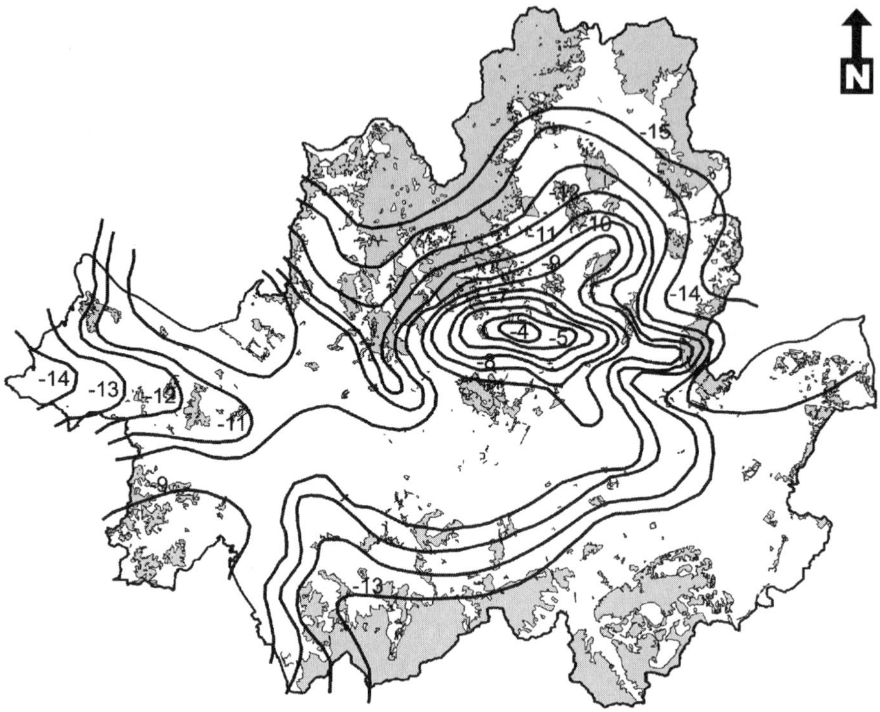

Fig. 25.15 Heat islands in Seoul (January 22, 1983; Lee 1985). Gray parts indicate vegetated area

The Effects of Urban Mesoclimate on Soil Physicochemical Properties

Air temperature inversions trap polluted urban air masses over cities for extended periods (Chiras, 2001; Bonan, 2002). Moreover, microcurrents of air due to temperature differences between urban and more rural areas can transport light gaseous air pollutants from the urban center to the urban fringe (Miller, 1998; Bonan, 2002). We found that differences in soil properties among sites in Seoul are likely to be explained in relation to such transport of pollutants by air microcurrents (Fig. 25.5). Soil acidification was more severe, and soil sulfate (SO_4) concentrations were higher in the urban fringe compared with soils in the urban center (Fig. 25.5). Soil acidification in those sites was due to deposition of acid precipitates, such as SO_x and NO_x (Rhyu and Kim, 1994). Gaseous SO_x and NO_x are transformed to sulfuric acid (H_2SO_4) and nitric acid (HNO_3) as they interact chemically with water in the air and soil, and are deposited in dry and wet form on soil (Kennedy, 1986; Reuss and Johnson, 1986).

As expected, the acidified soils of the urban periphery contained lower concentrations of basic cations, such as Ca^{2+} and Mg^{2+}, than soils in the urban center, because they had been leached through cation exchange mechanisms (Ulrich, 1980). But higher concentrations of Ca^{2+} and Mg^{2+} in soils in the urban center are also related to heavy particulate deposition probably from building materials (e.g., concrete; Lovett et al., 2000), or to direct applications of calcium chloride ($CaCl_2$) used for melting snow. Furthermore, acidified soil releases the ion Al^{3+}, which inhibits plant cell division and consequently retards plant growth (Wong and Swift, 2003). These changes in soil chemistry are known to cause forest decline (Ulrich, 1980).

The Effects of Urban Mesoclimate on Vegetation Structure

The Mongolian oak forests in the more urbanized areas of Seoul had a different species composition, lower richness, and retrogressive successional trends compared to those in suburbs and beyond (Figs. 25.6, 25.7, and 25.9). These differences are likely due to the development of thin canopy crowns in overstory Mongolian oaks, which have been exposed to severe air pollution stress over many years (Kim, 1994). By increasing the supply of light and precipitation to the forest floor below, thin crowns of canopy trees cause dense growth of subcanopy trees, such as the Korean mountain ash, *Sorbus alnifolia*. Therefore, over time vegetation structure and successional trends change (Smith, 1974; McClenahan, 1978; Scale, 1980; Lee et al., 2000). Once the subcanopy layer becomes denser, light again decreases and species richness can be expected to decline, a pattern we observed in our more urban stands (Figs. 25.7, 25.8, and 25.9).

Retrogressive succession, signs of which appeared in our urban oak communities, is usually caused by frequent or intense disturbance (Runkle, 1985). Although such situations have been frequently observed around industrial complexes exposed to severe air pollution (Kozlowski, 1985; Shugart and McLaughlin, 1985; Lee, 1993; Moravčik, 1994; Freedman, 1995; Gunn, 1995), it is a very rare phenomenon in urban areas. Retrogressive succession would be expected where pollution damage to trees is usually intense and acute. However, pollution in most urban areas, though chronic, is less severe than near industrial sites (Freedman, 1995). Although we could observe signs of severe air pollution damage from observing the vegetation structure in Seoul, several pollution indices do not spatially match these damage patterns to vegetation at least at the scales measured (Seoul City, 1997a; Ministry of Environment, 2005). Therefore, our results in Seoul could likely be explained as resulting from synergistic interactions between chronic air pollution and urban climate, rather than resulting solely from severe pollution (Olson and Sharpe, 1985). Atmospheric temperature inversions, occurrence of microcurrents of air due to local temperature differences, and soil acidification due to air pollutants transported by such microcurrents all interact to affect vegetation damage and community change. Therefore, we can recognize a new type of forest decline in Mongolian oak stands as a general phenomenon in the upper slopes on the mountainous ridges surrounding the Seoul basin (Seoul City, 1997b, 1998).

Necessity and Effects of Improving Green-Space Networks in Cities

As encroaching urbanization and sprawl threaten green space, biodiversity and environmental problems become more immediate issues to solve. Seoul has experienced rapid urbanization during the past several decades. With that urbanization has come a surge in and development and loss of open space. Moreover, the patterns of land development are typically ecologically or geographically imbalanced, particularly near urban centers. This situation has caused serious environmental problems, such as forest decline, and raises a serious challenge to local communities and their public officials and planners. Is it possible to preserve open space, while accommodating the need for well-connected and balanced land development?

Greenway networks offer exactly such a mechanism by preserving open space while at the same time creating a "green infrastructure" for linking people (and other species) and places (Ahern, 1995; Fabos, 1995). Greenway planning is a relatively new tool for planners, giving their respective communities the opportunity to create a new set of innovative design standards for urban growth (Ahern, 1995; Fabos, 1995; Diamond and Nooan, 1996). Greenways serve multiple purposes. They buffer environmental stresses, protect the natural environment, offer recreational opportunities, and provide alternative transportation routes (Landsberg, 1981; Upmanis et al., 1998; Conine et al., 2004). Among these options, our study and GIS exercise focused on increasing environmental protection and biodiversity conservation. When green networks are located along streams, they can aid in buffering surface waters from nonpoint pollution. In addition, they can ameliorate urban climate and reduce air pollution. Furthermore, green riparian networks can provide ecological corridors for wildlife movement in the city and beyond the city's boundaries. Riparian greenways would also serve recreational functions for people (Smith and Hellmund, 1993). As alternative transportation corridors, greenways can link origins and designations along the landscape, providing a means for pedestrians and cyclists to travel to and from places they want to go, thus encouraging less fossil fuel consumption during commutes to work and for errand-running within the city.

In our study, the most important goal for a greenway network was abating environmental stresses, such as severe air pollution and the heat-island effect. Urban vegetation interacts with air pollutants and has been found to reduce them significantly (Smith and Dochinger, 1976; Smith, 1978; Olson and Sharpe, 1985; Nowak et al., 1996). In addition, greenway planning can also contribute to environmental protection. When places along rivers and streams, greenways function as effective water quality buffers, trapping sediment and pollutants from urban and agricultural storm-water runoff (Arendt, 1994). Maintaining a natural wooded zone along streams allows for nutrient uptake by riparian vegetation, preventing many potentially degrading substances from entering the waterway (Lowrance et al., 1984; Peterjohn and Correll, 1984). Eroded soil may also be trapped by streamside vegetation and litter, thus reducing the amount of particles that would otherwise end up in the stream (Smith and Hellmund, 1993). By designating the flood plain as a greenway, development in the riparian zone is minimized and water quality protection is enhanced. This also restricts development in near the stream and permits only those activities that are compatible with the natural landscape (Fabos, 1985). These vegetated buffers are commonly recognized as best management practices (Conine et al., 2004).

The protection of green spaces is another environmental benefit served by greenways. While greenways may be established within developed areas, they may also be created in areas where there is the threat of future development, preserving land in its current natural state. These great benefits in all the three spheres of environmental protection, recreation, and alternative transportation cannot be realized unless the greenway planners take a systematic approach to the delineation of greenway paths. Such an approach has been advocated by many prominent landscape planners, including Fabos (1985), Hendrix et al. (1988), and McHarg (1969). Once it is delineated, mapped, and given a public identity, a greenway system can be protected, ensuring its conservation and public value for the future (Arendt, 1994).

The goal of this study was to systematically identify future greenway corridors within Seoul City that will best serve the multiple objectives of environmental protection, recreation, and alternative transportation. Ultimately, the alternative greenway scenarios generated through this analysis will be used by the local communities, public officials, and planners as a basis for discussing land-use priorities as they pursue a more balanced approach to land development in the city and create the city's comprehensive land-use plan.

Conclusion

Seoul, the capital of South Korea, is located on a basin surrounded by many mountains, and including Mt. Boukhansan National Park. Seoul has a long history of urbanization—one that has been occurring for over 600 years. Its population is now over 10 million and its mean density approximately 17,000/km^2. Although the climate is a continental type, the negative impacts of altered urban climate are increasing. Both flat plains and hilly terrain are being covered with residential, commercial, and public facilities, and thereby natural vegetation is now restricted to the urban fringe, where the terrain is mountainous. Those mountainous areas with natural vegetation are only passively managed by designating them as green belts, and so their continued ecological health is probably threatened. Except for a few natural parks, vegetation in the urban center is usually composed of artificial plantations introduced for erosion control or for landscape architectural purposes. Therefore, the center of Seoul endures quantitative as well as qualitative deficits of green-space vegetation, and what vegetation remains there is under severe environmental stress. Furthermore, the geographically unbalanced distribution of green space transports environmental stress to the urban boundary by altering atmospheric circulation patterns (Bonan, 2002). This has likely resulted in soil acidification of the mountainous areas in the urban boundary where most of the remnant natural vegetation exists.

We suggest that revegetating riparian areas along rivers and streams and creating extra green space alongside sidewalks is an initial locations for expanding the city's green spaces, because they cross the urban center and connect with core green spaces along the urban boundary. In addition, this approach promotes a more even spatial distribution for these green areas and improves connectivity among them. Governments in Seoul and its surrounding metropolitan area cities can promote these objectives through diverse programs. Projects for restoring Cheonggye stream, which passes through the urban center and Seoul forest, are examples of such programs that can acquire an enthusiastic public support base for expanding such efforts.

Acknowledgments This study was partly supported by fund of Long Term Ecological Research (LTER) provided from the Ministry of Environment of Korea.

References

Ahern, J. (1995) Greenways as a planning strategy. Landscape and Urban Planning 33:131–155.

Akbari, H., Davis, S.D., Huang, J., and Winnett S. (eds.) (1992) Cooling our communities. The Lawrence Berkeley National Laboratory Report 31587, Berkeley, California.

Akbari, H., Rosenfeld, A.H., and Taha, H.G. (1990) Summer heat islands, urban trees and white surfaces. 1990 The American Society of Heating, Refrigerating and Air Conditioning Engineers Transactions (Atlanta, GA, January 1990). LBL Report 28308, Berkeley, California.

Allen, S.E., Grimshaw, H.M., and Rowland, A.P. (1986) Chemical analysis. In: Moore, P.D., and Chapman, S.B. (eds.) Methods in Plant Ecology. Blackwell, Oxford, pp. 285–344.

Arendt, R. (1994) Rural by Design. American Planning Association, Chicago.

Barbour, M.G., Burk, J.H., Pitts, W.D., Gilliam, F.S., Schwartz, M.W. (1999) Terrestrial Plant Ecology, 3rd ed. Benjamin/Cummings Co., Menlo Park.

Bardsley, C.E., and Lancaster, J.D. (1960) Determination of reserve sulfur and soluble sulfates in soils. Proceedings of Soil Science Society of America 24:265–268.

Bonan, G. (2002) Ecological Climatology: Concepts and Applications. Cambridge University Press, Cambridge.

Braun-Blanquet, J. (1964) Pflanzensoziologie. Grundze der Vegetaionskunde. Springer-Verlag, Wien.

Chiras, D.A. (2001) Environmental Science: Creating a Sustainable Future. Jones and Bartlett, Sudbury.

Conine, A., Xiang, W.N., Young, J., and Whitley, D. (2004) Planning for multi-purpose greenways in Concord, North Carolina. Landscape and Urban Planning 68:271–287.

Diamond, H., and Noonan, P. (1996) Land Use in America. Island Press, Washington, DC.

Environmental System Research Institute (ESRI). (1996) ArcView GIS. Environmental System Research Institute, Inc. New York.

Fabos, J.G. (1985) Land Use Planning: From Global to Local Challenge. Chapman and Hall, New York.

Fabos, J.G. (1995) Introduction and overview: the greenway movement, uses and potentials of greenways. Landscape and Urban Planning 33:1–13.

Freedman, B. (1995) Environmental Ecology: The Ecological Effects of Pollution, Disturbance, and Other Stresses, 2nd ed. Academic Press, San Diego.

Gilbert, O.L. (1991) The Ecology of Urban Habitats. Chapman & Hall, London.

Grey, G.W., and Deneke, F.J. (1986) Urban Forestry. John Wiley and Sons, New York.

Gunn, J.M. (ed.). (1995) Restoration and Recovery of an Industrial Region. Springer-Verlag, New York.

Hendrix, W.G., Fabos, J.G., and Price, J.E. (1988) An ecological approach to landscape planning using geographic information system technology. Landscape and Urban Planning 15:211–225.

Henry, J.A., and Dicks, S.E. (1987) Association of urban temperatures with land use and surface materials. Landscape and Urban Planning 14:21–29.

Hiesler, G.M., and Herrington, L.P. (1976) Selection of trees for modifying metropolitan climates. In: U.S. Department of Agriculture. Better Trees for Metropolitan Landscapes. General Technical Report NE-22. USDA Forest Service, Washington, DC, pp. 31–37.

Hill, M.O. (1979) DECORANA—a Fortran Program for Detrended Correspondence Analysis and Reciprocal Averaging. Cornell University Ithaca, New York.

Innes, J.L., and Oleksyn, J. (ed.) (2000) Forest dynamics in heavily polluted regions. Report No. 1 of the The International Union of Forestry Research Organisations task force on environmental change. CABI, Vienna.

Kennedy, I.R. (1986) Acid Soil and Acid Rain. The Impact on the Environment of Nitrogen and Sulphur Cycling. Research Studies Press Ltd., Letchworth and John Wiley & Sons, New York.

Kent, M., and Cocker, P. (1992) Vegetation Description and Analysis: A Practical Approach. CRC Press, Boca Raton, FL.

Kim, J.C., and Choe, S.C. (1997) Seoul—The Making of a Metropolis. Wiley, New York.

Kim, J.H. (1991) Forest decline syndrome affected by air pollution. In: Proceedings of the Symposium on Human, Forest, and Environment. The Korean Society of Landscape Architecture and the Forestry Research Institute of Korea, Seoul, pp. 3–25 (In Korean).

Kim, J.W. (1994) Ecophysiological responses of *Quercus mongolica* stand to air pollutants and a model for its productivity affected by ambient ozone. Ph.D. thesis, Seoul National University (in Korean with English abstract).

Korea Metrological Administration (KMA). (1990) Annual Climatological Report. Korea Metrological Administration, Seoul (in Korean).

Korea Plant Name Index. (2004) http://www.koreaplants.go.kr:9090/.Kozlowski, T.T. (1985) SO_2 effects on plant community. In: Winner, W.E., Mooney, H.A., and Goldstein, R.A. (eds.) Sulfur Dioxide and Vegetation. Stanford University Press, Stanford, California, pp. 431–453.

Krestov, P.V., Song, J.S., Nakamura, Y., and Verkholat, V.P. (2006) A phytosociological survey of the deciduous temperate forests of mainland Northeast Asia. Phytocoenologia 36(1):77–150.

Landsberg, H.E. (1981). The Urban Climate. Academic Press, New York.

Lee, C.S. (1992) A study on restoration of vegetation destroyed by air pollution. A research report for Post-Doctoral Fellowship Program funded by Scientific Research Foundation of Korea, Seoul (in Korean with English abstract).

Lee, C.S. (1993) Regeneration of *Pinus densiflora* community around the Yeocheon industrial complex disturbed by air pollution. Korean Journal of Ecology 16(3):305–316 (in Korean with English summary).

Lee, C.S., Lee, A.N., Cho, Y.C., Shin, H.C., and Woo, H.S. (2004) Effects of partial restoration practiced by a method suitable for the riverine environment in Korea. In: Proceedings of 16th International Conference, Society for Ecological Restoration International, August 24–26, 2004, Victoria, Canada, pp. 1–10.

Lee, C.S., Hong, S.K., You, Y.H., Kim, J.E., and Moon, J.S. (2000) Landscape structure in the green belt around the metropolitan area of Seoul, Korea. Proceedings of the The International Association of Vegetation Science Symposium, pp. 230–233.

Lee, C.S., Lee, W.S., Cho, H.J., and Rhim, S.J. (2002a) Ecology of Seoul. Dangdae, Seoul.

Lee, C.S., Moon, J.S., and Cho, Y.C. (2007) Effects of soil amelioration and tree planting on restoration of an air-pollution damaged forest in South Korea. Water, Air, and Soil Pollution 179(1–4): 239–254.

Lee, C.S., You, Y.H., and Robinson, G.R. (2002b) Secondary succession and natural habitat restoration in abandoned rice fields of central Korea. Restoration Ecology 10:306–314.

Lee, E.B. (1997) Actual vegetation and potential natural vegetation of Pukhansan National Park, mid-western Korea. Korean Journal of Ecology 20(6): 439–450.

Lee, H.Y. (1985) A study on the urban temperature in Seoul. Ehwa Bulletin of Geography 1:104 (in Korean).

Lovett, G.M., Traynor, M.M., Pouyat, R.V., Carreiro, M.M., Zhu, W.X., and Baxter, J.W. (2000) Atmospheric deposition to oak forests along an urban-rural gradient. Environmental Science and Technology 34:4294–4300.

Lowrance, R., Todd, R., Fail, L. Jr., Leonard, R. Jr., and Asmussen, L. (1984) Riparian forests as nutrient filters in agricultural watersheds. Biological Science 34:374–378.

Magurran, A.E. (2004) Measuring Biological Diversity. Blackwell Publishing, Boston.

McCarigal K., and Marks, B.J. (1995) FRAGSTATS: spatial pattern analysis program for quantifying landscape structure. General Technical Report PNW-GTR-351. USDA Forest Service, Pacific Northwest Station, Portland, Oregon.

McClenahan, J.R. (1978) Community changes in a deciduous forest exposed to air pollution. Canadian Journal of Forest Restoration 8:432–438.

McHarg, I. (1969) Design with Nature. Natural History Press, New York.

Miller, R.W. (1998) Urban Forestry: Planning and Managing Urban Green Spaces. Prentice Hall, Upper Saddle River, NJ.

Ministry of Environment. (2005) A White Book on Environment. Ministry of Environment, Seoul (in Korean).

Moravčik, P. (1994) Development of new forest stands after a large scale forest decline in the Kršněhory Mountains. Ecological Engineering 3:57–69.

Nakagoshi, N., Kamada, M., and Hong, S.K. (1992) Map of actual vegetation map in Miwacho, Hiroshima Prefecture. Bulletin of the Biological Society of Hiroshima University 58:3–6.

National Institute of Environmental Research (NIER). (1981) Studies on the Ecosystem Around the Industrial Complexes. NIER, Seoul.

Nichol, J.E. (1996) High resolution surface temperature patterns related to urban morphology in a tropical city: a satellite-based study. Journal of Applied Meteorology 35:135–146.

Noss, R.F. (1991) Wilderness recovery: thinking big in restoration ecology. The Environmental Professional 13:225–234.

Nowak, D.J., Rowntree, R.A., McPherson, E.G., Sisinni, S.M., Kerkmann, E.R., and Stevens, J.C. (1996) Measuring and analyzing urban tree cover. Landscape and Urban Planning 36:49–57.

Olson, R.L. Jr., and Sharpe, P.J.H. (1985) Toward a mechanistic understanding of plant response to SO2. In: Winner, W.E., Mooney, H.A., and Goldstein, R.A. (eds.), Sulfur Dioxide and Vegetation. Stanford University Press, Stanford, CA, pp. 507–520.

Parker, J. (1989) The impact of vegetation on air conditioning consumption. Controlling summer heat islands. In: Proceedings of the Workshop on Saving Energy and Reducing Atmospheric Pollution by Controlling Summer Heat Islands. University of California, Berkeley, pp. 45–52.

Peterjohn, W.T., and Correll, D.L. (1984) Nutrient dynamics in an agricultural watershed; observations on the role of the riparian forest. Ecology 65:1466–1475.

Reuss, J.O., and Johnson, D.W. (1986) Acid Deposition and the Acidification of Soils and Waters. Springer-Verlag, New York.

Rhyu, T.C. (1994) Mechanism of forest decline syndrome of *Pinus rigida* stands by acid pollutants and its recovery in Seoul Metropolitan region. Ph.D. thesis, Seoul National University.

Rhyu, T.C., and Kim, J.H. (1994) Growth decline of pitch pine caused by soil acidification in Seoul metropolitan area. Korean Journal of Ecology 17:287–297.

Runkle, J.R. (1985). Disturbance regimes in temperate forests. In: Pickett, S.T.A., and White, P.S. (eds.) The Ecology of Natural Disturbance and Patch Dynamics. Academic Press, New York, pp. 17–33.

Scale, P.R. (1980) Changes in plant communities with distance from an SO_2 source. In: Miller, P.R. (ed.) Effects of Air Pollutants on Mediterranean and Temperate Forest Ecosystems. General Technical Report PSW-43. U.S. Forest Service, Berkeley, CA, pp. 1–248.
Seoul City. (1997a) Environment of Seoul. Seoul City, Seoul.
Seoul City. (1997b) A Report of a Detailed Survey on the Forest Ecosystems in Seoul, 1st report. Seoul City, Seoul.
Seoul City. (1998) A Report of Detailed Survey on the Forest Ecosystems in Seoul, 2nd Report. Seoul City, Seoul.
Seoul City. (2000a) Seoul Metropolitan Biotop Map. Seoul City, Seoul.
Seoul City. (2000b) A Study on Urban Planning Technique Based on Meteorological Characteristics of Seoul City. Seoul City, Seoul.
Seoul City. (2005) Statistics of Seoul. http://www.seoul.go.kr/seoul/summary/statistics/statisti.
Shugart, H.H., and McLaughlin, S.B. (1985) Modeling SO2 effects on forest growth and community dynamics. In: Winner, W.E., Mooney, H.A., and Goldstein, R.A. (eds.) Sulfur Dioxide and Vegetation. Stanford University Press, Stanford, California, pp. 478–491.
Smith, D.S., and Hellmund, P. (1993) Ecology of Greenways. University of Minnesota Press, Minneapolis, MN.
Smith, W.H. (1974) Air pollution: effects on the structure and function of the temperate forest ecosystem. Environmental Pollution 6:111–129.
Smith, W.H. (1978) Urban vegetation and air quality. In: Proceedings of the National Urban Forest Conference, ESF Pub. 80-003. SUNY, Syracuse, pp. 284–305.
Smith, W.H. (1990) Air Pollution and Forests, 2nd ed. Springer-Verlag, New York.
Smith, W.H., and Dochinger, S. (1976) Capacity of metropolitan trees to reduce atmospheric contaminants. In: Better Trees for Metropolitan Landscape. General Technical Report NE-22. USDA Forest Service, Washington, DC, pp. 49–59.
Souch, C.A., and Souch, C. (1993) The effect of trees on summertime below-canopy urban climates: a case study Bloomington, Indiana. Journal of Arboriculture 11:1–12.
Taoda, H. (1979) Effect of urbanization on the green broad-leaved forest in Tokyo, Japan. In: Miyawaki, A., and Okuda, S. (eds.) Vegetation and Landschaft Japans. The Yokohama Phytosociological Society, Yokohama, pp. 161–165.
Ulrich, B. (1980) Die Wälder in Mitteleuropa: Meβergebnisse ihrer Umweltbelastung. Theorie einter Gefährdung. Prognose ihrer Entwicklung. Allgemeine Forst Zeitschrift 35:1198–1202.
Upmanis, H., Eliasson, I., and Lindqvist, S. (1998) The influence of green areas on nocturnal temperatures in a high latitude city (Goteborg, Sweden). International Journal of Climatology 18:681–700.
Wong, M.T.F., and Swift, R.S. (2003) Role of organic matter in alleviating soil acidity. In: Rengel, Z. (ed.) Handbook of Soil Acidity. Marcel Dekker, New York, pp. 337–358.
Yim, Y.J. (1989) Urban environment and green belt. Nature Conservation 67:1251–1255.

26
The Construction of Near-Natural Forests in the Urban Areas of Shanghai

Liang-Jun Da and Yong-Chang Song

Throughout the ages, China has had a high regard for afforestation and forest protection. Old growth forests, secondary forests, Fengshui forests (which can be good luck for forest owners, such as monastery or temple forests, cemetery forests, and forests around houses), all of various areas and patterns, are seen across the nation. While forests have been planted throughout the nation since 1950 to improve environmental quality, the pace of forest construction has accelerated since 1980. As a result, China is now among the top nations with the greatest area of human-made forests in the world. Because the history of human activities in China has been long and often violent, indigenous vegetation has been severely damaged. This has made forest construction and reconstruction difficult by decreasing the survival rate of these forests and prolonging the time it takes for their restoration and establishment. Although this situation has been partially caused by problems such as policies and lack of funding, the lack of proper theory to support operational methods that can guide forest construction projects has also created setbacks. At present, there are discrepancies and inconsistencies in theories and methods for informing the process of restoring and reconstructing urban forest ecosystems (Hobbs, 1996; Allen et al., 1997; Palmer et al., 1997; Miyawaki, 1998; Bao and Chen, 1999; Zhang and Xu, 1999). Empirical studies that scientifically experiment with various methods and test different theories are needed both to advance the discipline of urban forestry and to determine best management techniques and approaches in different urban contexts.

The city is an artificial, constructed ecosystem where human occupation is dominant. Cities are complex ecosystems consisting of social, economic, and natural subsystems. The natural ecosystem within urban areas is in a precipitous state of decline due to intense human activities of long duration. The discipline of urban forestry, which arose in Western nations in the latter half of the 20th century, is a type of forestry that serves to improve the quality of human life in the city. Therefore, in its narrowest sense, the objective of urban afforestation is to make the city green, beautiful, and healthy for people. Near-natural forests is a new urban afforestation method developed and carried out by Miyawaki (1998, 1999a,b; see also Chapter 12). This recent restoration approach is based on the theories of potential natural vegetation and community succession. Native tree and shrub species are selected from the natural vegetation in a local area and their seeds germinated and

M.M. Carreiro et al. (eds.), *Ecology, Planning, and Management of Urban Forests: International Perspectives.*

grown in containers for subsequent out-planting. Within a relatively short time, near-natural forests grow to a climax stage at low cost. These forests should be characterized by the following: a complex community structure with high species diversity, multilayer canopies and high biomass, a static state, and less maintenance during later stages of succession (Da, 2004). This chapter describes experiments in the establishment of near-natural forests in China, especially in Shanghai, and makes recommendations for improving methods for constructing these forests.

Near-Natural Method

Theoretical Basis for Forest Construction Using the Near-Natural Method

At present, there are three types of urban greening activities typically conducted in China: (1) planting stands of monocultures of the same species and age class; (2) small-scale, combination woody tree and floral plantings mostly in residential and commercial areas; and (3) transplantation of mature trees in grassy gardens and parks for rapid visual impact. Each of these forests has weaknesses in terms of sustainability, diversity maintenance, and promotion of ecological functions in the urban landscape. Plantation monocultures consisting of conifers or other rapidly growing pioneer trees are used for afforesting wilderness areas, and creating timber and shelter-belt forests. The weakness of this kind of afforestation involves the extreme homogeneity of species in the forest community. Due to low-density planting in creating these monotypic stands, the resources for the trees are not limiting, and so competition and interdependence between individual trees are fairly low. In many cases, seedling and sapling recruitment is generally low, so forest regeneration becomes an issue over time. As compared with multispecies stands, monocultures are also more vulnerable to plant diseases and insect pests (Gibson and Jones, 1977; Ciesla and Donaubauer, 1994; Gadgil and Bain, 1999; Zhou, 2004). These forests also have a simple vertical structure and low diversity due to maintenance practices that cut down undergrowth. Hence, such monoculture plantations harbor lower animal diversity as well (Kloor, 2000).

Small-scale forests (actually wooded gardens) are built in small courtyards, or as street tree and hedgerow groupings in a largely impervious, human-made landscape. The main objectives of this category of plantings are usually decorative and aesthetic. Often many mature, peculiarly shaped, or even very ancient trees are transplanted, with a great deal of consideration given to being unconventional or unorthodox. Therefore, despite the area being small, the cost of building these wooded green spaces is high. Also, because most of the nursery trees are already mature, their survival rate is low, and replacement costs high.

A third urban greening category is the planting of mature trees into lawns, particularly in parks. This method is simple and fast. However, the planting and maintenance costs are high and the ecological functioning of such a system is not complex, contributing little to local biodiversity.

To avoid the disadvantages of these three methods, the natural image and diversity of forests should be restored (Kloor, 2000). The Miyawaki method (Wang and Chen, 1999; Wang et al., 2002) promotes the philosophy of constructing near-natural forests by determining what the potential native vegetation of the region would be at various stages in the process of succession. Hence, unlike much of the planting that has existed in China up until now, this method emphasizes constructing and restoring forests using native trees.

The near-natural method is based on community succession theory in plant ecology, and aims to restore the local potential vegetation (Fujiwara, 1997; Miyawaki, 1998; Wang et al., 2002; Da, 2004). Succession theory refers to the process in which one plant community is replaced by another plant community via such mechanisms as competition, tolerance, and facilitation as habitat conditions change. After a pioneer plant community is established on a bare open area, succession proceeds until a climax community is established, unless disturbance sets the community back to an earlier state. The climax community creates conditions that are conducive to continued reestablishment of the same species, and hence is relatively stable in species composition over long periods. In natural ecosystems, the successional process, from the establishment of the first plant community in an open area to a forested climax community, can take hundreds of years. However, by reforming the soil, controlling moisture conditions, collecting the seeds of native trees, and raising seedlings in the plastic pots, the near-natural method can accelerate succession and create a climax community suitable for the local climate in a much shorter time (Fig. 26.1).

Near-Natural Method and Its Procedure

Construction of vegetated patches by the near-natural method can be divided into five parts (Fig. 26.2) (Miyawaki, 1998):

1. Vegetation survey and charting: A vegetation survey is the foundation for constructing and restoring forests for protecting the environment. From these surveys, the actual local vegetation is determined, mapped, and used to deduce the potential natural, climax vegetation for the region. Habitat characteristics, including the climate, geology, land form, and soil types, are simultaneously determined, since these allow us to estimate the potential climax vegetation for a site.
2. Plant species selection: Plant species are selected from the list of actual and potential vegetation determined for an area. These are mainly canopy trees species and species that can colonize the understory, such as shrub species. Usually, there are at least 10 to 20 species of woody plants selected.
3. Seed collection and seedling cultivation: After selecting the species, seeds are collected when the fruits ripen. They are either collected by directly picking

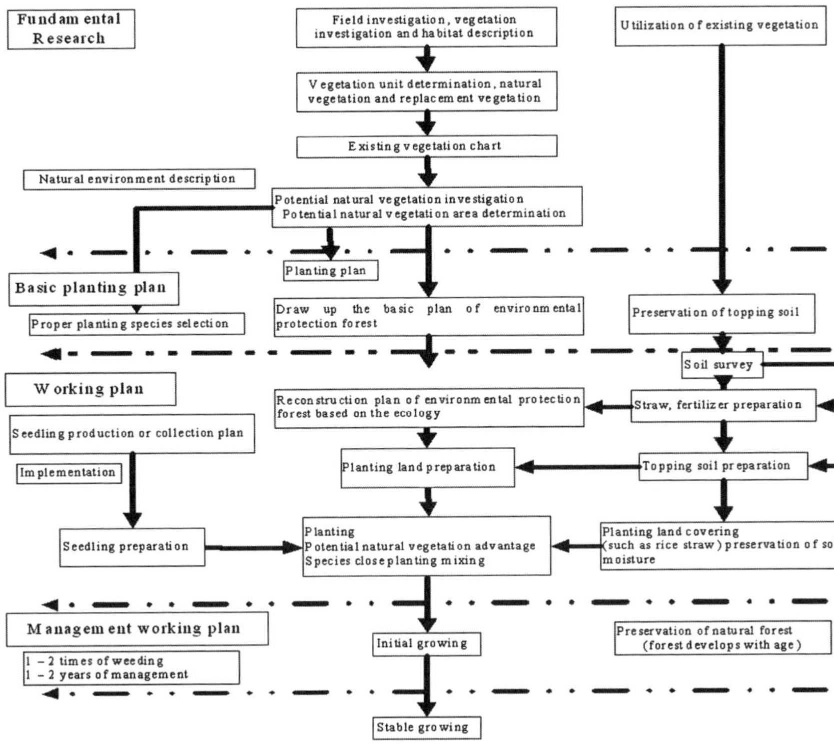

Fig. 26.1 Successional time line for reaching climax communities in slower natural ecosystems versus the accelerated rate achieved using the near-natural method that bypasses earlier series

fruits from the trees or after seeds fall to the ground. Seeds that are immature or greatly damaged by insects are removed. Seeds are soaked in water overnight to accelerate imbibition and smother some insect larvae, and sown in a seedbed consisting of soil and pearl stone. Once seedlings have two to six true leaves, they are transplanted to plastic pots (10 to 12 cm in diameter, 10 cm in height). Sometimes seedlings that germinated naturally in the seed production plots are transplanted into pots. The pots contain soil similar to that of a healthy, natural woodland, and is rich in organic matter, so that soil is loose and well aerated. After 2 to 3 years, when seedlings are 30 to 50 cm high with well-developed root systems, they are ready for transplantation in the field.

4. Site preparation and planting: Generally, soil conditions in the reforestation area are rather poor, often barren and dry, and so it is necessary to prepare the land manually. In China, a layer of soil, about 50 to 100 cm deep, should be added. On flat land, topography may need to be created and baffle plates are added (such as bamboo plate, timber plate, ribbon plate, iron nets) to reduce soil erosion. Additionally, in areas where the soil layer is too thin, rock exposed, or there are new road cuts, a V-shaped trench should be dug, and soil added to increase the soil thickness for later rooting by plants. Piles and baffle plates may also be needed.

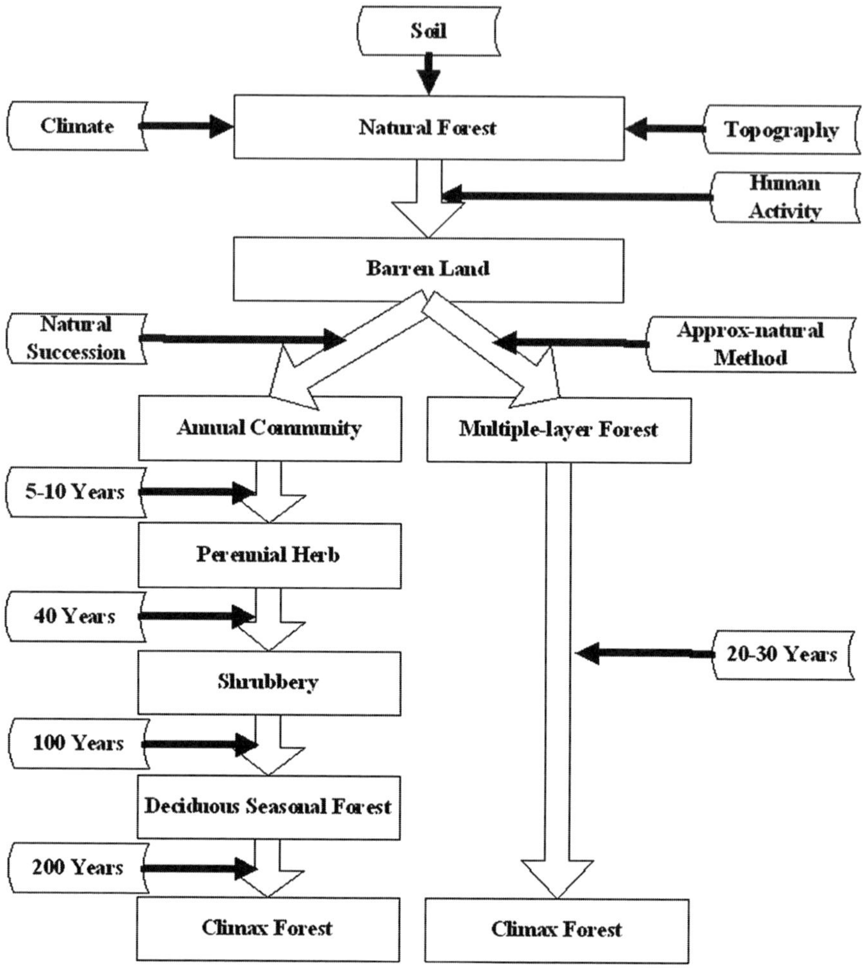

Fig. 26.2 Flow chart for restoration and creation of native forests using near-natural method

Once the site is prepared, seedlings or small saplings are ready for transplanting. Before out-planting, the pots with seedlings are soaked in water for 15 to 30 seconds, until no air bubbles stop escaping. A hole about 1.5 times the pot diameter is dug, and the seedling planted with the hole backfilled with soil and gently tamped down and compacted. Planting density should be about three to four stems per square meter. When planting, attention should be paid to species mixing and close planting, thus mimicking natural conditions for seedlings in forests. Dense planting helps maintain a proper microclimate for seedling growth, particularly in small areas where conditions external to the patch might penetrate into the patch interior. Natural thinning processes will occur over time. After planting, seedlings should be covered with rice straw or rotten straw and secured with hay ropes to avoid negative effects of wind and dryness. This mulch will also reduce weed growth. If possible, the area should be sprayed with water to soak roots, reduce fire risk, promote decomposition, and release soil nutrients.

5. Maintenance: One to three years after planting, simple maintenance procedures such as weeding, watering, and fertilizer application, should be done, and the seedlings allowed to thin via natural processes of competition and selection. Generally, 33% to 50% of the original seedlings do not reach the tree stratum, or die. After 15 to 50 years (depending on soil conditions and precipitation), a forest will emerge that is close in composition and vertical structure to that that would naturally develop without human assistance, hence the term near-natural forest.

Application of the Near-Natural Method in China

There have been over 600 successful examples in Japan of restoring environmental protection forests using the near-natural method. Since 1990, the near-natural method has also been used for restoring tropical rain forests, evergreen broad-leaved forests, and deciduous broad-leaved forest in Malaysia, Brazil, Chile, and Thailand (Miyawaki, 1998, 1999a,b). In China, this method has also been used successfully in Beijing, Shanghai, Xuzhou, Qingdao, and Inner Mongolia to rebuild evergreen broad-leaved forests and deciduous broad-leaved forests (Fujiwara et al., 2000; Da and Xu, 2003).

Construction of Near-Natural Forests in Shanghai

Goals for Constructing Near-Natural Forests in Shanghai

As a large metropolis characterized by a high-density population, concentrated industries, and resource shortages, Shanghai has realized the importance of environmental quality and reliable natural resources for supporting its large-scale development program in the new century. Therefore, there is a sense of urgency in Shanghai to accelerate its plans for environmental protection and construction. The greening process, one of the five environmental conservation foci in Shanghai, relates directly to attaining many of Shanghai's environmental goals, such as controlling and reducing air pollution, improving its urban image, landscaping residential areas, and promoting greater quality of life in general.

To improve the quality of this urban ecosystem and pursue a strategic goal of sustainable development, Shanghai is accelerating its green-space development. However, this speed may compromise the city's long-term goals, if more thought is not given to planning and discussing such issues as finding the proper combination and ratios of green space to impervious cover in different sectors of the city, and where near-natural forests can be most effectively planted. For example, finding locations where construction and postconstruction maintenance costs can be reduced can maximize the positive environmental effects even if the forest is small. Trying to find sites where multiple ecological benefits would ensue would also help prioritize where these forests should be built. These benefits include ameliorating the heat-island effect and local microclimate, and reducing air pollutants and local flooding. Improvements in species diversity from neighborhood to landscape scales should also influence the locations where these forests are constructed. In addition, these forests can provide recreational areas for local residents, and simultaneously serve as laboratories for scientific research and a school for environmental education for the public (Yan, 1998; Zhang, 2001).

Near-Natural Forest Construction in Shanghai

Since it combines natural growth with artificial buildings and surfaces, the concept of near-natural forests has been embraced and promoted as an essential element for transforming Shanghai into an eco-city. The first model forest stand was established in the Pudong New Area of Shanghai between Yingchun Road and Zhangjiabang River in June 2000. This 3000-m^2 forest lies just east of the office of the Mitsubishi Corporation, the forest sponsor, north of the Shanghai Science and Technology Museum across Yingchun Road, and west of the Pudong green belt. Three additional near-natural forests were built in 2002, 2004, 2005.

The Selection of Tree Species and Planting for the Pudong Forest

The tree species chosen for the forest were selected from a list of major dominant species comprising Shanghai's potential vegetation (Table 26.1). This included four species of broad-leaved evergreen trees, two deciduous species, and five

Table 26.1 Species of woody plants used to construct the Pudong forest stand

Species names	Life form	Seedling number
Cyclobalanopsis glauca	Evergreen broad-leaved tree	2500
C. myrsinaefolia	Evergreen broad-leaved tree	2500
Castanopsis sclerophylla	Evergreen broad-leaved tree	2500
Machilus thunbergii	Evergreen broad-leaved tree	2830
Ligustrum quihoui	Evergreen broad-leaved shrub	370
Distylium racemosum	Evergreen broad-leaved shrub	150
Pittosporum tobira	Evergreen broad-leaved shrub	400
Fatisia japonica	Evergreen broad-leaved shrub	300
Aucuba japonica	Evergreen broad-leaved shrub	300
Liquidambar formosana	Deciduous broad-leaved tree	300
Cornus alba	Deciduous broad-leaved shrub	200

broad-leaved evergreen shrubs to occupy the forest understory. The seed collection was carried out in the peripheral regions of Shanghai, which belongs to the same vegetation zone. Seedbed and container breeding were used to raise seedlings in greenhouses where lighting intensity and other environmental factors could be controlled. Within 1 or 2 years, the container seedlings developed a strong root system and reached a height of 20 to 40 cm.

Procedure for Reforming Topography and Seedling Maintenance

Due to the high water table and high soil salinities in Shanghai, some topographic locations are unfavorable for planting. Therefore, soil from other locations was imported to the area to create greater microtopography before transplanting seedlings. Two sloping fields, one large and another small, were built. The small field adjacent to the Zhangjiabang River is triangle-shaped and lies in the southeast corner of the model area. It is 1.7 m higher than its surroundings and has a 10° slope in the meridional direction, and about a 2° slope in the transmeridional direction. The larger field parallel to the road, is over 90 m long, 30 m wide, and 1.6 m higher than its surroundings. This area has a slope of 5° in the meridional direction.

When planting, mixing different tree species is preferred. Seedlings are planting at densities as high as three to four stems per square meter, since high-density planting will favor natural selection of the best seedlings that match their microsite conditions, and at the same time it will provide a source of excess additional seedling to meet greening needs in the future. After the seedlings are planted, only some general care is needed, such as weeding every 2 or 3 years. After this, the forest will grow naturally and no further maintenance is needed. After 5 years, it becomes readily evident that an incipient "forest" has begun to grow. Even when young and small the patch can provide an interesting green habitat for its surroundings. Within 10 to 20 years, a near-natural forest will have developed.

Plant Growth

With some maintenance, such as weeding, the seedling survival rate the first winter after out-planting is more than 90%. From 2000 to 2004 the total height increase per seedling across all species ranged from 94 to 409 cm, and vertical growth rates ranged from 381% to 775%. In 2004, four years after transplanting in the field, the largest seedling of *Liquidambar formosana* was 6.0 m tall, and *Cyclobalanopsis glauca* and *Ligustrum quihoui* were 5.6 m and 5.5 m, respectively. *C. myrsinaefolia* was 4.5 m, and *Distylium racemosum* and *Castanopsis sclerophylla* were 4.4 m and 4.1 m tall, respectively. In addition, native deciduous tree species (*Ulmus parvifolia, Sapium sebiferum, Salix matsudana, Broussonetia papyrifea*) invaded the stand, dispersed by wind and birds. Most of these were about 3.0 to 3.5 m tall (Table 26.2). In this area, a near-natural forest can be expected to develop within 10 years. Since both canopy and subcanopy species

Table 26.2 Growth rates of tree species planted in the Pudong site

Species names	Height of plants (cm)								Growth rate in each period (%)		
	2000		2001		2003		2004				
	Average	Max.	Average	Max.	Average	Max.	Average	Max.	2000–2001	2001–2003	2003–2004
Cyclobalanopsis glauca	42.6	94	53.8	115	200	410	287	560	26.3	271.7	43.5
C. myrsinaefolia	35.6	105	41.2	108	183	420	276	450	15.7	344.2	50.8
Castanopsis sclerophylla	33.3	74	39.0	83	90	190	127	410	17.1	130.8	41.1
Machilus thunbergii	23.8	40	32.4	58	123	200	223	400	36.1	279.6	81.3
Ligustrum quihoui	67.7	90	82.2	135	307	400	441	550	21.4	273.5	43.6
Distylium racemosum	60.0	107	70.4	125	183	350	253	440	17.3	159.9	38.3
Pittosporum tobira	50.0	72	68.5	100	187	250	311	400	37.0	173.0	66.3
Fatsia japonica	28.5	45	38.5	61	87	140	149	180	35.1	126.0	71.3
Liquidambar formosana	62.0	92	73.4	110	260	360	471	600	18.4	254.2	81.2

are planted, the forest has a vertical structure consisting of several vegetation layers of trees, shrubs, and herbaceous plants. This vertical structure itself fosters greater species diversity by providing more niches for organisms. At present, the area has grown into beautiful woodland and the landscape effect of this greening is remarkable (Fig. 26.3).

Fig. 26.3 **(A)** Initial planting (June 11, 2000). **(B)** Same site, 4 years later (June 8, 2004).

Economic Savings of Using the Near-Natural Method for Urban Greening

There are costs associated with obtaining seeds, growing seedlings in greenhouses, transplanting seedlings, and maintaining them in the early years after transplanting. There are also costs associated with preparing the site and soil. As the Miyawaki model suggests, corporate sponsorship can be sought for covering or defraying the public costs of building these forests. Detailed cost-benefit analyses should be conducted to guide future construction projects. In the meantime, we provide the following nonquantitative overview of relative costs and benefits for using the near-natural method over that of traditional methods for urban greening projects.

The cost of the growing of seedlings for the near-natural method is only a small fraction of that for current traditional greening projects using other methods like transplantation of mature trees. The high survival rate of seedlings after the first winter is greater than 90%. Seedlings are small and easily transported to the site. Large machinery is not needed for transplanting, as occurs when large, mature trees are moved and planted to provide "instant trees" for a site. Seedlings grown in pots in greenhouses are checked before transplanting to the field to make sure that they are healthy and have well-developed root systems. Therefore, their potential to grow into an adult tree is optimized. Since it is not necessary to trim the roots or canopy of small seedlings (as is done for large tree transplantations to reduce water stress), the vitality of small seedlings is high. On the contrary, large transplanted trees sustain higher mortality rates and hence larger costs for replacement.

The near-natural method fosters high biomass and an integrated forest structure. The forest community ultimately has a species composition consisting of both the planted native species and those that naturally invade the site. Therefore, species diversity becomes high without incurring additional costs. The variety of species will help resist pests, diseases, and different disasters. Since this particular greening process mimics natural rules, savings due to low-intensity human maintenance accrue over the long term. The near-natural method requires much less maintenance particularly in the later stages of forest growth than other forms of urban greening.

Recommendations for Improving the Near-Natural Forest Construction Method

Multiple-Layer Forest

In the actual application of the near-natural method, we find that some plants stay small and are likely to be washed away in strong rain events, because of their weak competitive ability in the stand. Therefore, we suggest ensuring that a multiple-layer forest will grow by using the following approach. In the initial stage of forestation, arrange some pioneer trees that are 2 to 3 m high in the upper story, such as *Chinese sweetgum* and *Pistacia chinensis*. These will shade the seedlings of the target climax community and fertilize them by creating a natural, forest floor litter layer with their fallen leaves. After the target tree species have grown to a certain size, some should be judiciously removed to reduce competition pressure. Seedlings removed can then be used for greening other areas. Constructing a multilayered forest is the optimal method for achieving short- and long-term landscape effects from both ecological and economic standpoints.

Advice for Adjusting the Near-Natural Method to Different Climatic Regions

The near-natural method favors forestation in wet regions where rainfall is abundant and more or less evenly distributed during the year. Therefore, when using the near-natural method in regions with less rain and shorter growing seasons, more attention must be paid to preserving soil moisture and watering seedlings at critical times to ensure success. Seedlings should be transplanted during rainy seasons. Additionally, the area of the holes dug for seedling transplantation should also be larger (30 to 50 cm^2). In the north, because rainfall is low and soil is poor, cultivated seedlings should be bigger when transplanted; we recommend 3 to 5 years of age or older. Consequently, pots for growing seedlings should also be bigger, 15 to 30 cm in diameter, so that the root system can become more developed and be more drought resistant.

Acknowledgments We thank Akira Miyawaki and Yan Ling-Zhang for their helpful suggestions and valuable comments. Thanks are also due to Zhu Ying-Rui, Chen Yan-Ping, Li Li-Na, Lin An-Guo, Fang He-Jun, Yang Yong-Chuan, Yang Tong-Hui, and Xia Ai-Mei, for their help with field work and data analysis. This research is supported by the project of National Natural Science Foundation of China (30130060, 30370255), the Major State Basic Research Development Plan of China (G2000046801), and the Key Academic Discipline Program of the Shanghai Government.

References

Allen, E.B., Covington, W., and Falk, D.A. (1997) Developing the conceptual basis for restoration ecology. Restoration Ecology 5:275–276

Bao, W.K., and Chen, Q.H. (1999) Process and characteristics of retrograded ecosystem. Chinese Journal of Applied Ecology 18(2):36–42 (in Chinese).

Ciesla, W.M., and Donaubauer, E. (1994) Decline and Dieback of Trees and Forests. A Global Overview. Food and Agriculture Organization of the United Nations (FAO) Forestry Paper 120. FAO, Rome.

Da, L. (2004). Construction of near-natural forests as ecological restoration of urban ecosystem in Shanghai. Chinese Forestry Science and Technology 3:60–64.

Da, L., and Xu, X. (2003) An attempt to build a "close-to-forest" in Shanghai. Journal of Chinese Urban Forestry 1(2):13–16 (in Chinese with English abstract).

Fujiwara, K. (1997) Methodology of vegetation analysis and vegetation mapping based on phytosociology and vegetation science. Bulletin of the Institute of Environmental Science and Technology, Yokohama National University 23:1–46.

Fujiwara, K., You, H.M., and Miyawaki, A. (2000) The potential natural vegetation and proposals for restoration of green environment in Xuzhou City. Bulletin of the Kanagawa Phytosociological Society 5:19–31.

Gadgil, P.D., and Bain, J. (1999) Vulnerability of planted forests to biotic and abiotic disturbances. New Forests 7:227–238.

Gibson, I.A.S., and Jones, T. (1977) Monoculture as the origin of major forest pests and diseases. In: Cherret, J.M., and Sagar, G.R. (eds.) Origins of Pest, Parasite, Disease and Weed Problems. Blackwell Scientific Publications, Oxford, pp. 139–161.

Hobbs, R.J. (1996) Towards a conceptual framework for restoration ecology. Restoration Ecology 4:93–110.

Kloor, K. (2000) Restoring American's forests to their "natural" roots. Science 287:573–575.

Miyawaki, A. (1998) Restoration of urban green environments based on the theories of vegetation ecology. Ecological Engineering 11:157–165.

Miyawaki, A. (1999a) Restoration of native forests by native trees. Plant Biotechnology 16:15–25.

Miyawaki, A. (1999b) Restoration of native forest. Dainoppon-Tosyo Press, Tokyo, pp. 112–156 (in Japanese).

Palmer, M.A., Ambrese, R.F., and Poff, N.L. (1997) Ecological theory and community restoration ecology. Restoration Ecology 5:291–300.

Wang, R.Q., Fujiwara, K., and You, H.M. (2002) Theory and practices for forest vegetation restoration: native forest with native trees—introduction of the Madawaska's method for reconstruction of "environmental protection forest (ecological method to reforestation)". Acta Phytoecologica Sinica 26(suppl):133–139(in Chinese).

Wang, X.-H., and Chen, X.-Y. (1999) Application of Miyawaki method to the construction of Shanghai urban ecological environment. Shanghai Environmental Sciences 18(2):100–104 (in Chinese).

Yan, L.-Zh. (1998) Consideration on bringing green space advantage into play. Chinese Landscape Architecture 17(74):44–46 (in Chinese).

Zhang, J.E., and Xu, Q. (1999) Major issues in restoration ecology researches. Chinese Journal of Applied Ecology 10:109–113 (in Chinese).

Zhang, Q.-F. (2001) Discussion on concept and principles of urban ecological greening. Chinese Landscape Architecture 17(76):34–36 (in Chinese).

Zhou, H. (2004) The protection forests of Yangtze Gorges in Chongqing Yunyang suffer a severe disease and insect damage. China News Service, June 17.

III
Synthesis and Directions for Future Research, Planning, and Implementation

27
Urban Forestry and the Eco-City: Today and Tomorrow

Margaret M. Carreiro and Wayne C. Zipperer

In 1990, the Chicago Academy of Sciences held a conference, Sustainable Cities: Preserving and Restoring Urban Biodiversity, which led to the publication of a book entitled *The Ecological City* (Platt et al., 1994). This symposium differed from others on cities at that time by focusing principally on cities as habitats for biodiversity. The thrust of the symposium was that interactions between people and nonhuman biological entities in urban landscapes had not received much scientific attention and warranted increased ecological investigation. More than a decade later in Shanghai, the International Meeting on Urban Forestry and Eco-Cities conference explored the role of urban forestry in creating more environmentally sound cities that enhance people's quality of life. During the interval between these two symposia, urban ecology has rapidly developed as an ecological discipline exploring the myriad elements that comprise an urban landscape. No longer are urban ecologists trying to convince the ecological community that urban landscapes are important and productive subjects for research, trying to convince planners that ecological concepts need to be incorporated into urban design, or trying to convince environmental managers that a multiple scale approach is needed to manage ecological goods and services and to restore habitats. However, this symposium also revealed that implementation of these principles can be difficult for a variety of reasons, not the least of which is that we still do not understand the nuances of the political and socioecological interactions that affect the structure and function of urban landscapes and how they can be influenced to improve environmental conditions citywide (e.g., Perkins et al., 2004).

The reality is that if our cities are to move in the direction of becoming *eco*-cities, a greater awareness of the ecosystem services provided by a city's urban forest (its entire green infrastructure) must be fostered not only among practitioners and scientists, but also among political leaders and the public. More opportunities should be created to formally and informally educate the public on the roles that urban nature plays in reducing a city's resource and energy use, in improving air and water quality, in decreasing flooding, and in maintaining our physical and psychological well-being. Such education provides the foundation for change. Making cities more comfortable places for people to live by incorporating more of the natural world into our daily lives, and by working *with* nature to prevent or

M.M. Carreiro et al. (eds.), *Ecology, Planning, and Management of Urban Forests: International Perspectives.*

mitigate problems that otherwise require costly engineered solutions are means of linking local quality of life for urban residents with global sustainability for the human species. Therefore, an environmentally educated populace with a greater shared vision of the future is essential, if the long-term goal of creating more ecologically sound and resource-efficient cities is to succeed.

The studies collected in this volume represent a global snapshot of many perspectives and activities of planners, managers, and environmental scientists centered on integrating more and better-planned green infrastructure into the hardscapes of our burgeoning cities. Such diverse experimentation is exciting and essential at this stage in the development of international urban forestry, if professionals are to assess which urban greening strategies are successful in their respective cities. Yet within all this diversity of approaches and opinion, shared ideas and needs have emerged. This concluding chapter highlights and reinforces the following major cross-cutting themes expressed by the international authors who contributed to this book:

- Defining the scope of the urban forest and the need for holistic management
- Quantifying the urban forest and its ecological services
- Expanding research in urban ecology and forestry
- Building partnerships for implementation, planning, and research
- Incorporating urban forestry into the vision of the eco-city

We recognize that most of the recommendations and issues described in this chapter and book have not benefited from the experiences of people in cities on all continents, but mainly reflect current urban forestry concerns in selected countries in Europe, Asia, and North America, where conference participants live and work. In addition, the urban forestry issues and studies included in this book deal primarily with cities in countries that have the economic capability of supporting an urban forestry program and mostly with cities in temperate climates. Therefore, this chapter also cites some potential contrasts with the urban forestry needs and challenges faced by cities in developing countries with fewer economic resources, many of which are in more tropical regions. By doing so, we hope to stimulate more international dialogue in identifying and articulating a spectrum of urban forestry goals that would match the varying needs of people in different cities throughout the world.

Defining the Urban Forest and the Need for Holistic Management

The simple act of defining the domain of urban forestry highlights its diversity rather than its unity. The urban forest is a mosaic of trees and other vegetation, some of which are managed intensively by different agencies or people, and others where natural successional forces, indirectly affected by urban conditions, determine species composition and regeneration (McDonnell et al., 1997; Zipperer et al., 1997; Silva Matos et al., 2002; also see Chapter 11). Therefore, the urban forest consists of street trees, remnant and emergent forest patches, tree plantations, and vegetation in parks, yards, highway verges, utility rights-of-way, and business and institutional campuses. A city's urban forest may also be considered to extend beyond its municipal boundary to encompass peri-urban agroforests or forested watersheds that provide a city's drinking water, such as the Catskill Mountain area does for New York City 160 kilometers away (Chichilnisky and Heal, 1998; Blaine et al., 2006). Such forests or plantations may often require complex management to sustain the multiple social, ecological, and economic services they provide for the urban public, a challenge described by Schulzke and Stoll (Chapter 18) and Jestaedt (Chapter 19). We also feel that urban forestry should expressly include consideration of the soil substrate, since soils, too, are within the purview and care of managers of these varied habitats and are critical determinants of long-term forest sustainability in all urban locales (Carreiro, 2005; also see Chapter 12).

Such an overarching definition enables the urban forest to be viewed more holistically, and provides a conceptual foundation for it to be managed for goods and services in a more integrative way. In addition, urban forestry can be studied and practiced from multiple perspectives that vary in focus over time and according to the developmental stage of urban forestry in different countries. For example, in developed countries, a prime focus in the past was management of the urban forest for aesthetic purposes (Howard, 1902; Pitt et al., 1979), whereas now, as urban populations have grown, intensified, and expanded, it has shifted to management for enhancing ecosystem services (e.g., Nowak and Dwyer, 2000). In developing countries, a more important focus may be managing vegetation to provide materials, such as firewood, fruit and timber, at very local scales (Carter, 1995). Over time, each city and region may manage its urban forest for an increasingly broader and more inclusive range of benefits. Hence, in defining the bounds of urban forestry as a discipline, it is important to consider the current developmental needs of a population as they establish urban forestry goals most suited to their city's social, economic, and geographic context. It may not be as important to rigorously define which vegetative elements are to be considered part of the urban forest as much as to identify the diverse contributions and functions of vegetation and unpaved soils, both within and outside the city, to a particular community's well-being.

Environmental professionals increasingly recognize the ecological reality that the different green areas in a city are, indeed, interacting in diverse ways with each other, with aquatic systems, with the built infrastructure and with people (see Chapter 7). However, the social reality is that the degree and type of management given to vegetation is compartmentalized, varies greatly depending on the group responsible for its care, and seldom involves interactive, coherent planning among these groups (see Chapters 9 and 16). One important dichotomy is the distinction between trees and other vegetation that occur in the public versus the private domain. Public trees can be managed by different agencies within a local government and are a primary means of providing a more socially equitable distribution of vegetation in a city. Hence, the contribution of public trees to total tree cover in lower income areas can be considerable (Heynan et al., 2006). In a recent paper, Escobedo et al. (2006) observed for the City of Santiago, Chile, that higher income *comunas* (an administrative unit approximately equivalent to a municipality) had a higher mean value of tree cover (33.4%) than lower income *comunas* (11.8%). But the higher income *comunas* had a lower mean value of public trees (29%) than lower income *comunas* (54%). Without public support, the lower income *comunas* would have little tree cover, since residents often cannot plant trees for a variety of reasons including financial limitation, land ownership, and lack of available planting space (Perkins et al., 2004).

Escobedo et al. (2006) and Heynan et al. (2006) also point out the importance of trees in the private sector, since they often comprise the greatest proportion of total canopy coverage in cities. Trees and other plants in the private domain are managed by homeowners, community associations, utility companies, and businesses. This socially diverse management can greatly affect the distribution of canopy coverage in cities, thereby creating urban inequities in apportionment of ecosystem services in different neighborhoods. This possible imbalance is a dimension of environmental justice that is not often considered and needs to be given more attention by the public and decision makers. Vegetation management in private lands may also contribute significantly to the total plant diversity found throughout human settlements (Rapoport, 1993; see also Chapter 16). The vertical complexity, species composition, health, and distribution patterns of this green urban mosaic will then reflect the variation in ownership patterns, professional training, aesthetic sensibilities and choices, perceived value of vegetation, funding levels, and education of these diverse managers. Understanding how such diversity affects the ecological functioning of the landscape as a whole remains an important challenge for urban ecologists and practitioners wishing to promote and distribute particular ecological functions at a citywide scale, while enhancing community well-being at the local level.

One means of achieving the goal of improved allocation of social benefits (Westphal, 2003), ecosystem services, and materials from urban vegetation is for the public and private sectors to work together. For example, at the neighborhood scale, homeowners and small businesses can collaborate with municipal government and developers to implement a plan that better meets local needs (e.g., Ames and Dewald, 2003; Wolf, 2003). Similarly, at the city scale professionals and stakeholders can collaborate to develop a comprehensive Urban Forest Master Plan. Establishing such a communication network among groups of individuals can reduce the negative impacts that fragmented responsibility and care engenders, while clarifying and prioritizing local and city-wide urban forestry goals. Throughout the process of developing a plan of action, site assessments need to include input from not only landowners and businesses, but also renters and even those employed but not living in the area (for related examples, see Wolf, 2003; Elmendorf et al., 2005; Yli-Pelkonen and Kohl, 2005). Furthermore, by working together in a cooperative way, this network can create a more unified political advocacy for urban and community forestry and other community needs as well.

Quantifying the Urban Forest and Its Ecological Services

Effective management and planning of urban forests for promoting ecological and social benefits depends on obtaining information and creating databases on the abundance and distribution of vegetation across the city in relation to such variables as social context and land use, both current and planned. In some cases, the primary focus may be to improve ecosystem services and biodiversity conservation at the city scale (Löfvenhaft et al., 2002). In others, urban forest management may be more focused on providing tangible commodities for residents, such as food or fuel (Carter, 1995). Through the use of geographic information systems (GIS), spatial overlays of current and planned development or management together with environmental maps can identify locations for new plantings that enhance social and ecological benefits, and identify opportunities for linking isolated forest components to provide greater ecological and social connectivity.

Several authors contributing to this book highlighted the critical need to take stock of our urban forest resource as an essential first step in creating an effective urban forestry program (see Chapters 15, 16, 17, and 22). Inventories can simply be lists of trees by species and their locations, or contain detailed information such as tree size, vertical structure and health in relation to site conditions, land use, distribution of canopy cover and vegetation, and cultural importance. Such inventories should also be updated regularly so they can be used to determine change in characteristics important to management, such as mortality and growth rates of trees, species composition, and distribution of canopy cover in private and public sectors (Nowak et al., 2004). The spatial analysis capacity afforded by using GIS can also permit comparison of these urban forest attributes over time to assess policy efficacy and inform adaptive management decision making for the future (e.g., Dwyer et al., 2000). Surprisingly, despite the economic value of trees and the expense of their maintenance, the proportion of cities with organized urban forestry programs or an urban forestry master plan is still quite low even in developed countries, as indicated by Kielbaso (see Chapter 15), Kielbaso (1990), and Elmendorf et al. (2003) for the United States. Therefore, professional and stakeholder support should be sought to urge municipal government agencies to create a position of urban forester to lead efforts to inventory the city's tree resources. However, even with the creation of such a position, municipalities need to support the position with additional resources so that management objectives can be carried out effectively.

Different tools and statistical design approaches have been developed to assist managers and scientists in obtaining data on urban forest structure. For example, several sampling strategies were described by Chen and Jim (Chapter 16) and Wu et al. (Chapter 17). Recently, the U.S. Forest Service developed i-Tree (http://www.itreetools.org), an inventory software package to assist urban forest managers in caring for the different components of the urban forest. Although published methods and software now exist to assist urban ecologists and foresters in obtaining and archiving inventory data, the purpose or objectives of the inventory, and not the capabilities of the software itself, should guide decisions as to which variables need to be collected and which methods or protocols should be used. Because of the complexities of these issues, the data needs of a local neighborhood wishing to plant fruit-bearing trees along its streets may be quite different from the data needs of a citywide analysis for air pollution removal by vegetation, for example. Both management activities require knowledge of species, species performance, and site conditions. However, for the citywide analysis above, information such as air pollution sources, meteorological patterns, and areas of greater human susceptibility to pollution (e.g., schools and hospitals) are also needed if management activities for air pollution removal are to be more effective in improving human health and comfort. Furthermore, other factors, such as available funds and personnel, dictate which variables receive priority for collection and analysis. Finally, it should be noted that many tools linked to ecosystem service models (e.g., Urban Forest Effects model [UFORE];, Nowak and Crane, 2000) were developed for a specific region and, therefore, need to be parameterized to local conditions. Nonetheless, judicious use of these tools and models can greatly assist management activities at the neighborhood and city scales once specific goals and objectives are defined.

If the science and management of urban areas are to benefit fully from the landscape ecology perspective described by Wu (Chapter 2), then information from remote-sensing images should also be obtained and integrated more regularly into the planning and management assessments of urban forest distribution (e.g., Löfvenhaft et al., 2002; Freeman and Buck, 2003). The areal extent and resolution of satellite and other aerial images are well matched to the citywide and regional scales needed for urban planning. Information on the spatial distribution of different types of vegetation patches and their canopy coverage and condition can also be determined remotely using multispectral scanning imagery. Once validated by sampling on the ground, additional attributes of the forest can then be measured remotely (Waring and Running, 1998; Kerr and Ostrovsky, 2003), like dominant taxa and species richness (Martin et al., 1998; Gould, 2000), leaf area index (see Chapter 21), productivity (Smith et al., 2002), degree of moisture stress (Zarco-Tejada et al., 2003), and infection by pathogens or pests (Nilsson, 1995; Xiao and McPherson, 2005). After spatial patterns of distribution are identified, appropriate management responses can be deployed more efficiently and at the local scale. Remote images from different points in time can be used to determine where trees and forest patches have accrued or been lost over the interval. The effectiveness of different planting or land-use policies in increasing and distributing forest canopy cover in ways that are socially equitable, improve ecosystem service delivery, or meet conservation goals can also be gauged over time by using remote sensing, as described in Yang et al. (Chapter 22).

Sophistication in computer modeling of ecosystem services has grown since the 1990 conference in Chicago. A recent modeling effort, UFORE (Nowak and Crane, 2000), has been used in Canada, Chile, China, and the United States to quantify ecosystem benefits of the urban forest at the city scale and by land use (http://www.fs.fed.us/ne/syracuse/Data/Internation/data_inter.htm). The model uses tree species composition and detailed measurements of canopy structure and condition, diameter at breast height, and tree position to estimate air pollution removal and carbon sequestration. By stratifying sampling plots according to land use, one can begin to examine how land use affects ecosystem goods and services. However, one limitation of this approach is that it does not capture the spatial heterogeneity of vegetation or built infrastructure *within* a land use, and so limits our understanding of how finer-grained variation in built and vegetated land-cover types influences the movement of materials, energy, organisms, and water throughout a city, that is, the actual ecosystem processes that influence the goods and services being estimated. High-resolution models linking land use and land cover to environmental quality are, therefore, needed to improve planning for greater urban sustainability.

To address this need, Pauleit and Duhme (2000) developed a spatially explicit model at scales useful for planning to quantify the effects of different urban land covers on urban climate, energy use, CO_2 emissions, and water flow in Munich, Germany. They accomplished this by developing a typology that delineated distinct configurations (generally of 4.6 ha or less) of built-up infrastructure, other physical features, and vegetation. Although Pauleit and Duhme were able to capture the spatial heterogeneity within a land use and to evaluate how it influenced urban hydrology at small scales, such intensive effort may be beyond the capabilities of many cities and towns in both developed and developing nations. Therefore, there is a need to develop spatially explicit models for estimating ecosystem services that capture not only the heterogeneity of a land use, but are also more user-friendly for managers and planners. For example, Heidt and Neef (Chapter 6) maintain that quantitative models, like that of Bruse (1999), are useful for evaluating the relative benefits of small-scale structural changes of buildings and vegetation for relieving heat stress caused by stagnant air at the street level. In this way, urban greening can become a more readily appreciated strategy for infilling and improving environmental conditions in dense urban neighborhoods, an important need also addressed by Jim (Chapter 9).

In contrast to managing urban forests for ecosystem services at the broad-city scale, management for material services, such as fruit tree and fuel wood production, often occurs at the local, finer scale of a neighborhood. Such management for material services may seem to be in conflict with the goals of holistically managing the urban forest, because decisions are often made on a piecemeal basis with neighborhoods making decisions independent of each other rather than optimizing resources in a synergistic way. But they are not. For example, biotope and ecotope mapping of a city, a GIS-based approach that can provide information on the diversity, abundance, and distribution of a city's available resources in relation to existing neighborhoods, can be used to link disparate resources with planning and management activities (Sukopp and Weiler, 1988). Biotope mapping, for example, can assist planners with spatially explicit information on a city's natural resources and provide a basis for evaluating how any particular management action taken by residents to supply specific goods may affect adjacent areas. In South Africa, biotope mapping is used to identify areas within neighborhoods for afforestation and small agricultural plots that supplement people's diet and income (Sarel Cilliers, personal communication). Over time, these patches of vegetation, managed for materials and food, may coalesce to form an urban forest in locations where a forest did not exist before and be linked with existing vegetation in other portions of the city to create additional citywide or even regional benefits, such as producing vegetation corridors important for movement of organisms and people (Zipperer et al., 2000; Löfvenhaft et al., 2002) or air and water pollution buffer strips. Therefore, as both management and the forest evolve, a shift in management philosophy may occur from one that emphasizes providing specific goods to one providing an array of ecosystem services, thus enhancing material and environmental quality benefits at both the neighborhood and broader city scales.

If the ecosystem services provided by green infrastructure and unpaved soils are to become a more integral part of cost-benefit analyses in urban planning, then service quantification (for example, tons of pollutant removed, or degrees of cooling) must be translated into monetary units and those values incorporated systematically in municipal tree and shrub value appraisals. Currently, tree appraisal by municipal arborists does not normally incorporate ecosystem services in these valuations (Council of Tree and Landscape Appraisers, 2000; Watson, 2002). However, several cost-benefit analyses that do include noncommodity values and ecosystem services have been conducted for public trees in different communities in the United States (e.g., McPherson et al., 1997, 2006). Such analyses are especially important in urban and suburban areas in more developed countries where logging and farming activities are less likely to occur and add market value to trees. Chen and Jim (Chapter 16) and others (Farber et al. 2002; Chaudry, 2006) have also observed that the value to society of ecosystem services and other nonmarket benefits, which natural areas and vegetation contribute, needs to be incorporated more regularly into land-use planning processes and legal land-use regulatory frameworks (Arnold, in press). Although scientific research that estimates the ecosystem services provided by natural ecosystems has been increasing over the last 15 years, there is a need for more research in ecological economics to develop improved and generally agreed upon methods for converting ecosystem services to monetary units so that trade-offs of different land uses or other changes to the natural components of the environment can be evaluated. Such methods should also include weighting factors that allow the social and ecological context of the parcel and the parcel type's rarity to contribute to the value outcome (e.g., Duever and Noss, 1990). This is especially important in urban areas, where the value of a plot of natural land or a particular tree can be greater than in equivalent rural areas due to the larger human population benefiting from that plot or tree's services (Farber, 2005).

Expanding Research in Urban Forestry and Urban Ecology

While the ability to acquire tools, staff, and adequate funding probably constitutes a major bottleneck to managing our urban forests, the knowledge base for managing the forest more sustainably does exist, but in a limited context and for a limited number of biomes, principally temperate forests. This knowledge base must be continually expanded through applied and basic scientific research, and greater information exchange between the academic and practitioner communities. Furthermore, there is a critical need for multidisciplinary research within and among the social, physical, and natural sciences to understand the interactions and feedbacks between green infrastructure and its social and physical context (e.g., Alberti et al., 2003). The Urban Long-term Ecological Research sites in Baltimore (http://www.beslter.org/) and Phoenix (http://caplter.asu.edu/), funded by the National Science Foundation, are examples of programs addressing such research needs. Increasing the hierarchical scales of scientific inquiry can then parallel the disciplinary, multidisciplinary, and transdisciplinary research that can improve not only our management of urban green environments, but also our understanding of how a city functions as an ecosystem (see Chapter 2). Incentives for promoting networks of academics and practitioners to perform research at these larger scales would not only inform policy making, but in time increase our ability to understand the ecology *of* the city as an ecosystem and not simply the responses of green ecological units *in* cities, an important distinction made by Grimm et al. (2000) and Wu (Chapter 2). Positive signs that such networks are, indeed, being rapidly created and formalized into academic, governmental, or "think-tank" centers and institutes can be appreciated simply by searching the Internet using the key words *center (centre), institute, urban, ecology,* and *sustainability*.

While complex multi- and transdisciplinary research is at the pioneering edge of science, contributions at the disciplinary and interdisciplinary levels are still needed to lay the foundations for a more holistic understanding of the reciprocal impacts of the sociophysical city environment and its urban forests (e.g., Stewart et al., 2004). For example, greater practical and scientific understanding of the biological and ecological responses of native and exotic vegetation to varying and often stressful conditions needs to be gained from the scale of individual species and cultivars to that of communities in natural patches. This knowledge can then be applied in many ways, including improving site matching for planting of street trees, increasing the native species palette at nurseries for public and private use, improving restoration techniques for deteriorating natural areas (see Chapters 12 and 24), and improving reclamation strategies for unvegetated and derelict sites, such as landfills (Robinson and Handel, 2000) and former mining areas (see Chapter 23). In addition, comparative ecological research among cities (e.g., Globenet et al., 2000) would lay a foundation for distinguishing common urban effects and responses from those specific to a particular city or group of cities due to variation in factors such as geography, climate, soils, urban morphology, cultural values, and political and economic systems.

Climate change, biological species invasions, pests, diseases, and regional pollution threaten urban vegetation, as well as natural ecosystems throughout the world. Some urban natural areas, such as forest remnants, can be used as laboratories for basic ecological research to understand species and ecosystem responses not only to climate change, but also to invasive species, altered community trophic structures and disturbances, and elevated air pollutants including CO_2, (Carreiro and Tripler, 2005; also see Chapter 11). This information would be particularly pertinent for predicting the health and regeneration of urban forest patches where successional forces, rather than direct human planting and management, determine future species composition (Zipperer, 2002; Lugo, 2004; Lugo and Helmer, 2004). For instance, negative effects of urban land use on seedling regeneration could compromise the future ability of these forested patches to provide the ecosystem services of air pollution reduction, microclimate mediation, carbon sequestration, and flood control. Plant demographic research, coupled with successional trajectory modeling, (e.g., Pacala et al., 1996; Meurk and Hall, 2006) could inform timely mitigation interventions to prevent or reduce undesirable outcomes. Basic research is also needed on the effects of varying the abundance and distribution of urban vegetation patches on landscape connectivity, a factor important for maintaining meta-population and ecosystem processes at the landscape level (Byers and Mitchell, 2005; Ray, 2005; Reice, 2005; Sanjayan and Crooks, 2005). Such studies could then contribute to species conservation efforts from local to regional scales as well as to estimation of ecosystem services.

Since cities are human-dominated ecosystems, flows of information among and within groups of professionals, policy-makers, and the public are paramount for understanding how urban systems function ecologically as well as socially. Human activities engender responses from the socioeconomic and natural components of cities, some of which may require technical "translation" by experts before they can be perceived by policy makers and the public. Human and institutional reactions (or lack thereof) to these environmental responses then constitute feedback circuits that either perpetuate the same conditions or change them. Researchers in the social, economic, and natural sciences create and use aggregative indices as a means of measuring and communicating the multiple responses of their respective systems to internal and external forces, either human or natural. Examples in these disciplines include the Index of Social Health (http://iisp.vassar.edu/ish.html), the gross domestic product, and the Index of Biotic Integrity (http://www.epa.gov/bioindicators/ html/ibi-hist.html). One of the reasons for acquiring and creating such information is to provide early warning of undesirable change before the system itself "informs" us after reaching a more observable tipping point where corrective action becomes more costly. As Zhang et al. (Chapter 4) point out, while index development for measuring environmental sustainability at the national and regional levels is progressing (e.g., Heinz Center, 2002, http://www.heinzctr.org/ecosystems/report.html), sustainability indicator development at the city scale is still in its early stages (e.g., Urban Quality Index of Song and Gao, Chapter 6; Menegat, 2002). One of the research issues involved is the construction of indicators that are sensitive enough to capture the most important interactions among the social, ecological, and economic components of cities, and yet are simple enough for communicating to the public and policy makers. Among these are the interactions between people and the natural habitats in cities. The ecosystem services concept is proving valuable for communicating the important roles that nature plays in supporting human societies (Millennium Ecosystem Assessment, 2005, http://www.millenniumassessment.

org/en/Index.aspx), but the importance of ecosystems services in contributing to human well-being in urban landscapes is perhaps less publicly appreciated. However, as discussed earlier in this chapter, research and tools for converting nature's services into monetary terms would greatly assist the planning and management communities in evaluating different development options for urban and urbanizing areas. The construction of urban sustainability indices and the valuation of ecosystem services will be critical particularly in the near-term, if we are to prevent undesirable trajectories and gauge the efficacy of our collective actions in creating more ecologically sound cities.

Building Partnerships for Implementation, Planning, and Research

As cities grow and competition for space intensifies, the need for integrative planning and management of green infrastructure becomes more apparent. Indeed, the need for a more holistic approach to urban forest planning and management was perhaps the most recurring point made by the contributors to this book. Building partnerships to conserve, restore, and manage urban forests was advocated as one means of achieving this goal. Assembling a diverse expertise base with multiple viewpoints into partnerships to address a city's urban forestry issues can inform plans and their implementation at the outset, thereby avoiding some costly problems during and after project completion (Ames and Dewald, 2003). The perceived benefits of integration through partnerships include improving delivery of ecosystem services and materials to the most appropriate locations, reducing vegetation care and maintenance costs, distributing the health and recreation benefits of trees and parks in a more socially equitable manner, and providing habitat for wildlife in the most suitable sites.

It is not surprising that creating and maintaining a healthy diversity of vegetation and adequate levels of ecosystem services for people requires greater planning and integration of human effort, particularly in an ecosystem that is, after all, human-dominated and dynamic. More simply stated, "it takes more than an understanding of trees to sustain a successful urban forest" (Jones, Chapter 8). Partnerships among governmental and nongovernmental agencies, academic researchers, educators, and businesses can provide opportunities for stimulating public awareness and involvement in supporting a city's green infrastructure (Johnson, 2002), thereby providing social, environmental, and economic benefits for settlements large and small (e.g., African Conservation Trust's Manukelana Project, http://www.projectafrica.com/ manukelana.htm). People's involvement in planting and growing trees in their neighborhoods, schools, and public places is generally thought to promote the long-term success of urban greening programs. Participation of people in various greening activities in cities can build a sense of ownership that helps prevent problems like vandalism and may create a greater appreciation of a city's local biotic legacy and uniqueness (e.g., http://www.olmstedparks.org/conservancy/volunteer.html).

Greening activities may also provide social benefits to individuals and entire communities. However, claims of success or failure in the accrual of social improvements due to urban greening projects should be evaluated more rigorously than is often done, so that future activities can benefit from past insights (Westphal, 2003).

In some cases, bottom-up demand and follow-through from the public has affected forestry restoration and reclamation at a regional scale. As Jones (Chapter 8) describes, urban forestry in England arose in the early 20th century from the efforts of a volunteer community association that planted trees on lands badly despoiled by coal mining and metal smelting. Today, these plantings are an important part of the green infrastructure of some cities in the British Midlands. The current Urban Forest program in this "Black Country" of England involves partnerships among public, private, and volunteer organizations, and such partnerships have provided important models for successful restoration and greening activities elsewhere in the United Kingdom. Miyawaki (Chapter 12) has codified his philosophy (known internationally as the Miyawaki method) to restoring and constructing new urban forest patches, one that depends on partnerships. Miyawaki's approach relies on knowledge from basic and applied vegetation and soil science for selecting and growing native trees and shrubs, relies on government and private businesses for funding planting endeavors, and uses public volunteers as labor for the initial plantings. As he states (Chapter 12), "Reforestation can be viewed as analogous to dramas: vegetation ecologists write play scenarios, government and private companies work as producers and directors, and citizens, including school children, play the part of leading characters on the stage. Everyone has the opportunity to play a role in reforesting their region." The success of his method over the last 30 years is attested to by his estimate of having planted 30 million trees in over 1200 sites in Asia and Brazil.

Partnerships also inform the planning process. For example, university researchers Secco and Zulian (Chapter 20) offer urban planners a quantitative modeling tool, sensitive to social context, for making decisions about the location and equipment needs of urban recreational parks that best match neighborhood demographics and available transportation. Linking ecological and social systems provides decision makers with information for developing comprehensive management plans for the urban forest that also improve ecosystem and material benefits for urban residents (Yli-Pelkonen and Niemelä, 2005). Decision-making tools, especially those with scenario-building capacity, are needed to assist planners and decision makers with these complex assessments. For example, Keith Jones (Chapter 13) has described the development and use of the GIS-based Public Benefit Recording System that ranks different patches in a city using four criteria of public benefit: social, public access, economic, and environmental. These multiple dimensions of benefit can also provide a basis for fostering partnerships between the public and private sectors when positive synergies among the four categories are identified.

Perhaps the experience of the citizens of Porto Alegre, Brazil, best illustrates the benefits of widespread and continuous public involvement in urban planning. The city has a broad-based participatory budgeting and planning process, one that has directly involved approximately 150,000 residents (Menegat, 2002). This evolving social and political experiment begun in 1989 has led to resident-driven, environmental management plans and programs in a city of 1.3 million, which now boasts the highest standard of living and the highest amount of green space per inhabitant in Brazil (14 m^2/person). As part of this process, the need to understand the city's environmental setting and biotic resources for planning and management purposes was identified and resulted in the publication of the *Environmental Atlas of Porto Alegre* (Menegat et al., 1998, http://www6.ufrgs.br/gaia/gb/atlas/atlasframe.html). Environmental management and planning in Porto Alegre is based on six principles, three of which are as follows: (1) the city is an integral part of its natural ecosystems, (2) the watershed is the unit of environmental management, and (3) education and communication with citizens about the city's green environments is essential to secure long-term societal commitment to increasing and maintaining environmental quality and green space allocation. To meet the objectives of this last principle, parts of the atlas were published in a series of inserts in the local newspaper in order to disseminate that knowledge more broadly to the public. The atlas was also freely distributed to all municipal schools in the city, and as a consequence triggered the construction of urban environmental intelligence laboratories in the schools (Alexandre Ruszczyk, personal communication).

Through participatory research, partnerships among academic researchers, environmental managers, other practitioners, and stakeholders not only improve the implementation of management plans and practices, but also expand the breadth of research questions and research opportunities. Thus both the management and scientific knowledge base in urban ecology and urban forestry is increased at local and global scales. For example, ecological restoration of natural habitats in cities, a major management activity in urban environments, is one way of achieving this goal at the same time as it improves urban forest quality (e.g., Silva Matos et al., 2002). Often, managers do not have the time or possibly the resources to document restoration activities. By partnering with the academic community, a more rigorous evaluation of a restoration's efficacy can be conducted using proper statistical designs and analyses (Giardina et al., 2007). A properly designed project would include setting benchmarks for determining success before the restoration is initiated, replicating procedures or treatments at proper scales, using reference sites or treatment controls, and collecting pre- and post-treatment data to establish baseline and document the range of variability in habitat responses (see Chapters 23, 24, and 26). Additional benefits of partnering on a restoration project might also occur, and include: (1) the opportunity to build in long-term commitments for project evaluation, (2) the ability to determine the ecological mechanisms that underlie a project's success or failure, and (3) the opportunity to train future managers. Higgs (1997) further argues for partnerships with the broader local community to increase the democratization of restoration projects and to identify the unique cultural and ethical contexts of project sites. Such multifaceted discussions that capture the needs and understanding of many individuals at the local scale by their very nature require diverse partnerships and benefit long-term restoration success by promoting what Higgs calls "place awareness" and "authentic engagements between people and ecosystems" (e.g., Primack et al., 2000).

Incorporating Urban Forestry Into the Vision of the Eco-City

Over the last decade it has become increasingly apparent that the ability of our planet to provide people with resources for supporting current population levels without compromising future generations and other species has become strained. The fact that the planet's economic "metabolism" is now large enough to affect our planet's "metabolism" and climate regulatory system has become ever more accepted and mainstream (Stern, 2007). Climate uncertainty further complicates our ability to predict our planet's capacity to provide food, water, materials, and ecosystem services for our exponentially growing populations (Intergovernmental Panel on Climate Change (IPCC) report, http://www.ipcc.ch/). We are reaching, or perhaps have already reached, a critical threshold that requires bold and widespread responses from the human community to avoid a downturn in our collective quality of life. Finding ways to partner with the natural world in solving environmental problems, instead of viewing nature primarily as a commodity or amenity, must become an integral component of our adaptation to changing global conditions.

Progress in addressing these global challenges is increasing. In the early phase of this international awareness, the United Nations convened the Earth Summit in Rio de Janeiro, Brazil in 1992. Among other accomplishments, delegates to this conference provided a declaration of principles and a roadmap for promoting human sustainability known as the Agenda for the 21st Century (Agenda 21, for short; http://www.un.org/esa/sustdev/documents/agenda21/index.htm). Since a growing proportion of humanity was and still is becoming urban, delegates also realized that the solutions to many global problems lay in changing the activities and resource consumption patterns of people in cities. These challenges were addressed in Chapter 28 of Agenda 21, known as Local Agenda 21. This document created the impetus for subsequent conferences where policy and implementation frameworks for achieving sustainability goals consonant with Agenda 21 principles were produced. One of the better known of these was the first European Conference on Sustainable Cities and Towns held in Aalborg, Denmark, in 1994, which resulted in the Charter of European Cities and Towns Towards Sustainability (the Aalborg Charter; http://www.aalborgplus10.dk/ default.aspx?m=2&i=371). As of the Aalborg-Plus 10 meeting in 2004 (http://www.aalborgplus10.dk/default.aspx?m=2&i=308), 497 European cities and towns have committed to charter goals as full signatories and 531 additional cities have declared their intention to sign, indicating a groundswell of support from leaders and the public for the realization that we must learn to live within the bounds of earth's carrying capacity for our species.

Progress in creating more support for a sustainable cities movement has been made by other groups as well. The eco-cities movement (http://www.ecocitybuilders.org/), now almost 20 years old, has provided a venue for supporting projects, creating networks, and accelerating transdisciplinary exchange of information on urban sustainability, and has hosted six international conferences since 1990, with a seventh planned for 2008. While cities in the U.S. have not explicitly adopted the United Nations's Local Agenda 21, many have become more engaged in their commitment to urban sustainability planning and implementation (Sustain Lane, http://www.sustainlane.com/us-city-rankings/). So far in the U.S., the impetus for change has come mostly from the bottom up, as evidenced by 185 U.S. cities joining a total of 627 cities in 67 countries worldwide as members of the International Council for Local Environmental Initiatives–Local Governments for Sustainability (http://www.iclei.org/index.php?id=772). Mayors from 600 U.S. cities in all 50 states have also demonstrated leadership in committing to reductions in greenhouse gas emissions by signing a climate protection agreement (http://usmayors.org/climateprotection/). There are also encouraging signs that in addition to governmental and nongovernmental organizations, businesses are more willing to respond to the complex challenges imposed by climate change and urban sustainability (The Climate Group, http://theclimategroup.org/index.php/ reducing_emissions/case_studies/). And recently, the William J. Clinton Foundation in an alliance with several banks is financing green building technology in major cities worldwide to reduce urban energy use and CO_2 emissions (http://www.lintonfoundation.org/cf-pgm-cci-home.htm).

As a result of this rapid increase in awareness of sustainability issues, concepts such as "ecological footprint," "green technology," "cyclic economies," and "sustainability" are heard more often in the public parlance. They are no longer terms used only by academics and environmentalists, but are increasingly discussed by policy makers, businesses, and the public. However, even as we use a term that represents the color of plant life, ironically people seem not to think first about "greening" in terms of enhancing vegetation cover in their surroundings, but instead apply "green" more reflexively to items and processes that are human engineered. Perhaps this is due to the fact that most people and their leadership live in urban centers, where the built infrastructure dominates, and plants are often viewed as ornamental "extras" and not as integral contributors to the health, comfort, safety, and material needs of a city's people. This is also probably indicative of how far urban forestry professionals and advocates have yet to go in pressing home the fact that a city's plant life and soils are vital urban infrastructure, requiring and deserving as much deliberate, scientifically informed management and long-term commitment to care as our built infrastructure.

How can professionals and the public work with nature to move their cities closer to the eco-city ideal? Increasing the "amount and kind of nature … through conservation and restoration activities" is one of the five principles listed by Wittig (Chapter 3) for guiding this transformational process. Many opportunities exist in urban areas for increasing and integrating nature into human settlements. One of the most successful has been the development of greenway or greenbelt plantings in cities throughout the world (Fábos and Ryan, 2006). In many cases, strategies for creating corridor networks begin with identifying, from aerial photographs or other forms of remote sensing, linear vegetation features that are already part of the landscape. Most linear arrangements of trees and other vegetation occur along rivers, streams, canals, highways, and other transportation corridors and can serve as nucleating sites for restoration projects aimed at increasing the connectivity of green elements across the landscape. The motivations for creating greenways have varied over time and have reflected changing and varied societal needs from local to national scales (Fábos and Ryan, 2006). In some countries greenways are constructed primarily to preserve air and water quality, and reduce flooding. In others, initial reasons were to provide shelter-belts for agriculture and urban protection from storms (Yu et al., 2006), but are now expanding to include additional network functions, such as recreation, escape routes for disasters, and the conservation of biological communities and historic and cultural features (Bryant, 2006; Fábos and Ryan, 2006; see also Chapters 10, 14, and 25). However, while creating greater landscape connectivity through use of greenways is a common planning goal, certain caveats should be heeded, particularly in urban areas, since unintended negative consequences can sometimes occur after an ecological patch of high quality becomes linked to one of low quality (Simberloff and Cox, 1987; Environmental Law Institute, 2003). Also, political structures, which vary in their top-down versus bottom-up approaches to planning and implementation of environmental projects, may also influence the type, extent, and success of greenway plantings, as noted by Yu et al. (2006) in their comparison of greenway projects in China with those in Europe and North America.

Opportunities for greening cities as part of a path toward developing into an eco-city will also vary with economic status and changing demographics of cities. For example, the needs and opportunities for tree and vegetation planting differ greatly between the rapidly growing cities of developing nations and postindustrial shrinking cities in more developed nations. In developing cities, urbanization and the rapid influx of rural migrants often occur without benefit of government planning, infrastructure, and services. Consequently, supplying people's fundamental needs such as sanitation and potable water is grossly inadequate (Carter, 1995). In addition, food, energy, and materials for housing construction may also be insufficient. By improving soil stability, mitigating flooding, reducing air and water pollution, and providing fuel wood and shade, urban and periurban forestry projects, if integrated with economic and health policies and programs, have the potential to ameliorate many of the negative consequences of crowded and polluted environments (Konijnendijk et al., 2004). Trees in developing cities can also supply food, honey, fodder, spices, medicine, and craft supplies—all of which supplement diet or incomes (Carter, 1995). However, the difficulties in promoting greening can be formidable under these circumstances. For example, since other forms of energy are expensive, the urban poor in developing countries often rely on wood for fuel. This can result in the stripping of trees in streets and local parks and the creation of zones of desertification around a city (Olembo and de Rham, 1987; Carter, 1995). Livestock browsing in these cities can also destroy saplings, which then require extra protection strategies after planting. As a consequence, the United Nations Food and Agricultural Organization (FAO) has provided resources for the establishment of periurban agroforests for multiple purposes throughout the world. Over the years these experiments have met with mixed success (Haque, 1987; Konijnendijk et al., 2004). If the potential benefits of urban forestry are to be achieved in these difficult circumstances, then planning for the types and locations of greening must meet the direct material and environmental needs of people and cannot occur without considerable public support and partnership in their continued management and protection (Kuchelmeister and Braatz, 1993; Carter, 1995).

The recent phenomenon of shrinking cities also creates new opportunities to rethink urban planning and green space distribution. In Europe and especially in the United States, urban planning has long been focused on dealing with economic growth and areal expansion of cities. However, since the mid-20th century, a combination of forces including suburbanization, the expansion of global markets, and shifts from industrial to service and information-based economies in these countries has resulted in the simultaneous decline in population and the economy in urban centers. Embracing urban contraction as an opportunity for improving the quality of life of the remaining residents is still novel and difficult for politicians and planners alike, given the many cascading social problems that follow the shrinking of cities. Yet, midsized cities in the rustbelt of the midwestern U.S., such as Youngstown and Cleveland, Ohio, and Detroit and Flint, Michigan, are rising to the challenge (http://www.governing.com/articles/11cities.htm). Residents in areas that were only partially developed or in declining neighborhoods that are now mostly abandoned are given incentives to leave so that these locations can be turned into woodland, wetland, prairie, parks, or community gardens. This not only increases the environmental value of the land and the ecosystem services delivered, but may also provide economic value for a city, due to compensatory wetland mitigation laws applied to developers who drain and build on wetlands in other locations. Cities in Eastern Europe have also been grappling with these difficulties, prompting recent discussions, exhibits, and special issues of professional journals focused on the shrinking cities phenomenon (Müller, 2004).

In both growing and shrinking cities, planners, designers, decision makers, urban foresters, and residents have recognized the link between the urban forest and community well-being and livability. Many cities have started to move toward becoming eco-cities, cities where inhabitants not only realize the importance of reducing their ecological footprint, but also of improving their urban forests. This has resulted in communities creating policies to protect, conserve, and manage their urban forests to optimize ecosystem services, materials, and social benefits, and in so doing also reduce the rate at which planetary wide global warming occurs. Although the eco-city will not stop global climate disruption by itself, it may create the realization among decision makers and the public that if our cities are to remain livable during these changing and uncertain times, then improving green infrastructure is equally as important as improving the gray. If greater allocation and improved siting of green infrastructure is not planned and implemented, then more costly engineered solutions become our only adaptive alternatives. Only through a comprehensive, broad-scale approach to planning and management can the urban forest be conserved during urbanization and maintained in a healthy condition in settled areas (LaGro, 2001). To achieve this, green infrastructure cannot be an afterthought in the development process and cannot be "last in line" for municipal budgeting. Likewise, only through working with local residents can managers identify collective needs and how best to afforest a neighborhood, and thereby contribute to the evolution of an eco-city.

Conclusion

We have provided a broad overview of the varied ways in which urban forestry professionals, policy makers, and citizens throughout the world are working to incorporate more trees and other green spaces into their cities, thereby realizing the potential of urban forests to contribute to their community's well-being and sense of place. For the most part in this chapter, we have stressed the utilitarian functions of trees, forests, and other green areas within dense human settlements, since these ecosystem and socioeconomic benefits are compelling and motivating reasons for increasing green space allocation for people in cities. However, by treating the incorporation of nature in cities as a purely pragmatic exercise in engineering for addressing our physical needs, we overlook other powerful reasons that many of us have for greening our homes and communities—the solace, pleasure, excitement, and joy that we experience by being part of a greater natural world (Kellert and Wilson, 1993). This in essence was another common theme expressed by many of the authors who contributed to this book—our shared desire to bring more grace into our lives and to live more harmoniously with nature. As our species enters its urban century, we must be proactive in assuring that the everyday environment for the greatest number of our descendants will contain places of natural beauty where we and they can regain and retain our humanity.

References

Alberti, M., Marzluff, M. J., Shulenberger, E., Bradley, G., Ryan, C., and Zumbrunnen, C. (2003) Integrating humans into ecology: opportunities and challenges for studying urban ecosystems. BioScience 53:1169–1179.

Ames, B. and Dewald, S. (2003) Working proactively with developers to preserve urban trees. Cities 20:95–100.

Arnold, C. A. (In press) The structure of the land use regulatory system in the United States. The Journal of Land Use and Environmental Law 22.

Blaine, J., Sweeney, B., and Arscott, D. (2006) Enhanced source-water monitoring for New York City: Historical framework, political context and project design. Journal of the North American Benthological Society 25:851–866.

Bryant, M. M. (2006) Urban landscape conservation and the role of ecological greenways at local and metropolitan scales. Landscape and Urban Planning 76:23–44.

Bruse, M. (1999) Modelling and strategies for improved urban climates. Invited Paper. In: Proceedings International Conference on Urban Climatology & International Congress of Biometeorology, Sydney, Australia, November 8–12. (http://klima.geographie.ruhr-uni-bochum.de/bruse/paper63_00.pdf)

Byers, D. and Mitchell, J. (2005) Sprawl and species with limited dispersal activities. In: Johnson, E. A., and Klemens, M. W. (eds.) Nature in Fragments: The Legacy of Sprawl. Columbia University Press, New York, pp. 157–180.

Carreiro, M. M. (2005) Effects of urbanization on decomposer communities and soil processes in forest remnants. In: Johnson, E. A., and Klemens, M. W. (eds.) Nature in Fragments: The Legacy of Sprawl. Columbia University Press, New York, pp. 125–143.

Carreiro, M. M. and Tripler, C. E. (2005) Forest remnants along urban-rural gradients: examining their potential for global change research. Ecosystems 8:568–582.

Carter, J. (1995) The potential of urban forestry in developing countries: a concept paper. Food and Agriculture Organization, Rome, Italy. http://www.fao.org/docrep/005/t1680e/T1680E00.htm.

Chaudhry, P. (2006) Recreational use value of Chandigarh city's urban forestry. Current Science 91:1440–1441.

Chichilnisky, G. and Heal, G. (1998) Economic returns from the biosphere. Nature 391:629–630.

Council of Tree and Landscape Appraisers. (2000) Guide for Plant Appraisal (9th ed.). International Society of Arboriculture, Champaign, IL.

Duever, L. C., and Noss, R. F. (1990) A computerized method of priority ranking for natural areas. In: Mitchell, R. S., Sheviak, C. J., and Leopold, D. J. (eds.) Ecosystem management: Rare species and significant habitats. New York State Museum, Albany, pp. 22–33.

Dwyer, J., Novak, D., Noble, M., and Sisinni. S. (2000) Connecting people with ecosystems in the 21st century. Technical Report PNW-GTR-490. USDA Forest Service.

Elmendorf, W. R., Cotrone, V. J., and Mullen, J. T. (2003) Trends in urban forestry practices, programs and sustainability: Contrasting a Pennsylvania, U.S., study. Journal of Arboriculture 29:237–248.

Elmendorf, W. R., Willits, F. K., Sasidharan, V., and Godbey, G. (2005) Urban park and forest participation and landscape preference: a comparison between blacks and whites in Philadelphia and Atlanta, U. S. Journal of Arboriculture 31: 318–326.

Environmental Law Institute (2003) Conservation Thresholds for Land Use Planners. Environmental Law Institute, Washington, DC.

Escobedo, F. J., Nowak, D. J., Wagner, J. E., et al. (2006) The socioeconomics of management of Santiago de Chile's public urban forests. Urban Forestry and Urban Greening 4:105–114.

Fábos, J. G., and Ryan, R. L. (2006) An introduction to greenway planning around the world. Landscape and Urban Planning 76:1–6.

Farber, S. C. (2005) The economics of biodiversity in urbanizing ecosystems. In: Johnson, E. A., and Klemens, M. W. (eds.) Nature in Fragments: The Legacy of Sprawl. Columbia University Press, New York, pp. 263–283.

Farber, S. C., Costanza, R., and Wilson, M. A. (2002) Economic and ecological concepts for valuing ecosystem services. Ecological Economics 41:375–392.

Freeman, C. and Buck, O. (2003) Development of an ecological mapping methodology for urban areas in New Zealand. Landscape and Urban Planning 63:161–173.

Giardina, C., Litton, C., Thaxton, J., Cordell, S., Hadway, L., and Sandquist. D. (2007) Science driven restoration: a candle in a demon haunted world—response to Cabin (2007). Restoration Ecology 15:171–176.

Gould, W. (2000) Remote sensing of vegetation, plant species richness, and regional biodiversity hotspots. Ecological Applications 10:1861–1870.

Grimm, N., Grove, J. M., Pickett, S. T. A. and Redman, C. (2000) Integrated approaches to long-term studies of urban ecological systems. BioScience 50:571–584.

Haque, F. (1987) Thirteen city profiles. Unasylva 39. Food and Agriculture Organization. Rome, Italy. http://www.fao.org/docrep/s1930e/s1930e00.htm#Contents

Heinz Center for Science, Economics and the Environment (2002) The State of the Nation's Ecosystems. Cambridge University Press, Cambridge, UK. http://www.heinzctr.org/ecosystems/index.shtml

Heynen, N., Perkins, H. A., and Roy, P. (2006) The political ecology of uneven urban green space. Urban Affairs Review 42:3–25.

Higgs, E. (1997) What is good ecological restoration? Conservation Biology 11:338–348.

Howard, E. (1902) Garden Cities of Tomorrow. Faber and Faber, London, UK.

Johnson, L. (2002) Cities in nature: Case studies of urban greening partnerships, Evergreen, Canada. http://www.evergreen.ca/en/cg/cg-case1.pdf

Kellert, S. R. and Wilson, E. O. (eds.) (1993) The Biophilia Hypothesis. Island Press, Washington, D.C.

Kerr, J. T. and Ostrovsky, M. (2003) From space to species: ecological applications for remote sensing. Trends in Ecology and Evolution 18:299–305.

Kielbaso, J. (1990) Trends and issues in city forests. Journal of Arboriculture 16:69–76.

Konijnenddijk, C., Sadio, S., Randrup, T., and Schipperijn, J. (2004) Urban and peri-urban forestry in a development context– strategy and implementation. Journal of Arboriculture 30:269–276.

Kuchelmeister, G. and Braatz. S. (1993) Urban forestry revisited. Unasylva 44. http://www.fao.org/docrep/u9300e/u9300e00.htm#Contents

LaGro, J. A. Jr. (2001) Site Analyses: Linking Program and Concept in Land Planning and Design. Wiley and Sons, New York.

Löfvenhaft, K., Björn, C., and Ihse, M. (2002) Biotope patterns in urban areas: a conceptual model integrating biodiversity issues in spatial planning. Landscape and Urban Planning 58:223–240.

Lugo, A. E. (2004) The outcome of alien tree invasion in Puerto Rico. Frontiers in Ecology and the Environment 2:265–273.

Lugo, A. E., and Helmer, E. (2004) Emerging forest on abandoned land: Puerto Rico's new forests. Forest Ecology and Management 190:145–161.

Martin, M., Newman, S., Aber, J., and Congalton, R. (1998) Determining forest species composition using high spectral resolution remote sensing data. Remote Sensing of Environment 65:249–254.

McDonnell, M., Pickett, S., Groffman, P., et al. (1997) Ecosystem processes along an urban-to-rural gradient. Urban Ecosystems 1:21–36.

McPherson, E. G., Nowak, D., Heisler, G., et al. (1997) Quantifying urban forest structure, function, and value, the Chicago Urban Forest Climate Project. Urban Ecosystems 1:49–61.

McPherson, E. G., Simpson, J. R., Peper, P. J., et al. (2006) Midwest community tree guide: Benefits, costs, and strategic planting. USDA Forest Service General Technical Report PSW-GTR-199, Berkeley, CA.

Menegat, R. (2002) Participatory democracy and sustainable development: integrated urban environmental management in Porto Alegre, Brazil. Environment and Urbanization 14:181–206.

Menegat, R., Porto, M., Carraro, C., and Fernandes, L. (coordinators) (1998) Atlas ambientale de Porto Alegre. Edufrgs, Porto Alegre, Brazil.

Meurk, C. and Hall, G. (2006) Options for enhancing forest biodiversity across New Zealand's managed landscapes based on ecosystem modeling and spatial design. New Zealand Journal of Ecology 30:131–146.

Millennium Ecosystem Assessment. (2005) Ecosystems and Human Well-being: Synthesis. Island Press, Washington, DC, USA.

Müller, B. (2004) Demographic change and its consequences for cities: Introduction and overview. German Journal of Urban Studies 44. http://www.difu.de/index.shtml?/publikationen/dfk/ .

Niemalä, J., Kotze, J., Ashworth, A., et al. (2000) The search for common anthropogenic impacts on biodiversity: a global network. Journal of Insect Conservation 4:3–9.

Nilsson, H. -E. (1995) Remote sensing and image analysis in plant pathology. Annual Review of Phytopathology 15:489–527.

Nowak, D. J., and Crane, D. E. (2000) The urban forest effect (UFORE) model: Quantifying urban forest structure and functions. In: Hansen, M., and Burk, T. (eds.) Integrated Tools for Natural Resources Inventories in the 21st Century: Proceedings of the IUFRO Conference. USDA Forest Service, North Central Research Station, St. Paul, MN, Boise, ID, pp. 714–720.

Nowak, D. J. and Dwyer, J. F. (2000) Understanding the benefits and costs of urban forest ecosystems. In: Kuser, J. E. (ed.) Handbook of Urban and Community Forestry in the Northeast. Kluwer Academic/Plenum Publishers, New York, pp. 11–25.

Nowak, D. J., Kuroda, M., and Crane, D. E. (2004) Tree mortality rates and tree population projections in Baltimore, Maryland, USA. Urban Forestry and Urban Greening 2:139–147.

Olembo, R. and de Rham, P. (1987) In two different worlds. Unasylva 39:26–35. http://www.fao.org/docrep/s1930e/s1930e00.htm#Contents

Pacala, S., Canham, C., Saponara, J., Silander, J. Jr., Kobe, R., and Ribbens, E. (1996) Forest models defined by field measurements: estimation, error analysis and dynamics. Ecological Monographs 66:1–43.

Pauleit, S. and Duhme, F. (2000) Assessing the environmental performance of land cover types for urban planning. Landscape and Urban Planning 52:1–20.

Perkins, H. A., Heynen, N., and Wilson, J. (2004) Inequitable access to urban reforestation: the impact of urban political economy on housing tenure and urban forests. Cities 21:291–299.

Pitt, D., Soergell, K. II, and Zube, E. (1979) Trees in the city. In: Laurie, I. (ed.) Nature in Cities: The Natural Environment in the Design and Development of Urban Green Space. Wiley, Chichester, England, pp. 205–229.

Platt, R. H., Rowntree, R. A., and Muick, P. C. (eds.) (1994) The Ecological City. The University of Massachusetts Press, Amherst, MA.

Primack, R., Kobori, H., and Mori, S. (2000) Dragonfly pond restoration promotes conservation awareness in Japan. Conservation Biology 14:1553–1554.

Rapoport, E. (1993) The process of plant colonization in small settlements and large cities. In: McDonnell, M. J., and Pickett, S. T. A. (eds.) Humans as Components of Ecosystems. Springer-Verlag, New York, pp. 190–207.

Ray, J. (2005) Sprawl and highly mobile or wide-ranging species. In: Johnson, E. A. and Klemens, M. W. (eds.) Nature in Fragments: The Legacy of Sprawl. Columbia University Press, New York, pp. 181–205.

Reice, S. (2005) Ecosystems, disturbance and the impact of sprawl. In: Johnson, E. A. and Klemens, M. W. (eds.) Nature in Fragments: The Legacy of Sprawl. Columbia University Press, New York, pp. 90–108.

Robinson, G. and Handel, S. (2000) Directing spatial patterns of recruitment during an experimental urban woodland reclamation. Ecological Applications 10:174–188.

Sanjayan, M. and Crooks, K. (2005) Maintaining connectivity in urbanizing landscapes. In: Johnson, E. A., and Klemens, M. W. (eds.) Nature in Fragments: The Legacy of Sprawl. Columbia University Press, New York, pp. 239–262.

Silva Matos, D., Santos, C., De R. Chevalier, D. (2002) Fire and restoration of the largest urban forest of the world in Rio de Janeiro City, Brazil. Urban Ecosystems 6:151–161.

Simberloff, D. and Cox, J. (1987) Consequences and costs of conservation corridors. Conservation Biology 1:63–71.

Smith, M-L., Ollinger, S., Martin, M., Aber, J., Hallett, R., and Goodale, C. (2002) Direct estimation of aboveground forest productivity through hyperspectral remote sensing of canopy nitrogen. Ecological Applications 12:1286–1302.

Stern, N. 2007. The Economics of Climate Change: the Stern Review. Cambridge University Press, Cambridge, UK. http://www.hm-treasury.gov.uk/independent_reviews/independent_reviews_index.cfm

Stewart, G., Ignatieva, M., Meurk, C., and Earl, R. (2004) The re-emergence of indigenous forest in an urban environment, Christchurch, New Zealand. Urban Forestry and Urban Greening 2:149–158.

Sukopp, H. and Weiler, S. (1988) Biotope mapping and nature conservation strategies in urban areas of the Federal Republic of Germany. Landscape and Urban Planning 15:39–58.

Waring, R. H. and Running, S. W. (1998) Forest Ecosystems: Analysis at Multiple Scales. Academic Press, San Diego, CA.

Watson, G. (2002) Comparing formula methods of tree appraisal. Journal of Arboriculture 28:11–18.

Westphal, L. (2003) Urban greening and social benefits: a study of empowerment outcomes. Journal of Arboriculture 29:137–147.

Wolf, K. L. (2003) Public response to the urban forest in inner-city business districts. Journal of Arboriculture 29:117–126.

Xiao, Q. and McPherson, E. G. (2005) Tree health mapping with multispectral remote sensing data at UC Davis, California. Urban Ecosystems 8: 349–361.

Yli-Pelkonen, V. and Kohl, J. (2005) The role of local ecological knowledge in sustainable urban planning: perspectives from Finland. Sustainability: Science, Practice, and Policy 1:3–14.

Yli-Pelkonen, V., and Niemelä, J. (2005) Linking ecological and social systems in cities: urban planning in Finland as a case. Biodiversity and Conservation 14:1947–1967.

Yu, K., Dihua, L. and Nuyu, L. (2006) The evolution of greenways in China. Landscape and Urban Planning 76: 223–239.

Zarco-Tejada, P. J., Rueda, C., and Ustin, S. (2003) Water content estimation in vegetation with MODIS reflectance data and model inversion methods. Remote Sensing of Environment 85:109–124.

Zipperer, W. C. (2002) Species composition and structure of regenerated and remnant forest patches within an urban landscape. Urban Ecosystems 6:271–290.

Zipperer, W. C., Foresman, T. W., Sisinni, S. M., and Pouyat, R. V. (1997) Urban tree cover: an ecological perspective. Urban Ecosystems 1:229–246.

Zipperer, W. C., Wu, J., Pouyat, R. V., and Pickett, S. T. A. (2000) The application of ecological principles to urban and urbanizing landscapes. Ecological Applications 10:685–688.

Index

Printed in the United States of America.